OPTICAL PHYSICS

S. G. LIPSON, Ph.D.
Associate Professor of Physics, Israel Institute of Technology, Haifa

AND

H. LIPSON, F.R.S.
Emeritus Professor of Physics in the Faculty of Technology
University of Manchester

SECOND EDITION

CAMBRIDGE UNIVERSITY PRESS
Cambridge
London New York New Rochelle
Melbourne Sydney

Published by the Press Syndicate of the University of Cambridge
The Pitt Building, Trumpington Street, Cambridge CB2 1RP
32 East 57th Street, New York, NY 10022, USA
296 Beaconsfield Parade, Middle Park, Melbourne 3206, Australia

© Cambridge University Press 1969, 1981

First published 1969
Second edition 1981

Printed in the United States of America
Typeset by J. W. Arrowsmith Ltd, Bristol, England
Printed and bound by The Book Press, Brattleboro, Vermont

British Library cataloguing in publication data

Lipson, Stephen Geoffrey
Optical physics.–2nd ed.

1. Optics
I. Title II. Lipson, Henry
535 QC355.2 80-40153
Second edition
ISBN 0 521 22630 9 hard covers
ISBN 0 521 29584 X paperback
(First edition ISBN 0 521 06926 2)

OPTICAL PHYSICS

To my mother
S.G.L.

To my wife
H.L.

Contents

Preface to the second edition

Since the writing of the first edition the subject of Optics, as studied in universities, has grown greatly both in popularity and scope, and both we and the publishers thought that the time had arrived for a new edition of *Optical Physics.*

In preparing the new edition we have made substantial changes in several directions. First, we have attempted to correct all the mistakes and misconceptions that have been pointed out to us during the nine years the book has been in use. Secondly, we have made one important change in the subject matter: we have absorbed the chapter on Quantum Optics into the rest of the book. During the years, there have appeared many books devoted to laser physics, and it now seems impracticable for a book on physical optics to cover the subject at all satisfactorily in one chapter. However, since some knowledge of the principles of the laser is necessary for the understanding of physical optics today, particularly when coherence is being discussed, we have covered what we feel to be the necessary minimum as parts of Chapters 7 and 8.

In addition to the above changes in the subject matter, we have increased the number of exercises offered to the reader, organized them according to chapter, and provided solutions. We have also included a few suggestions, based on our experience, for student projects illustrating the material in the book.

We are, of course, most grateful to all those who have pointed out to us errors and room for improvement. But in particular we must thank D. S. Tannhauser, I. Senitsky and M. Neugarten who have helped us considerably by reading and criticizing in detail parts of the revised manuscript. We are also extremely grateful to the staff of Cambridge University Press, who have contributed considerably to the elimination of faults and inaccuracies in our manuscript.

S. G. Lipson

May 1979 H. Lipson

Preface to the first edition

There are two sorts of textbooks. On the one hand, there are works of reference to which students can turn for the clarification of some obscure point or for the intimate details of some important experiment. On the other hand, there are explanatory books which deal mainly with principles and which help in the understanding of the first type.

We have tried to produce a textbook of the second sort. It deals essentially with the principles of optics, but wherever possible we have emphasized the relevance of these principles to other branches of physics – hence the rather unusual title. We have omitted descriptions of many of the classical experiments in optics – such as Foucault's determination of the velocity of light – because they are now dealt with excellently in most school textbooks. In addition, we have tried not to duplicate approaches, and since we think that the graphical approach to Fraunhofer interference and diffraction problems is entirely covered by the complex-wave approach, we have not introduced the former.

For these reasons, it will be seen that the book will not serve as an introductory textbook, but we hope that it will be useful to university students at all levels. The earlier chapters are reasonably elementary, and it is hoped that by the time those chapters which involve a knowledge of vector calculus and complex-number theory are reached, the student will have acquired the necessary mathematics.

The use of Fourier series is emphasized; in particular, the Fourier transform – which plays such an important part in so many branches of physics – is treated in considerable detail. In addition, we have given some prominence – both theoretical and experimental – to the operation of convolution, with which we think that every physicist should be conversant.

We would like to thank the considerable number of people who have helped to put this book into shape. Professor C. A. Taylor and Professor A. B. Pippard had considerable influence upon its final shape – perhaps more than they realize. Dr I. G. Edmunds and Mr T. Ashworth have read through the complete text, and it is thanks to them that the inconsistencies are not more numerous than they are. (We cannot believe that they are zero!) Dr G. L. Squires and Mr T. Blaney have given us some helpful advice about particular parts of the book. Mr F. Kirkman and his assistants – Mr A. Pennington and Mr R. McQuade – have shown exemplary patience in producing some of our more exacting photographic illustrations, and in providing beautifully finished prints for the press. Mr L. Spero gave us considerable help in putting the finishing touches to our manuscript.

And finally we should like to thank the three ladies who produced the final manuscript for the press – Miss M. Allen, Mrs E. Midgley and Mrs K. Beanland. They have shown extreme forbearance in tolerating our last-minute changes, and their ready help has done much to lighten our work.

S.G.L.
H.L.

1

History of ideas

1.1 Importance of history

Why should a textbook on physics begin with history? Why not start with what is known now and refrain from all the distractions of out-of-date material? These questions would be justifiable if physics were a complete and finished subject; only the final state would then matter and the process of arrival at this state would be irrelevant. But physics is not such a subject, and optics in particular is very much alive and constantly changing. It is important for the student to understand the past as a guide to the future. To study only the present is equivalent to trying to draw a graph with only one point.

Moreover, by studying the past we can sometimes gain some insight – however slight – into the minds and methods of the great physicists. No textbook can, of course, reconstruct completely the workings of these minds, but even to glimpse some of the difficulties that they overcame is worthwhile. What seemed great problems to them may seem trivial to us merely because we now have generations of experience to guide us; or, more likely, we have hidden them by cloaking them with words. For example, to the end of his life Newton found the idea of 'action at a distance' repugnant in spite of the great use that he made of it; we now accept it as natural, but have we come any nearer than Newton to understanding it? By being brought back occasionally to such fundamental problems the physicist is bound to have his wits sharpened; no amount of modern knowledge can produce the same effect. The way to study physics is to ask questions, as the geniuses of the past asked them. The ordinary physics student will find someone to answer them; the good physics student will answer them himself.

1.2 The nature of light

1.2.1 *The first ideas.* Some odd ideas about the nature of light were put forward by the ancients, who wanted the sense of sight to be somehow similar to the sense that they knew best – that of touch. But this idea did not make much headway, and it was not until Galileo (1564–1642) introduced the experimental method that progress really began. Galileo was the first scientist effectively to propagate the idea of testing theories by experiment and, as an example, he tried to measure the speed of light. He failed, but even to have thought of the concept was an intellectual triumph.

1.2.2 *The basic facts.* What was known about light in the seventeenth century? First of all, it travelled in straight lines. Secondly, it was reflected off smooth surfaces and the laws of reflexion were known. Thirdly, it changed direction when it passed from one medium to another (refraction); the laws for this phenomenon were not so obvious, but by the year 1600 they had been established by Snell (1591–1626) and were later confirmed by Descartes (1596–1650). Fourthly, what we now call Fresnel diffraction had been discovered by Grimaldi (1618–63) and by Hooke (1635–1703). Finally, double refraction had been discovered by Bartholinus (1625–98). It was on the basis of these phenomena that a theory of light had to be constructed.

The last two facts were particularly puzzling. Why did shadows reach a limiting sharpness as the size of the source became small, and why did fringes appear on the light side of the shadow (Fresnel diffraction)? And why did light passing through a crystal of Iceland spar produce two images when light passing through most other transparent materials produced only one?

1.2.3 *The wave–corpuscle controversy.* As usual in science when there is inadequate evidence, controversy resulted. Newton threw his authority on the theory that light is corpuscular, mainly because his first law of motion said that if no force acts on a particle it will travel in a straight line; he therefore postulated that light corpuscles are not acted upon by ordinary forces such as gravity. Double refraction he explained by some asymmetry in the corpuscles, so that their directions depended upon whether they passed through the crystal forwards or sideways. He envisaged the corpuscles as resembling magnets and the word 'polarization' is still used (Chapter 5) although this explanation has long been discarded.

Diffraction, however, was difficult. Newton realized its importance and carried out some crucial experiments in the subject; he showed that the fringes formed in red light were separated more than those formed in blue light. But when he found that the corpuscular theory could not be made to fit in, he weakly dropped his experiments, saying that he was rather busy!

Newton was also puzzled by the fact that light was partly transmitted and partly reflected by a glass surface; how could his corpuscles sometimes go through and sometimes be reflected? He answered this question by propounding the idea of 'fits of reflexion' and 'fits of transmission'; in a train of corpuscles some would go one way and some the other. He even worked out the lengths of these 'fits' (which came close to what we now know as half the wavelength). But the idea was very cumbersome and was not really satisfying.

His opponent, Huygens (1629–95) was a supporter of the wave theory. With it he could account for diffraction and for the behaviour of two sets of waves in a crystal, without explaining *how* the two sets arose. Both he and Newton had a common misconception – that light waves, if they existed, must be like sound waves, which are longitudinal. It is surprising that the two greatest minds in science should have had this blind spot. If they had thought of transverse waves, the difficulties of explaining double refraction would have disappeared.

1.2.4 *Triumph of wave theory.* Newton's authority kept the corpuscular theory going until the end of the eighteenth century, but by then ideas were coming forward that could not be suppressed. In 1801 Young (1773–1829) produced his double-slit fringes (Fig. 1.1) – an experiment

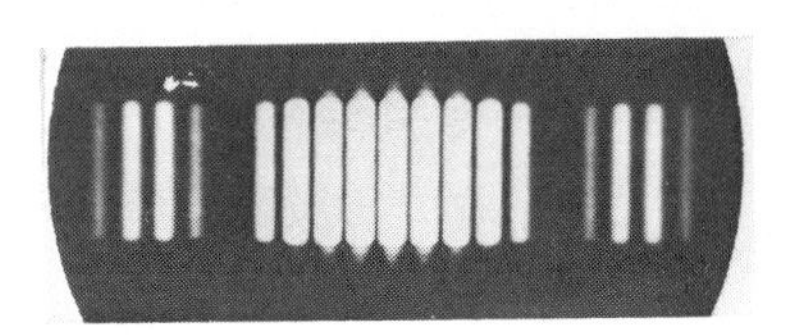

Fig. 1.1. Young's fringes.

so simple to carry out and so simple to interpret that the results were incontrovertible; in 1815 Fresnel (1788–1827) worked out the theory of the Grimaldi–Hooke fringes; and in 1821 Fraunhofer (1787–1826) produced diffraction patterns in parallel light which were much more amenable to theoretical treatment than were the Grimaldi–Hooke fringes. These three men laid the foundation of the wave theory that is still the basis of what is now called physical optics.

Nevertheless the corpuscularists were not quite defeated. In 1818 Poisson (1781–1840) produced an argument that seemed to invalidate the wave theory; he used the device of *reductio ad absurdum*. Suppose that a shadow of a perfectly round object is cast by a point source: at the periphery all the waves will be in phase, and therefore the waves should also be in phase at the centre of the shadow; there should therefore be a bright spot at this point. Absurd! Fresnel and Arago (1786–1853) carried out the experiment and found that there really *was* a bright spot at the centre (Fig. 1.2). Users of *reductio ad absurdum* should make sure of the absurdity of the result they are criticizing.

The triumph of the wave theory seemed complete.

Fig. 1.2. The bright spot at the centre of the shadow of a disc.

1.3 Speed of light

1.3.1 *Measurement.* The methods that Galileo used to measure the speed of light were far too crude to be successful. In 1678 Römer (1644–1710) realized that an anomaly in the times of successive eclipses of the moons of Jupiter could be accounted for by a finite speed of light, and deduced that it must be about 3×10^8 m s^{-1}. In 1726 Bradley (1693–1762) made the same deduction from observations of the small ellipses that the stars describe in the heavens; since these ellipses have a period of one year they must be associated with the movement of the Earth.

It was not, however, until 1850 that direct measurements were made, by Fizeau (1819–96) and Foucault (1819–68); their experiments are fully described in elementary textbooks, and their results confirmed those of Römer and Bradley. In the hands of Michelson (1852–1931) their methods achieved a high degree of accuracy – about 0.03 per cent. This, then, was one important problem completely disposed of.

1.3.2 *Refractive index.* Foucault's method was the more versatile in that it required a relatively short path, and the speed of light could be measured in media other than air – water, for example. This enabled another important result to be obtained. According to the corpuscular theory (§ 1.2.3) the speed of light should be greater in water than in air because the corpuscles must be attracted towards the water to account for the changed direction of the refracted light; according to the wave theory, the waves must travel more slowly in water and 'slew' round to give the new direction (Fig. 1.3). Foucault's method confirmed the wave theory completely, but of course gave no indication of the nature of the waves.

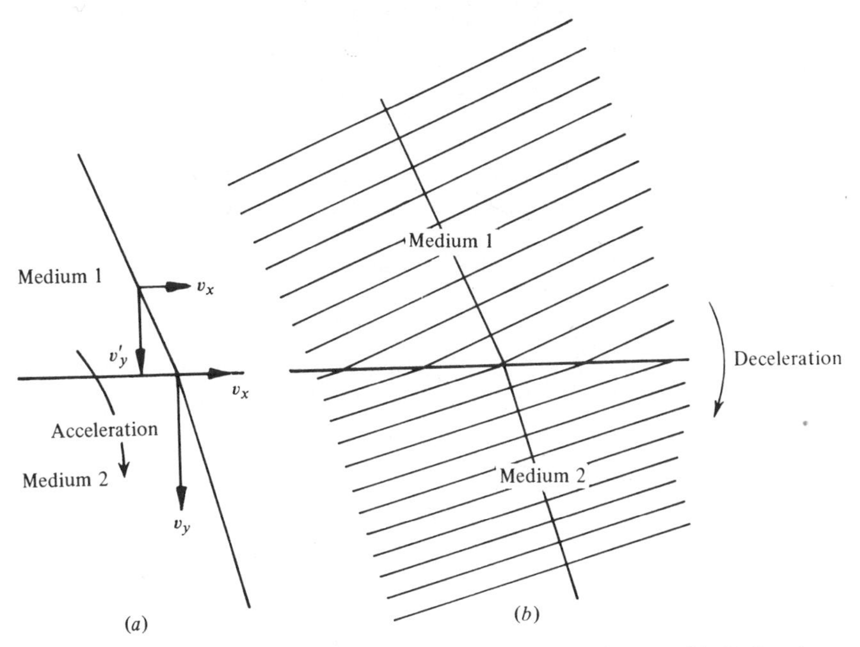

Fig. 1.3. (*a*) Refraction according to the corpuscular theory; (*b*) Refraction according to the wave theory.

1.4 Transverse or longitudinal waves?

1.4.1 *Polarization.* The distinction between transverse and longitudinal waves had been appreciated early in the history of physics; sound waves were found to be longitudinal and water waves were obviously transverse.

The phenomenon that enabled a decision to be made was that of double refraction in Iceland spar (§ 1.2.2). Huygens pointed out that this property, which is illustrated in Fig. 1.4, must mean that the orientation of the crystal must be related somehow to some direction in the wave.

Fig. 1.4. Double refraction in an Iceland spar crystal.

We now know that this is the correct solution. We shall discuss the results of these ideas in more detail in Chapter 4.

1.4.2 *Nature of light.* These experiments, of course, tell us nothing about the nature of light; they are all concerned with its behaviour. The greatest step towards understanding came from a completely different direction – the theoretical study of magnetism and electricity.

In the first half of the nineteenth century the relationship between magnetism and electricity had been worked out fairly thoroughly, by men such as Oersted (1777–1851), Ampère (1775–1836) and Faraday (1791–1867). In 1864 Clerk Maxwell (1831–79) was inspired to put their results in mathematical form, and in manipulating the equations he found that they could assume the form of a wave equation. The velocity of the wave could be derived from the known magnetic and electric constants. Evaluation of this velocity showed that it was equal to the velocity of light, and thus light was established as an electromagnetic disturbance. This was one of the most brilliant episodes in physics, bringing together several different branches of physics and showing their relationship to each other.

1.5 Instruments

1.5.1 *The telescope.* Although single lenses had been known from time immemorial, it was not until the beginning of the seventeenth century that optical instruments as we know them came into being. Hans Lippershey (d. 1619) in 1608 discovered, probably accidentally, that two separated lenses could produce a clear enlarged image of a distant object.

Galileo seized upon the discovery, made his own telesope, and began to make a series of discoveries – such as Jupiter's moons and Saturn's rings – that completely altered the subject of astronomy. Newton, dissatisfied with the colour defects in the image, invented the reflecting telescope (Fig. 1.5). Since then the telescope has not changed in essence, but it has changed in size and cost.

Fig. 1.5. Newton's reflecting telescope. (From Lodge, 1920.)

1.5.2 *The microscope.* The story of the microscope is quite different. Its origin is uncertain; many people contributed to its early development; new ways of using it are still being found; and it is possible that further developments may still be forthcoming.

The microscope originated from the magnifying glass. In the sixteenth and seventeenth centuries considerable ingenuity was exercised in making high-powered lenses; a drop of water or honey could produce wonderful results in the hands of an enthusiast. Hooke (1635–1703) played perhaps the greatest part in developing the compound microscope, and some of his instruments (Fig. 1.6) already showed signs of future trends in design. One can imagine the delight of such an able experimenter in having the privilege of developing a new instrument and of using it to examine the world of the very small, as described in his *Micrographia* (1665).

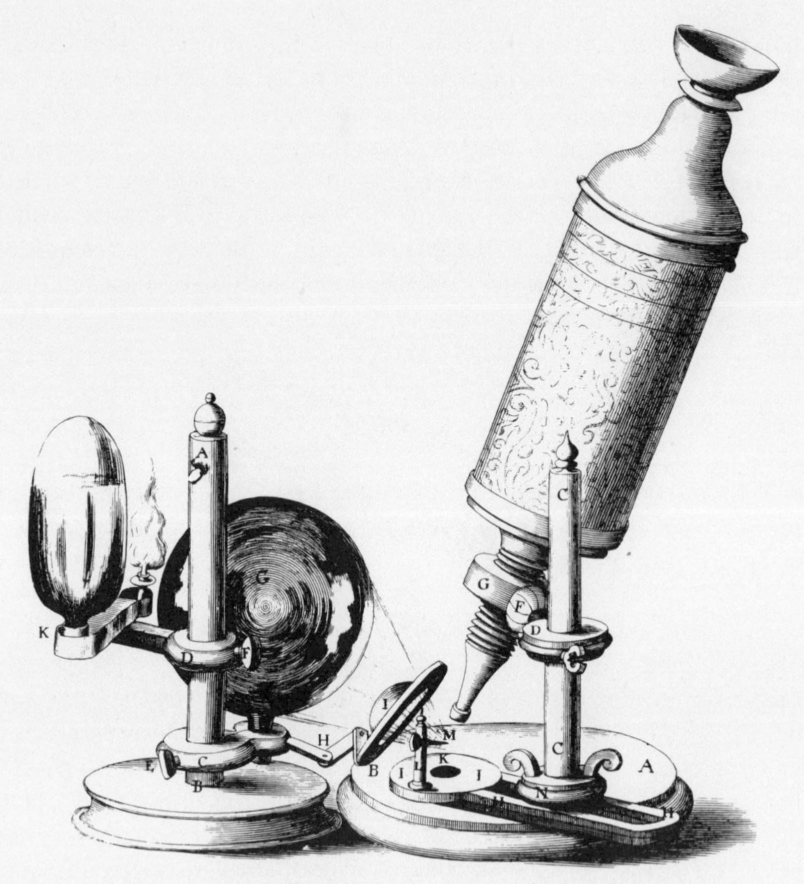

Fig. 1.6. Hooke's microscope, from his *Micrographia*.

1.5.3 *The spectrometer and diffraction grating.* Once the principles of the subject were known, instruments other than the obvious telescope and microscope arose. The most important were the spectrometer, which Fraunhofer invented for precise measurement of the optical properties of glass, and the diffraction grating. The invention of the latter was apparently accidental; Fraunhofer was using a spectrometer to look at the diffraction pattern of a slit and was dissatisfied with the intensity. He therefore conceived the idea of increasing the intensity by using a number of slits. The intensity certainly increased, but so also did the number of patterns. Dissatisfaction can be a powerful ally to a physicist!

The diffraction grating, allied to the spectrometer, has proved to be one of the most useful tools in physics. It added, as it were, a new dimension to astronomy, for example. Lives have been devoted to making better and larger gratings, and their uses are considerable. Again, a chance discovery opened up new avenues in physics.

1.5.4 *Geometrical optics.* In order to put the theory of optical instruments on a sound basis, a branch of optics – called geometrical optics – was founded. It was based entirely on the concept of rays of light, which travelled in straight lines in uniform media and bent abruptly on passing from one medium to another, obeying Snell's law.

Based upon geometrical optics, rules for improving the performances of the microscope and telescope were found. Lenses were improved by the skilful shaping of surfaces and by combining lenses so that their errors cancelled. Newton knew how to correct for chromatic aberration, but glasses were not available to put the correction into effect; Huygens devised a compound eyepiece that is still named after him. Progress seemed to be limited only by the ingenuity of designers and the skill of craftsmen. The end of the story will be told in Chapter 9.

1.6 Quantum theory

1.6.1 *The origins.* With the marriage of geometrical optics and wave theory (physical optics) it seemed, up to the end of the last century, that no further rules about the behaviour of light were necessary. Nevertheless there were some basic problems, as the study of the light emitted by hot bodies indicated. Why do such bodies become red-hot at about 600 °C and become whiter as the temperature increases? The great physicists such as Kelvin (1824–1907) were well aware of this problem, but it was not until 1900 that Planck (1858–1947) put forward, very tentatively, an *ad hoc* solution, now known as the quantum theory.

The idea, which we shall deal with in more detail in Chapter 8, is that wave energy is divided into packets (quanta) whose energy content is proportional to the frequency; the lower frequencies, such as those of red light, are then more easily produced than higher frequencies. The idea was not liked, but gradually scepticism was overcome as more and more experimental evidence in its favour was produced. By about 1920 it was generally accepted, largely on the basis of Einstein's (1879–1955) study of the photoelectric effect in 1905 and of Compton's (1892–1962) discovery of the Compton effect in the scattering of X-rays in 1923.

1.6.2 *Wave–particle duality.* But the problem had not really been solved. We still cannot conceive the equivalence of waves and particles of energy. The energy of a wave is distributed through space; the energy of a particle is concentrated. Perhaps one of the closest analogies is that of the surf rider carried along and guided by a wave. If he had to pass through a narrow opening he would either miss it and be completely stopped or he would go completely through. The wave guiding him, however, would be curtailed and could no longer guide him with the same accuracy; this is equivalent to diffraction.

But no physical picture is quite adequate; we can accept Planck's and Einstein's ideas and we can use them, but we do not understand them. This is not a depressing state of affairs; physics has developed in this way for three centuries, the difficulties of one generation being the stepping stones for the next. Perhaps someone will one day find a convincing physical picture – for those of us who like physical pictures – or perhaps we shall just accustom ourselves to the idea of the equivalence of waves and quanta, as we have accustomed ourselves to action at a distance. For those who do not need physical pictures, there is already no trouble.

1.6.3 *Corpuscular waves.* This history, however, is still not completely written; as usual in physics one idea leads to another and in 1924 a new idea occurred to de Broglie (born 1892), based upon the principle of symmetry. Faraday had used this principle in his discovery of electromagnetism; if electricity produces magnetism, does magnetism produce electricity? De Broglie asked, 'If waves are corpuscles, are corpuscles waves?' By now physicists had learnt not to be sceptical, and within three years his question had been answered. Davisson (1881–1958) and Germer (born 1896) by ionization methods and G. P. Thomson (1892–1975) by photographic methods, showed that fast-moving electrons could be diffracted by matter similarly to X-rays. Since then other particles such as neutrons and protons have also been diffracted. Schroedinger (1887–1961) in 1928 produced a general wave theory of matter, and the subject of wave mechanics is now a basic part of physics.

1.7 Resolving power

1.7.1 *Wave limitations.* The view that progress in optical instruments depended only upon the skill of their makers was suddenly brought to an end by Abbe (1840–1905) in 1873. He showed that the geometrical theory – useful though it was in developing optical instruments – was incomplete in that it took no account of the wave properties of light. Geometrically, the main condition that is necessary to produce a perfect image is that the rays from any point in the object should be so refracted that they meet together at a point on the image. Abbe showed that this condition is necessarily only approximate; waves spread because of diffraction and so cannot intersect in a point.

He put forward another view of image formation – that an image is formed by two processes of diffraction. The first occurs when light is scattered by the object; the second occurs when the scattered beams are brought together again and combine to form an image. Limitations occur at both stages – at the first because light is diffracted through a finite angle (less than 180°) and at the second because only a finite part of the diffracted light passes through the instrument. The first is the fundamental limitation; in fact one cannot conceive of making use of radiation diffracted through more than 90°. The second is a practical limitation and places emphasis on making instruments which can accept as wide a cone of light as possible.

Abbe showed that one cannot resolve detail less than about half a wavelength, even with a perfectly corrected instrument. This simple result was greeted by microscopists with disbelief; many of them observed detail less than this with good rigidly-mounted instruments. Abbe's theory, however, proves that such detail is erroneous; it is a function of the instrument rather than of the object. Improving lenses further is not the right way to improve microscopes.

1.7.2 *Resolving-power challenge.* Any conclusion of this sort must not be considered as depressing; it must be regarded as a challenge. Until difficulties are clearly exposed no real progress is possible. Now that it was known where the limitations of optical instruments lay, it was possible to concentrate upon *them* rather than upon lens design. Since resolving power is a function of wavelength, we need to consider new radiations with shorter wavelengths rather than new lens combinations.

Ultra-violet light is an obvious choice, but the experimental difficulties are too great to justify a gain of perhaps a factor of two. The radiations that *have* been effective are X-rays and electron waves; these have wavelengths about 10^{-3} or 10^{-4} of those of visible light. They have produced such revolutionary results that separate sections need to be devoted to them.

1.7.3 *X-ray diffraction.* X-rays were discovered in 1895, but for seventeen years no one knew whether they were particles or waves. Then, in 1912, a brilliant idea of von Laue (1879–1960) solved the problem; he envisaged the possibility of using a crystal as a (three-dimensional) diffraction grating and the experiment of passing a fine beam of X-rays on

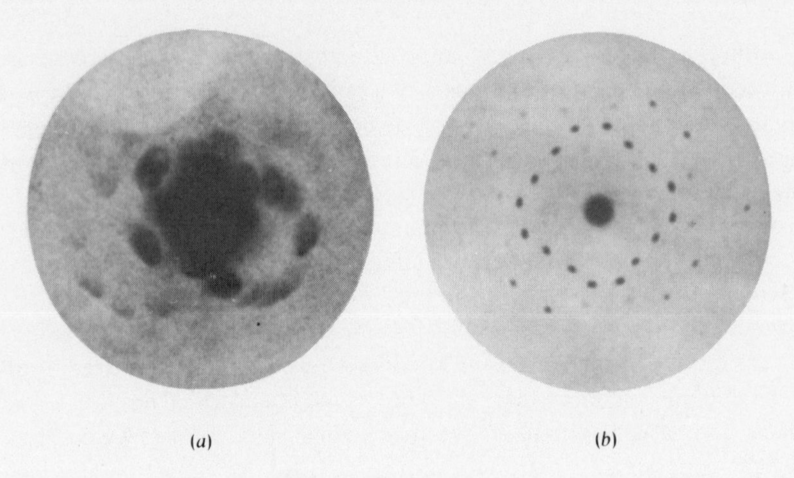

Fig. 1.7. X-ray diffraction patterns produced by a crystal. (From Ewald, 1962.)

to a crystal of $CuSO_45H_2O$ (Fig. 1.7(*a*)) showed definite indications of diffraction. A more symmetric crystal gave still more positive results (Fig. 1.7(*b*)). In addition to solving the problem, a new subject – X-ray crystallography – was born.

1.7.4 *Electron microscopy.* The realization that moving particles also have wave properties (§ 1.6.3) introduced a new factor into the subject. If such particles are charged they can be deflected electrostatically or magnetically, and so refraction can be simulated. It was found theoretically that suitably shaped fields could produce focused beams so that image formation was possible.

Electrons have been used with much success for this work, and instruments with magnetic (or more rarely electrostatic) 'lenses' are available for producing images with very high magnifications; such instruments are called electron microscopes.

By using accelerating voltages of the order of 100 kV, wavelengths of the order of 0.1 Å can be produced, and thus a limit of resolution far better than that with X-rays should be obtainable. In practice, however, electron lenses are quite crude by optical standards and thus only small apertures are possible; the electron microscope developed rapidly in the 1930s, and the limit of resolution soon reached and then surpassed that of the light microscope. But now a limit of about 10 Å has been reached and does not seem likely to be easily surpassed.

This is a remarkable improvement on light microscopy. It has opened up new fields in many branches of science; biology, for example, has been revolutionized. For physicists the limit of resolution is tantalizingly low; the electron microscope will show molecular detail (Fig. 1.8), but it will not reach down to atoms. If the limit of resolution could be taken down by another order of magnitude an enormous range of new information would become available.

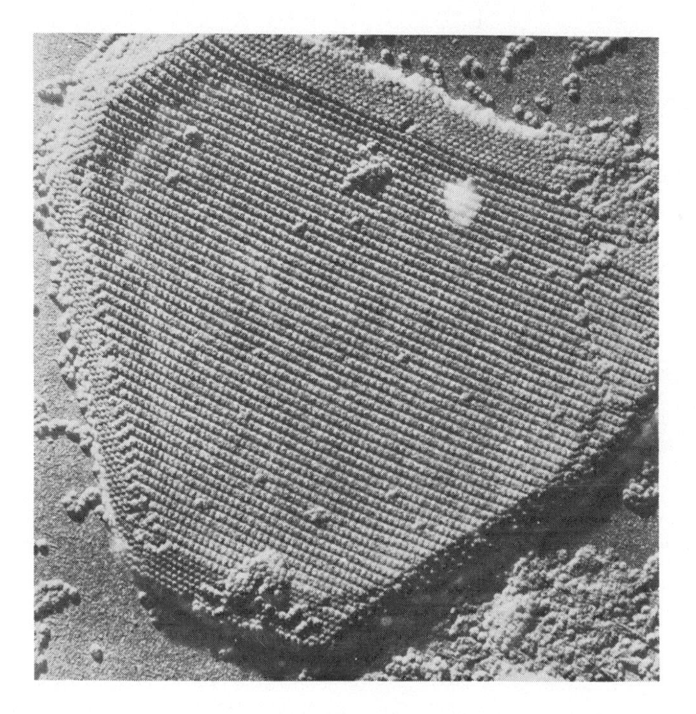

Fig. 1.8. Structure of a crystal. (From Lipson, 1962.)

1.8 Recent developments

This history has taken us to about the 1930s. At this stage there was a lull in the subject and some physicists began to say that no more advances could be expected, that it was now possible to regard optics as a closed book, and that it was hardly worth teaching. This attitude was, of course, mistaken; even if optics were finished, it would still form a necessary part of physics teaching for it permeates the whole of the rest of the subject.

One man largely kept the optics flag flying. This was Fritz Zernike (1888–1966), of the Netherlands, who chose to apply his talents to this subject in spite of the blandishments of other branches which appeared to be more exciting. He appreciated the importance of the concept of coherence and showed that there was much more to it than the two extremes of complete coherence and incoherence. He introduced the idea of partial coherence of a single beam of light. He invented the phase-contrast microscope, a device that revitalized biology by eliminating the need for staining and for which he was awarded the Nobel Prize in 1953.

It seemed, however, that perfect coherence – like any other perfection in physics – would remain an unattainable idea, until Townes, of the USA, and Basov and Prokharov, of the USSR, invented the laser, a device for which they were awarded the Nobel Prize in 1964. They managed to arrange that, in effect, all the atoms in a system should vibrate in phase and so produce a completely coherent beam. Clearly, such a beam can be coherent in only one direction and so the laser produces a finely directed beam in which all the energy is concentrated; even with powers as small as 1 mW an extremely intense beam is produced.

With the practical invention of the laser and the theoretical concept of coherence, new optical procedures became possible. Chief amongst these was holography whereby a diffraction pattern can be recorded photographically both in amplitude and phase; from this pattern an image can be reproduced three-dimensionally. This idea had been suggested by Gabor in 1940 for improving images obtained by the electron microscope; when its importance was realized he was awarded the Nobel Prize in 1971. Holography has turned out to be of great value in applied physics; for example, vibrations of objects can be followed in detail by forming holograms at different instants of time (§ 9.7.3).

Moreover, the intense beam of the laser enabled new phenomena to be observed. Previously, in a beam of light the electric and magnetic fields were extremely weak and could produce only linear effects on any atom on which they fell. But now, with extremely powerful lasers, non-linear effects could be produced and a new branch of the subject opened up – non-linear optics. The study of matter is bound to be affected by this new discovery.

All this activity tends to support our thesis that no part of physics can be written off. Students must learn what has gone on in the past so that they can be ready for new ideas when they arise in the future.

1.9 Summary

In this chapter we have tried to trace the development of optics from the first hesitant concepts to the very considerable body of knowledge that the subject now comprises. In addition to giving the reader an overall view of the way optics has advanced, we have also laid down a pattern for the succeeding chapters.

We start with waves in general (Chapter 2) and the extremely important treatment in terms of Fourier analysis (Chapter 3). We then deal with electromagnetic waves in particular (Chapter 4), and the effects – of which polarization is of outstanding importance – produced when they pass through anisotropic media (Chapter 5). The interaction of waves leads to the phenomena of diffraction and interference (Chapters 6 and 7). We continue with a discussion of coherence (Chapter 8) and on the basis of these chapters we are able to deal in detail with what is the most important application of optics – image formation (Chapter 9). We then consider scattering and dispersion in terms of classical theory (Chapter 10) and in Chapter 11 we conclude by taking a selection of topics and treating them in much more detail in terms of the concepts laid down in the intervening parts of the book.

The appendices contain matter – very briefly outlined – that is necessary for the understanding of the various chapters, but which we felt would interrupt the flow of ideas if it were inserted in appropriate places in the text.

We have included a number of questions that are designed to test the understanding of the textual matter, rather than to serve as examples of examination-type questions.

2

Waves

2.1 Introduction

Optics, as discussed in this book, is the study of wave propagation, and wave propagation is described mathematically in terms of wave equations. The exact definition of a *wave equation* is somewhat subject to personal taste, but for the purposes of this book it is defined as any equation which leads to a harmonic wave as a solution. This wave may be dispersive or non-dispersive; it may be progressive, standing, attenuated or evanescent. Because of the variety of waves which can be involved, there is accordingly a large variety of wave equations. This chapter is devoted to the study of such equations, and how they arise as the result of the properties of media.

2.2 Non-stationary disturbances

To see the origin of a wave in a way which will lead to the understanding of wave equations, let us consider a simple experiment. We have a large stretched sheet of rubber and we press down on it at a certain point. An equilibrium state is reached, in which the elastic forces at the pressure-point exactly balance the applied force, and at all neighbouring points they exactly cancel out. Now remove the applied pressure. Instantaneously only the pressure-point is not in equilibrium; all the neighbouring points are unchanged. But as the rubber at the pressure-point moves to restore equilibrium, that at the neighbouring points is disturbed, and as a result they in turn move, and so the disturbance propagates outwards. To produce a quantitative theory of such an effect we therefore need to consider

(a) the equations governing the setting-up of an equilibrium state under the application of a steady applied force, and

(b) the equation of motion at a point in the medium when it is not in equilibrium.

2.2.1 *The form of wave equations.* Thinking in one dimension, we might expect that a wave should be basically a disturbance of some sort which can travel with a fixed velocity and be unchanged in form from point to point. In mathematical terms, the waveform at time t and position x is the same as that at time 0 and position $x-vt$, v being the velocity:

$$f(x,t)=f(x-vt,0)=f(\xi_-,0). \tag{2.1}$$

Alternatively, the wave can travel in the opposite direction with the same velocity giving

$$f(x,t)=f(x+vt,0)=f(\xi_+,0),$$

where

$$\xi_\pm = x\pm vt. \tag{2.2}$$

Differentiating (2.1) by x and t respectively, we have

$$\frac{\partial f}{\partial x}=\frac{\mathrm{d}f}{\mathrm{d}\xi_-}; \qquad \frac{\partial f}{\partial t}=-v\frac{\mathrm{d}f}{\mathrm{d}\xi_-}, \tag{2.3}$$

and for (2.2)

$$\frac{\partial f}{\partial x}=\frac{\mathrm{d}f}{\mathrm{d}\xi_+}; \qquad \frac{\partial f}{\partial t}=v\frac{\mathrm{d}f}{\mathrm{d}\xi_+}. \tag{2.4}$$

Equations (2.3) and (2.4) can be reconciled to a single equation by a second similar differentiation followed by eliminating f'' between the pairs; either equation gives

$$\frac{\partial^2 f}{\partial x^2}=\frac{\mathrm{d}^2 f}{\mathrm{d}\xi^2}; \qquad \frac{\partial^2 f}{\partial t^2}=v^2\frac{\mathrm{d}^2 f}{\mathrm{d}\xi^2},$$

whence

$$\frac{\partial^2 f}{\partial x^2}=\frac{1}{v^2}\frac{\partial^2 f}{\partial t^2}, \tag{2.5}$$

of which equations (2.1) and (2.2) are the most general solutions. Equation (2.5) is known as the *non-dispersive wave equation,* because the solutions imply that the form of the wave does not alter as it progresses. This property, as the examples which follow in later sections will show, is not common to all waves; in practice it is quite rare.

Although equation (2.5) has a general solution given by (2.1) and (2.2), there is a particular solution to it which is more important because it is general for a large class of equations known as *wave equations.* This solution is a simple-harmonic wave of amplitude A in its complex exponential form:

$$f(x,t)=A\exp\left\{2\pi\mathrm{i}\left(vt-\frac{x}{\lambda}\right)\right\},$$

where ν is the frequency in cycles per unit time and λ is the wavelength. A more tidy expression can be written in terms of

the wave number $k = 2\pi/\lambda$,

the angular frequency $\omega = 2\pi\nu$,

which give

$$f(x, t) = A \exp\{i(\omega t - kx)\}. \tag{2.6}$$

It is easy to verify that this waveform satisfies equation (2.1), provided that the velocity is given by $v = \omega/k$. The importance of the waveform (2.6) is that it is by definition a solution of all wave equations; if a disturbance begins simple-harmonically in a medium satisfying such an equation, it will continue simple-harmonically and will not be distorted.

We shall define a general wave equation as that which has equation (2.6) as its solution. It can be produced by differentiating (2.6); we have in general:

$$\frac{\partial^n f}{\partial x^n} = (-ik)^n A \exp\{i(\omega t - kx)\} = (-ik)^n f \tag{2.7}$$

and

$$\frac{\partial^n f}{\partial t^n} = (i\omega)^n A \exp\{i(\omega t - kx)\} = (i\omega)^n f. \tag{2.8}$$

Notice that the use of the complex exponential in equation (2.6) has avoided the untidy alternation between cosine and sine which is involved if a cosine or sine function is used. The function f in equations (2.7) and (2.8) can always be eliminated; therefore, if there exists a polynomial relation between differentials of f, we can use the operator $(\partial/\partial x)^n$ to replace $\partial^n/\partial x^n$ (and similarly for t), when the general wave equation becomes:

$$g\left(\frac{\partial}{\partial x}, \frac{\partial}{\partial t}\right) f = 0, \tag{2.9}$$

where g is a polynomial function of the operators $(\partial/\partial x)$ and $(\partial/\partial t)$. For example, if

$$g = \left(\frac{\partial}{\partial x}\right)^2 - \frac{1}{v^2}\left(\frac{\partial}{\partial t}\right)^2$$

the equation becomes

$$\frac{\partial^2 f}{\partial x^2} - \frac{1}{v^2}\frac{\partial^2 f}{\partial t^2} = 0,$$

which is the non-dispersive equation (2.5).

Once the wave equation has been decided from the physics of the problem, we can immediately work out the dispersive properties of the wave – i.e. the relationship between the angular frequency ω, the wave number k, and their quotient, the wave velocity ω/k. This is simply done. From equation (2.7) it will immediately be clear that the operation $\partial/\partial x$ is exactly equivalent to multiplying by $(-\mathrm{i}k)$; equation (2.8) shows that $\partial/\partial t$ is equivalent to $\mathrm{i}\omega$. Thus an equation relating ω and k can be obtained from the wave equation by these substitutions. This is called a *dispersion equation*; for example, equation (2.5) gives:

$$(-\mathrm{i}k)^2 f = v^{-2}(\mathrm{i}\omega)^2 f$$

or

$$\frac{\omega^2}{k^2} = v^2; \qquad \frac{\omega}{k} = \pm v. \tag{2.10}$$

The velocity is $\pm v$, as expected. Thus we can see that any medium whose properties can be expressed by a polynomial equation in time and space derivatives

$$g\left(\frac{\partial}{\partial x}, \frac{\partial}{\partial t}\right) f = 0$$

must always have a solution of the form

$$f = A \exp\{\mathrm{i}(\omega t - kx)\}.$$

The relationship between ω and k is entirely determined by the polynomial, which contains the physics of the problem.

2.2.2 *Examples of wave equations.* The more common forms of wave equation should already be familiar to the reader. As examples we shall quote the following.

(a) Mechanical waves: for example, longitudinal sound waves in a compressible fluid. If the fluid has compressibility K and density ρ, the equilibrium-state equation ((a) in § 2.2) is Hooke's law:

$$P = K\frac{\partial \eta}{\partial x}, \tag{2.11}$$

where P is the stress, the local pressure, and η the local displacement from equilibrium. The differential $\partial\eta/\partial x$ is thus the strain. The dynamic equation ((b) in § 2.2) relates the deviation from the equilibrium state (constant P) to the local acceleration:

$$\rho\frac{\partial^2 \eta}{\partial t^2} = \frac{\partial P}{\partial x}. \tag{2.12}$$

Equations (2.11) and (2.12) lead to a wave equation

$$\frac{\partial^2 \eta}{\partial x^2} = \frac{\rho}{K} \frac{\partial^2 \eta}{\partial t^2}. \tag{2.13}$$

Thus the waves are non-dispersive, with wave velocity

$$v = \left(\frac{K}{\rho}\right)^{\frac{1}{2}}.$$

The relationship between ω and k is (from (2.10))

$$\omega = \pm vk. \tag{2.14}$$

(b) The diffusion equation. Heat diffuses under steady-state conditions according to the equation

$$q = -\kappa \frac{\partial \theta}{\partial x}, \tag{2.15}$$

where q is the heat flow per unit area, κ the thermal conductivity and θ the local temperature. The dynamic equation represents the rate of rise of temperature when heat flows into a region of specific heat s per unit volume:

$$s \frac{\partial \theta}{\partial t} = -\frac{\partial q}{\partial x}, \tag{2.16}$$

whence the wave equation:

$$\frac{\partial \theta}{\partial t} = \frac{\kappa}{s} \frac{\partial^2 \theta}{\partial x^2} \tag{2.17}$$

and the dispersion relation

$$\omega = \mathrm{i} \frac{\kappa}{s} k^2. \tag{2.18}$$

(c) The Schroedinger wave equation. One of the most important wave equations in the history of modern physics has been derived essentially from a knowledge of its dispersion relation. For a moving particle, we know that the sum of its kinetic energy T and potential energy V is equal to a constant total energy E:

$$E = T + V. \tag{2.19}$$

Now the kinetic energy can be expressed in terms of the momentum p and mass m as

$$T = \frac{p^2}{2m} \tag{2.20}$$

and thus

$$E = \frac{p^2}{2m} + V. \tag{2.21}$$

We now resort to two hypotheses concerned with the diffraction of particle beams and the quantization of the energy in light waves. The first of these is due to de Broglie (§ 1.6.3), who suggested that an electron has a wavelength related to its momentum by Planck's constant h:

$$\lambda = hp^{-1}, \tag{2.22}$$

which can be written:

$$p = \hbar k, \tag{2.23}$$

where $\hbar$ is $h/2\pi$. The second hypothesis was invoked by Planck to explain the absence of the so-called ultra-violet catastrophe in radiation from a hot body (§ 1.6.1). He found it possible to explain the observed spectrum of a hot black body only by assuming that the total energy of radiation is quantized in units of $h\nu$, or equivalently $\hbar\omega$. Thus, extending the result to all particles,

$$E = \hbar\omega. \tag{2.24}$$

Substituting equation (2.23) and (2.24) in (2.21) we have

$$\hbar\omega = \frac{\hbar^2 k^2}{2m} + V, \tag{2.25}$$

which is a dispersion relation, and arises from the one-dimensional wave equation

$$\mathrm{i}\hbar \frac{\partial \psi}{\partial t} = -\frac{\hbar^2}{2m} \frac{\partial^2 \psi}{\partial x^2} + V\psi. \tag{2.26}$$

This is Schroedinger's equation.

2.3 Complex quantities: attenuation

2.3.1 *Complex velocity of a wave.* The reader may have noticed that the dispersion relations (2.18) and (2.25) sometimes give rise to imaginary or complex velocities. Let us see what a complex velocity might mean by substituting it in equation (2.6). As the velocity is ω/k we can replace ω by vk, and since the complex velocity is

$$v = v_1 + \mathrm{i}v_2 \quad (v_1 \text{ and } v_2 \text{ both real}), \tag{2.27}$$

$$\begin{aligned} f &= A \exp\{\mathrm{i}(\omega t - kx)\} \\ &= A \exp\{\mathrm{i}k(v_1 t + \mathrm{i}v_2 t - x)\} \\ &= A \exp\{(-v_2 kt)\} \exp\{\mathrm{i}k(v_1 t - x)\}. \end{aligned} \tag{2.28}$$

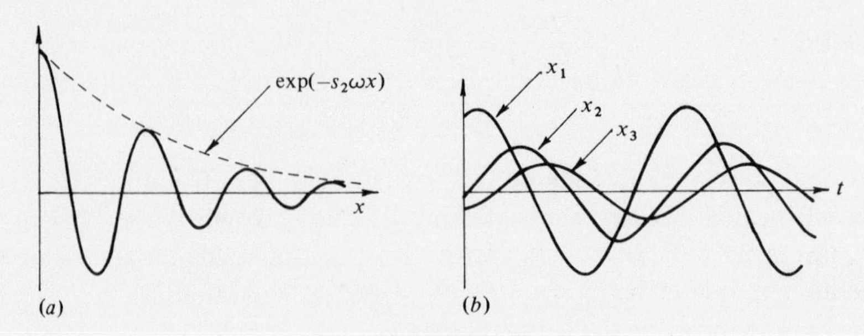

Fig. 2.1. An attenuated wave (k a complex function of ω): (a) as a function of distance at a fixed instant; (b) as a function of time at three positions $x_1 < x_2 < x_3$.

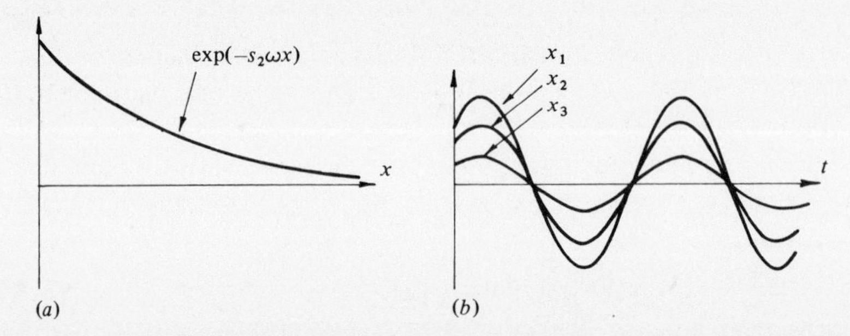

Fig. 2.2. An evanescent wave (k an imaginary function of ω): (a) as a function of distance at a given instant; (b) as a function of time at three positions $x_1 < x_2 < x_3$.

In other words, the real part v_1 represents a true wave velocity – i.e. a phase progression with distance – whereas the imaginary part v_2 implies attenuation – a decay with time. It is more common to make ω real, when we have

$$f = A \exp(-s_2 \omega x) \exp\{i\omega(t - s_1 x)\}, \tag{2.29}$$

where we define

$$s = s_1 - is_2 = v^{-1} = k/\omega. \tag{2.30}$$

A wave emitted at $x = 0$ with frequency ω and amplitude A is thus attenuated to amplitude $A \exp(-s_2 \omega x)$ at a distance x away from the source. The complex velocity thus has a practical meaning (Fig. 2.1).

2.3.2 *Imaginary velocity.* Sometimes the velocity is calculated to be purely imaginary. The effect of this is to make s_1 or v_1 zero in equations (2.27) and (2.30). In the latter expression the wave clearly has no harmonic space-dependence at all; it is a purely exponential function of x, but still oscillating in time with frequency ω (Fig. 2.2). It is then called an *evanescent wave.*

2.3.3 *The diffusion equation.* The diffusion equation leads to attenuated and evanescent waves and will be used as an example. We had (equation (2.17))

$$\frac{\partial \theta}{\partial t} = D\frac{\partial^2 \theta}{\partial x^2} \qquad \left(D = \frac{\kappa}{s}\right),$$

for the evolution of the temperature distribution in a one-dimensional medium – say a straight bar. The $\omega(k)$ dispersion relation is

$$\mathrm{i}\omega = -Dk^2$$

giving

$$k = \left(\frac{\omega}{2D}\right)^{\frac{1}{2}}(1-\mathrm{i}). \tag{2.31}$$

If one end of the bar is subjected to alternating heating so that its temperature θ varies as

$$\theta = \theta_0 \exp(\mathrm{i}\omega t) \qquad (\omega \text{ real}),$$

the wave is propagated along the bar in the form

$$\begin{aligned}\theta(x, t) &= \theta_0 \exp\left[\mathrm{i}\left\{\omega t - \left(\frac{\omega}{2D}\right)^{\frac{1}{2}}(1-\mathrm{i})x\right\}\right]\\ &= \theta_0 \exp\left\{-\left(\frac{\omega}{2D}\right)^{\frac{1}{2}}x\right\}\exp\left[\mathrm{i}\left\{\omega t - \left(\frac{\omega}{2D}\right)^{\frac{1}{2}}x\right\}\right],\end{aligned} \tag{2.32}$$

which is attenuated in distance along the bar with characteristic decay distance $(2D/\omega)^{\frac{1}{2}}$. This quantity is the distance in which θ decays by a factor e^{-1}. The propagated disturbance is still a wave, however; the phase of oscillation progresses regularly along the bar with a velocity $(2D\omega)^{\frac{1}{2}}$. Now suppose that the same bar has an initial temperature distribution

$$\theta = \theta_0 \exp(-\mathrm{i}kx) \qquad (k \text{ real})$$

impressed upon it at time $t = 0$, and the temperature distribution is left to its own devices. We should expect, intuitively, that heat would flow from hot to cold regions until the uneven distribution disappeared; we should not expect the temperature at any point to overshoot and oscillate, but simply to decay to its final value. To show that intuition is correct, we write down the subsequent temperature distribution:

$$\begin{aligned}\theta(x, t) &= \theta_0 \exp\{\mathrm{i}(\omega t - kx)\}\\ &= \theta_0 \exp\{-(Dk^2 t)\}\exp(-\mathrm{i}kx) \qquad \text{(from (2.32))}.\end{aligned}$$

There is no oscillatory time-dependence; the phase of the distribution $\exp(-\mathrm{i}kx)$ remains unchanged, but its amplitude decays to zero with time-constant $(Dk^2)^{-1}$. This is a wave evanescent in time. Thus the heat-diffusion equation illustrates both types of behaviour; it supports a wave attenuated in distance for a real frequency, or evanescent in time for a real wavelength.

To sum up, we have discovered three types of solution to wave equations, all of which can be represented by velocities of various types:

(a) real velocity: the wave is propagated at constant amplitude,
(b) complex velocity: the wave is attenuated, either in distance or time,
(c) imaginary velocity: the wave is evanescent.

2.4 Group velocity

The non-dispersive wave equation (2.5) is unique in that it leads to the propagation of disturbances of all frequencies with the same velocity, and as a result any wave travels unchanged. In general, provided that the dispersion is not very large, a disturbance which is very like a wave of a single frequency ω_0 will travel with little distortion, but the velocity at which it moves is not the same as the velocity ω_0/k_0. Such a disturbance is called a *wave-group* (Fig. 2.3) and is discussed fully in Chapter 8; for the

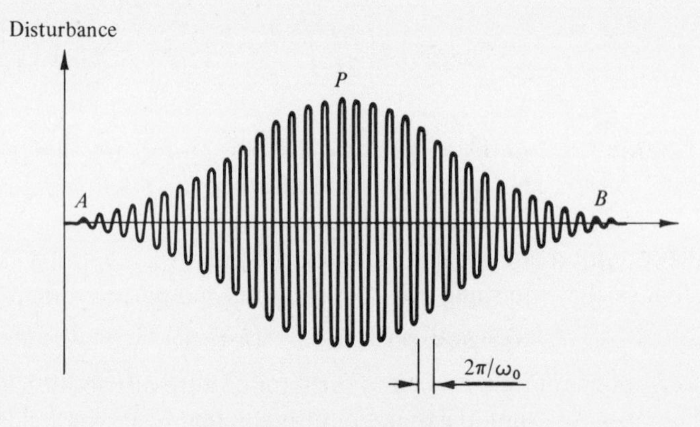

Fig. 2.3. A wave-group.

moment we need just accept the fact that it can be built up from waves of frequency close to ω_0. The obvious property we shall use, however, is that the position of the group is defined not by the individual waves in it, but by the envelope, which has its maximum point at *P*, despite the fact that the amplitude of the wave may be zero there at a particular moment. The *group velocity* is the velocity at which the envelope moves. The simplest and most usual method of deriving the group velocity considers the periodic wave-groups or beats produced by two waves of almost equal frequency. As this method is not obviously extendable to a single group consisting of waves of many close frequencies, we shall use a more general one. We assume that the maximum in the envelope occurs at the point where all the components add together in phase; in other words, that the variation of the argument $(\omega t - kx) \equiv \phi$ with frequency is zero:

$$\frac{\partial \phi}{\partial \omega} = t - x\frac{\partial k}{\partial \omega} = 0. \tag{2.33}$$

The group velocity v_g is the relationship between the positions x and time t for which this is satisfied, i.e.

$$v_g = \frac{x}{t} = \frac{\partial \omega}{\partial k}. \tag{2.34}$$

This is the most useful expression for the group velocity. Notice that it has the same dimensions as the wave velocity ω/k. The equation (2.34) can be written in many other forms, one example being as follows. The wave velocity is given by $v = \omega/k$ whence the group velocity

$$\begin{aligned} v_g &= \frac{\partial \omega}{\partial k} = \frac{\partial (kv)}{\partial k} \\ &= v + k\frac{\partial v}{\partial k} \\ &= v - \lambda\frac{\partial v}{\partial \lambda}. \end{aligned} \tag{2.35}$$

Other forms of the expression will be derived when necessary.

2.4.1 *The group velocity and energy transfer.* Since the wave amplitude outside the wave-group is zero (at points A and B in Fig. 2.3, for example) it is clear that any energy attached to the waves travels with the group, because a wave of zero amplitude cannot store any energy. If, however, a wave-group becomes distorted on its passage through a medium, it becomes difficult to define exactly what velocity corresponds to the transfer of energy. Such conditions correspond to a highly frequency-dependent group velocity, and are discussed more fully in § 10.3.2, where they are considered as relevant to anomalous dispersion. The velocity of transfer of energy is called the *signal velocity*, and under conditions where a wave-group does not become distorted the group velocity and signal velocity are equal.

The main importance of the group velocity, then, is that it is the velocity at which energy is transported by the waves. As a result of the theory of relativity we know that energy cannot be transported at a velocity greater than that of light in free space; this means that in no medium can the signal velocity be greater than the velocity of light. There is no such restriction on the wave velocity, however, so that a dispersion relation such as that for X-rays in a medium (§ 10.3.4)

$$v = \frac{\omega}{k} = c\left(1 - \frac{1}{2}\frac{\Omega^2}{\omega^2}\right)^{-1} > c \tag{2.36}$$

is quite permissible since the group velocity

$$v_g = \frac{\partial \omega}{\partial k} = c\left(1 + \frac{1}{2}\frac{\Omega^2}{\omega^2}\right)^{-1} \tag{2.37}$$

is then less than c. In fact, the relationship which occurs approximately between equations (2.36) and (2.37),

$$vv_g = c^2, \tag{2.38}$$

is quite common. If (2.38) does hold, but v is less than c, it will follow that v_g is greater than c and the relativistic prediction will be violated. Fortunately, under such conditions there is an absorption of the wave resulting from its dispersion, but we shall postpone further mention of this point until § 10.3.3.

2.5 Boundary conditions

The problem of boundaries is not a difficult one in principle. If the properties of a medium change abruptly, an incident wave may be partially reflected at the boundary and partially transmitted. We need then to consider the boundary conditions, which are the relationships between quantities on opposite sides of the boundary. In the transmission of sound waves, for example, the displacement of the medium must remain continuous across the boundary (if it is well glued!), otherwise either a crack or a region of infinite compression will occur. Similarly the pressure must be continuous otherwise the boundary layer, which is a mathematical plane containing zero mass of material, would accelerate at infinite rate under the action of a finite force. Since there exist three waves at the boundary – incident, reflected and transmitted – their relative amplitudes can be calculated consistently with these two boundary conditions. We shall not concern ourselves with more detail about the calculation here, as it is adequately illustrated by the discussion of reflexion and refraction characteristics for electromagnetic waves under many conditions in Chapter 4.

2.6 Waves in three dimensions

In three dimensions there are several types of wave having sinusoidal time-dependence. The simplest form is the *plane wave*; *cylindrical* and *spherical* waves having position-dependent amplitude are also possible solutions of the various wave equations.

2.6.1 *Plane waves.* The one-dimensional argument presented earlier in this chapter can be rewritten in terms of position vector **r**; the wave amplitude **f**(**r**, t) and the quantity **k** are now vectors so that (2.6) becomes

$$\mathbf{f}(\mathbf{r}, t) = \mathbf{A} \exp\{i(\omega t - \mathbf{k}\cdot\mathbf{r})\}. \tag{2.39}$$

This function can easily be shown to be a solution of the three-dimensional generalized wave equation (compare (2.9))

$$g\left(\frac{\partial}{\partial x}, \frac{\partial}{\partial y}, \frac{\partial}{\partial z}, \frac{\partial}{\partial t}\right)\mathbf{f} = 0, \tag{2.40}$$

giving a dispersion equation

$$g(-ik_x, -ik_y, -ik_z, i\omega) = g(-i\mathbf{k}, i\omega) = 0. \tag{2.41}$$

The vectors **f** and **A** represent the *polarization* of the wave, i.e. the direction in which the displacement takes place; the vector **k** represents the reciprocal of the wavelength along the direction of travel of the wave. For this reason it is known as the *wave-vector*, and it is **k**, and not λ, which can be resolved into components along different directions (Fig. 2.4). The

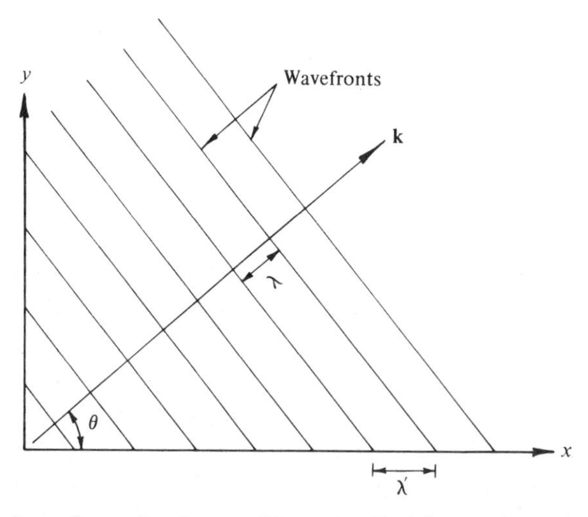

Fig. 2.4. A three-dimensional wave. The wave disturbance along the x-axis has wavelength $\lambda' = \lambda/\cos\theta$. Thus $k' = k\cos\theta$ and so **k**, not λ, is a vector.

dispersion equation is then a relationship between ω and the vector **k**, and the propagation properties come as a result of the equation.

A three-dimensional wave of the form (2.39) has a *wavefront*, which is a plane of constant phase

$$\mathbf{k} \cdot \mathbf{r} = \text{constant} \tag{2.42}$$

and is clearly normal to the wave-vector **k**. The wave velocity is parallel to **k** and is therefore also normal to the wavefronts. However, the group velocity is not necessarily parallel to **k** in an anisotropic medium, but we shall leave such considerations until Chapter 5, where the dependence of the propagation on both the magnitude and direction of **k** is of fundamental importance.

2.6.2 *Spherical and cylindrical waves.* Other possible waves in three dimensions are the *spherical wave*

$$\mathbf{f}(\mathbf{r}, t) = \mathbf{A}(\mathbf{r}) \exp\{i(\omega t - kr)\} \tag{2.43}$$

and the *cylindrical wave*:

$$\mathbf{f}(\mathbf{r}, t) = \mathbf{A}(\mathbf{r}) \exp[i\{\omega t - k(x^2 + y^2)^{\frac{1}{2}}\}]. \tag{2.44}$$

Problems concerning polarization and wave-vectors arise with these waveforms, and in this book their use will be limited to the scalar wave theory (§ 6.1.3).

2.7 Two important principles

2.7.1 *Huygens' construction.* As we have seen, Huygens was a staunch advocate of the wave theory and introduced several ideas which have stood the test of time. One of these was the wavefront which we have already described as a surface of constant phase (§ 2.6.1). Huygens considered the wave more as a transient phenomenon, which was emitted from a point at a certain instant. Then the wavefront is defined as the surface which the wave disturbance has reached at a given time. In principle, a wavefront can have almost any shape, but only plane, spherical or occasionally ellipsoidal wavefronts have any analytical importance.

Huygens pointed out that if one knows the wavefront at a certain moment, the wavefront at a later time can be deduced by considering each point on the first one as the source of a new disturbance. The new disturbance is a spherical wave. At the later time, the spherical waves will have grown to a certain radius, and the new wavefront is the envelope of all the new disturbances (Fig. 2.5). The mathematical embodiment of this

Fig. 2.5. Huygens' construction.

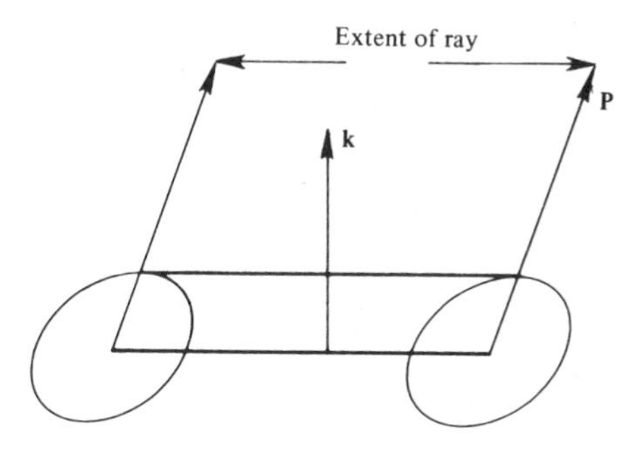

Fig. 2.6. Huygens' construction in an anisotropic medium.

idea is the Kirchhoff diffraction integral, which will be discussed in § 6.5. The new disturbances are known as *Huygens' wavelets*. On the basis of this principle, one can easily see that a spherical wave will remain spherical, but any other shape will distort.

The construction can be applied to anisotropic materials (Chapter 5) by realizing that, if the wave velocity is a function of direction, the wavelets are ellipsoids and not spheres. It then follows, by considering the progress of a wavefront of limited dimensions, that the direction of energy flow (**P**, the *ray direction*) is given by joining the origin of each wavelet to the point at which it touches the envelope; this direction does not always coincide with that of the wave-vector **k** (Fig. 2.6).

It is sometimes argued that Huygens' principle is deficient in that it predicts a backward wave as well as a forward one. This is not so. If we wish to find out how a wavefront A progresses during a time t, we can divide t into two parts, say t_1 and t_2. The wavefront A will develop into B after t_1 and B into C after t_2. If we take backward waves into account the wavefront C is accompanied by C' and C'' which lie at positions depending on the exact values of t_1 and t_2. All such waves will interfere destructively and only the forward wave, which depends on $t = t_1 + t_2$ alone, remains (Fig. 2.7).

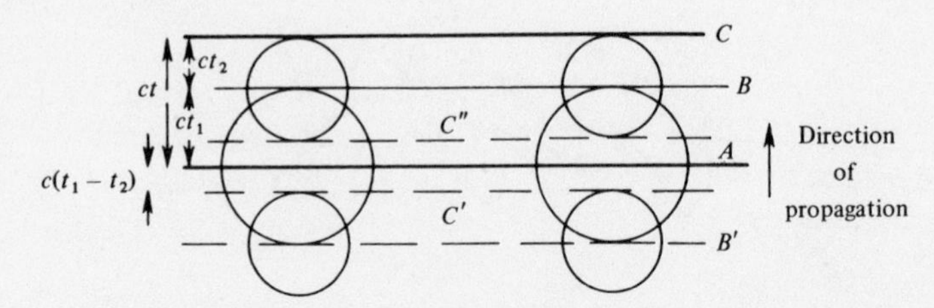

Fig. 2.7. Using combinations of backward and forward waves, wavefronts such as C' and C'' can be deduced, whose positions depend on the choice of t_1.

2.7.2 *Fermat's principle.* Fermat originally stated that light travels from point A to point B by the path that takes the minimum time, thus illustrating Nature's concern with economy! The law of rectilinear propagation in a homogeneous medium is an obvious result, and the laws of reflexion and refraction can also be deduced (Appendix I, 1). In using this principle it is convenient to define the *optical path* from A to B as

$$\overline{AB} = \int_A^B \mu(x)\,\mathrm{d}x, \tag{2.45}$$

where $\mu(x)$ is the refractive index. The time taken for light to travel from A to B is then $\overline{AB}/c$. When two points O and I are conjugate, i.e. they are object and image through some optical system, Fermat's principle leads us to the conclusion that all optical paths $\overline{OI}$ through the system must be equal in length. Waves travelling through the system by various routes will then arrive in phase at the point I and will therefore interfere constructively.

Fermat's principle, as he stated it, is not really correct. Sometimes profligate Nature takes the longest path! Without going into details, it is fairly easy to show that the time taken by light from the object O to a point *further than* the image I is a maximum. Fermat's principle is therefore properly called the *principle of extreme path.*

As we remarked earlier, in Chapter 6 we shall return to these subjects in a more formal manner, but for the present they should be accepted as intuitive ideas which have been extremely fruitful in advancing the understanding of wave optics.

3

Fourier theory

3.1 Introduction

J. B. J. Fourier (1763–1830) was one of the French scientists of the time of Napoleon who raised French science to extraordinary heights. He was essentially an applied scientist, and the work by which his name is now known was his contribution to the theory of heat transmission. He was faced with the problem of solving the one-dimensional heat-diffusion equation (2.15) for the development of the temperature distribution $\theta(x, t)$ in a body

$$\frac{\partial \theta}{\partial t}=D\frac{\partial^2 \theta}{\partial x^2} \tag{3.1}$$

for which he knew some initial conditions – the temperature as a function $\theta(x, 0)$ of position x at $t=0$. This equation, which is a wave equation of the general type dealt with in § 2.2.2, has analytic solutions if θ is a sinusoidal function of x,

$$\theta(x, 0)=\theta_0 \sin kx, \tag{3.2}$$

the solution then being an exponential with characteristic time related to k:

$$\theta(x, t)=\theta_0 \exp(-Dk^2 t) \sin kx. \tag{3.3}$$

As a sinusoidal initial distribution is a very artificial example, Fourier devised a method of expressing any periodic function, or any non-periodic function restricted to a body of regular shape, as the sum of a series of sinusoidal terms of various wavelengths for each of which the equation (3.1) could be solved individually. As the equation is homogeneous in θ, the solutions can then be added.

This principle of expressing an arbitrary function as the sum of a set of sinusoidal terms is called *Fourier theory* and has found applications far beyond the boundaries of heat-transmission theory. Since optics is concerned with light waves and their interactions with obstacles which represent regions in which the wave equations must be solved under different conditions from their surroundings, Fourier theory is obviously a basic tool which we shall use repeatedly. The intention of this chapter is to make the theory familiar to readers, and to derive some of the more important results and ideas so that they can be used later without breaking the thread of an argument.

3.2 Analysis of periodic functions

3.2.1 *Fourier's theorem.* Fourier's theorem states that any periodic function $f(x)$ can be expressed as the sum of a series of sinusoidal functions which have wavelengths which are integral sub-multiples of the wavelength λ of $f(x)$. To make this statement complete, zero is counted as an integer, giving a constant leading term to the series:

$$f(x)=\tfrac{1}{2}C_0+C_1\cos\left(\frac{2\pi x}{\lambda}+\alpha_1\right)+C_2\cos\left(\frac{2\pi x}{\lambda/2}+\alpha_2\right)+\dots$$
$$+C_n\cos\left(\frac{2\pi x}{\lambda/n}+\alpha_n\right)+\dots. \quad (3.4)$$

The ns are called the *orders* of the terms, which are harmonics. The following argument demonstrates the theorem as reasonable. If we cut off the series after the first term, the equation is satisfied only at a discrete number of points – at least two per wavelength. If we add a second term the number of points of agreement will increase; as we continue adding terms the number of intersections between the synthetic function and the original can be made to increase without limit (Fig. 3.1). This does not

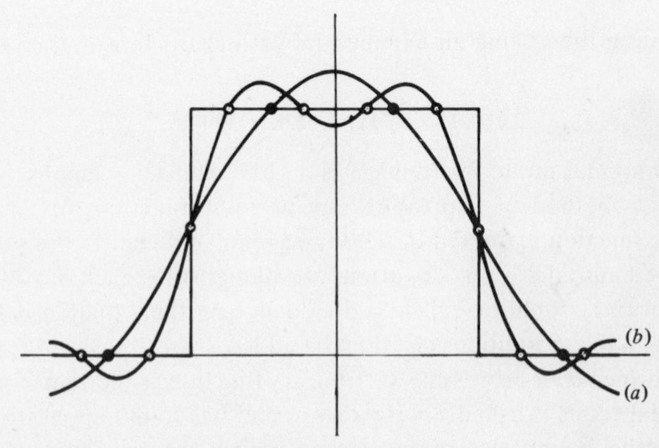

Fig. 3.1. Intersection between a square wave and its series terminated after (a) the first and (b) the third term.

prove that the functions *must* be identical when the number of terms becomes infinite; there are examples which do not converge to the required function, but the regions of error must become vanishingly small.

This reasoning would, of course, apply to basic functions other than sine waves. The sine curve, however, being the solution of all wave equations, is of particular importance in physics, and hence gives Fourier's theorem its fundamental significance.

3.2.2 *Fourier coefficients.* Each term in the series (3.4) has an *amplitude* C_n and a *phase angle* α_n. The latter quantity provides the degree of freedom necessary for relative displacements of the terms of the series along the x-axis. The determination of these quantities for each term of the series is called *Fourier analysis.*

It is, however, not always convenient to specify an amplitude and phase; we can express each term in the form:

$$C_n \cos(nk_0x+\alpha_n) = A_n \cos nk_0x + B_n \sin nk_0x, \tag{3.5}$$

where $A_n = C_n \cos \alpha_n$ and $B_n = -C_n \sin \alpha_n$, and $k_0 = 2\pi/\lambda$. The series (3.4) is then written as

$$f(x) = \tfrac{1}{2}A_0 + \sum_1^\infty A_n \cos nk_0x + \sum_1^\infty B_n \sin nk_0x. \tag{3.6}$$

The process of Fourier analysis consists of evaluating the pairs (A_n, B_n) for each value of n.

3.2.3 *Complex Fourier coefficients.* The real functions $\cos\theta$ and $\sin\theta$ can be regarded as real and imaginary parts of the complex exponential $\exp \mathrm{i}\theta$. Algebraically, there are many advantages in using the complex exponential, and in this book we shall use it almost without exception.

We can write (3.6) in the form

$$f(x) = \frac{A_0}{2} + \sum a_n \exp(\mathrm{i}nk_0x), \tag{3.7}$$

where the range of summation is as yet unspecified. Now let us equate (3.7) and (3.6). We then have:

$$\sum a_n\{\cos(nk_0x) + \mathrm{i}\sin(nk_0x)\}$$

$$= \sum_1^\infty \{A_n \cos(nk_0x) + B_n \sin(nk_0x)\}. \tag{3.8}$$

When we equate equivalent cosine and sine terms independently (assuming the ranges of the summations to be identical) we get:

$$a_n = A_n; \qquad \mathrm{i}a_n = B_n. \tag{3.9}$$

These equations are *not independent*, and cannot be generally true. We have to carry out the complex summation from $n = -\infty$ to $+\infty$ in order to solve the problem. There are then two independent complex coefficients, a_n and a_{-n}, corresponding to the pair A_n, B_n, and we then have, on comparing terms in (3.8):

$$a_n + a_{-n} = A_n; \qquad \mathrm{i}(a_n - a_{-n}) = B_n, \tag{3.10}$$

whence

$$a_n = \tfrac{1}{2}(A_n - \mathrm{i}B_n) = \tfrac{1}{2}C_n \exp(\mathrm{i}\alpha_n), \tag{3.11}$$

$$a_{-n} = \tfrac{1}{2}(A_n + \mathrm{i}B_n) = \tfrac{1}{2}C_n \exp(-\mathrm{i}\alpha_n). \tag{3.12}$$

The Fourier series is therefore written in complex notation as:

$$f(x) = \sum_{-\infty}^{\infty} a_n \exp(\mathrm{i}nk_0x), \tag{3.13}$$

where $a_0 = A_0/2$.

No assumptions have been made as to the reality or complexity of A_n and B_n, and the above formula is therefore true also for complex functions $f(x)$, for which A_n and B_n themselves are complex. However, for a real function, when A_n and B_n are real quantities, it follows immediately from (3.11) and (3.12) that a_n and a_{-n} are complex conjugates:

$$a_n = a^*_{-n}. \tag{3.14}$$

3.3 Fourier analysis

For certain functions Fourier analysis can be carried out analytically by a process that depends on an obvious property of a sinusoidal function – that its integral over a complete number of wavelengths is zero. Consequently, the integral of the product of two sinusoidal functions with integrally-related wavelengths over a complete number of cycles is also zero with one exception: when the two wavelengths are equal the product always has constant sign and the integral is finite. Therefore, if we integrate the product of $f(x)$ (wavelength λ) with a sine function of wavelength λ/m, the result will be zero for all the Fourier coefficients of $f(x)$ except the mth, which has wavelength λ/m, and the value of the integral will then give the amplitude of the coefficient a_m.

To express this mathematically let us find the mth Fourier coefficient by multiplying the function $f(x)$ by $\exp(imk_0x)$ and integrating over a complete wavelength λ. It is convenient to take x in angular measure as $\theta = k_0x$ and then to take the integral I_m over the range $-\pi \leqslant \theta \leqslant \pi$, which is one wavelength. Then

$$I_m = \int_{-\pi}^{\pi} f(\theta) \exp(im\theta)\, \mathrm{d}\theta$$

$$= \int_{-\pi}^{\pi} \sum_{-\infty}^{\infty} a_n \exp(in\theta) \exp(im\theta)\, \mathrm{d}\theta. \tag{3.15}$$

Every term in the summation is sinusoidal, with wavelength $\lambda/(m+n)$, with the exception of the one for which $n = -m$. The sinusoidal terms, being integrated over $(m+n)$ wavelengths, do not contribute; so that

$$I_m = \int_{-\pi}^{\pi} a_{-m}\, \mathrm{d}\theta = 2\pi a_{-m}. \tag{3.16}$$

Thus we have a general expression for the mth Fourier coefficient:

$$a_m = \frac{1}{2\pi} \int_{-\pi}^{\pi} f(\theta) \exp(-im\theta)\, \mathrm{d}\theta. \tag{3.17}$$

Note that it includes the zero term, the mean value of $f(\theta)$:

$$a_0 = \frac{1}{2\pi} \int_{-\pi}^{\pi} f(\theta)\, \mathrm{d}\theta. \tag{3.18}$$

3.3.1 *Even and odd functions.* A function is said to be *even* or *symmetric* if $f(\theta) = f(-\theta)$, and *odd* or *antisymmetric* if $f(\theta) = -f(-\theta)$ (see Fig. 3.2). Let us return for a moment to the formulation (3.6) of the Fourier series in terms of the sine and cosine functions. Now a periodic even function must be expressed as a sum of cosine functions only, since the sine terms make contributions of opposite sign at $+\theta$ and $-\theta$. Thus $B_n = 0$ and it follows from (3.11) and (3.12) that

$$a_n = a_{-n} \qquad \text{(even function)}. \tag{3.19}$$

If, in addition, the function is real, so that (3.14) is true, we find

$$a_n = a_{-n}^* = a_{-n} \qquad \text{(real even function)}, \tag{3.20}$$

implying that a_n is real.

Similarly, for an odd function, we must have coefficients $A_n = 0$ and

$$a_n = -a_{-n} \qquad \text{(odd function)}, \tag{3.21}$$

$$a_n = a_{-n}^* = -a_{-n} \qquad \text{(real odd function)}, \tag{3.22}$$

implying that a_n is purely imaginary in the latter case.

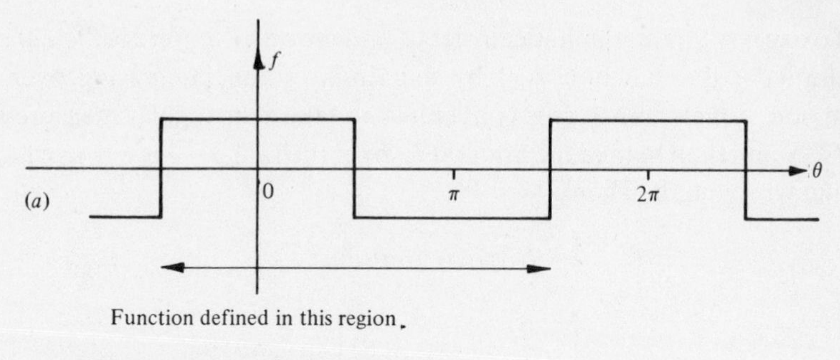

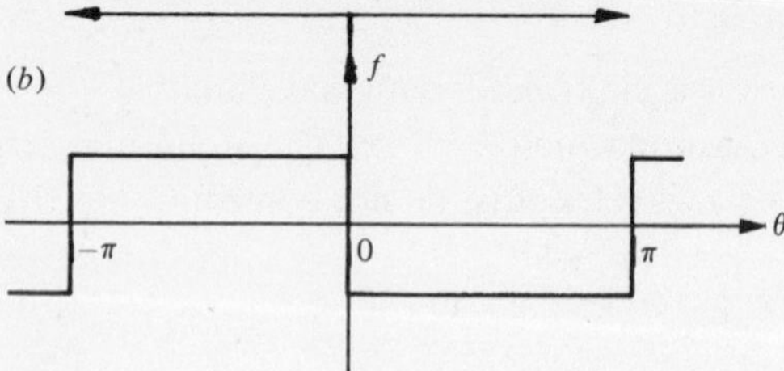

Fig. 3.2. A square wave (*a*) as an even function, (*b*) as an odd function.

3.3.2 *The square wave.* Mathematical textbooks contain many examples of Fourier analysis, and we shall therefore deal with only one here – the *square wave.* This has constant value over half its period ($-\pi/2$ to $\pi/2$) and the opposite sign over the other half ($\pi/2$ to $3\pi/2$) (Fig. 3.2(*a*)). The function as defined above is even; a_n is therefore real. If possible, it is often worthwhile choosing the position of the origin to make a function even, as the mathematics is usually simpler; if we had chosen to make the function a positive constant from $-\pi$ to 0 and an equal negative constant for 0 to π it would have been odd and its coefficients all imaginary (Fig. 3.2(*b*)). This effect – the altering of the phase of all coefficients together by a shift of origin – is often important; the form of the function determines the *relative* phases of the coefficients only.

For the even function (Fig. 3.2)

$$f(\theta)=1\ (-\pi/2\leq\theta\leq\pi/2);\qquad f(\theta)=-1\ (\pi/2\leq\theta\leq 3\pi/2), \tag{3.23}$$

$$\begin{aligned} a_n &= \frac{1}{2\pi}\int_{-\pi}^{\pi} f(\theta)\exp(-in\theta)\,\mathrm{d}\theta \\ &= \frac{1}{2\pi}\int_{-\pi/2}^{\pi/2}\exp(-in\theta)\,\mathrm{d}\theta-\frac{1}{2\pi}\int_{\pi/2}^{3\pi/2}\exp(-in\theta)\,\mathrm{d}\theta \\ &= \frac{1}{n\pi}\sin\frac{n\pi}{2}\{1-\exp(-in\pi)\}. \end{aligned} \tag{3.24}$$

Thus we have, evaluating a_0 from (3.18),

$$a_0=0,\qquad a_1=\frac{2}{\pi},\qquad a_2=0,\qquad a_3=-\frac{2}{3\pi},$$

$$a_4=0,\qquad a_5=\frac{2}{5\pi}\ldots.$$

3.3.3 *Reciprocal space in one dimension.* We can think of the Fourier coefficients a_n as a function $a(n)$ of n. As $a(n)$ exists only for integral values of n, the function can be considered as being defined for non-integral values but as having zero value there; the function which represents the series for a square wave can therefore be drawn as in Fig. 3.3. Given this drawing, we could simply reconstruct the original square

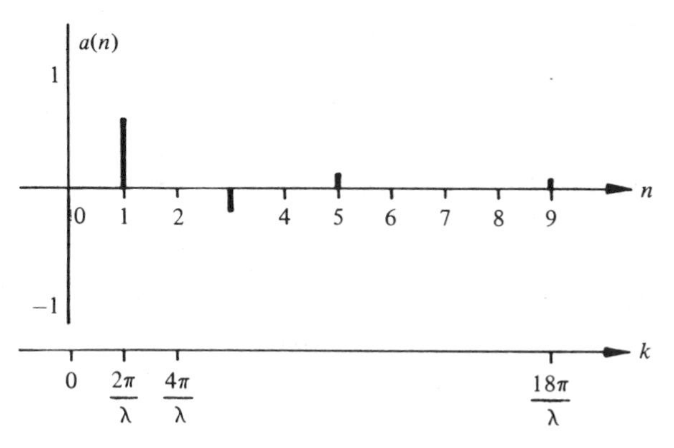

Fig. 3.3. Functions $a(n)$ and $a(k)$ for a square wave.

wave by summing the series it represents, except that it gives no information about the wavelength λ of the original wave. This defect can be simply remedied. Written in terms of x, the expression for a_n is

$$a_n=\frac{1}{\lambda}\int_{\text{one wavelength}} f(x)\exp(-\mathrm{i}nk_0x)\,\mathrm{d}x. \tag{3.25}$$

Information about the wavelength λ can be included in (3.25) by using the variable nk_0 rather than n; this is conventionally called k, and corresponds to a harmonic of wavelength λ/n. The extra information about the wavelength is included by relabelling the abscissa in Fig. 3.3 (lower axis) and the function (3.25) becomes

$$a(k)=\frac{1}{\lambda}\int_{\text{one wavelength}} f(x)\exp(-\mathrm{i}kx)\,\mathrm{d}x. \tag{3.26}$$

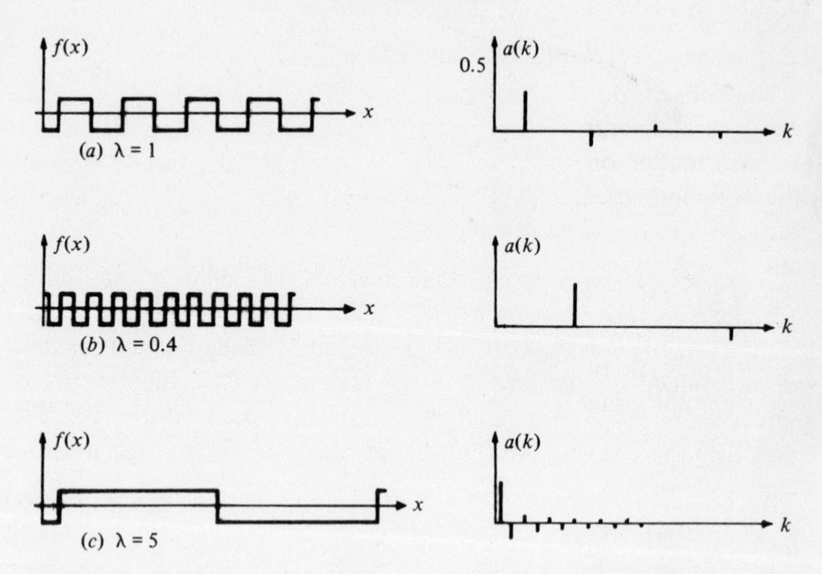

Fig. 3.4. Square waves of different scales and their Fourier coefficients $a(k)$.

It is useful now to compare the functions $a(k)$ as λ changes. In Fig. 3.4 this comparison is carried out, the scales of k and x being the same in *a*, *b*, *c*. Clearly the scale of $a(k)$ is inversely proportional to that of $f(x)$. For this reason (k proportional to $1/\lambda$) the space whose coordinates are measured by k is called reciprocal space; real space has coordinates measured by x and reciprocal space by x^{-1}. So far, of course, we have discussed a purely one-dimensional space; the extension to two and three dimensions is simple, and will be discussed in Chapter 7.

3.3.4 *Analysis of a general function.* In general, the integration involved in equation (3.16) is much more difficult than the example quoted. In many cases, even when $f(x)$ is a simple analytic function, the integral cannot be evaluated analytically and approximate or numerical methods must be used. The process, however, is the same in principle; the integral

$$\frac{1}{2\pi}\int_{-\pi}^{\pi} f(\theta)\exp(-in\theta)\,d\theta \tag{3.27}$$

has to be evaluated for a series of values of n, and because of the repetition involved techniques for reducing the work to a minimum are very valuable.

A very efficient computer technique, called the *fast Fourier transform*, is now generally used, and is described in essence in Appendix V.

3.4 Non-periodic functions

Periodic objects do not usually occur naturally; although crystals, which have accurately repeated sets of atoms in three dimensions, are entirely periodic, matter on the macroscopic scale is usually not so. Natural objects sometimes simulate periodicity in their growth, but this is never precise and most objects which we have to deal with optically (i.e. on a scale greater than the wavelength of light) are completely non-periodic.

Since this book is concerned with light and real objects we may therefore ask why Fourier methods are of any importance, since they apply to periodic functions only. The answer is that the theory has an extension, not visualized by Fourier himself, to non-periodic functions. The extension is based upon the concept of the *Fourier transform.*

3.4.1 *Fourier transform.* We have seen in § 3.2.1 that a periodic function can be analysed into harmonics of wavelengths ∞, λ, $\lambda/2$, $\lambda/3, \ldots$, and we have shown by Fig. 3.4 how the form of the function $a(k)$ depends on the scale of λ. When our interest turns to non-periodic functions we can proceed as follows. Construct a wave of wavelength λ in which each unit consists of our non-periodic function (Fig. 3.5). We can always make λ so large that an insignificant amount of the function lies outside the one-wavelength unit. Now allow λ to increase without limit, so that the repeats of the non-periodic function separate further and further. What happens to the function $a(k)$? The spikes approach one another as λ increases, but if one works through a particular example one finds that the envelope of the tips of the spikes remains invariant; it is determined only by the unit, which is our non-periodic function. In the limit of $\lambda \to \infty$ the spikes are infinitely close to one another, and the function $a(k)$ has just become the envelope. This envelope is called the Fourier transform of the non-periodic function. The limiting process is illustrated in Fig. 3.5.

Admittedly, this demonstrates, rather than proves, that the Fourier series for a non-periodic function is a continuous function, rather than a set of spikes at discrete frequencies. The argument does not show that in the limit $\lambda \to \infty$ the set of spikes, now infinitely close together, does really become a continuous function, but physically the difference is unimportant. From the mathematical point of view it is better to work in reverse. We define the Fourier transform of a function $f(x)$ as

$$a(k) = \int_{-\infty}^{\infty} f(x) \exp(-\mathrm{i}kx) \, \mathrm{d}x \qquad (3.28)$$

which is a continuous function, and then show that if $f(x)$ is periodic, the transform $a(k)$ is non-zero at discrete and periodic values of k only. Proof

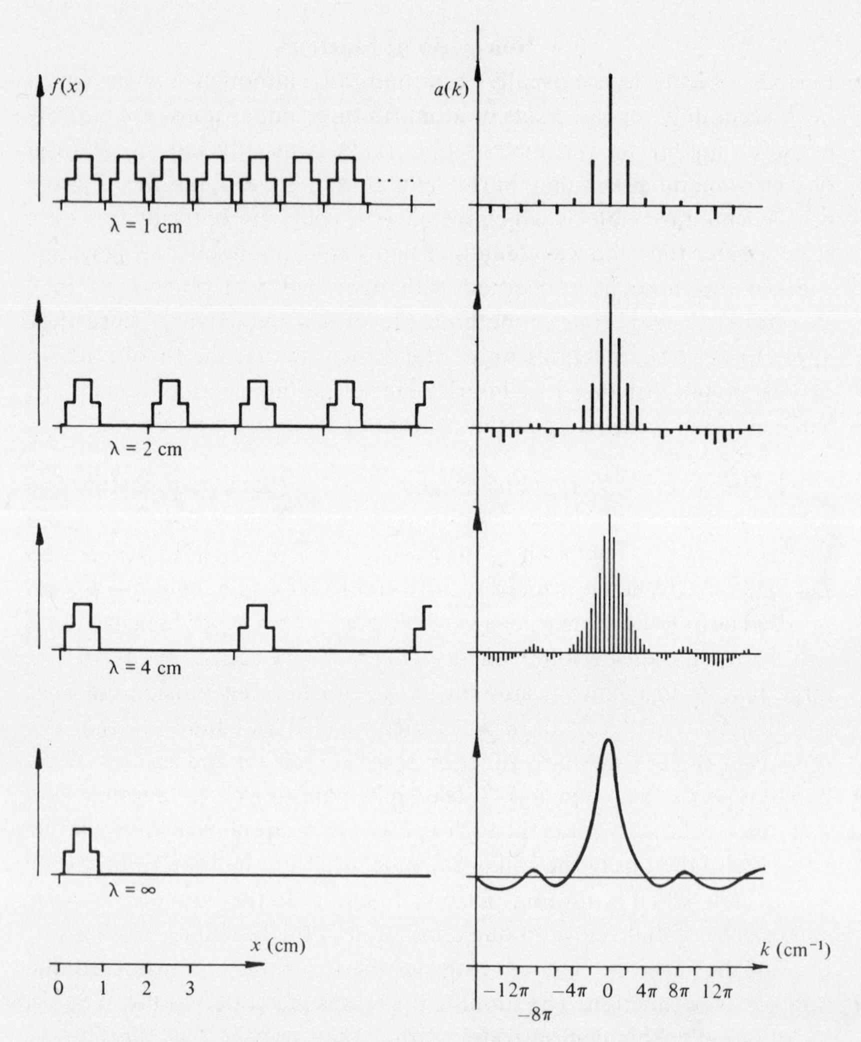

Fig. 3.5. Illustrating the progression from Fourier series to transform.

of this statement will become trivially easy once we have introduced the concept of convolution (§ 3.6).

An important outcome of this reasoning is that the set of orders of a periodic function can be regarded as equally-spaced ordinates of the Fourier transform of the unit. As the spacing is changed by altering the wavelength λ, the orders sweep through the transform (Fig. 3.5). This process is called *sampling* the transform and is a very fruitful idea in understanding diffraction gratings.

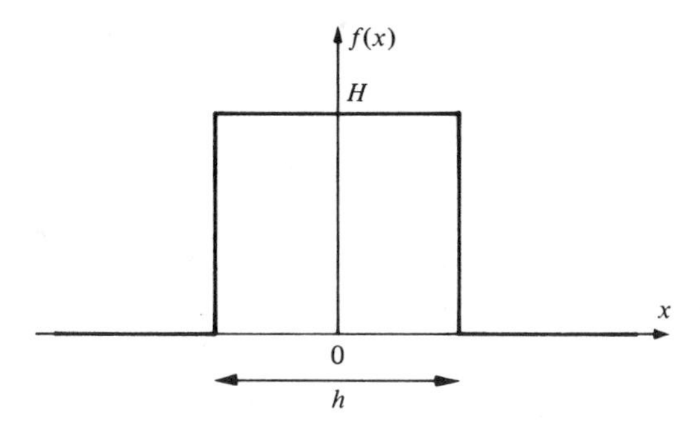

Fig. 3.6. A square pulse.

3.4.2 *Fourier transform of a square pulse.* We can illustrate the calculation of a Fourier transform by using the equivalent example to that in § 3.3.2, a single square pulse. This is one of the simplest, and optically most useful, functions. We define it to have height H and width h (Fig. 3.6) and the integral (3.18) becomes

$$\begin{aligned} a(k) &= \int_{-h/2}^{h/2} H \exp(-\mathrm{i}kx)\,\mathrm{d}x \\ &= \frac{H}{-\mathrm{i}k}\left\{\exp\left(\frac{-\mathrm{i}kh}{2}\right) - \exp\left(\frac{\mathrm{i}kh}{2}\right)\right\} \\ &= Hh\frac{\sin(kh/2)}{kh/2}. \end{aligned} \tag{3.29}$$

The function $\sin\theta/\theta$ appears very frequently in Fourier transform theory, and has therefore been given the name 'sinc (θ)'. Equation (3.29) can thus be written:

$$a(k) = Hh \operatorname{sinc}(kh/2). \tag{3.30}$$

The transform is illustrated in Fig. 3.7. It has a value Hh (the area under the pulse) at $k = 0$ and decreases as k increases, reaching zero when $kh = 2\pi$. It then alternates between positive and negative values, being zero at $kh = 2n\pi$ ($n \neq 0$). It should be noted that the transform is always real since we have defined the function to be symmetrical about the origin; otherwise the transform would be complex, although the magnitude of its value would be unaffected by the displacement (§ 3.3.1).

We can see the reciprocal property of the transform mentioned in § 3.3.3. As h is increased, the value of k at which the transform becomes zero decreases and the interval between successive zeros also decreases; the coarser the function, the finer is the detail of its transform. Conversely, as h decreases the transform spreads out and when h reaches

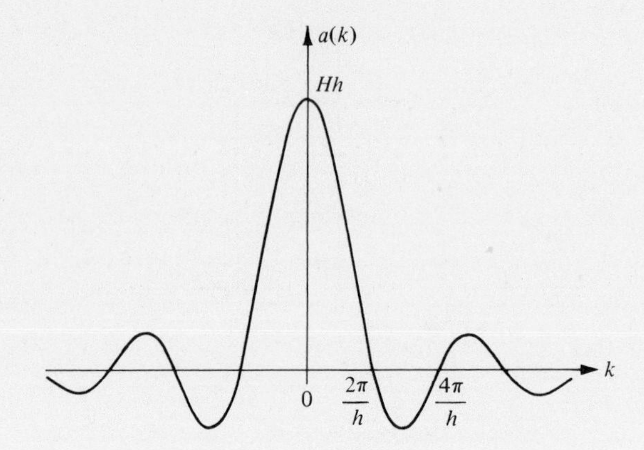

Fig. 3.7. Transform of a square pulse.

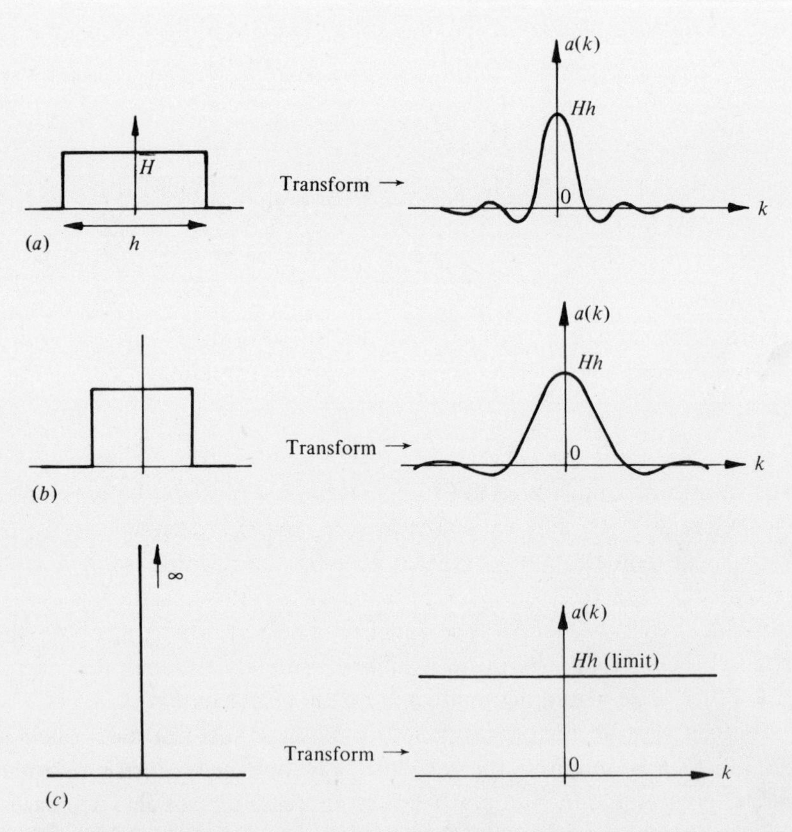

Fig. 3.8. Progression from a square pulse, (a) and (b), to a δ-function (c). The area Hh remains constant throughout.

zero there is no detail at all in the transform, which has become a constant, Hh (see Fig. 3.8).

3.4.3 *The δ-function.* The limiting process above has introduced a new and very useful function, the δ-function. It is the limit of a square pulse as its width h goes to zero but its enclosed area Hh remains at unity. It is therefore zero everywhere except at $x = 0$, when it has infinite value, the particular value of infinity being $\lim_{h\to 0} 1/h$. The transform of the δ-function can be found by the limiting process above; we start with a square pulse of width h and height h^{-1}, which has transform

$$a(k) = \operatorname{sinc}(kh/2) \tag{3.31}$$

and see that as $h \to 0$ the transform becomes unity for all values of k. The transform of a δ-function at the origin in one dimension is a constant.

3.4.4 *A shift of origin.* To illustrate the change in phase, but not in amplitude, which occurs when the origin is shifted, we can calculate the transform of the δ-function at $x = x_0$:

$$f(x) = \delta(x - x_0) = \delta(x') \tag{3.32}$$

$$\begin{aligned} a(k) &= \int_{-\infty}^{\infty} \delta(x - x_0) \exp(-\mathrm{i}kx)\, \mathrm{d}x \\ &= \int_{-\infty}^{\infty} \delta(x') \exp\{-\mathrm{i}k(x' + x_0)\}\, \mathrm{d}x' \\ &= \exp(-\mathrm{i}kx_0) \int_{-\infty}^{\infty} \delta(x') \exp(-\mathrm{i}kx')\, \mathrm{d}x' \\ &= \exp(-\mathrm{i}kx_0). \end{aligned} \tag{3.33}$$

This has constant amplitude, unity, but steadily changing phase.

3.4.5 *Multiple δ-function.* A collection of δ-functions at various values of x is another useful function:

$$f(x) = \sum_n \delta(x - x_n). \tag{3.34}$$

Its transform is clearly

$$a(k) = \sum_n \exp(-\mathrm{i}kx_n). \tag{3.35}$$

If there are two δ-functions, for example, at $x_n = \pm\alpha/2$ we have a transform

$$a(k) = 2\cos(k\alpha/2) \tag{3.36}$$

which is entirely real (even function) and is oscillatory (Fig. 3.9). Its importance in discussing the optical experiment of Young's fringes will be evident in § 7.4.1.

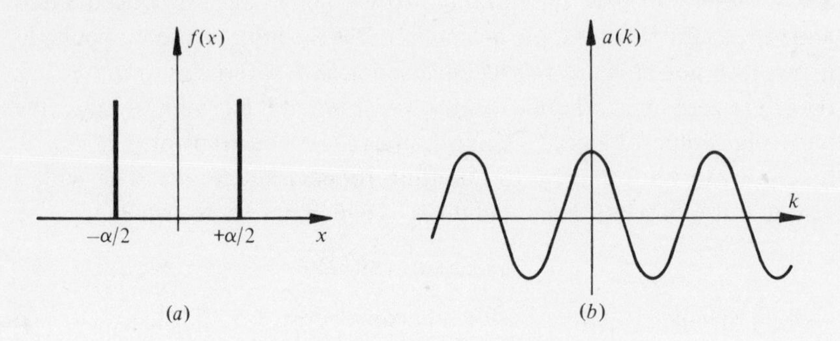

Fig. 3.9. (*a*) Two δ-functions at $\pm\alpha/2$ and (*b*) their transform.

The transform of a regular array of δ-functions is particularly important:

$$f(x)=\sum_{-\infty}^{\infty}\delta(x-n\alpha). \tag{3.37}$$

It follows from (3.35) that

$$a(k)=\sum_{-\infty}^{\infty}\exp(-\mathrm{i}kn\alpha). \tag{3.38}$$

Now the function $f(x)$ extends to infinity in both the positive and negative directions, and its integral is infinite. As a result it can be shown that from the mathematical point of view it does not have a Fourier transform. However, we know that the mathematics only represents a real physical entity, which must itself be finite in extent. From the point of view of the physics we shall therefore introduce no serious error by fading out the series after a large but finite number of terms. (It is always better to fade out than to cut the series off abruptly, as will be demonstrated in § 9.4.5 and § 9.6.6.) It is easiest to do this by multiplying the series by the factor $\exp(-s|n|)$, where s is a number small enough not to affect the series appreciably within any specified region of interest. But now (3.38) can be summed; we write it as the sum of two geometric series, less a single term which appears twice:

$$a(k)=\sum_{0}^{\infty}\exp(-\mathrm{i}kn\alpha-sn)+\sum_{-\infty}^{0}\exp(-\mathrm{i}kn\alpha+sn)-1, \tag{3.39}$$

in which s can be made as small as required. Equation (3.39) gives:

$$a(k)=\{1-\exp(-\mathrm{i}k\alpha-s)\}^{-1}+\{1-\exp(\mathrm{i}k\alpha-s)\}^{-1}-1. \quad (3.40)$$

This function has periodic peaks of value $2(1-\mathrm{e}^{-s})\approx 2/s$ whenever $k\alpha=2\pi m$, for any integer value of m. These peaks approximate to δ-functions as $s\to 0$. Essentially, the function (3.40) can be considered as the periodic series of δ-functions:

$$a(k)=\sum_{m=-\infty}^{\infty}\delta(k-2\pi m/\alpha). \quad (3.41)$$

3.4.6 *The Gaussian function.* Another function whose Fourier transform is particularly useful in optics is the *Gaussian*:

$$f(x)=\exp(-x^2/2\sigma^2). \quad (3.42)$$

From the definition of the transform, (3.28), we have

$$\begin{aligned} a(k)&=\int_{-\infty}^{\infty}\exp(-x^2/2\sigma^2)\exp(-\mathrm{i}kx)\,\mathrm{d}x \\ &=\int_{-\infty}^{\infty}\exp\left[-\left\{\frac{x}{(2\sigma^2)^{\frac{1}{2}}}+\mathrm{i}k\left(\frac{\sigma^2}{2}\right)^{\frac{1}{2}}\right\}^2\right]\exp\left\{-k^2\left(\frac{\sigma^2}{2}\right)\right\}\mathrm{d}x \end{aligned} \quad (3.43)$$

by completing the square in the exponent. The integral is standard (it occurs frequently in statistical theory) and has the value

$$\int_{-\infty}^{\infty}\exp\frac{-\xi^2}{2\sigma^2}\,\mathrm{d}\xi=(2\pi\sigma^2)^{\frac{1}{2}} \quad (3.44)$$

and therefore

$$a(k)=(2\pi\sigma^2)^{\frac{1}{2}}\exp\left\{-k^2\left(\frac{\sigma^2}{2}\right)\right\}. \quad (3.45)$$

The original function (3.42) was a Gaussian with variance σ; the transform is also a Gaussian, but with variance σ^{-1}. The *half-peak width* of the Gaussian is its width at half its maximum height; it can be shown to be equal to 1.94σ. This example particularly emphasizes the reciprocal relationship between the scales of the function and its transform.

3.4.7 *The Fourier transform in two and three dimensions.* All that has been said so far about Fourier transforms and series in one dimension also applies to larger dimensionalities. In particular, two-dimensional functions (screens) are very important in optics. The transform is defined in terms of two spatial frequency components, k_x and k_y, by a double integral:

$$a(k_x, k_y) = \iint_{-\infty}^{\infty} f(x, y) \exp\{-\mathrm{i}(xk_x + yk_y)\}\, \mathrm{d}x\, \mathrm{d}y. \tag{3.46}$$

If the function $f(x, y)$ can be written as the product $f_1(x)f_2(y)$ the integral (3.46) can be factorized into two one-dimensional transforms:

$$\begin{aligned} a(k_x, k_y) &= \int_{-\infty}^{\infty} f_1(x) \exp(-\mathrm{i}xk_x)\, \mathrm{d}x \int_{-\infty}^{\infty} f_2(y) \exp(-\mathrm{i}yk_y)\, \mathrm{d}y \\ &= a_1(k_x)a_2(k_y). \end{aligned} \tag{3.47}$$

In the same way as the components (x, y) form a vector $\mathbf{r}$ in direct space, the components (k_x, k_y) form a vector $\mathbf{k}$ in reciprocal space. Three-dimensional analogues of (3.46) and (3.47) can be written down with no trouble.

3.5 The Fourier inversion theorem

One very useful property of Fourier transforms is that the processes of transforming and untransforming are identical. This property is not trivial, and will be proved below. Another way of stating the property is to say that the Fourier transform of the Fourier transform is the original function again – which is true except for some minor details. This property is known as the *Fourier inversion theorem.*

If the original function is $f(x)$, the Fourier transform $f_1(x')$ of its Fourier transform can be written down directly as a double integral:

$$f_1(x') = \iint_{-\infty}^{\infty} f(x) \exp(-\mathrm{i}kx)\, \mathrm{d}x \exp(-\mathrm{i}kx')\, \mathrm{d}k, \tag{3.48}$$

which can be evaluated as follows.

$$\begin{aligned} f_1(x') &= \iint_{-\infty}^{\infty} f(x) \exp\{-\mathrm{i}k(x + x')\}\, \mathrm{d}x\, \mathrm{d}k \\ &= \int_{-\infty}^{\infty} f(x) \left[\frac{\exp\{-\mathrm{i}k(x + x')\}}{-\mathrm{i}(x + x')}\right]_{k=-\infty}^{k=\infty} \mathrm{d}x. \end{aligned} \tag{3.49}$$

The function with the square brackets can be written as the limit:

$$\lim_{k\to\infty} \frac{2 \sin ky}{y}, \qquad \text{where } y = (x + x'), \tag{3.50}$$

which can be shown† to be equal to $2\pi\delta(y)$. The transform $f_1(x')$ is thus:

$$f_1(x') = \int_{-\infty}^{\infty} 2\pi\delta(x+x')f(x)\,\mathrm{d}x = 2\pi f(-x'). \tag{3.51}$$

On retransforming the transform we have therefore recovered the original function, intact except for inversion through the origin (x has become $-x$) and multiplied by a factor 2π. In the two-dimensional transform of a function $f(x, y)$ (§ 3.4.7), the result of retransforming the transform is to invert *both* axes, which is equivalent to a rotation of 180° about the origin.

It is *conventional* to redefine the inverse transform in a way which 'corrects' the above two deficiencies, so that the transform of the transform comes out exactly equal to the original function. One defines the forward transform, $f(x)$ to $a(k)$, as before (3.28):

$$a(k) = \int_{-\infty}^{\infty} f(x) \exp(-ikx)\,\mathrm{d}x \tag{3.52}$$

and the inverse transform, $a(k)$ to $f(x)$, as

$$f(x) = \frac{1}{2\pi}\int_{-\infty}^{\infty} a(k) \exp(ikx)\,\mathrm{d}k. \tag{3.53}$$

Then the inverse transform of the forward transform is exactly identical to the original function. Of course, physical systems are ignorant of such conventions. If we carry out the transform and its inverse experimentally, as in an imaging system (§ 9.3.3), the image is indeed inverted!

3.5.1 *Examples.* The Fourier inversion theorem can be illustrated by any function which can itself be transformed analytically, and whose transform can also be transformed analytically. In § 3.4 we have already introduced the periodic array of δ-functions (§ 3.4.5) and the Gaussian (§ 3.4.6) which both transform into themselves. Another example from § 3.4.5 is the pair of δ-functions which transforms into a cosine (3.36):

$$\delta(x+\tfrac{1}{2}\alpha) + \delta(x-\tfrac{1}{2}\alpha) \rightarrow 2\cos(\tfrac{1}{2}k\alpha). \tag{3.54}$$

† That the function

$$\lim_{k\to\infty} (2\sin ky)/y$$

is a δ-function, can be justified by drawing the function out for a few values of k. That it has a value of 2π follows from the standard definite integral:

$$\int_{-\infty}^{\infty} \frac{\sin ky}{y}\,\mathrm{d}y = \pi.$$

The magnitude of a δ-function is equal to the area underneath it.

The transform of the cosine can be evaluated from (3.53):

$$\frac{1}{2\pi}\times 2\int_{-\infty}^{\infty} \cos\left(\tfrac{1}{2}k\alpha\right)\exp\left(\mathrm{i}kx\right)\mathrm{d}k$$

$$=\frac{1}{2\pi}\int_{-\infty}^{\infty}\left[\exp\{\mathrm{i}k(x+\tfrac{1}{2}\alpha)\}+\exp\{\mathrm{i}k(x-\tfrac{1}{2}\alpha)\}\right]\mathrm{d}k,$$

$$=\delta(x+\tfrac{1}{2}\alpha)+\delta(x-\tfrac{1}{2}\alpha), \tag{3.55}$$

which is the original function. The Fourier inversion theorem is particularly useful, of course, when the transform in one direction only can be carried out analytically.

3.6 Convolution

An operation which appears very frequently in optics – and indeed in physics in general – is called *convolution*, or folding. The convolution of two real functions is defined mathematically as:

$$F(x)=\int_{-\infty}^{\infty} f(x')g(x-x')\,\mathrm{d}x', \tag{3.56}$$

where F is the convolution of f and g, and x' is a dummy variable.

The convolution is often represented by the symbol $*$ so that (3.56) is written:

$$F(x)=f(x)*g(x). \tag{3.57}$$

3.6.1 *Illustration by means of an out-of-focus camera.* The convolution function is best illustrated by that most simple of optical instruments, the camera. Suppose we consider the photograph of a plane object taken with an out-of-focus camera. Any one bright point on the object will produce a blurred spot in the image-plane, centred at the point x' where the image would come if focusing were exact. In one dimension this blurred spot would be described as a function of position x, $g(x-x')$ centred on the point x'. The intensity of the blurred spot is proportional to the intensity of the original point, which can be written as the intensity $f(x')$ that the sharp image would have at x'. The intensity at point x is therefore

$$f(x')g(x-x') \tag{3.58}$$

Fig. 3.10. The convolution of two-dimensional functions by the out-of-focus camera method described in § 3.6.1. The camera lens has been masked by three apertures, which can be observed by convoluting them with a δ-function, a pinhole, in (*b*), (*d*) and (*f*). The function (*a*) has been photographed through these apertures to give: (*c*) as the convolution of (*a*) and (*b*): (*e*) as the convolution of (*a*) and (*d*); (*g*) as the convolution of (*a*) and (*f*).

(a)

(c)

(b)

(e)

(d)

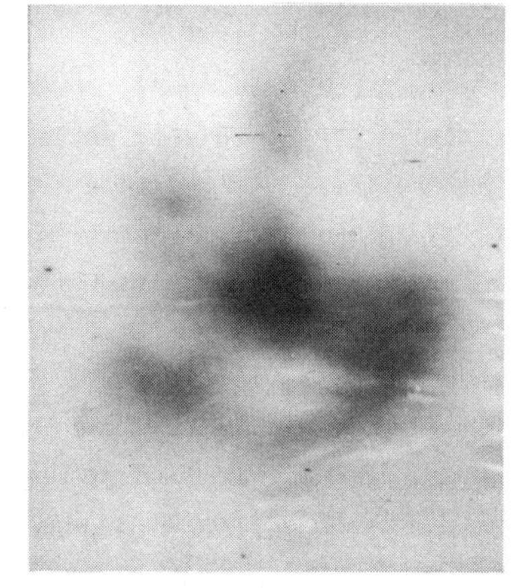

(f)

(g)

and for the complete blurred image the total intensity observed at x is the integral

$$F(x)=\int_{-\infty}^{\infty} f(x')g(x-x')\,\mathrm{d}x'. \qquad (3.59)$$

This process, called *convoluting* the functions f and g, is of great importance in optics and Fourier analysis. It is illustrated by the means described above in Fig. 3.10, where two dimensions have been employed and the function is written

$$F(x, y)=\iint_{-\infty}^{\infty} f(x', y')g(x-x', y-y')\,\mathrm{d}x'\,\mathrm{d}y'. \qquad (3.60)$$

3.6.2 *Convolution with an array of δ-functions.* One of the most important applications of convolution in physical optics occurs when one of the functions is an array, regular or otherwise, of δ-functions. It can be illustrated in two dimensions by an infinite sheet of postage stamps (Fig. 3.11), and it will be seen that under these conditions the idea of convolution becomes particularly simple. Basically a sheet of stamps can be described as a rectangular lattice of points (Fig. 3.11(*a*)) with separation such as 2.5 cm along the y-axis and 2.0 cm along the x-axis, and at each lattice point is placed an identical unit, that of one postage stamp. We define a particular point on the postage stamp as the origin, and describe one stamp as a function $f(x, y)$ referred to this origin; we then describe the lattice by a series of δ-functions at the lattice points:

$$g(x, y)=\sum_{l,m} \delta(x-la)\delta(y-mb), \qquad (3.61)$$

where $a=2$ cm, $b=2.5$ cm and l and m are integers. The sheet of postage stamps can then be written as a composite function representing the density of ink at all points in the (x, y) plane (Fig. 3.11(*b*)):

$$F(x, y)=\iint g(x', y')f(x-x', y-y')\,\mathrm{d}x'\,\mathrm{d}y'. \qquad (3.62)$$

3.6.3 *Convolution in optics.* The convolution function is important in optics because it can be used to represent two important objects: a crystal and a diffraction grating. A crystal clearly consists of a unit cell, containing a definite arrangement of atoms, convoluted with the crystal lattice, which is a set of δ-functions at lattice points.

A diffraction grating with arbitrary line-shape is described as the convolution of a function representing the line-shape $g(x')$ with a regular series of δ-functions, one for each line. This is equivalent to a one-dimensional crystal.

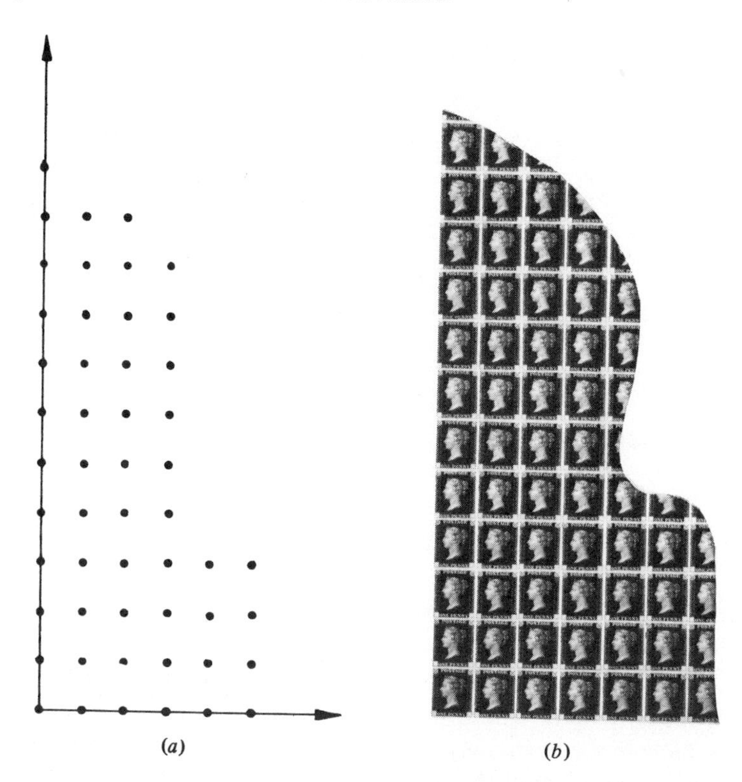

Fig. 3.11. Sheet of postage stamps, representing convolution of one stamp with a two-dimensional lattice of functions. (By kind permission of HM Postmaster-General.)

A third important use of convolution occurs when the two functions f and g are identical. This is called self-convolution, and will be discussed further in § 3.6.7.

3.6.4 *Fourier transform of a convolution.* Not only does the convolution operation occur frequently in physics, but its Fourier transform is particularly simple. This fact makes it very convenient to use. We shall now demonstrate the *convolution theorem*, which states that the *Fourier transform of the convolution of two functions is the product of the transforms of the original functions.*

Consider the convolution $F(x)$ of the functions $f(x)$ and $g(x)$,

$$F(x) = \int_{-\infty}^{\infty} f(x')g(x-x')\,\mathrm{d}x'. \tag{3.59}$$

Its Fourier transform is

$$A(k)=\int_{-\infty}^{\infty}\left\{\int_{-\infty}^{\infty} f(x')g(x-x')\,\mathrm{d}x'\right\}\exp(-\mathrm{i}kx)\,\mathrm{d}x$$
$$=\iint_{-\infty}^{\infty} f(x')g(x-x')\exp(-\mathrm{i}kx)\,\mathrm{d}x'\,\mathrm{d}x. \tag{3.63}$$

By writing $y=x-x'$, we can rewrite this as

$$A(k)=\iint_{-\infty}^{\infty} f(x')g(y)\exp\{-\mathrm{i}k(x'+y)\}\,\mathrm{d}x'\,\mathrm{d}y, \tag{3.64}$$

which separates into two factors:

$$\int_{-\infty}^{\infty} f(x')\exp(-\mathrm{i}kx')\,\mathrm{d}x'\int_{-\infty}^{\infty} g(y)\exp(-\mathrm{i}ky)\,\mathrm{d}y=a(k)b(k), \tag{3.65}$$

where $b(k)$ is the transform of $g(x)$.

This is the required result.

We can now invoke the Fourier inversion theorem (§ 3.5) and deduce immediately that the *Fourier transform of the product of two functions must be equal to the convolution of their individual transforms*, which is an alternative statement of the convolution theorem.

3.6.5 *Transforms and convolutions of complex functions.* In § 3.3.1 we discussed the relationships between a_n and a_{-n} for functions having various symmetry properties. We included in the discussion the possibility that $f(x)$ was complex, and since complex functions form the backbone of wave optics we must extend our discussion of transforms to include them.

If the function $f(x)$ is complex, and has transform $a(k)$ defined in the usual manner, we can write down the transform of its complex conjugate $f(x)^*$ as:

$$\int_{-\infty}^{\infty} f(x)^*\exp(-\mathrm{i}kx)\,\mathrm{d}x=\left\{\int_{-\infty}^{\infty} f(x)\exp(\mathrm{i}kx)\,\mathrm{d}x\right\}^*=a(-k)^*. \tag{3.66}$$

Thus the transform of $f(x)^*$ is $a(-k)^*$. Other rules can then be formulated. For example, if $f(x)$ is real, then $f(x)=f(x)^*$ and so

$$a(-k)^*=a(k) \qquad \text{(real function, cf. (3.14).} \tag{3.67}$$

For an even, or symmetric, function, $f(x)=f(-x)$ and we can write

$$a(-k)=\int_{-\infty}^{\infty} f(x)\exp(\mathrm{i}kx)\,\mathrm{d}x$$

$$=-\int_{\infty}^{-\infty} f(-x)\exp\{-\mathrm{i}k(-x)\}\,\mathrm{d}(-x)$$

$$=a(k) \qquad \text{(even function).} \tag{3.68}$$

This corresponds exactly to (3.19) for the Fourier series, when the index n is replaced by the variable k. All the equations (3.20)–(3.22) have equivalents:

$$a(k)=-a(-k) \qquad \text{(odd function).} \tag{3.69}$$

In addition, $a(k)$ must be real for a real even function and imaginary for a real odd function.

The definition of convolution can be accepted as an extension of (3.56) to include complex functions, and the convolution theorem follows with no change.

3.6.6 *An example of the convolution theorem in use.* There are many examples of functions which can most conveniently be Fourier-transformed after they have been broken down into a convolution or a product, and the reader will meet many of them in the succeeding chapters. We shall just give one simple example here, which will be used with time and frequency variables in § 8.2.2.

A *Gaussian wave-group* is a travelling wave, of the form $A\exp(\mathrm{i}k_0x)$ modified by a Gaussian envelope (§ 3.4.6) having variance σ (Fig. 3.12). It can be written in the form:

$$f(x)=A\exp(\mathrm{i}k_0x)\exp(-x^2/2\sigma^2). \tag{3.70}$$

This function will immediately be recognized as the *product* of the complex exponential $\exp(\mathrm{i}k_0x)$ and the Gaussian (3.42). Its transform is therefore the *convolution* of the transforms of these two functions which are, respectively:

$$2\pi A\delta(k-k_0)$$

and (3.45), namely,

$$(2\pi\sigma^2)^{\frac{1}{2}}\exp(-k^2\sigma^2/2).$$

Now the first of these transforms is a δ-function at the point $k=k_0$, and on convoluting the latter transform with it, we simply shift the origin of the transform Gaussian to that point, getting:

$$a(k)=(2\pi)^{\frac{3}{2}}\sigma A\exp\{-(k-k_0)^2\sigma^2/2\}. \tag{3.71}$$

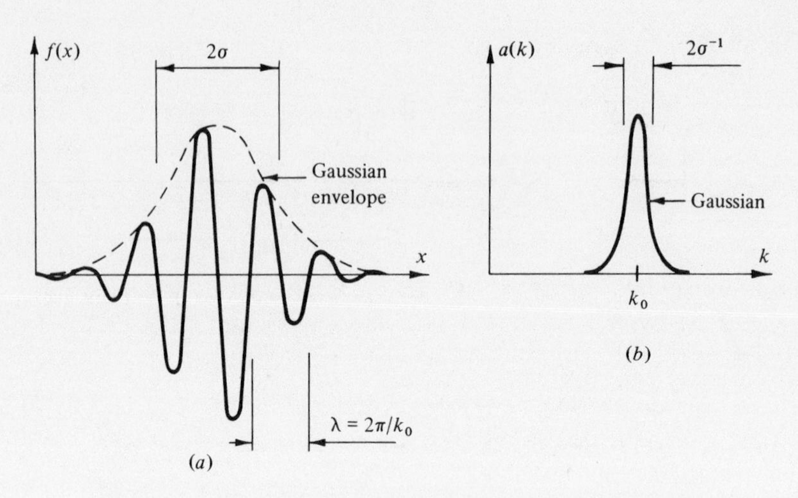

Fig. 3.12. (*a*) A wave-group; (*b*) transform of a wave-group.

3.6.7 *Self-convolution and auto-correlation.* Two convolutions which can be generated from a single function, and are of considerable importance in optics, are the convolution of a function with its own complex conjugate, and its convolution with the inverse of its conjugate.

The first of these, called the *self-convolution*, is

$$F_S(x) = f(x) * f(x)^* = \int_{-\infty}^{\infty} f(x')f(x-x')^* \, dx' \tag{3.72}$$

and its transform follows from (3.56) and (3.66) to be

$$A_S(k) = a(k)a(-k)^*. \tag{3.73}$$

The second function, the *auto-correlation* function, is important in statistics and in communication theory as well as in optics, and is defined as

$$F_C(x) = f(x) * f(-x)^* = \int_{-\infty}^{\infty} f(x')f(x'-x)^* \, dx', \tag{3.74}$$

whence its transform

$$A_C(k) = a(k)a(k)^* = |a(k)|^2. \tag{3.75}$$

This is the intensity of the function $a(k)$, and is always real and positive.

It is instructive to see the way in which the auto-correlation function is built up in two dimensions (Fig. 3.13). On each point of the function $f(x', y')$ we put the origin of the inverse function $f(-x, -y)^*$, which is just $f(x, y)^*$ rotated by 180°. We immediately see a strong point developing at the origin. This strong point at the origin, which is intrinsic to the

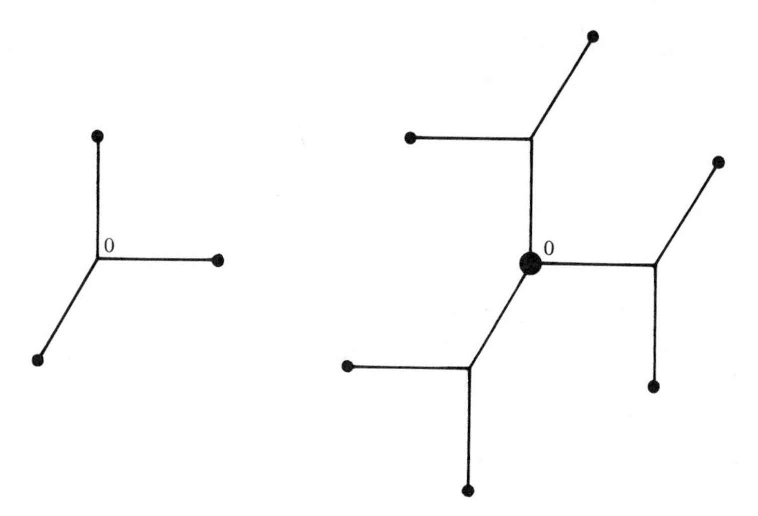

Fig. 3.13. Auto-correlation of a two-dimensional function consisting of three δ-functions (shown on left). The lines are inserted to guide the eye, and are not part of the function.

auto-correlation function, is responsible for the all-real transform $|a(k)|^2$ and is a recurrent theme in diffraction theory (§ 7.4.6 and §§ 11.4.1–11.4.3) and in holography (§ 9.7). Experimental methods of determining spatial auto-correlation functions will be discussed briefly in § 9.8.

3.6.8 *Energy conservation: Parseval's theorem.* The process of Fourier transformation essentially takes a certain function $f(x)$ and represents it as the superposition of a set of waves. We shall see later, in Chapter 7, that in optics Fraunhofer diffraction is described by a Fourier transform, where $f(x)$ represents the amplitude distribution leaving the diffracting obstacle and $a(k)$ represents the amplitude distribution in the diffraction pattern. No light energy need be lost in this process, and it would therefore seem necessary that the total power leaving the object be equal, or at least proportional, to that arriving at the diffraction pattern. In mathematical terms, we expect that

$$\int_{-\infty}^{\infty} |f(x)|^2 \,\mathrm{d}x = C\int_{-\infty}^{\infty} |a(k)|^2 \,\mathrm{d}k. \qquad (3.76)$$

This is called *Parseval's theorem.* It can be deduced easily from our discussion of the auto-correlation function in § 3.6.7. From (3.74) and (3.75) we see that the transform of the object auto-correlation function $F_C(x)$ is the transform intensity $|a(k)|^2$. From the Fourier inversion theorem, the transform of $|a(k)|^2$ must therefore be equal to $F_C(x)$. Writing this out explicitly, from (3.53) we have

$$\frac{1}{2\pi}\int_{-\infty}^{\infty}|a(k)|^2 \exp(\mathrm{i}kx)\,\mathrm{d}k = \int_{-\infty}^{\infty} f(x')f(x'-x)^*\,\mathrm{d}x'. \qquad (3.77)$$

Now let $x = 0$ in this equation. This gives:

$$\frac{1}{2\pi}\int_{-\infty}^{\infty}|a(k)|^2\,\mathrm{d}k = \int_{-\infty}^{\infty}|f(x')|^2\,\mathrm{d}x' \qquad (3.78)$$

which is Parseval's theorem. We should remind the reader that the factor $1/2\pi$, which somewhat detracts from the physical intuition implicit in the theorem, has arisen solely because of the *convention* that the same factor be introduced into the definition (3.53) of the inverse transform.

4

Electromagnetic waves

4.1 Electromagnetism

In the previous chapters we have dealt with some of the general properties of waves without considering in detail any particular medium. This chapter will use Chapter 2 as a basis for understanding wave-like disturbances propagated in the electromagnetic field, both in free space and in the presence of material media. We shall not consider disturbances other than simple-harmonic waves, however, but we shall leave it to the reader to make his own synthesis of Chapters 3 and 4 when he needs it; he should find little difficulty in principle, although the mathematics may be complicated.

This chapter starts with a derivation of the electromagnetic wave equation from Maxwell's equations, and continues with some of the applications of the wave equation which are important in optics. The reader's understanding of many of the elementary aspects of the electromagnetic field will be assumed (see e.g. Grant & Phillips, 1975). As is customary now, SI units will be generally used. We should emphasize the distinction made in this book between μ (magnetic permeability) and $\mathrm{\mu}$ (refractive index).

In Chapter 2 the subject of boundary conditions was introduced but left without example. Here one of the most important sets of boundary conditions is used to deduce the reflexion and refraction coefficients that occur at plane boundaries between media, both dielectric and conducting. Following this discussion of sharp boundaries, we continue with consideration of very blurred boundaries in which the properties of the medium change gradually as the wave progresses. And finally we discuss two aspects of the theory of radiation, which will be important in some of our later work.

4.1.1 *Maxwell's equations.* There are four equations derived by Maxwell which describe in a formal manner the behaviour of electromagnetic fields. Vector differential operators give a very simple formulation of the equations (although we should remark that Maxwell's original paper, in which he deduced the existence of electromagnetic waves from them, was hardly believed on its publication because the mathematics, in terms of components, was too complicated to follow):

$$\nabla \cdot \mathbf{D} = \rho, \tag{4.1}$$

$$\nabla \cdot \mathbf{B} = 0, \tag{4.2}$$

$$\nabla \times \mathbf{H} = \frac{\partial \mathbf{D}}{\partial t} + \mathbf{j}, \tag{4.3}$$

$$\nabla \times \mathbf{E} = -\frac{\partial \mathbf{B}}{\partial t}, \tag{4.4}$$

where **D** is the electric displacement, **B** the magnetic induction, **H** the magnetic field and **E** the electric field.

If we were pedantic, we should point out at this stage that the equations are valid only in the conditions under which they have been experimentally verified, which are static or quasi-static. It is therefore completely unjustifiable to do what we shall proceed to do – i.e. to apply them to very-high-frequency conditions; $10^{14}\ \mathrm{s}^{-1}$ is a frequency so far removed from the stationary conditions of Gauss's theorem, or the fast withdrawal of a magnet from a coil implied in Faraday's law, that we have no right to expect Maxwell's equations to be accurate for light and X-rays. The only justification for their accuracy at such frequencies is that they do predict the subject of optics almost exactly. Quantization of the energy of the waves is the only aspect not included in Maxwell's equations.

4.2 Electromagnetic waves

4.2.1 *Maxwell's wave equation.* Maxwell deduced the existence of electromagnetic waves from the equations (4.1)–(4.4). In a uniform isotropic dielectric medium in which there are no space-charges ρ or currents $\mathbf{j}$, we can write these equations in the form:

$$\nabla \cdot \mathbf{D} = \epsilon\epsilon_0 \nabla \cdot \mathbf{E} = 0, \tag{4.5}$$

$$\nabla \cdot \mathbf{B} = \mu\mu_0 \nabla \cdot \mathbf{H} = 0, \tag{4.6}$$

$$\nabla \times \mathbf{H} = \frac{\partial \mathbf{D}}{\partial t} = \epsilon\epsilon_0 \frac{\partial \mathbf{E}}{\partial t}, \tag{4.7}$$

$$\nabla \times \mathbf{E} = -\frac{\partial \mathbf{B}}{\partial t} = -\mu\mu_0 \frac{\partial \mathbf{H}}{\partial t}. \tag{4.8}$$

Taking ($\nabla\times$) of both sides of (4.8) and replacing the result by the right-hand side of (4.7), we have

$$\nabla\times(\nabla\times\mathbf{E}) = -\mu\mu_0 \frac{\partial}{\partial t}(\nabla\times\mathbf{H}) = -\mu\mu_0\epsilon\epsilon_0 \frac{\partial^2\mathbf{E}}{\partial t^2}. \tag{4.9}$$

The quantity $\nabla\times(\nabla\times\mathbf{E})$ can be expanded:

$$\nabla\times(\nabla\times\mathbf{E}) = \nabla(\nabla\cdot\mathbf{E}) - \nabla^2\mathbf{E} \tag{4.10}$$

by virtue of which, and of (4.5), we write (4.9) as

$$\nabla^2\mathbf{E} = \epsilon\mu\epsilon_0\mu_0 \frac{\partial^2\mathbf{E}}{\partial t^2}, \tag{4.11}$$

$\nabla^2\mathbf{E}$ being the vector $\nabla\cdot(\nabla\mathbf{E})$,

$$\left(\frac{\partial^2 E_x}{\partial x^2}, \frac{\partial^2 E_y}{\partial y^2}, \frac{\partial^2 E_z}{\partial z^2}\right)$$

in Cartesian coordinates, and equivalent expressions in other coordinates.

4.2.2 *Velocity of the waves.* This wave equation, which is of the non-dispersive type (equation (2.5)), gives rise to a well-defined and unique wave velocity:

$$v = (\epsilon_o\mu_0\epsilon\mu)^{-\frac{1}{2}}. \tag{4.12}$$

In free space, where ϵ and μ take on the values unity, we have the velocity in free space

$$c = (\epsilon_0\mu_0)^{-\frac{1}{2}}, \tag{4.13}$$

which is an important physical constant.

The question of why electromagnetic waves, in contrast to almost all other types of wave, can be propagated in the absence of a medium, has historically been a much-argued question; it is probably fair to say that we are no nearer an answer to it today than were our predecessors one hundred years ago.

The numerical values of ϵ_0 and μ_0 in SI units are $10^7/4\pi c^2$ and $4\pi\times10^{-7}$ H m^{-1} respectively, so that (4.13) is automatically satisfied. (In our opinion, this is one of the disappointing features of the SI units – that (4.13) contains no surprises, since the value of ϵ_0 has been 'fudged' in advance.) However, it remains possible to make a static determination of ϵ_0 by measuring the capacitance C of a capacitor of simple geometry. C can also be calculated in terms of ϵ_0 and so the latter can be determined experimentally. An example of an experiment to do this is illustrated in

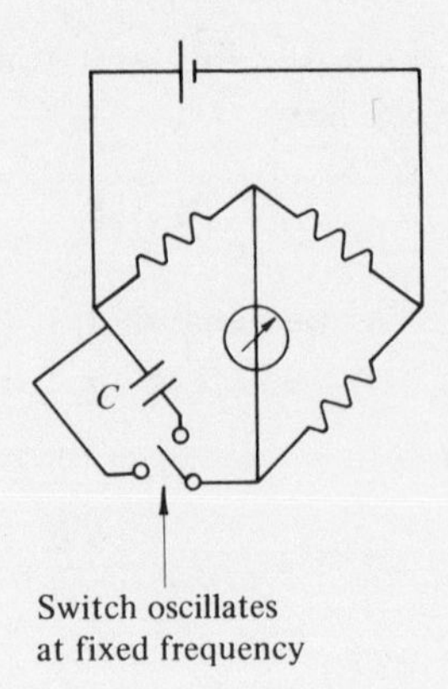

Fig. 4.1. Low-frequency experiment to determine the product $\epsilon_0\mu_0$.

Fig. 4.1, where C is measured in a bridge circuit against a standard resistor. The switch oscillates at frequency n thus charging and discharging the condenser n times per second and passing a current

$$I = nCV \tag{4.14}$$

through the arm of the bridge. It can be seen that the capacitor behaves as a resistor of magnitude $1/nC$, and C can thus be measured. This measurement, in terms of a resistor, gives C in farads. Provided that the capacitor is geometrically simple, its capacity can be calculated from its dimensions, and values for ϵ_0, and hence C, deduced. The estimate is essentially a static determination of C, since the frequency n is only of the order of 10 Hz. The value from this experiment and the observed velocity of light agree within experimental error.

For optical purposes it is rarely necessary to consider media in which the magnetic permeability μ differs significantly from unity. As a result of this, we can combine equations (4.12) and (4.13) to give

$$\frac{c}{v} = \epsilon^{\frac{1}{2}}. \tag{4.15}$$

This quantity is, by definition, the refractive index μ. There follows the important relationship

$$\mu = \epsilon^{\frac{1}{2}}. \tag{4.16}$$

4.2.3 *Solutions of the wave equation.* As pointed out in Chapter 2, the basic solution of the wave equation is sinusoidal. There are only two sinusoidal solutions of any importance:

(a) A radially propagating wave (equation (2.43))

$$\mathbf{E} = \mathbf{E}_0 \frac{a}{r} \exp\{\mathrm{i}(\omega t - kr)\} \tag{4.17}$$

which is a solution over a restricted solid angle. This solution is used in §§ 6.2.1 and 6.5, in which its vector form is not important.

(b) A plane wave (equation (2.39))

$$\mathbf{E} = \mathbf{E}_0 \exp\{\mathrm{i}(\omega t - \mathbf{k}\cdot\mathbf{r})\} \tag{4.18}$$

is the most important solution. A useful way of treating wave fields of this form consists of replacing the operators ∇ and $\partial/\partial t$ in Maxwell's equations by $-\mathrm{i}\mathbf{k}$ and $\mathrm{i}\omega$ respectively (§ 2.2.1). As a result of this, equations (4.5)–(4.8) become

$$\mathbf{k}\cdot\mathbf{D} = 0, \tag{4.19}$$

$$\mathbf{k}\cdot\mathbf{B} = 0, \tag{4.20}$$

$$\mathbf{k}\times\mathbf{H} = -\omega\mathbf{D} = -\omega\epsilon\epsilon_0\mathbf{E}, \tag{4.21}$$

$$\mathbf{k}\times\mathbf{E} = \omega\mathbf{B} = \omega\mu\mu_0\mathbf{H}. \tag{4.22}$$

Thus the vectors **E**, **k** and **H** are mutually orthogonal (Fig. 4.2) and the wave is transversely polarized. The relative magnitude of **E** and **H** follow

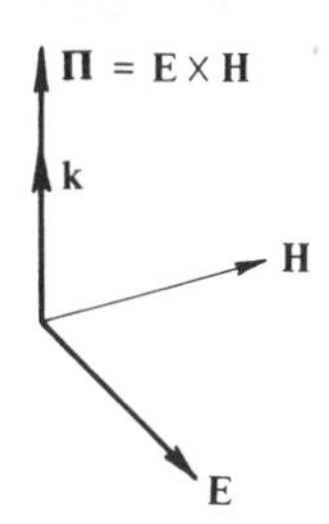

Fig. 4.2. Orthogonality of **E**, **H** and **k** in an electromagnetic wave.

from (4.21) or (4.22); the wave velocity v is, of course, ω/k (§ 2.2.1) and therefore

$$\frac{E}{H} = \frac{k}{\epsilon\epsilon_0\omega} = \left(\frac{\mu\mu_0}{\epsilon\epsilon_0}\right)^{\frac{1}{2}} = Z. \tag{4.23}$$

This ratio is the *impedance* of the medium, and is further discussed in § 4.2.5.

For a transverse plane wave, the *plane of polarization* is defined as the plane containing **E** and **k**. We shall see in Chapter 5 that in an anisotropic medium the propagation depends on the orientation of this plane with respect to the crystal axes.

4.2.4 *Flow of energy in an electromagnetic wave.* It is an important feature of electromagnetic waves that they are able to transport energy. The vector describing the flow of energy is called the *Poynting vector* which can be shown (e.g. see Grant & Phillips, 1975) to be

$$\mathbf{\Pi} = \mathbf{E} \times \mathbf{H}. \tag{4.24}$$

Reassuringly, the direction of this vector turns out to be along the direction of travel of the wave. The Poynting vector has dimensions of power per unit area, and its magnitude is equal to the intensity of the wave. It was shown above that in the simplest solution of Maxwell's equation **E** and **H** are transverse and normal to each other. The Poynting vector is therefore non-zero and along the direction of the wave (Fig. 4.2). The average value of $\mathbf{\Pi}$, $\langle \mathbf{\Pi} \rangle$, will depend on the phase of **E** and **H**. If they are in phase,

$$\langle \mathbf{\Pi} \rangle = \langle E_0 \sin \omega t \cdot H_0 \sin \omega t \rangle = \tfrac{1}{2} E_0 H_0 \tag{4.25}$$

but if they have components in quadrature, those components lead to the result that

$$\langle \mathbf{\Pi} \rangle = \langle E_0 \sin \omega t \cdot H_0 \cos \omega t \rangle = 0. \tag{4.26}$$

Thus electric and magnetic fields in quadrature lead to no transfer of energy. We shall shortly find that evanescent waves form an example of such behaviour.

4.2.5 *Impedance of media to electromagnetic waves.* In § 4.2.3 we showed that the ratio of E to H is the impedance, Z. It clearly has the right dimensions; in SI units E is measured in V m^{-1} and H in A m^{-1} so that the ratio E/H has the dimensions of ohms. The concept of impedance is useful when we consider the transfer of electromagnetic waves from medium to medium; the calculations become analogous to those for the transfer of a.c. signals through networks.

We have already deduced an expression (4.23) for E/H as

$$\frac{E}{H} = Z = \pm \left(\frac{\mu \mu_0}{\epsilon \epsilon_0} \right)^{\frac{1}{2}},$$

the sign depending on the direction of the energy flow. Clearly in non-magnetic media, where $\mu = 1$, we have:

$$Z = Z_0 / \epsilon^{\frac{1}{2}}, \tag{4.27}$$

where Z_0 is the impedance of free space:

$$(\mu_0 / \epsilon_0)^{\frac{1}{2}} = 4\pi c \times 10^{-7} = 377\ \Omega. \tag{4.28}$$

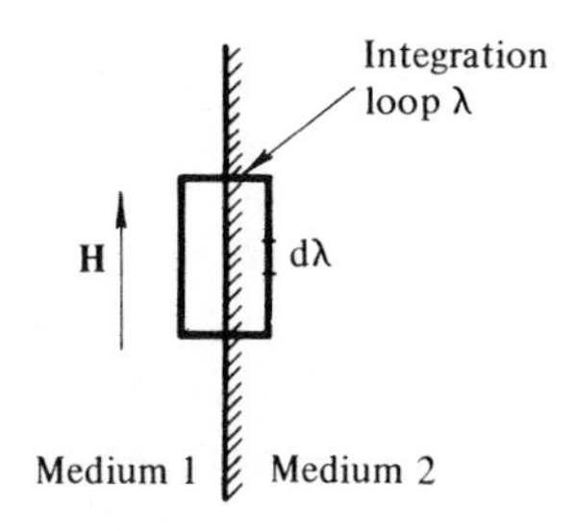

Fig. 4.3. Continuity of **H**.

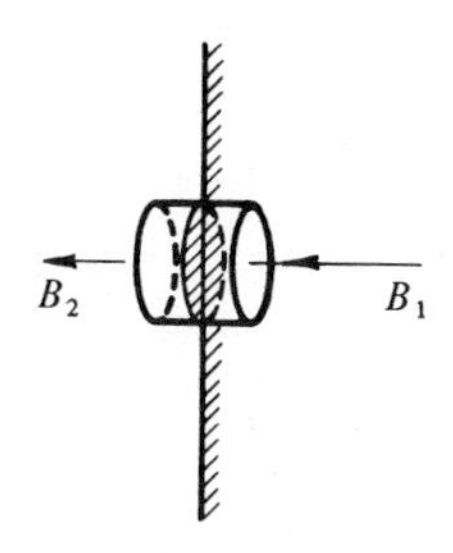

Fig. 4.4. Continuity of **B**.

In terms of refractive index μ the impedance is written

$$Z = Z_0/\mu. \tag{4.29}$$

4.3 Reflexion and refraction

To determine the reflective and refractive properties of electromagnetic waves we need first to decide on the boundary conditions. These are fairly simple; we can immediately see that for real media

(a) **E** parallel to the boundary surface must be continuous, because the potential V cannot have a discontinuity without causing breakdown;

(b) **H** parallel to the surface must be continuous, unless there exists a surface current (Fig. 4.3);

(c) **B** normal to the surface must be continuous, because $\nabla \cdot \mathbf{B} = 0$ (Fig. 4.4);

(d) **D** normal to the surface must be continuous, similarly, unless there is a surface charge.

4.3.1 *Reflexion and refraction at a boundary.* We can now solve the problem of reflexion of a plane-polarized wave at the plane boundary between two media of impedances Z_1 and Z_2. Without loss of generality we need treat only the two polarizations with the field **E** lying in, and perpendicular to, the plane of incidence. Any other plane of polarization can be resolved into two components. (These two cases are often referred to as the TM and TE modes respectively, following waveguide usage.) In three-dimensional calculations of this sort it is very necessary to establish a rigorous sign-convention. We shall do this by allowing the light to be incident from the left on the plane $z=0$, z being measured from left to right. Angles will be measured anticlockwise from the relevant normal, so that the reflected wave occurs at angle $\hat{j}=-\hat{\imath}$. Fig. 4.5 shows the details. We shall first consider the plane of polarization

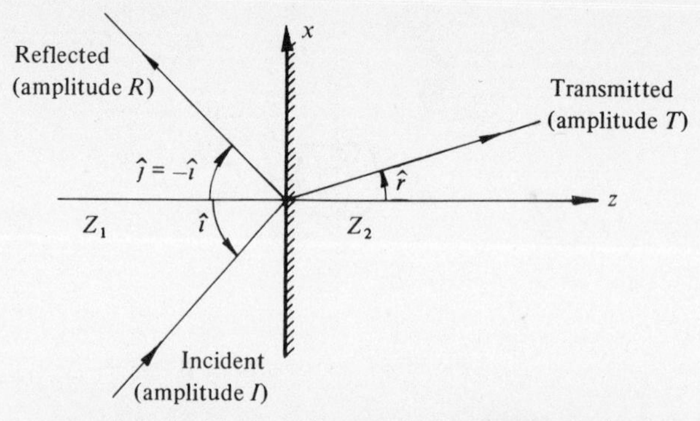

Fig. 4.5. Incident, reflected and transmitted rays.

normal to the plane of incidence, so that $\mathbf{E}=E_y$ only. We denote the wave numbers $2\pi/\lambda$ in the two media by k_1 and k_2. At instant $t=0$,

$$\left.\begin{aligned}&\text{incident wave} && E_y=E_{y\mathrm{I}}=I\exp\{-\mathrm{i}(k_1z\cos\hat{\imath}+k_1x\sin\hat{\imath})\},\\&\text{reflected wave} && E_y=E_{y\mathrm{R}}=R\exp\{-\mathrm{i}(k_1z\cos\hat{j}+k_1x\sin\hat{j})\},\\&\text{transmitted wave} && E_y=E_{y\mathrm{T}}=T\exp\{-\mathrm{i}(k_2z\cos\hat{r}+k_2x\sin\hat{r})\}.\end{aligned}\right\}\qquad(4.30)$$

Any changes of phase occurring on reflexion and transmission will be indicated by negative or complex values of R and T. The magnetic fields are related by impedance $Z=E/H$ and are perpendicular to $\mathbf{k}$ and $\mathbf{E}$. The fact that the reflected wave travels in the opposite z-direction to the others is represented by making its impedance equal to $-Z$ so that the Poynting vector, the energy flow, is in the correct direction.

$$\left.\begin{aligned}&\text{Incident wave} && H_z=E_{y\mathrm{I}}Z_1^{-1}\sin\hat{\imath},\\& && H_x=-E_{y\mathrm{I}}Z_1^{-1}\cos\hat{\imath}.\\&\text{Reflected wave} && H_z=-E_{y\mathrm{R}}Z_1^{-1}\sin\hat{j},\\& && H_x=E_{y\mathrm{R}}Z_1^{-1}\cos\hat{j}.\\&\text{Transmitted wave} && H_z=E_{y\mathrm{T}}Z_2^{-1}\sin\hat{r},\\& && H_x=-E_{y\mathrm{T}}Z_2^{-1}\cos\hat{r}.\end{aligned}\right\}\qquad(4.31)$$

The boundary conditions can then be applied. E_y is itself the parallel component, so that at the point $x=0$ and $z=0$ we have

$$I+R=T \tag{4.32}$$

and in general for all points in the plane $z=0$ we then have

$$k_1 \sin \hat{\imath} = -k_1 \sin \hat{\jmath} = k_2 \sin \hat{r}, \tag{4.33}$$

whence $\hat{\imath} = -\hat{\jmath}$ and

$$\begin{aligned} \sin \hat{\imath} &= \frac{k_2}{k_1} \sin \hat{r} \\ &= \frac{\mu_2}{\mu_1} \sin \hat{r} \quad \text{(Snell's law)}, \\ &= \mu_r \sin \hat{r}, \end{aligned} \tag{4.34}$$

where μ_r is the relative refractive index between the two media. Continuity of the parallel component H_x of $\mathbf{H}$ at $(x, z) = (0, 0)$ gives

$$-IZ_1^{-1} \cos \hat{\imath} + RZ_1^{-1} \cos \hat{\jmath} = -TZ_2^{-1} \cos \hat{r}. \tag{4.35}$$

Whence, for this polarization (denoted by subscript $\perp$)

$$\left.\begin{aligned} R_\perp &= \frac{Z_2 \cos \hat{\imath} - Z_1 \cos \hat{r}}{Z_2 \cos \hat{\imath} + Z_1 \cos \hat{r}} = \frac{\cos \hat{\imath} - \mu_r \cos \hat{r}}{\cos \hat{\imath} + \mu_r \cos \hat{r}}, \\ T_\perp &= \frac{2Z_2 \cos \hat{\imath}}{Z_2 \cos \hat{\imath} + Z_1 \cos \hat{r}} = \frac{2 \cos \hat{\imath}}{\cos \hat{\imath} + \mu_r \cos \hat{r}}. \end{aligned}\right\} \tag{4.36}$$

The coefficients for the other plane of polarization (denoted by $\|$, since $\mathbf{E}$ is parallel to the plane of polarization) can be worked out similarly. When R and T refer to the component E_x, we find:†

$$\left.\begin{aligned} R_\| &= \frac{Z_2 \cos \hat{r} - Z_1 \cos \hat{\imath}}{Z_2 \cos \hat{r} + Z_1 \cos \hat{\imath}} = \frac{\cos \hat{r} - \mu_r \cos \hat{\imath}}{\cos \hat{r} + \mu_r \cos \hat{\imath}}, \\ T_\| &= \frac{2Z_2 \cos \hat{r}}{Z_2 \cos \hat{r} + Z_1 \cos \hat{\imath}} = \frac{2 \cos \hat{r}}{\cos \hat{r} + \mu_r \cos \hat{\imath}}. \end{aligned}\right\} \tag{4.37}$$

The functions $R_\perp$ and $R_\|$ are illustrated in Fig. 4.6. The reflexion coefficients for normal incidence $\hat{\imath} = \hat{r} = 0$ are equal:

$$R_\| = R_\perp = \frac{Z_2 - Z_1}{Z_2 + Z_1} = \frac{1 - \mu_r}{1 + \mu_r} = \frac{\mu_1 - \mu_2}{\mu_1 + \mu_2}, \tag{4.38a}$$

† There is sometimes some confusion about the sign of $R_\|$. If the coefficient were to refer to the component E_z (which might be more convenient at angles near grazing incidence), the sign of $R_\|$ would be reversed. To avoid any confusion, we shall only refer to the coefficient to E_x in this book.

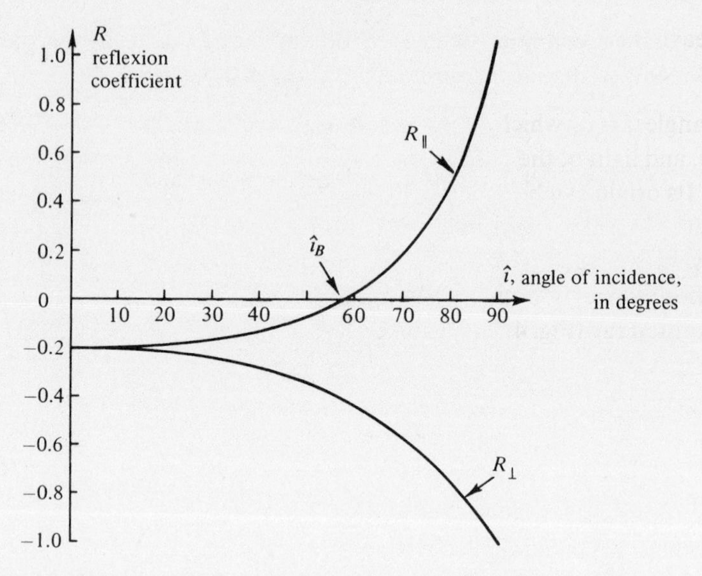

Fig. 4.6. Reflexion coefficient at the surface of a medium of refractive index $\mu = 1.5$ for waves polarized parallel (‖) and perpendicular ($\perp$) to the plane of incidence.

which is negative when $\mu_r > 1$. Similarly

$$T_\parallel = T_\perp = \frac{2Z_2}{Z_2 + Z_1} = \frac{2}{1 + \mu_r} = \frac{2\mu_1}{\mu_1 + \mu_2}. \tag{4.38b}$$

If incidence is in the dense medium so that $\mu_r < 1$, it is clear that $T > 1$; this does not violate the principle of conservation of energy because the energy flow per unit area is $E^2 Z^{-1}$ and the proportion transmitted is thus

$$\left(\frac{2Z_2}{Z_1 + Z_2}\right)^2 \frac{Z_1}{Z_2} = \frac{4Z_2 Z_1}{(Z_2 + Z_1)^2} = \frac{4\mu_r}{(1 + \mu_r)^2}, \tag{4.39}$$

which reaches a maximum value of 1 when $\mu_r = 1$. At non-normal incidence the fact that the areas of transmitted and reflected beams are in the ratio $\cos \hat{\imath} : \cos \hat{r}$ must also be taken into account when calculating total energy flows.

For the polarization-plane parallel to the incidence plane, Fig. 4.6 indicates that the reflexion coefficient is zero at a particular angle $\hat{\imath}_B$. For this condition we have

$$Z_2 \cos \hat{r} - Z_1 \cos \hat{\imath} = 0,$$

$$\frac{\cos \hat{r}}{\cos \hat{\imath}} = \frac{Z_1}{Z_2} = \mu_r = \frac{\sin \hat{\imath}}{\sin \hat{r}}. \tag{4.40}$$

We leave it to the reader to show that this equation can be rewritten as

$$\tan \hat{\imath} = \cot \hat{r} = \mu_r. \tag{4.41}$$

The angle $\hat{\imath} = \hat{\imath}_B$ which is the solution of this equation is the *Brewster angle*, and light of the parallel polarization is not reflected when incident at $\hat{\imath}_B$. Its origin can be seen physically as a result of the transverse nature of light. Consider the reflected ray of light as re-radiated by a plane of atoms in the surface layer of the medium, which are vibrating in a direction which is both in the plane of incidence and normal to the transmitted ray (Fig. 4.7). It is clear that no light can be re-radiated along

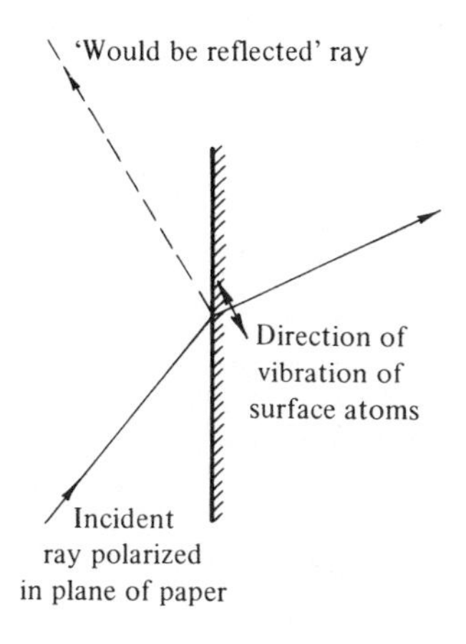

Fig. 4.7. Origin of the Brewster angle.

the direction of vibration since light is a transverse and not a longitudinal wave, and when this direction coincides with the direction of the reflected ray it must have zero amplitude. Under these conditions, which are that the reflected and transmitted ray are perpendicular to each other, the reflexion coefficient is zero for this plane of polarization. This argument cannot be reproduced when light is incident *in the medium* on its interface with a vacuum.

4.3.2 *Incidence in the denser medium.* Implicitly we have assumed $\mu_r > 1$ so far, in that we have drawn Fig. 4.6 with a real angle of refraction for all angles of incidence. If incidence occurs in the denser medium, so that the equation

$$\frac{\sin \hat{\imath}}{\sin \hat{r}} = \mu_r \qquad (\mu_r < 1) \tag{4.42}$$

cannot always be satisfied by a real angle $\hat{r}$, total internal reflexion occurs when $\hat{\imath}$ exceeds the critical angle $\hat{\imath}_c = \sin^{-1}\mu_r$. We can investigate this situation more closely as follows.

We postulate that a solution of equation (4.42) always exists, even if it leads to unreality. We shall then calculate the reflexion and transmission coefficients for all angles $\hat{\imath}$ and show that total reflexion does occur, and the disturbance on the far side of the boundary is an evanescent (§ 2.3.2) rather than a progressive wave. For the equation

$$\sin\hat{r} = \frac{1}{\mu_r}\sin\hat{\imath} > 1$$

we have

$$\cos\hat{r} = (1-\sin^2\hat{r})^{\frac{1}{2}} = \pm \mathrm{i}\beta \qquad (\beta \text{ real and positive}). \tag{4.43}$$

Of the two signs for $\cos\hat{r}$, the lower one will shortly be shown to be physically unacceptable. Substituting in the equations for R and T ((4.36), (4.37)) we obtain:

$$\left.\begin{aligned} R_\perp &= \frac{\cos\hat{\imath} \mp \mathrm{i}\mu_r\beta}{\cos\hat{\imath} \pm \mathrm{i}\mu_r\beta}, \qquad & T_\perp &= \frac{2\cos\hat{\imath}}{\cos\hat{\imath} \pm \mathrm{i}\mu_r\beta},\\ R_\parallel &= \frac{\pm\mathrm{i}\beta - \mu_r\cos\hat{\imath}}{\pm\mathrm{i}\beta + \mu_r\cos\hat{\imath}}, \qquad & T_\parallel &= \frac{2\cos\hat{\imath}}{\mu_r\cos\hat{\imath} \pm \mathrm{i}\beta}. \end{aligned}\right\} \tag{4.44}$$

As the reflexion coefficients are both of the form

$$R = \frac{u-\mathrm{i}v}{u+\mathrm{i}v} = \exp\left\{-2\mathrm{i}\tan^{-1}\left(\frac{v}{u}\right)\right\} = \exp(-\mathrm{i}\alpha) \tag{4.45}$$

it is clear that they represent complete reflexion ($|R| = 1$) but with a phase change α

$$\alpha_\perp = \pm 2\tan^{-1}\frac{\mu_r\beta}{\cos\hat{\imath}}, \qquad \alpha_\parallel = \mp 2\tan^{-1}\frac{\mu_r\cos\hat{\imath}}{\beta}. \tag{4.46}$$

Neither of the transmission coefficients is zero, however, and so we must investigate the transmitted wave more closely.

We shall write the space-dependent part of the transmitted wave in full:

$$\begin{aligned} E &= E_0\exp\{-\mathrm{i}(kz\cos\hat{r} + kx\sin\hat{r})\}\\ &= E_0\exp(\mp k\beta z)\exp\{-\mathrm{i}kx(1+\beta^2)^{\frac{1}{2}}\}. \end{aligned} \tag{4.47}$$

The z-dependence shows that the wave is evanescent and decays rapidly to zero as z increases provided that the upper signs in (4.43)–(4.47) are chosen. The characteristic decay distance is $(k\beta)^{-1}$. For a critical angle of 45° ($\mu_r = 2^{-\frac{1}{2}}$) at an incident angle of 46°, for example,

$$\beta = (2\sin^2 46° - 1)^{\frac{1}{2}} = 0.19$$

and the decay distance is thus

$$\frac{\lambda}{2\pi\beta} \approx 0.8\lambda.$$

The evanescent disturbance can be observed by putting a second parallel surface within the decay distance of the first boundary. The evanescent wave then gives rise to a real wave in the second medium (Fig. 4.8) which can be observed. To calculate the amplitude of the observed wave we need to solve the complete problem satisfying boundary conditions at the second surface as well; as this is complicated, although not difficult, we shall not go into the problem further.

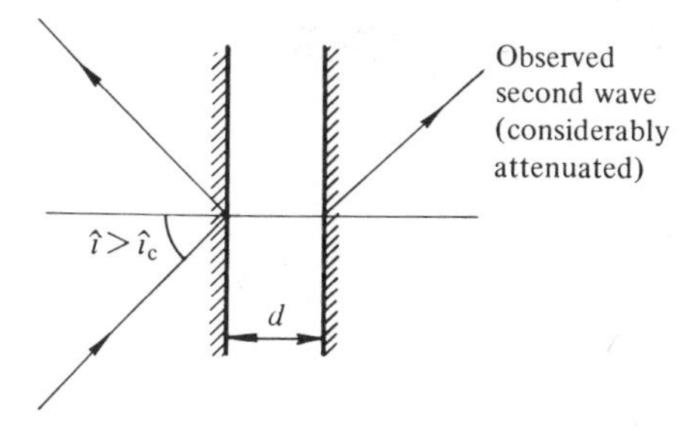

Fig. 4.8. 'Tunnelling' of an electromagnetic wave incident at an angle greater than the critical angle. The spacing d must be of the order of $(k\beta)^{-1}$.

The phenomena of total internal reflexion and the consequent evanescent wave have several important uses. By total internal reflexion at an angle of incidence of 45° (less than the critical angle in glass) various types of prism are used in optical instruments to reflect light rays with or without inversion of an image. An important application is found in the common design of field glasses.

In optical fibres, repeated total internal reflexion at the wall is used to transfer light energy along the length of the fibre, with negligible loss. In addition, the existence of the evanescent wave outside the fibre gives rise to one of the ways in which energy can be extracted without any mechanical disturbance. Optical fibres will be discussed in more detail in § 11.9.

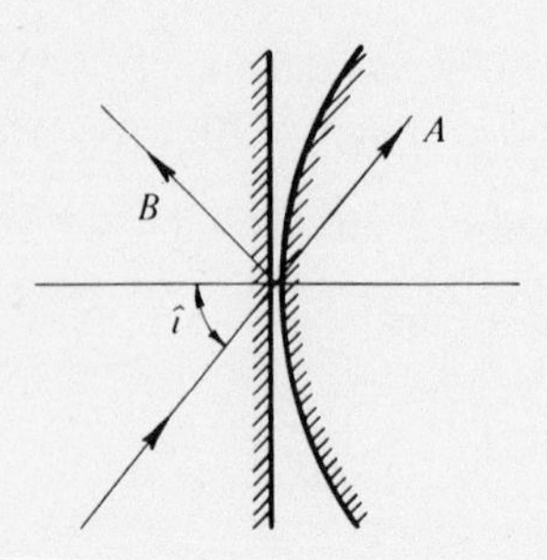

Fig. 4.9. Use of curved surface to observe tunnelling as a function of separation.

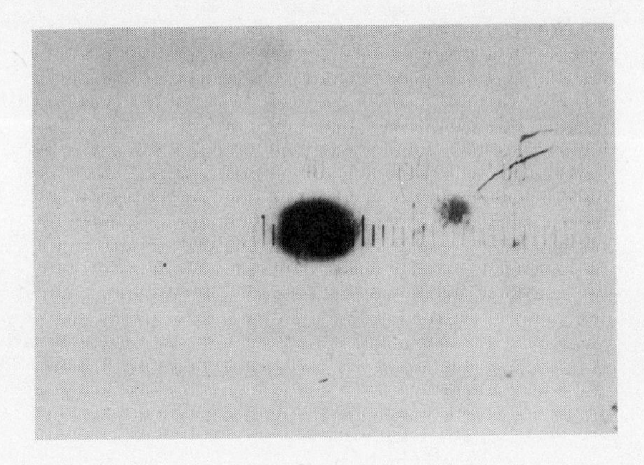

Fig. 4.10. Observations at B (Fig. 4.9) with incidence just greater than critical angle ($\hat{\imath} > \hat{\imath}_c$).

4.3.3 *An experiment to illustrate the evanescent wave.* To demonstrate clearly the *frustration* of total reflexion by a second surface we make the separation d (Fig. 4.8) vary from zero upwards by replacing the second plane surface by a spherical one, that of a biconvex lens of focal length about 20 cm being adequate. The two surfaces are almost in contact at one point (Fig. 4.9). The intensity of the wave observed at A drops off – but not discontinuously – as we move away from the point of contact. Alternatively the intensity at B can be observed, in which case a dark spot is observed around the point of contact (Fig. 4.10). To show that transmission occurs up to a separation of the order of one wavelength, the direction of illumination is altered to be non-critical so that a set of Newton's rings is observed at either A or B (Fig. 4.11), and the separation between the surfaces can be estimated in terms of them. The observations

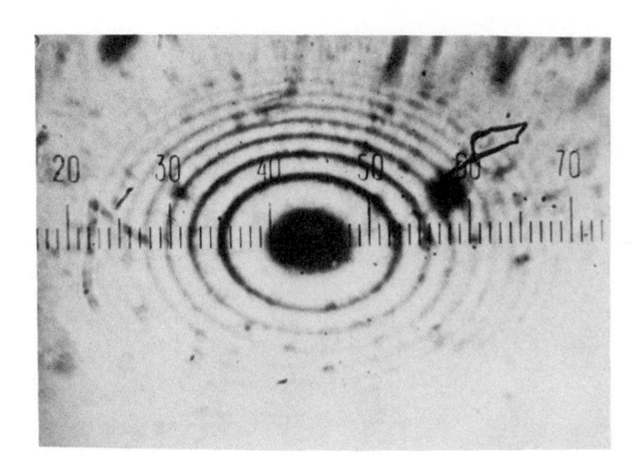

Fig. 4.11. As Fig. 4.10, but with incidence just below critical angle ($\hat{\imath} < \hat{\imath}_c$).

shown in Figs. 4.10 and 4.11 indicate transmission over a region where the gap is up to about $\frac{3}{4}\lambda$.

4.3.4 *Energy flow in the evanescent wave.* Because the amplitude of the evanescent wave decays with z no energy can be transported away from the surface. The wave which was incident with perpendicular polarization has transmitted components which have space-variation

$$\left.\begin{aligned} E_{yT} &\sim \exp(-k\beta z)\exp\{-\mathrm{i}kx(1+\beta^2)^{\frac{1}{2}}\}, \\ H_{zT} &\sim Z_2^{-1}(1+\beta^2)^{\frac{1}{2}} E_{yT}, \\ H_{xT} &\sim Z_2^{-1}\mathrm{i}\beta E_{yT}. \end{aligned}\right\} \tag{4.48}$$

The Poynting vector $\mathbf{\Pi} = \mathbf{E}\times\mathbf{H}$ thus has components

$$\left.\begin{aligned} \mathbf{\Pi}_x &\sim (E_{yT})^2 Z_2^{-1}(1+\beta^2)^{\frac{1}{2}}, \\ \mathbf{\Pi}_z &\sim (E_{yT})^2 Z_2^{-1}\mathrm{i}\beta. \end{aligned}\right\} \tag{4.49}$$

The former, being real, represents a real flow of energy along the x direction, parallel to the surface, but restricted to a layer of half the decay distance. The latter represents a wave in which **E** and **H** oscillate in quadrature with one another (the imaginary iβ indicates this) and the Poynting vector has zero time average. The evanescent wave thus transports no energy away from the surface.

4.3.5 *Phase changes on critical reflexion.* The phase changes (4.46) for the two polarizations

$$\alpha_\perp = 2\tan^{-1}\frac{\mu_r\beta}{\cos\hat{\imath}}, \qquad \alpha_\| = -2\tan^{-1}\frac{\mu_r\cos\hat{\imath}}{\beta}, \tag{4.50}$$

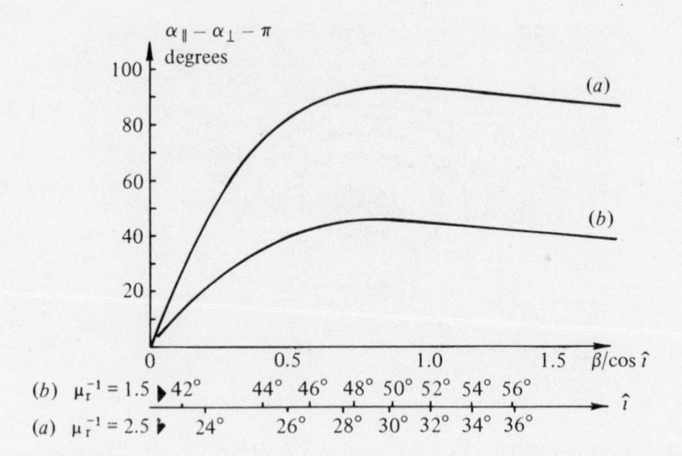

Fig. 4.12. The difference in phase change produced on internal reflexion of waves polarized in and normal to the plane of incidence. Here $\alpha_\parallel - \alpha_\perp - \pi$, is plotted as a function of $\beta/\cos \hat{\imath}$ for the examples (*a*) $\mu_r^{-1} = 2.5$ and (*b*) $\mu_r^{-1} = 1.5$.

have somewhat different dependence on the angle $\hat{\imath}$ in the region between $\hat{\imath}_c$ and $\pi/2$. They have the values 0 and π respectively at $\hat{\imath} = \hat{\imath}_c$ ($\beta = 0$); at $\hat{\imath} = \pi/2$ they are $-\pi$ and 0. The difference $\alpha_\parallel - \alpha_\perp - \pi$ can be evaluated for any particular value of μ_r. It is shown in Fig. 4.12, plotted as a function of $\beta/\cos t$ for the two values of μ_r

$$\mu_r = 1.5^{-1},$$
$$\mu_r = 2.5^{-1}.$$

For $\beta/\cos \hat{\imath} = 1$ the phase difference has its maximum value, $\pi + 46°$ and $\pi + 94°$ respectively for the two values of μ_r. This calculation becomes quite important both in the study of fibre optics (§ 11.9) and in understanding the Fresnel rhomb (§ 5.2.4).

As a result of the calculations we have carried out it is now possible to work out the reflexion coefficient over the whole range of $\hat{\imath}$ from zero to $\pi/2$ in the case where $\mu_r < 1$. These are illustrated in Fig. 4.13 for $\mu = 1/1.5$.

4.3.6 *Electromagnetic waves incident on a conductor.* The media we have discussed so far have all been insulators, as a result of which we have been able to neglect the current term in the equation (4.3),

$$\nabla \times \mathbf{H} = \frac{\partial \mathbf{D}}{\partial t} + \mathbf{j}. \tag{4.51}$$

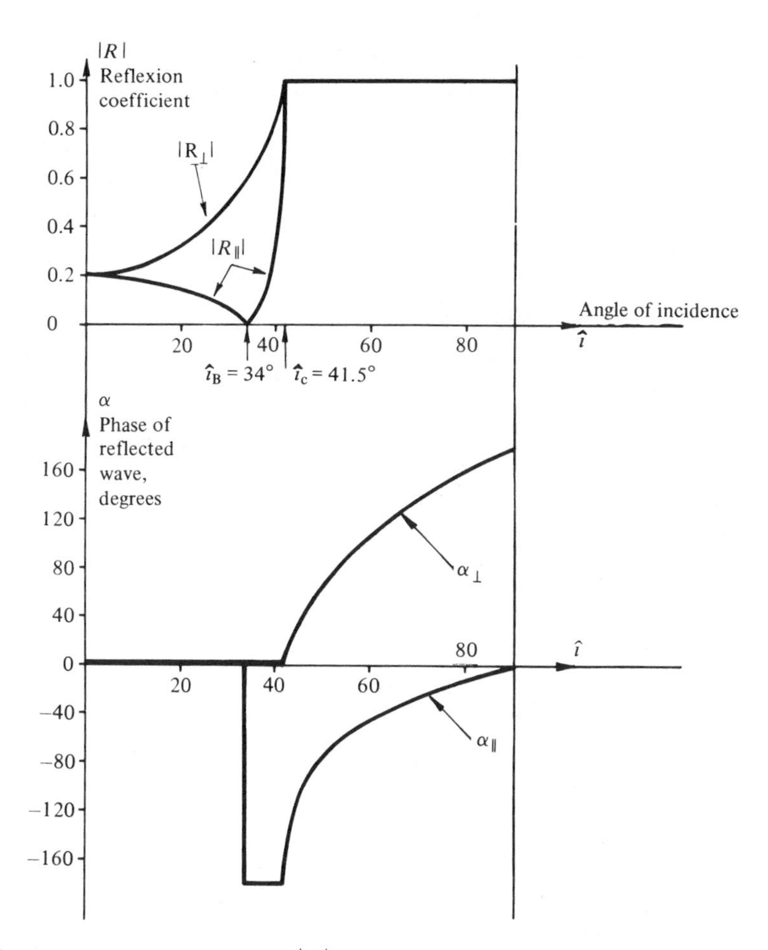

Fig. 4.13. Reflexion coefficient $|R|$ and phase change α occurring with incidence in the denser medium, calculated for $\mu_r^{-1} = 1.5$. The complete reflexion coefficient is $R = |R| \exp(\mathrm{i}\alpha)$.

If we wish to consider what happens to an electromagnetic wave incident on a conductor we must bring this term into play, as the electric field $\mathbf{E}$ will induce a non-zero current density $\mathbf{j}$ if the conductivity σ is appreciable:

$$\mathbf{j} = \sigma \mathbf{E}. \tag{4.52}$$

We can substitute (4.52) into (4.51) at the same time replacing $\mathbf{D}$ by $\epsilon\epsilon_0\mathbf{E}$ to give:

$$\nabla \times \mathbf{H} = \epsilon_0 \epsilon \frac{\partial \mathbf{E}}{\partial t} + \sigma \mathbf{E}. \tag{4.53}$$

Now, remembering that the wave is oscillatory with frequency ω, we replace the operator $\partial/\partial t$ by $\mathrm{i}\omega$ (§ 2.2.1) and thus obtain the equation

$$\nabla \times \mathbf{H} = \epsilon_0 \mathbf{E} \mathrm{i}\omega\left(\epsilon + \frac{\sigma}{\mathrm{i}\omega\epsilon_0}\right). \tag{4.54}$$

The conductivity term can be absorbed into the equation if the real dielectric constant is replaced by a complex one

$$\epsilon_\mathrm{c} = \epsilon + \frac{\sigma}{\mathrm{i}\omega\epsilon_0}. \tag{4.55}$$

This is an important result; propagation in a conductor can be treated formally as propagation in an insulator with a complex dielectric constant. The reason is easy to see. In an insulator the electric field produces displacement current $\partial\mathbf{D}/\partial t$ in quadrature with itself; in a conductor the real current density is in phase with $\mathbf{E}$, and thus the net effect is a total current at an intermediate phase angle, which is represented by a complex ϵ_c.

As the mathematics is similar to that in § 4.2.2 for a real dielectric, we shall take the standard result:

$$v = (\epsilon\mu)^{-\frac{1}{2}}c,$$

and substitute ϵ_c from equation (4.55) to give

$$v = \left\{\mu\left(\epsilon - \frac{\sigma \mathrm{i}}{\omega\epsilon_0}\right)\right\}^{-\frac{1}{2}} c. \tag{4.56}$$

If there are no unusual molecular polarization effects (such as ferroelectricity), one can assume ϵ to be of the order of unity. Substitution of any reasonable values for σ and ω then shows the imaginary term to be completely dominant. We therefore write:

$$c/v = \mathrm{\mu} = (-\mathrm{i}\mu\sigma/\epsilon_0\omega)^{\frac{1}{2}} = (\mu\sigma/2\epsilon_0\omega)^{\frac{1}{2}}(1-\mathrm{i}). \tag{4.57}$$

Following § 2.3.1 we can then write down the effect of applying a wave of frequency ω:

$$\mathbf{E} = \mathbf{E}_0 \exp\{\mathrm{i}(\omega t - kz)\} = \mathbf{E}_0 \exp\{\mathrm{i}\omega(t - \mathrm{\mu} z/c)\} \tag{4.58}$$

normally to the surface $z = 0$ of a conductor; at depth z we have from (4.57):

$$\mathbf{E}(z) = \mathbf{E}_0 \exp\{-(\mu\mu_0\sigma\omega/2)^{\frac{1}{2}}z\} \exp[\mathrm{i}\{\omega t - (\mu\mu_0\sigma\omega/2)^{\frac{1}{2}}z\}]. \tag{4.59}$$

This is an attenuated wave, with characteristic decay length l and wavelength λ inside the conductor given by

$$l = \lambda/2\pi = (\mu\mu_0\sigma\omega/2)^{\frac{1}{2}}. \tag{4.60}$$

The decay per wavelength is thus independent of the frequency. The existence of such a decay implies that a wave cannot travel more than a few wavelengths inside a conductor.

Table 4.1 shows some typical values of the *skin-depth* l at various frequencies; it is clear that at optical frequencies the penetration into copper is almost negligible, even on an atomic scale, although admittedly the theory above is not applicable at such frequencies, where the skin-depth is less than the electronic mean free path in the metal.

Table 4.1. *Skin-depths in copper at various frequencies at* 0 °C

Frequency (Hz)	Free-space wavelength	Skin-depth
50		10 mm
1000		2.4 mm
10^6	300 m	0.01 mm
0.6×10^{15}	5000 Å	17 Å

4.3.7 *Reflexion by a metal surface.* As a result of the above calculation one can calculate some of the properties of an ideal conductor, i.e. one in which Maxwell's equations describe the true conditions. Such calculations ignore atomic effects, which give rise to the characteristic colour of copper, for example, but might be expected to be true in the visible region for simpler metals such as aluminium. One can substitute the complex value for μ from (4.57) in equation (4.37) for the reflectivity, for example, and deduce results of which we shall just mention one.

Light is incident obliquely on a metal surface in the TM mode (**E** in the plane of incidence). The reflexion coefficient as a function of angle is

$$R_{\|} = \{\cos \hat{r} - s(1-\mathrm{i}) \cos \hat{\imath}\}/\{\cos \hat{r} + s(1-\mathrm{i}) \cos \hat{\imath}\}, \tag{4.61}$$

where

$$s = (\mu\sigma/2\epsilon_0\omega)^{\frac{1}{2}} \tag{4.62}$$

Since $s \gg 1$ we can assume $\cos \hat{r} = 1$, whence

$$R_{\|} = [1 - s(1-\mathrm{i}) \cos \hat{\imath}]/[1 + s(1-\mathrm{i}) \cos \hat{\imath}]. \tag{4.63}$$

For small angles of incidence $\hat{\imath}$, $R_{\|}$ has the value -1: perfect reflexion with a π phase change. As we approach glancing incidence, the phase of the reflected wave changes continuously to give $R_{\|}$ the value of $+1$ at $\hat{\imath} = \pi/2$. The phase change occurs around what might be described as a 'complex Brewster angle' at which the real and imaginary parts of R are comparable, i.e.

$$s \cos \hat{\imath} \approx 1. \tag{4.64}$$

Here the value of $|R|$ falls to a value somewhat less than unity. With an aluminium surface at room temperature, this angle is about 89° for visible light, and the intensity reflexion coefficient $|R|^2$ falls to a minimum of about 20 per cent. At smaller angles of incidence, the behaviour of real metals is indistinguishable from that of an ideal metal with $\sigma \to \infty$.

4.4. Effects in non-homogeneous media

The results that we have discussed in § 4.3 are all concerned with sudden changes in properties at a plane boundary between one medium and another. There are many situations where a gradual change of the properties of the medium occur in a distance of the order of, or greater than, the wavelength. One meets an important example in the study of the reflexion of radio waves from a charged ionosphere, which is the principal means of investigating the upper atmosphere (see § 11.7). And the same principles are involved in understanding the everyday phenomenon of the mirage.

4.4.1 *The Wentzel, Kramers and Brillouin* (*WKB*) *method.* The method of treatment is, of course, to try to solve Maxwell's equations under the given conditions; the difficulty arises in the mathematics because the value of μ, which has hitherto been a constant, is now a function of position. The easiest case to consider in general is the one in which the characteristic distance for a change of properties is very much larger than the wavelength, so that locally the disturbance can be well represented by a plane-wave of amplitude A:

$$E(\mathbf{r}) = A(\mathbf{r}) \exp\{\mathrm{i}(\omega t - \mathbf{k}(\mathbf{r}) \cdot \mathbf{r})\}. \tag{4.65}$$

We observe that the frequency must remain the same at all points in a time-independent system. The method of tackling the problem is quite adequately illustrated by a one-dimensional example. The refractive index is assumed to be a function of z only:

$$\mu = \epsilon^{\frac{1}{2}} = \mu(z) \tag{4.66}$$

and the wave equation for the electric field becomes

$$\frac{\partial^2 E}{\partial t^2} = \frac{c^2}{\mu^2(z)} \nabla^2 E = \frac{c^2}{\mu^2(z)} \frac{\partial^2 E}{\partial z^2}, \tag{4.67}$$

which has a local solution of the form (4.65) with $k(r) = [\mu(r)\omega]/c$. The method of solution of the equation to find $A(\mathbf{r})$ is called the WKB method after its originators, Wentzel, Kramers and Brillouin. In a uniform medium the phase of the wave is $(\omega t - kz)$; in the non-uniform medium it will clearly be the integral $(\omega t - \int_0^z k(z)\,\mathrm{d}z)$ as the phase is 2π times the number of cycles undergone by the wave along its path from $z = 0$. We thus take, for a trial solution of equation (4.67), the wave

$$E(z) = A(z) \exp\left\{\mathrm{i}\left(\omega t - \int_0^z \frac{\mu(z)\omega}{c}\,\mathrm{d}z\right)\right\}. \tag{4.68}$$

Substituting this into the wave equation (4.67) leads to

$$-\omega^2 A = \frac{c^2}{\mu^2}\left(\frac{\partial^2 A}{\partial z^2} + 2\frac{\mathrm{i}\omega\mu}{c}\frac{\partial A}{\partial z} - \frac{\omega^2\mu^2}{c^2}A + \frac{\mathrm{i}\omega}{c}\frac{\partial\mu}{\partial z}A\right), \tag{4.69}$$

$$0 = \frac{c^2}{\mu^2}\frac{\partial^2 A}{\partial z^2} + 2\frac{\mathrm{i}\omega c}{\mu}\frac{\partial A}{\partial z} + \frac{\mathrm{i}\omega c}{\mu^2}\frac{\partial\mu}{\partial z}A. \tag{4.70}$$

The mathematics so far is exact. To make the equation tractable in general (for particular functions $\mu(z)$ there will no doubt be exact solutions) we use the fact that conditions are only varying slowly – $\partial\mu/\partial z$ is small – and so $\partial A/\partial z$ is small and $\partial^2 A/\partial z^2$ smaller still. We then neglect the last of these in equation (4.70) giving

$$0 = \frac{2}{A}\frac{\partial A}{\partial z} + \frac{1}{\mu}\frac{\partial\mu}{\partial z}, \tag{4.71}$$

which has the simple solution

$$A = B\mu^{-\frac{1}{2}}, \tag{4.72}$$

where B is an arbitrary constant. The form of the wave is thus

$$E(z) = B\mu(z)^{-\frac{1}{2}} \exp\left\{\mathrm{i}\left(\omega t - \int_0^z \frac{\omega\mu(z)}{c}\,\mathrm{d}z\right)\right\}, \tag{4.73}$$

provided that the refractive index μ varies slowly enough with distance and the amplitude A also varies slowly.

An example of the solution for a linear change of μ is shown in Fig. 4.14.

4.4.2 *Energy flow*. The continuity of energy flow is consistent with the above calculation. H is given, as before, by

$$H = E/Z$$

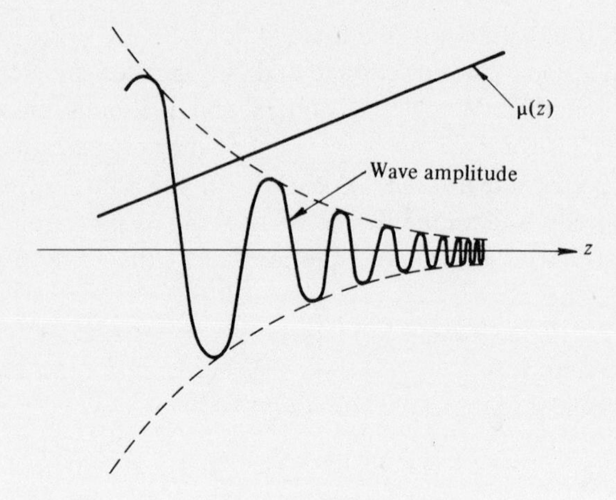

Fig. 4.14. Progression of a wave in a region where the refractive index μ is a linear function of position.

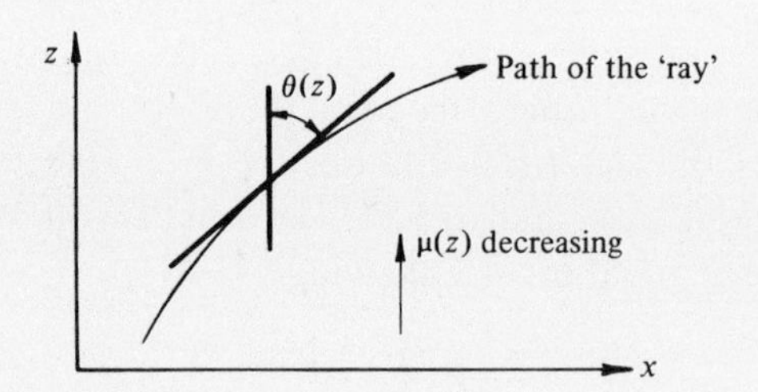

Fig. 4.15. Oblique ray in a region of changing μ.

and as Z is proportional to μ^{-1} we have

$$|\mathbf{\Pi}| = |\mathbf{E}\times\mathbf{H}| = \frac{|E|^2}{Z} \sim \mu^0, \tag{4.74}$$

which is independent of the position z. Thus energy is conserved.

4.4.3 *Reflexion of the wave. Mirages.* When the wave is incident at an angle to the normal, reflexion takes place at a certain value of μ. This is what occurs when light is reflected by a temperature gradient in air, for example. We consider the same refractive conditions $\mu(z)$, but allow the wave-vector to make an angle $\theta(z)$ with the z-axis (Fig. 4.15). We now write the wave as

$$E = A(z)\exp\left[\mathrm{i}\omega\left\{t - \int_0^{x,z}\frac{\mu(z)}{c}(\sin\theta(z)\,\mathrm{d}x + \cos\theta(z)\,\mathrm{d}z)\right\}\right]. \tag{4.75}$$

(Compare equation (4.68), obtained by putting $\theta = 0$.) We can eliminate $\theta(z)$ by using Snell's law of refraction which states that

$$\mu(z) \sin\theta = K, \quad \text{a constant.} \tag{4.76}$$

The values of K depends on the initial values of θ and μ at $z = 0$. Then (4.75) becomes

$$E = A(z)\exp\left[\mathrm{i}\omega\left\{t - \int_0^x K\,\mathrm{d}x/c - \int_0^z (\mu^2 - K^2)^{\frac{1}{2}}\,\mathrm{d}z/c\right\}\right]. \tag{4.77}$$

We can substitute this value of E into the two-dimensional form of the wave equation (4.67),

$$\frac{\partial^2 E}{\partial t^2} = \frac{c^2}{\mu^2(z)}\left(\frac{\partial^2 E}{\partial x^2} + \frac{\partial^2 E}{\partial z^2}\right)$$

to give, in an identical manner to the previous calculation,

$$A(z) = C(\mu^2 - K^2)^{-\frac{1}{4}}. \tag{4.78}$$

The amplitude of the wave thus appears to become infinite at the height where $\mu = K$; from (4.76) this corresponds to $\sin\theta = 1$, and is just the height at which critical reflexion (§ 4.3.2) occurs. The infinite magnitude of the result is not to be trusted, however, as the WKB approximation breaks down here because $\partial^2 A/\partial x^2$ can no longer be neglected. In fact the wave-disturbance travelling along the x-axis is progressive; the wave can clearly be written as the product

$$E = C(\mu^2 - K^2)^{-\frac{1}{4}} \exp(\mathrm{i}\omega t) \times \exp\left(-\frac{\mathrm{i}\omega K x}{c}\right) \cdot \exp\left\{\frac{\mathrm{i}\omega}{c}\int_0^z (\mu^2 - K^2)^{\frac{1}{2}}\,\mathrm{d}z\right\}. \tag{4.79}$$

Thus the light ray is bent to return downwards. Above the level where $\mu = K$ the wave becomes evanescent, as there $(\mu^2 - K^2)^{\frac{1}{2}}$ is imaginary. By comparing the values of the evanescent wave at the reflexion level and the necessarily standing wave below it we can in fact confirm that, to satisfy continuity of E, the amplitude of disturbance does not become infinite.

The practical results of this, as a mirage, is often observed above a black road during the summer. Because of the absorption of the sun's radiation by the road a layer of hot air is produced just next to the surface; the refractive index $\mu(z)$ thus increases with height, and a light ray proceeding downwards is turned round as in Fig. 4.16. At temperature T (°C) air has a refractive index

$$\mu(T) = 1.000\,291 - 1 \times 10^{-6}\,T,$$

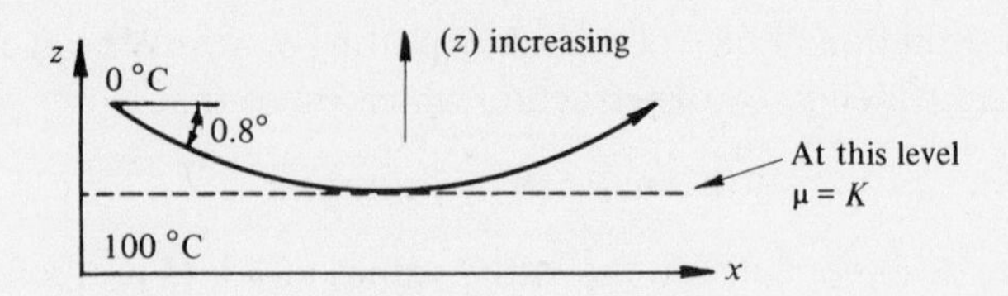

Fig. 4.16. A mirage.

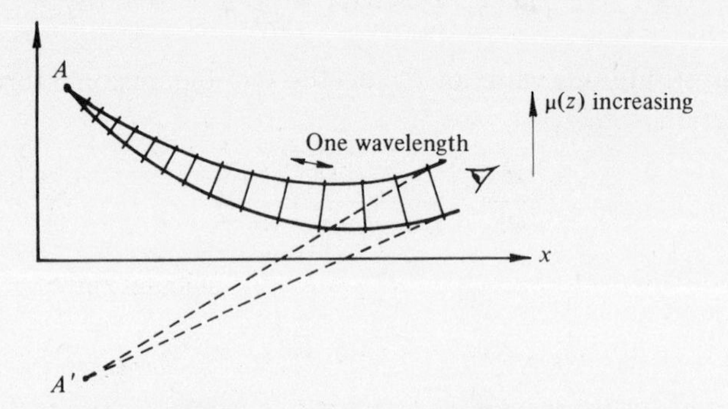

Fig. 4.17. Fermat's principle applied to the mirage.

so that if the road has a temperature of 100 °C and the observer 0 °C, the condition $\mu = K$ gives

$$\mu(100\,°\mathrm{C}) = K = \mu(0\,°\mathrm{C}) \sin\theta,$$

where θ is the angle of observation. This leads to $\pi/2 - \theta = 0.8°$; under these conditions mirages can be observed at angles smaller than this value. The phenomenon of a mirage can also be treated by means of Fermat's principle (§ 2.7.2) as we shall now show. Consider two light rays starting from A in slightly different directions. As the lower one is always in a region where the refractive index is lower, the velocity there is larger, and therefore the upper ray must travel a shorter path to take the same time. The direction of travel at any point in the path is given by the normal to the wavefront which, from the definition in § 2.6.1, is the surface of constant phase – or time. Fig. 4.17 then illustrates why the path is curved. By the use of Lagrange's theory of minimizing or maximizing pathlengths as a function of the exact path, this reasoning can be made quantitative.

4.5 Radiation

There are several aspects of the radiating properties of accelerating charge distributions which are of importance in optics, both classical and quantum. In this section we shall describe the radiation by multipole systems, and the phenomenon of radiation pressure. The results will be used later for the discussion of scattering (§ 10.2).

4.5.1 *Radiation by a small electric dipole.* A radiating atom is much smaller than one wavelength of light; so also are many classes of scattering particles. In this section we shall describe the radiation pattern of such an object, assuming that it behaves as an oscillating electric dipole. Scattering particles which are not small compared with the wavelength can be treated only approximately by rather complicated mathematics (Mie scattering) and are completely out of the scope of this book.

The problem we have to solve is to calculate the electric and magnetic field produced by a point dipole **p** which oscillates as a function of time:

$$\mathbf{p} = \mathbf{p}_0 \sin \omega t. \tag{4.80}$$

It clearly creates its own electrostatic field. In addition, the oscillation is equivalent to a periodic transfer of charge from one end of the elementary dipole to the other. This means that it has the properties of an oscillating current element, which creates a magnetic field. The magnetic field changes with time and therefore induces an additional electric field. The vector product of these electric and magnetic fields gives a non-zero Poynting vector, which indicates a flow of energy to or from the dipole.

The mathematical problem outlined above is discussed in detail in texts on electromagnetic theory (see e.g. Grant & Phillips, 1975) and it seems out of place to give the detailed derivation here. It is found that for a dipole $(0, 0, p_0)$ in the z-direction, the fields at distance r, large compared to a wavelength λ, are, in polar coordinates (r, θ, ϕ):

$$(E_r, E_\theta, E_\phi) = \left(0, \frac{-\omega^2 p_0 \sin\theta}{4\pi\epsilon_0 c^2} \frac{\sin\{\omega(t - r/c)\}}{r}, 0\right), \tag{4.81}$$

$$(H_r, H_\theta, H_\phi) = \left(0, 0, \frac{-\omega^2 p_0 \sin\theta}{4\pi c} \frac{\sin\{\omega(t - r/c)\}}{r}\right), \tag{4.82}$$

and the Poynting vector is thus radial,

$$(\Pi_r, \Pi_\theta, \Pi_\phi) = \left(\frac{\omega^4 p_0^2 \sin^2\theta}{16\pi^2\epsilon_0 c^3} \frac{\sin^2\{\omega(t - r/c)\}}{r^2}, 0, 0\right). \tag{4.83}$$

The directions of the vectors are illustrated in Fig. 4.18. One should notice the behaviour of these fields:

(a) The radiated wave is characteristic of a spherical wave emitted from a source at the origin (§ 2.6.2). Its amplitude E falls off as r^{-1}, and its intensity Π as r^{-2}, as would be expected.

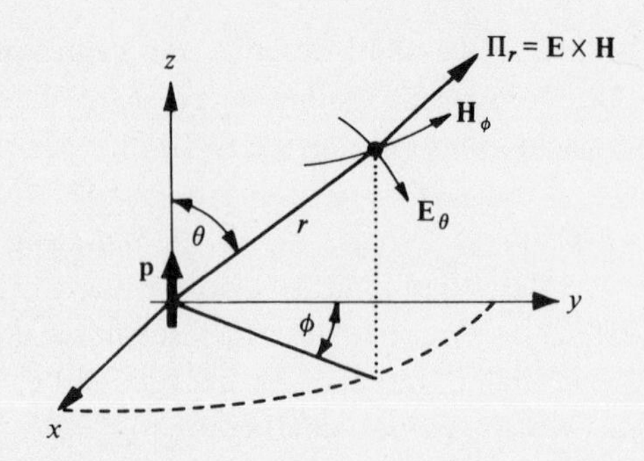

Fig. 4.18. Radiation from an oscillating electric dipole.

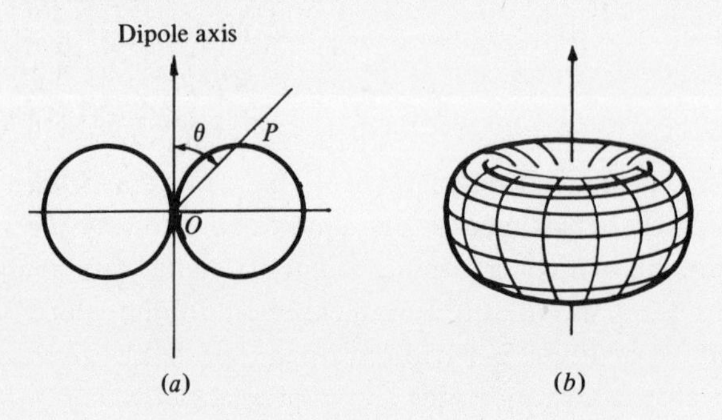

Fig. 4.19. Radiation polar-diagram for a dipole oscillator: (*a*) two-dimensional section; (*b*) three-dimensional sketch.

(b) The polarization of the wave is the direction of the vector **E** which always lies in the plane of $\mathbf{p}_0$ and **r**. In other words, the wave is plane-polarized parallel to the source dipole.

(c) The intensity of the wave is greatest in the direction normal to $\mathbf{p}_0$ $(\theta = \pi/2)$ and falls to zero as the axis of $\mathbf{p}_0$ is approached, the functional relationship to the angle θ being $\sin^2 \theta$. In fact one sees that the radiated amplitude in any direction **r** is just proportional to the component of $\mathbf{p}_0$ transverse to that direction, i.e. to $(p_0 \sin \theta)$. Radiation along the z-axis, in which direction the charge motion is longitudinal, is exactly zero.

The intensity as a function of angle is often described graphically by a *radiation polar-diagram*, in which $\Pi(\theta, \phi)$ is drawn as a function of θ and ϕ. For the dipole, Fig. 4.19 shows Π as a function of θ only, and as a function of (θ, ϕ) for later comparison with other multipoles.

The total radiated energy is given by integrating $\mathbf{\Pi}(r, \theta, \phi)$ over a surface enclosing the dipole. The result is independent of the choice of surface as a result of Gauss's law; choosing a sphere of radius r we have instantaneously radiated power

$$P_{\mathrm{d}}=\frac{\omega^4 p_0^2 \sin^2\{\omega(t-r/c)\}}{16\pi^2\epsilon_0 c^3}\int_0^{2\pi}\int_0^{\pi}\sin^2\theta\cdot 2\pi\sin\theta\,\mathrm{d}\theta\,\mathrm{d}\phi$$

$$=\frac{\omega^4 p_0^2 \sin^2\{\omega(t-r/c)\}}{6\pi\epsilon_0 c^3}. \tag{4.84}$$

The mean value of this power over many periods of the wave is

$$\overline{P_{\mathrm{d}}}=\omega^4 p_0^2/12\pi\epsilon_0 c^3. \tag{4.85}$$

This is the rate of radiation of energy by a classical oscillating dipole.

4.5.2 *Higher-order electric multipoles.* Calculation of the radiation by oscillating multipoles of higher order can be conveniently carried out by considering them as collections of dipoles. Multipoles are important when we consider the radiation properties of atomic systems.

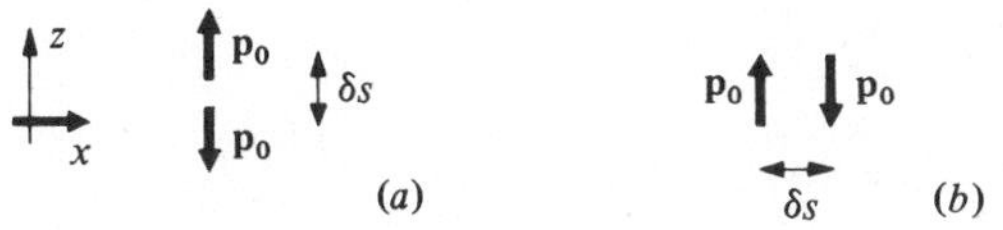

Fig. 4.20. Two types of quadrupole.

We shall only briefly discuss the quadrupole here; there are two types of quadrupole created by putting opposed dipoles either end-to-end or side-by-side (Fig. 4.20). The radiation pattern of the first type (Fig. 4.20(*a*)) is then given by superimposing the pattern of $\mathbf{p}_0$ centred on the z-axis at $z=\delta s/2$ on that of $-\mathbf{p}_0$ at $z=-\delta s/2$. The result is fields given by:

$$\mathbf{E},\mathbf{H}=\delta s\frac{\partial}{\partial z}(\mathbf{E},\mathbf{H}\text{ for the dipole }\mathbf{p}_0). \tag{4.86}$$

Likewise, the second type (Fig. 4.20(*b*)) gives us

$$\mathbf{E},\mathbf{H}=\delta s\frac{\partial}{\partial x}(\mathbf{E},\mathbf{H}\text{ for the dipole }\mathbf{p}_0).$$

The radiation polar diagrams are shown in Fig. 4.21 for the two quadrupoles. The first one radiates a total mean power of

$$\overline{P_{\mathrm{q}}}=\omega^6 p_0^2\delta s^2/20\pi\epsilon_0 c^5. \tag{4.87}$$

This is generally much smaller than $\overline{P_{\mathrm{d}}}$, implying that the quadrupole is a poor source of electromagnetic radiation compared with the dipole.

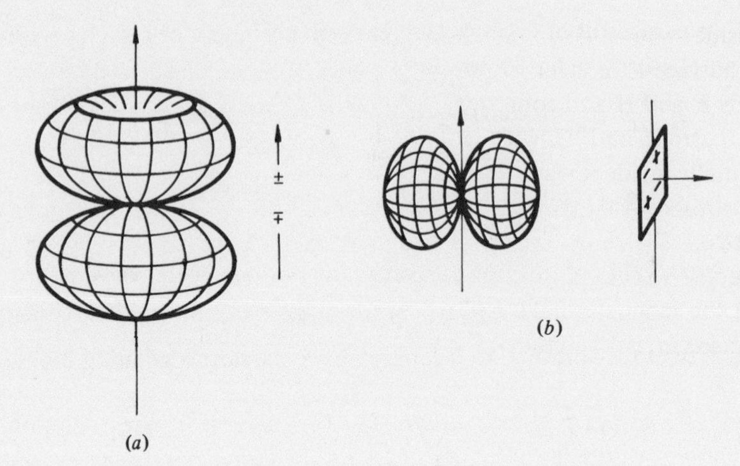

Fig. 4.21. Radiation polar-diagrams for two types of quadrupole: (*a*) *zz*; (*b*) *zy*.

However, if we write the ratio

$$\overline{P_{\mathrm{q}}}/\overline{P_{\mathrm{d}}} = 3\omega^2\,\delta s^2/5c^2 = \frac{3}{5}\left(\frac{2\pi}{\lambda}\right)^2\,\delta s^2, \qquad (4.88)$$

we see that the radiated powers may become comparable if

$$\delta s \approx \lambda/2\pi. \qquad (4.89)$$

In other words, when the wavelength of the radiation is comparable with the size of the radiator, radiation by multipoles may be comparable with that by dipoles. This result is important in the discussion of 'forbidden' atomic transitions.

4.5.3 *Radiation pressure.* Electromagnetic radiation conveys momentum as well as energy. When the direction or magnitude of the energy flow is altered by interaction with a body, there is consequently a reaction equal to the rate of change of vector momentum in the wave, and this reaction is termed *radiation pressure.*

In terms of photons, a concept dealt with in more detail in later sections, the magnitude of radiation pressure is easily related to the radiation-energy density. In this section we shall show that it is not purely a particle property, but can be explained as a result of Maxwell's equations.

4.5.4 *Momentum in an electromagnetic wave.* The electromechanical origin of radiation pressure as a surface effect is not difficult to see. If we consider reflexion by a metal surface, the wave induces electric currents in the surface-skin layer which are in phase with the electric and magnetic fields **E** and **H** and consequently there is a force of interaction between the current and **H** with non-zero time-average. The reflexion of a normally incident wave by a perfect conductor is particularly simple, and we can show that the radiation pressure is equal to the energy density of electromagnetic radiation at the surface, including both incident and reflected waves.

If the magnetic field in the combined waves at the surface is **H**, the surface current **J** in the metal is normal to **H**, and has magnitude

$$J = H \tag{4.90}$$

and is distributed throughout the skin depth. This arises simply from the Ampère circuit integral. The wave-disturbance within the metal is evanescent, and therefore maintains the same phase as at the surface. The force per unit area is then given by

$$\mathbf{R} = \mathbf{J} \times \mathbf{B}, \tag{4.91}$$

which has magnitude BH and is directed normal to the surface. The magnitude is twice the energy density $\mathscr{E}$ of the incident wave:

$$\mathscr{E} = \tfrac{1}{2}(\mathbf{B} \cdot \mathbf{H} + \mathbf{D} \cdot \mathbf{E}) \tag{4.92}$$

since for a perfect conductor $\mathbf{E} = 0$ on its surface. The radiation pressure is thus equal to the combined energy density of incident and reflected waves at the surface. The result can be written in terms of a momentum density in the wave:

$$\rho_{\mathrm{p}} = \frac{\mathscr{E}}{v} \quad \left(= \frac{\mathscr{E}}{c} \text{ in free space} \right), \tag{4.93}$$

which is reversed in the reflexion process. This formulation ties the result rather neatly to the concept of the photon as a relativistic particle.

4.5.5 *Radiation pressure in a dielectric.* In a dielectric a similar result holds as in a metal; the radiation pressure is equal to the *difference* between the electromagnetic energy densities at the *two* surfaces, since the wave is partially transmitted. In this case, the current density in the metal is replaced by the displacement-current density. Now the displacement current, $\partial \mathbf{D}/\partial t$, is in quadrature with the fields **E** and **H**, so at first sight the mechanical force would appear to be zero. However, we must take into account partial reflexion of the wave at the surfaces of the material, and the consequent interaction between the **B** of the wave in each direction with the $\partial \mathbf{D}/\partial t$ of the wave in the other direction. These fields are not necessarily in quadrature. For some thicknesses of material the total force just comes out to be zero, and these thicknesses correspond to the condition for complete transmission of the incident wave as a result of interference effects (§ 7.5.1).

4.5.6 *Order of magnitude of radiation pressure.* Radiation pressure is certainly not a large effect. The intensity of sunlight, for example, is about 10^3 W m^{-2}. These 10^3 watts are in fact all the energy in the prism of light 1 m^2 in cross-section and c metres long, which arrives in one second on that square metre. The energy density is thus $10^3/c$ m^{-2}, i.e.

$$\mathscr{E} = 10^3/3\times 10^8 \approx 3\times 10^{-6}\ \mathrm{J\ m^{-3}}.$$

This is the pressure in N m^{-2}; it is about 2×10^{-8} mm Hg. Only when one considers the effects of high-energy lasers does radiation pressure produce easily observable effects.

5

Polarization and anisotropic media

5.1. Introduction

Wave propagation in anisotropic media is usually considered as a specialized topic, mathematically too complicated to be studied by the general physicist. To a certain extent this is true; a complete analysis of crystal optics usually involves a great deal of algebra and the gains in understanding seem very small compared to the cost.

Our aim in this chapter is to present the basic physics of electromagnetic-wave propagation in anisotropic materials in an unconventional manner, by the use of a geometrical method. The advantage of this method is that it is directly applicable to a number of studies other than optics; the method is in fact based on that used to study the dynamics of electrons in crystals. We shall see it applied in this chapter to light propagation in birefringent and optically active materials, and in § 11.7 to wave propagation in the ionosphere where the anisotropy is induced by the Earth's magnetic field.

The chapter begins with a formal discussion of polarized light in terms of vectors and goes on to discuss the general formation of the problem of wave propagation in any medium. It shows that the complete propagation characteristics can be deduced in a straightforward manner by the construction of a refractive-index surface. Following this, we shall use the geometrical method mentioned above to calculate the refractive-index surface for a dielectric crystal and hence to deduce its wave properties. In the final section we shall discuss some practical uses of crystal optics.

5.2 Polarized light

We shall discuss the main types of polarized radiation. All forms of polarization can be described as elliptical polarization of various degrees, and for some purposes this is a convenient specification. However, we shall here proceed in the usual fashion by describing linearly and circularly polarized light first and then absorbing the two into the more general description of elliptically polarized light.

5.2.1 *Plane-polarized light.* In § 4.2.1 we showed that the simplest basic solution to Maxwell's wave equation was a plane-polarized transverse wave. If the wave-vector is **k**, the electric and magnetic field vectors **E** and **H** are mutually perpendicular and lie in a plane normal to **k**; they both oscillate in phase, so that:

$$\mathbf{E} = \mathbf{E}_0 \exp\{\mathrm{i}(\omega t - \mathbf{k}\cdot\mathbf{r})\},$$
$$\mathbf{H} = \mathbf{H}_0 \exp\{\mathrm{i}(\omega t - \mathbf{k}\cdot\mathbf{r})\},$$
$$\mathbf{E}\cdot\mathbf{H} = 0; \qquad \mathbf{E}\cdot\mathbf{k} = 0; \qquad \mathbf{H}\cdot\mathbf{k} = 0. \tag{5.1}$$

This solution was obtained for an isotropic medium, and in this section we shall restrict ourselves to that condition. It follows trivially from the orthogonality of **E**, **H** and **k** that the direction of energy flow **Π** is parallel to **k**;

$$\mathbf{\Pi} = (\mathbf{E}\times\mathbf{H}),$$
$$\mathbf{\Pi}\times\mathbf{k} = \{(\mathbf{E}\times\mathbf{H})\times\mathbf{k}\} = \{\mathbf{H}(\mathbf{E}\cdot\mathbf{k}) - \mathbf{E}(\mathbf{H}\cdot\mathbf{k})\} = 0. \tag{5.2}$$

Such a state of affairs is indicated by Fig. 5.1. In an anisotropic medium it will soon appear (§ 5.3.4) that matters are not quite so simple.

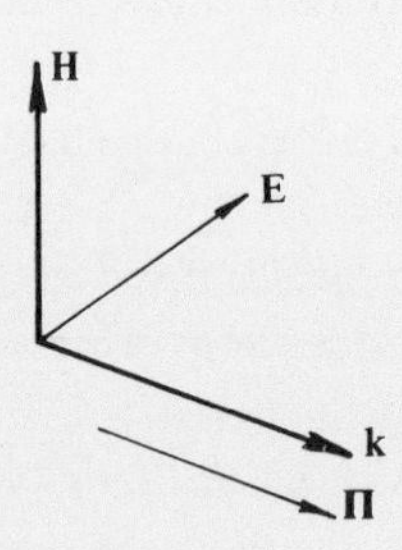

Fig. 5.1. Orthogonality of **E**, **H** and **k** in an electromagnetic wave.

5.2.2 *Circularly polarized light.* There is a second form of polarized light that is just as important as plane-polarized light. It is the result of adding two equal plane-polarized waves with the same **k** but perpendicular electric vectors and a $\pi/2$ phase difference between their oscillations. The two waves are respectively (suffices 1 and 2, which can represent x and y,

k being along z)

$$\left.\begin{aligned} \mathbf{E}_1 &= \mathbf{E}_{10} \exp\{\mathrm{i}(\omega t - \mathbf{k}\cdot\mathbf{r})\}, \\ \mathbf{E}_2 &= \mathbf{E}_{20} \exp\left\{\mathrm{i}\left(\omega t - \mathbf{k}\cdot\mathbf{r} - \frac{\pi}{2}\right)\right\}, \\ \mathbf{H}_1 &= \mathbf{H}_{10} \exp\left\{\mathrm{i}\left(\omega t - \mathbf{k}\cdot\mathbf{r} - \frac{\pi}{2}\right)\right\}, \\ \mathbf{H}_2 &= \mathbf{H}_{20} \exp\{\mathrm{i}(\omega t - \mathbf{k}\cdot\mathbf{r})\}, \end{aligned}\right\} \tag{5.3}$$

$$\left.\begin{aligned} &\mathbf{E}_1\cdot\mathbf{E}_2 = 0 = \mathbf{H}_1\cdot\mathbf{H}_2, \\ &E_{10} = E_{20} = E_0, \qquad H_{10} = H_{20} = E_0/Z. \end{aligned}\right\} \tag{5.4}$$

The combined wave represented by (5.3) and (5.4) consists of an electric field of magnitude E_{10} which rotates with frequency ω in the plane normal to **k**. This is most easily shown by noting that the observed electric field is the real part of the waveform in (5.3). We then have

$$E_1 = E_0 \cos(\omega t - \mathbf{k}\cdot\mathbf{r}), \qquad E_2 = E_0 \sin(\omega t - \mathbf{k}\cdot\mathbf{r}) \tag{5.5}$$

whence $\mathbf{E} = \mathbf{E}_1 + \mathbf{E}_2$ has modulus E given by

$$\begin{aligned} E^2 = |\mathbf{E}_1 + \mathbf{E}_2|^2 &= E_1^2 + E_2^2 + 2|\mathbf{E}_1\cdot\mathbf{E}_2| \\ &= \mathbf{E}_0^2\{\cos^2(\omega t - \mathbf{k}\cdot\mathbf{r}) + \sin^2(\omega t - \mathbf{k}\cdot\mathbf{r})\} \\ &= \mathbf{E}_0^2 \quad \text{since} \quad \mathbf{E}_1\cdot\mathbf{E}_2 = 0. \end{aligned} \tag{5.6}$$

The angle θ at which this resultant lies in the plane normal to **k** is given by

$$\tan\theta = \frac{|E_2|}{|E_1|} = \frac{\sin(\omega t - \mathbf{k}\cdot\mathbf{r})}{\cos(\omega t - \mathbf{k}\cdot\mathbf{r})} = \tan(\omega t - \mathbf{k}\cdot\mathbf{r});$$

therefore

$$\theta = (\omega t - \mathbf{k}\cdot\mathbf{r}). \tag{5.7}$$

The field is thus of magnitude E_0 and lies at an angle $(\omega t - \mathbf{k}\cdot\mathbf{r})$ to the axis of $\mathbf{E}_1$.

If we look at the source of the light and observe the electric field vector in a wavefront ($\mathbf{k}\cdot\mathbf{r}$ = constant), we see it rotating in an *anticlockwise* sense ($\theta = \omega t$). This is conventionally known as *left-hand circularly polarized light.* Alternatively we can freeze the same field at a given instant of time, and the vector lies at angle $\theta = -\mathbf{k}\cdot\mathbf{r}$ which traces out the helix of a *left-handed screw.*

When the vector $\mathbf{E}_2$ leads in phase, we write for (5.5):

$$E_1 = E_0 \cos(\omega t - \mathbf{k}\cdot\mathbf{r}), \qquad E_2 = -E_0 \sin(\omega t - \mathbf{k}\cdot\mathbf{r}) \tag{5.8}$$

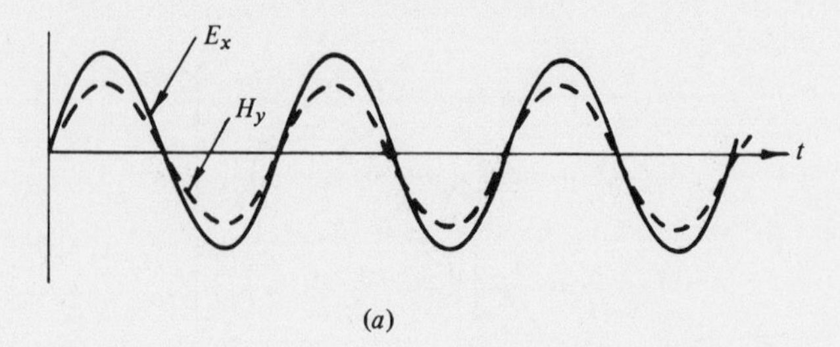

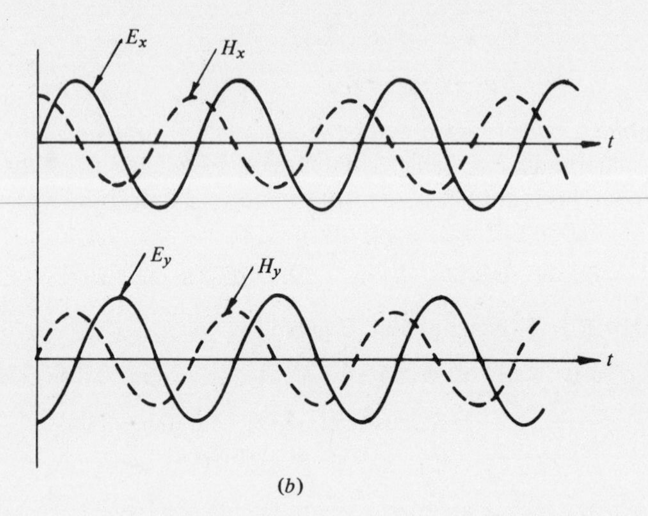

Fig. 5.2. (*a*) Plane-polarized waves. (*b*) Circularly polarized waves.

whence

$$\theta = -(\omega t - \mathbf{k} \cdot \mathbf{r}), \tag{5.9}$$

which is right-handedly polarized, the electric vector rotating in a clockwise sense. Fig. 5.2 compares for plane- and circularly-polarized light the progress of **E** and **H** with time in a given plane.

5.2.3 *Elliptically polarized light.* Both circularly- and plane-polarized light can be described as special cases of *elliptically polarized* light in which the electric vector rotates with a periodic change in length as the wave progresses, and can be thought of as describing an ellipse. Two quantities are necessary to describe elliptically polarized waves completely; these are the orientation of the axes of the ellipse and its eccentricity. The latter is defined as $(1 - a_2^2/a_1^2)^{\frac{1}{2}}$, where a_1 and a_2 are

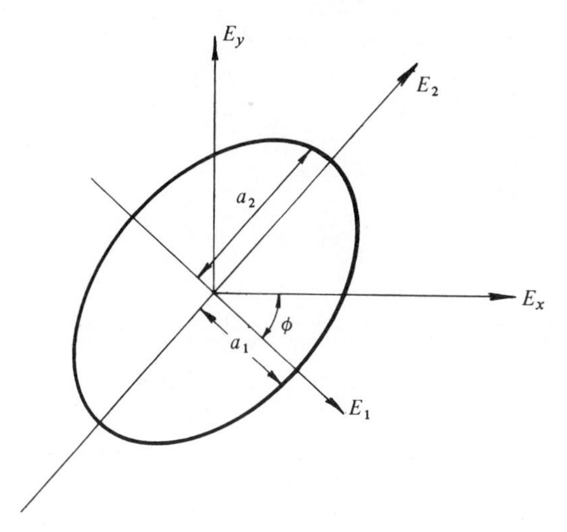

Fig. 5.3. Rotation of coordinates for elliptically polarized light.

major and minor axes respectively. Plane-polarized light then corresponds to an eccentricity of unity, whereas circularly polarized light has an eccentricity of zero. In this latter case the ellipse axes are degenerate.

We can combine the two quantities into one single complex measure of the degree of elliptical polarization, which introduces some simplifications into theories such as that of ionosphere propagation (§ 11.7). We must choose two reference axes, x and y, in a plane perpendicular to the wave normal. We can then define

$$R = \frac{E_x}{E_y}, \tag{5.10}$$

where E_x and E_y are the time-dependent components of $\mathbf{E}$ resolved along the two axes.

The electric vector rotating round an ellipse of semi-axes a_1 and a_2 can be written in terms of its principal components:

$$(E_1, E_2) = (a_1, \mathrm{i}a_2) \exp \{\mathrm{i}(\omega t - kz)\}. \tag{5.11}$$

The observed field is again the real part of the above complex vector. Rotating this to axes (E_x, E_y) at angle ϕ to the principal axes (Fig. 5.3) we have

$$\begin{aligned}(E_x, E_y) &= \begin{pmatrix} \cos\phi & \sin\phi \\ -\sin\phi & \cos\phi \end{pmatrix} \begin{pmatrix} E_1 \\ E_2 \end{pmatrix} \\ &= (a_1 \cos\phi + \mathrm{i}a_2 \sin\phi, -a_1 \sin\phi + \mathrm{i}a_2 \cos\phi) \exp \mathrm{i}(\omega t - kz),\end{aligned} \tag{5.12}$$

whence

$$R=\frac{E_x}{E_y}=\frac{a_1\cos\phi+\mathrm{i}a_2\sin\phi}{-a_1\sin\phi+\mathrm{i}a_2\cos\phi}. \tag{5.13}$$

We can now look at two special cases.

(a) *Plane-polarized light*

$$a_1=E_0,\qquad a_2=0,$$

$$R=-\cot\phi. \tag{5.14}$$

The perpendicular polarization has $\phi=\phi+\pi/2$, whence

$$R=-\cot\left(\phi+\frac{\pi}{2}\right)=\tan\phi. \tag{5.15}$$

(b) *Circularly polarized light.* If light is right-handedly polarized (5.8), its vectors can be represented by (5.11) if

$$a_1=a_2=E_0,$$

$$R=\frac{1+\mathrm{i}}{-1+\mathrm{i}}=-\mathrm{i}. \tag{5.16}$$

Similarly, for left-handedly polarized light,

$$a_1=-a_2=E_0,$$

$$R=\frac{1-\mathrm{i}}{-1-\mathrm{i}}=+\mathrm{i}. \tag{5.17}$$

The type of polarization in a wave can be recognized from its R-value. We have the general rules:

R is real – plane-polarized

R is complex – elliptically polarized

$R=\pm\mathrm{i}$ – circularly polarized.

Two modes of polarization, represented by R_1 and R_2, can be termed *orthogonal* if their R-values satisfy

$$R_1R_2^*=-1. \tag{5.18}$$

This is clearly satisfied for plane-polarized waves with orthogonal planes since, from (5.14) and (5.15),

$$R_1R_2^*=-\cot\phi\tan\phi=-1 \tag{5.19}$$

and also applies to left- and right-handedly polarized light;

$$R_1R_2^*=\mathrm{i}^2=-1, \tag{5.20}$$

from (5.16) and (5.17). But every elliptically polarized wave has an orthogonal conjugate too, and we shall see later in this chapter, and also in § 11.7, that pairs of characteristic waves are always orthogonal.

5.2.4 *Production of polarized light by reflexion.* The reflexion coefficient at the Brewster angle (§ 4.3.1)

$$\hat{i}_B = \tan^{-1} \mu$$

is zero for polarization in the plane of incidence but not for the perpendicular polarization, so that all the light reflected at this angle is polarized normal to the incidence plane. Although in principle perfectly polarized radiation is produced, the exact surface conditions are rather critical and in practice the light is not usually completely polarized. The most important property of light polarized by Brewster-angle reflexion is that its plane of polarization is exactly defined by the experimental arrangement (Fig. 5.4); polaroid materials and Nicol prisms need calibration (§ 5.7.3).

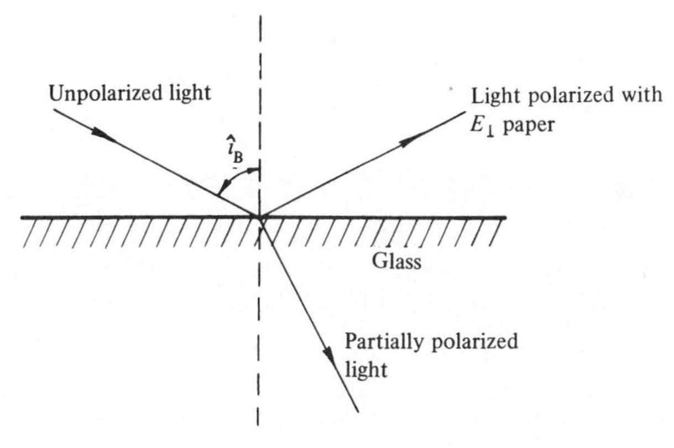

Fig. 5.4. Production of plane-polarized light by reflexion at the Brewster angle.

The *Fresnel rhomb* is a device for producing circularly polarized light from plane-polarized by using the phase changes on critical reflexion (§ 4.3.5). Light polarized at 45° to the plane of the paper is incident on the rhomb in Fig. 5.5 and is critically reflected twice. The angle of incidence $\hat{\imath}$ is arranged so that the two components of the light, polarized perpendicular to the paper and parallel to it, suffer phase changes $\alpha_\perp$ and $\alpha_\parallel$ differing by $\pi/4$, so that after two reflexions a difference of $\pi/2$ in their phases is achieved. The compound wave is thus circularly polarized. In principle it would be possible to achieve this condition after one reflexion; $\alpha_\perp - \alpha_\parallel - \pi$ has a maximum value of 94° in the example $\mu_r = 1/2.5$ illustrated in Fig. 4.12(a) but such a refractive index is uncommon in transparent materials and it is easier to aim at 45° per reflexion, for which the example of $\mu_r = 1/1.5$ is adequate (the maximum value of $\alpha_\perp - \alpha_\parallel - \pi$

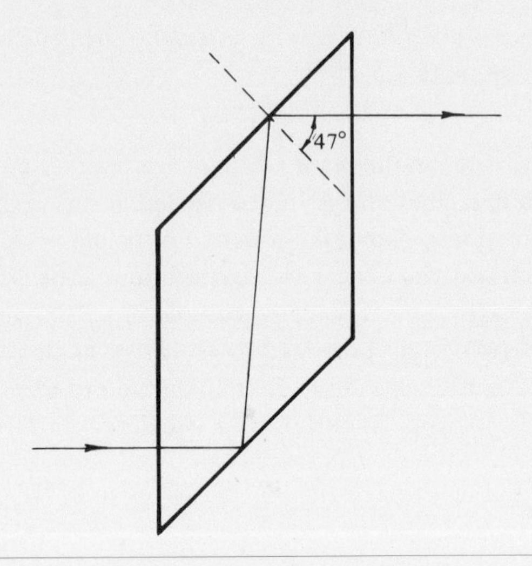

Fig. 5.5. The Fresnel rhomb.

is 47°). The incidence angle for this refractive index should be either 47° or 51°, but it is clear from Fig. 4.12(*b*) that gross errors in $\hat{\imath}$ affect the phase difference imperceptibly. By altering the plane of polarization of the incident light any degree of elliptical polarization can also be created.

We shall leave further discussion of the production of polarized light until § 5.7.3, since crystal propagation affords much more convenient methods of production of both plane- and circularly-polarized light.

5.3 General problems of wave propagation in anisotropic media

5.3.1 *The refractive-index surface.* Before considering any anisotropic medium in detail, we shall discuss a general formation of the problem which we can then apply to any medium. In § 2.2.1 we showed that the wave velocity was given by

$$v = \omega/k \tag{5.21}$$

and the group velocity, the velocity at which energy is transported, by the relationship

$$v_g = \frac{\partial \omega}{\partial k}. \tag{5.22}$$

The study of anisotropic media immediately poses the question: how do we write these expressions to show their dependence on **k** as a vector, rather than a scalar? We cannot immediately replace k in (5.21) by **k**, as

$(\mathbf{k})^{-1}$ is not a vector (see § 2.6.1); we can, however, make use of the refractive index μ instead of the velocity v, since

$$\mu = \frac{c}{v} = \frac{ck}{\omega} = \frac{k}{k_0}, \tag{5.23}$$

where k_0 is the value of k in free space, which is isotropic. Now let k be replaced by $\mathbf{k}$, representing both the magnitude and direction of k; μ becomes $\boldsymbol{\mu}$, which is the refractive index for propagation in a particular direction. We have, then

$$\boldsymbol{\mu} = \frac{1}{k_0}\mathbf{k}. \tag{5.24}$$

Variations of the index $\boldsymbol{\mu}$ can be represented by a surface plotted as a function of its direction, and the surface has the simple meaning that its radius vector represents the refractive index for propagation in that direction (Fig. 5.6). The surface is defined for a particular value of k_0,

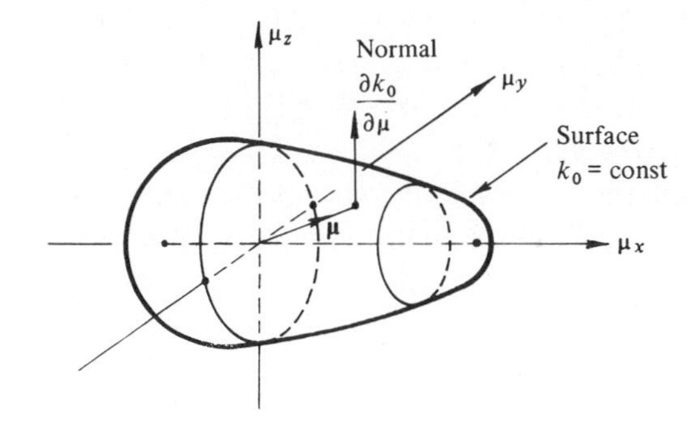

Fig. 5.6. Construction of a μ-surface.

corresponding to a particular frequency ω; because of dispersion the shape and dimensions will change slightly with ω, but we shall ignore this effect by restricting our attention to some fixed frequency. The surface is thus to be regarded as dependent upon the frequency. The group velocity can also be expressed geometrically in terms of the surface. Writing k as $\mathbf{k}$ in (5.22) we can interpret $\mathbf{v}_g$ as

$$\mathbf{v}_g = \frac{\partial \omega}{\partial \mathbf{k}} = \left(\frac{\partial \omega}{\partial k_x}, \frac{\partial \omega}{\partial k_y}, \frac{\partial \omega}{\partial k_z}\right) \tag{5.25}$$

and substituting

$$\frac{\omega}{c} = k_0, \qquad \boldsymbol{\mu} = \frac{\mathbf{k}}{k_0}, \tag{5.26}$$

we have

$$\mathbf{v}_g = \frac{1}{ck_0}\frac{\partial k_0}{\partial \boldsymbol{\mu}} = \frac{1}{ck_0}\left(\frac{\partial k_0}{\partial \mu_x}, \frac{\partial k_0}{\partial \mu_y}, \frac{\partial k_0}{\partial \mu_z}\right). \tag{5.27}$$

This vector, which can alternatively be written as

$$\mathbf{v}_g = \frac{1}{ck_0} \operatorname{grad}_\mu k_0, \tag{5.28}$$

is the normal to the μ-surface since $k_0 = \text{constant}$. This is analogous to the electrostatic theorem that field lines ($\mathbf{E} = -\nabla V$) are normal to equipotential surfaces ($V = \text{constant}$).

The general problem for anisotropic media can now be analysed. We must study the particular anisotropic system to calculate the form and features of the μ-surface, representing the refractive index as a function of the direction of the wave-vector. Then, given the wave-vector, we can find the direction of the associated energy flow by constructing the normal to the surface at the point corresponding to the wave-vector direction.

5.3.2 *An isotropic medium as a special case.* The application to an isotropic medium is simple. The refractive index μ is independent of direction and so the μ-surface is a sphere. The normal to the surface of the sphere is parallel to its radius vector, and therefore the energy is transported parallel to $\mathbf{k}$ (Fig. 5.7). In the terminology of crystal optics this is an *ordinary* ray; the ray vector (the direction along which energy travels, and which is therefore observed) is the same as the wave-vector $\mathbf{k}$. The ray vector is identical with the Poynting vector $\mathbf{\Pi}$, as both represent the energy flow.

5.3.3 *Warning about μ-surfaces and wave-surfaces.* We must give a warning about this approach. The μ-surfaces, which we shall deduce, are

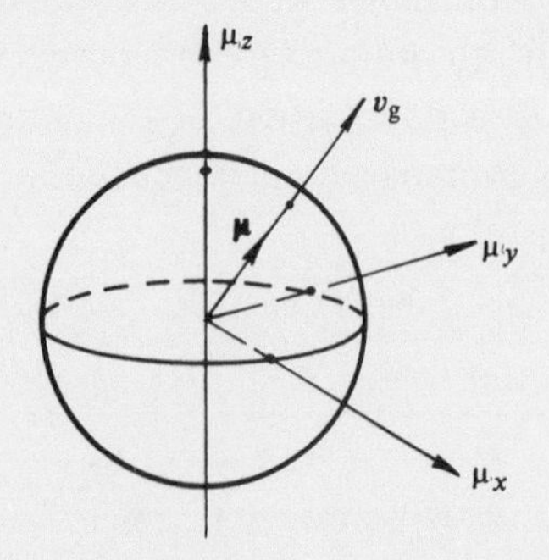

Fig. 5.7. The μ-surface for an isotropic medium.

the reciprocals of the velocity surfaces, or wave-surfaces, which are usually used in the theory of crystal optics. Superficially the two look similar. Both are fourth-order surfaces for a crystal, with two values corresponding to two orthogonal polarizations propagated along a particular direction. But the normal to the wave-surface is not the ray vector, and Huygens' construction is not necessary when using the μ-surface. The μ-surface approach is, however, the usual method of approach in magneto-ionic theory, to which § 11.7 is a simplified introduction, and also in most other problems of wave propagation in anisotropic materials. For example, the Fermi surface – the invaluable representation of the relationship between energy and **k** for electron waves in a metal – is the direct analogue of the μ-surface introduced above. It is with this general view in mind that a departure from the standard treatment has been made.

5.3.4. *Electromagnetic waves in an anisotropic medium.* The second general problem which we can discuss before approaching any particular medium is that of wave-propagation when the dielectric properties of the medium are anisotropic. We should expect Maxwell's equations to hold and, in view of their being applied to waves of the form

$$\mathbf{E} = \mathbf{E}_0 \exp\{i(\omega t - \mathbf{k}\cdot\mathbf{r})\} = \mathbf{E}_0 \exp\{i(\omega t - xk_x - yk_y - zk_z)\},$$

we can use the operator substitutions (§ 2.2.1):

$$\frac{\partial}{\partial t} = i\omega, \qquad \nabla \equiv \left(\frac{\partial}{\partial x}, \frac{\partial}{\partial y}, \frac{\partial}{\partial z}\right) \equiv -i\mathbf{k}. \tag{5.29}$$

Then we can write Maxwell's equations (4.1.1) in the form:

(a) $$\nabla\cdot\mathbf{B} = -i\mathbf{k}\cdot\mathbf{B} = 0. \tag{5.30}$$

This implies $i\mathbf{k}\cdot\mathbf{H} = 0$, since the medium is assumed to be magnetically isotropic.

(b) $$\nabla\cdot\mathbf{D} = i\mathbf{k}\cdot\mathbf{D} = 0 \quad \text{(for an uncharged region).} \tag{5.31}$$

This no longer implies $\nabla\cdot\mathbf{E}$ to be zero, since **D** and **E** are not parallel in an electrically anisotropic medium.

(c) $$\nabla\times\mathbf{H} = \frac{\partial\mathbf{D}}{\partial t} \quad \text{gives} \quad i\mathbf{k}\times\mathbf{H} = -i\omega\mathbf{D}. \tag{5.32}$$

(d) $$\nabla\times\mathbf{E} = -\frac{\partial\mathbf{B}}{\partial t} \quad \text{gives} \quad i\mathbf{k}\times\mathbf{E} = i\omega\mathbf{B} = i\omega\mu_0\mathbf{H}. \tag{5.33}$$

We have assumed here that the magnetic permeability differs negligibly from that of free space.

To derive a wave equation we proceed as before by operating $\nabla\times$, or $i\mathbf{k}\times$, on (5.33) giving

$$\mathbf{k}\times(\mathbf{k}\times\mathbf{E}) = \omega\mu_0(\mathbf{k}\times\mathbf{H}) = -\omega^2\mu_0\mathbf{D}. \quad (5.34)$$

We can check that this equation is reasonable by applying it to an isotropic medium. Then a scalar dielectric constant ϵ applies, giving

$$\mathbf{k}\times(\mathbf{k}\times\mathbf{E}) = \mathbf{k}(\mathbf{k}\cdot\mathbf{E}) - k^2\mathbf{E} = -\omega^2\mu_0\epsilon_0\epsilon\mathbf{E}. \quad (5.35)$$

In such a medium, however, (5.31) does imply $\mathbf{k}\cdot\mathbf{E}$ to be zero, so that we have

$$\frac{\omega^2}{k^2} = v^2 = \frac{1}{\mu_0\epsilon_0\epsilon} = \frac{c^2}{\epsilon},$$

which has already been deduced for an isotropic medium in § 4.2.2.

5.3.5 *Characteristic waves and their velocity.* Now, however, equation (5.34) cannot be treated so simply, although one important fact emerges with no trouble. Equation (5.34) cannot have a solution unless the vectors $\mathbf{k}\times(\mathbf{k}\times\mathbf{E})$ and $\mathbf{D}$ are parallel. The direction of the former expression is perpendicular to both $\mathbf{k}$ and $\mathbf{k}\times\mathbf{E}$, and thus lies in the plane of $\mathbf{k}$ and $\mathbf{E}$, normal to $\mathbf{k}$ (Fig. 5.8). From (5.31) $\mathbf{D}$ is also normal to $\mathbf{k}$, and so a

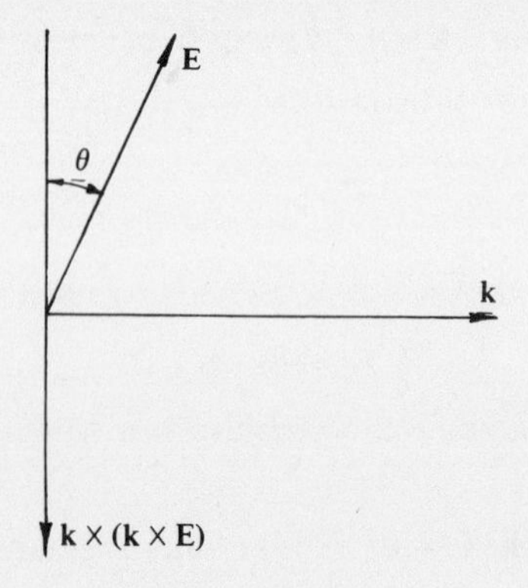

Fig. 5.8. The vectors $\mathbf{k}$, $\mathbf{E}$ and $\mathbf{k}\times(\mathbf{k}\times\mathbf{E})$.

condition for a solution of (5.34) to exist is that $\mathbf{k}$, $\mathbf{E}$ and $\mathbf{D}$ must be coplanar. It is this condition that defines the *characteristic waves*; if they are not coplanar the plane of polarization will rotate as the wave progresses, and the disturbance cannot be described by a single harmonic wave.

Provided that $\mathbf{D}$, $\mathbf{E}$ and $\mathbf{k}$ are coplanar, the equation (5.34) leads to an expression for the wave velocity. The vector $\mathbf{k}\times(\mathbf{k}\times\mathbf{E})$ has magnitude

$-k^2E\cos\theta$, **D** being defined as positive, and therefore we obtain

$$k^2E\cos\theta = \omega^2\mu_0 D$$

or

$$\frac{\omega^2}{k^2} = v^2 = \frac{E\cos\theta}{\mu_0 D}, \tag{5.36}$$

θ being the angle between **D** and **E**. Fig. 5.9 shows the relative disposition of the vectors associated with the wave.

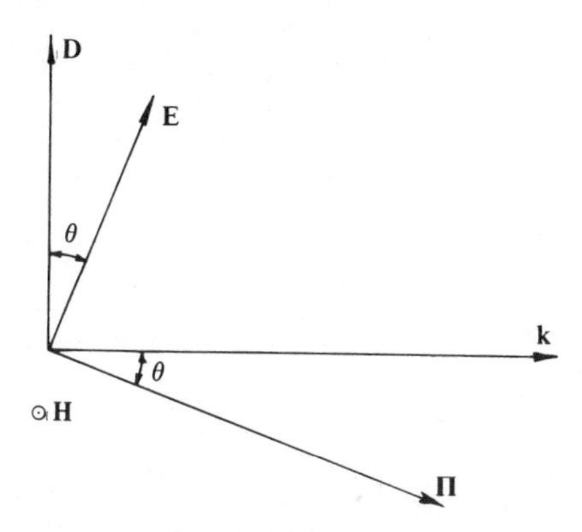

Fig. 5.9. The vectors **D**, **E**, **Π**, **k** and **H** for a wave in an anisotropic medium.

5.3.6 *Analysis of a particular medium.* To analyse the propagation properties of a medium the procedure is now straightforward in principle. Firstly we must find an expression for **D** as a function of **E**. As this depends on the direction of the vectors, it will be necessary to express the relationship by a tensor equation. Having done this, for every direction of **k** we must investigate all the transverse directions of **D** (since **D** and **k** are perpendicular, as a result of (5.31)) and choose those two for which **D**, **E** and **k** are coplanar as characteristic waves. Any arbitrary polarization can then be expressed as a linear combination of the two; it will turn out that their polarizations are orthogonal. Having chosen the characteristic waves, equation (5.36) will give a velocity for each one; the two velocities will generally be different, and can be used to construct a μ-surface having two sheets (which may intersect or touch). Finally, the wave-propagation characteristics can be studied from the μ-surface.

5.4 Crystal Optics

5.4.1 *The dielectric tensor.* Crystals are anisotropic because of their crystal structure. Instead of defining a single dielectric constant for the material as a whole it is necessary to take into account the fact that the dielectric constant depends on the direction of the electric field and also that the resulting electric displacement **D** may then not be parallel to **E**. The relationship between the two vectors will usually be a tensor equation:

$$\mathbf{D} = \boldsymbol{\epsilon} \cdot \mathbf{E} \tag{5.37}$$

and the tensor

$$\boldsymbol{\epsilon} = \begin{pmatrix} \epsilon_{11} & \epsilon_{12} & \epsilon_{13} \\ \epsilon_{21} & \epsilon_{22} & \epsilon_{23} \\ \epsilon_{31} & \epsilon_{32} & \epsilon_{33} \end{pmatrix} \tag{5.38}$$

is called the dielectric tensor. In transparent dielectric materials the tensor is symmetric:

$$\epsilon_{ji} = \epsilon_{ij} \tag{5.39}$$

so that the number of independent coefficients in (5.38) is six. The meaning of the tensor is straightforward. If an electric field

$$\mathbf{E} = (E_1, E_2, E_3)$$

is applied to the material, the resultant electric displacement is

$$\mathbf{D} = (D_1, D_2, D_3),$$

where

$$D_i = \epsilon_{1i} E_1 + \epsilon_{2i} E_2 + \epsilon_{3i} E_3 \qquad (i = 1, 2, 3). \tag{5.40}$$

For particular directions of field the vectors **D** and **E** do turn out to be parallel:

$$\frac{D_1}{E_1} = \frac{D_2}{E_2} = \frac{D_3}{E_3} = \epsilon \tag{5.41}$$

and such directions are called *principal directions*; there are in general three of them, each having a characteristic value of ϵ in (5.41). The three principal directions for a tensor satisfying (5.39) are real and mutually orthogonal, and if the tensor is referred to these three axes as x-, y- and z-axes it takes on the simpler form

$$\boldsymbol{\epsilon} = \begin{pmatrix} \epsilon_1 & 0 & 0 \\ 0 & \epsilon_2 & 0 \\ 0 & 0 & \epsilon_3 \end{pmatrix}. \tag{5.42}$$

The process of referring it to its principal axes is called *diagonalizing the tensor*, and the details are dealt with quite thoroughly in most books on tensor analysis. For some materials two of the principal values ϵ_i are equal; such materials are called *uniaxial crystals* (§ 5.5).

5.4.2 *Representation of the tensor by an ellipsoid.* The possibility of a geometrical approach to the mathematics of crystal optics geometrically rests on the fact that a tensor of the form (5.41) can be represented as an ellipsoid. An ellipsoid in Cartesian coordinates can be written as the solution of a matrix equation:

$$(x, y, z)\begin{pmatrix} a_{11} & a_{12} & a_{13} \\ a_{21} & a_{22} & a_{23} \\ a_{31} & a_{32} & a_{33} \end{pmatrix}\begin{pmatrix} x \\ y \\ z \end{pmatrix} = 1, \tag{5.43}$$

where $a_{ij} = a_{ji}$. This equation can be expanded to give:

$$a_{11}x^2 + xy(a_{12} + a_{21}) + a_{22}y^2 + yz(a_{23} + a_{32}) + a_{33}z^2 + zx(a_{31} + a_{13}) = 1$$

and is simplified in matrix notation to

$$\mathbf{x}\cdot\boldsymbol{A}\cdot\mathbf{x}' = 1. \tag{5.44}$$

By referring the tensor $\boldsymbol{A}$ to its principal axes we obtain the equation

$$(x, y, z)\begin{pmatrix} a_1 & 0 & 0 \\ 0 & a_2 & 0 \\ 0 & 0 & a_3 \end{pmatrix}\begin{pmatrix} x \\ y \\ z \end{pmatrix} = 1. \tag{5.45}$$

or

$$a_1x^2 + a_2y^2 + a_3z^2 = 1 \tag{5.46}$$

in which the x-, y-, and z-axes are clearly the symmetry axes of the ellipsoid. Thus the principal axes of the tensor are equivalent to the symmetry axes of the ellipsoid. The three semi-axes of the ellipsoid (5.46) have values $a_1^{-\frac{1}{2}}$, $a_2^{-\frac{1}{2}}$, $a_3^{-\frac{1}{2}}$.

To represent the dielectric tensor $\boldsymbol{\epsilon}$ (equation (5.38)) by an ellipsoid we shall use its reciprocal tensor $\boldsymbol{\epsilon}^{-1}$ referred to principal axes:

$$\boldsymbol{\xi} = \boldsymbol{\epsilon}^{-1} = \begin{pmatrix} \epsilon_1^{-1} & 0 & 0 \\ 0 & \epsilon_2^{-1} & 0 \\ 0 & 0 & \epsilon_3^{-1} \end{pmatrix} \tag{5.47}$$

and write the analogous equation to (5.44):

$$\mathbf{D}\cdot\boldsymbol{\xi}\cdot\mathbf{D}' = 1. \tag{5.48}$$

This equation (5.48) represents an ellipsoid called the *optical indicatrix* plotted on the axes D_1, D_2 and D_3 and clearly it has semi-axes $\epsilon_1^{\frac{1}{2}}$, $\epsilon_2^{\frac{1}{2}}$, $\epsilon_3^{\frac{1}{2}}$. What does it represent? The vector $\mathbf{D}$ from the origin to a point on the ellipsoid is a solution of (5.48), which is

$$\mathbf{D}\cdot\boldsymbol{\xi}\cdot\mathbf{D}' = \mathbf{D}\cdot\mathbf{E} = 1 \tag{5.49}$$

from the definition of $\boldsymbol{\xi}$. We therefore imagine both **D** and **E** to vary in magnitude and direction in such a way that (5.49) is always satisfied. For a given direction of **D**, the equation then determines both **D** and **E** completely, and the ellipsoid is the locus of the vector **D** drawn out from the origin (Fig. 5.10).

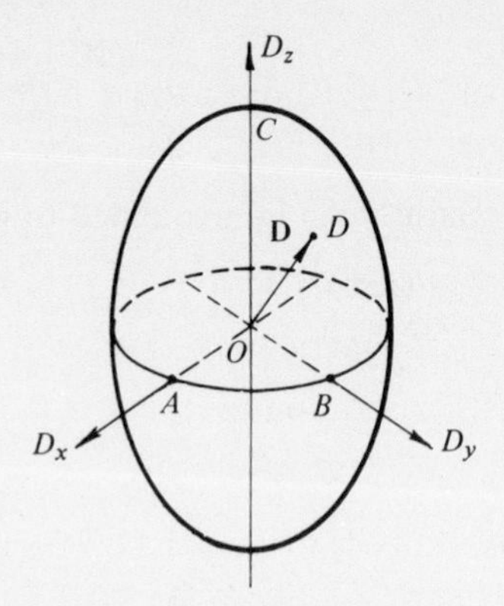

Fig. 5.10. The optical indicatrix.

5.4.3 *Refractive index.* The refractive index then follows very simply from (5.36). We have

$$v^2 = \frac{c^2}{\mu^2} = \frac{1}{\epsilon_0 \mu_0 \mu^2} = \frac{E \cos\theta}{\mu_0 D} = \frac{\mathbf{E}\cdot\mathbf{D}}{\mu_0 D^2}. \tag{5.50}$$

Since all vectors from the origin to the ellipsoid satisfy $\mathbf{E}\cdot\mathbf{D} = 1$ we have

$$\mu^2 = \frac{D^2}{\epsilon_0}$$

or

$$\mu = \frac{D}{\epsilon_0^{\frac{1}{2}}}. \tag{5.51}$$

The refractive index for polarization along the direction of **D** is thus given directly as the radius of the indicatrix in that direction. The indicatrix is the surface representing μ as a function of the polarization of the wave. To see that this is reasonable, consider the principal axes. It follows from (5.51) that the three principal refractive indices are $\epsilon_0^{-\frac{1}{2}} OA$, etc.; from the value of the semi-axes deduced above we then have

$$\mu_1 = \left(\frac{\epsilon_1}{\epsilon_0}\right)^{\frac{1}{2}}, \quad \text{etc.,}$$

which is clearly correct.

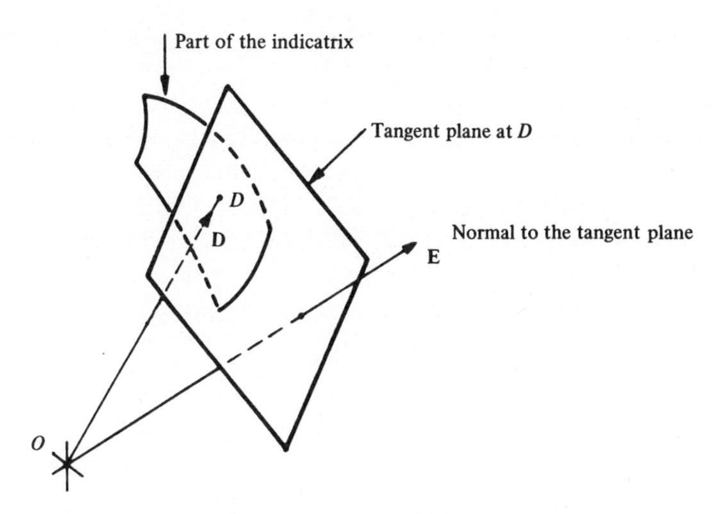

Fig. 5.11. Relationship between **D**, **E** and the optical indicatrix.

5.4.4 *Characteristic waves.* One part of the necessary calculation has thus been carried out. Provided a wave with polarization vector **D** is a characteristic wave (§ 5.3.5), it will have a refractive index given by the indicatrix in Fig. 5.10. This indicatrix, it should be emphasized, is *not* the μ-surface of § 5.3.1, as it represents μ as a function of the polarization vector **D** and not the wave-vector **k**. We shall come to the μ-surface presently. How do we decide what vector **E** corresponds to **D**, in order to judge the criterion for a characteristic wave, which is that **D**, **E** and **k** are coplanar? The construction to achieve the direction of **E** is to draw the tangent plane to the indicatrix at the end of the vector **D**; the direction of **E** is then given by the normal to this tangent plane (Fig. 5.11).† The

† This construction can easily be justified algebraically. The tangent to the ellipsoid

$$a_1x^2 + a_2y^2 + a_3z^2 = 1 \tag{5.46}$$

at (x_1, y_1, z_1) is

$$a_1x_1x + a_2y_1y + a_3z_1z = 1.$$

A vector normal to this plane is

$$(a_1x_1, a_2y_1, a_3z_1).$$

Translating this to the indicatrix (5.48) for which

$$a_1 = \epsilon_1^{-1}, \quad \text{etc.,}$$

the vector

$$\mathbf{D} \equiv (x_1, y_1, z_1)$$

and consequently the tangent-normal is

$$\left(\frac{x_1}{\epsilon_1}, \frac{y_1}{\epsilon_2}, \frac{z_1}{\epsilon_3}\right) = \left(\frac{D_1}{\epsilon_1}, \frac{D_2}{\epsilon_2}, \frac{D_3}{\epsilon_3}\right) \equiv \mathbf{E}.$$

construction is once again clearly correct for the principal axes, for which the plane-normal and **D** then coincide.

Now we can proceed to deduce the characteristic waves for a given wave-vector **k**. Since **k** is normal to **D** (equation (5.31)), it follows that all possible **D** vectors lie in a plane through O normal to **k** (Fig. 5.12(a)). This plane cuts an elliptical section $PQRS$ of the indicatrix. The characteristic polarizations are those **D** vectors whose associated **E** vectors lie in the (**D**, **k**) plane, and it is clear from the symmetry of the ellipse that they must correspond to the major and minor axes OP and OQ (Fig. 5.12(b)).

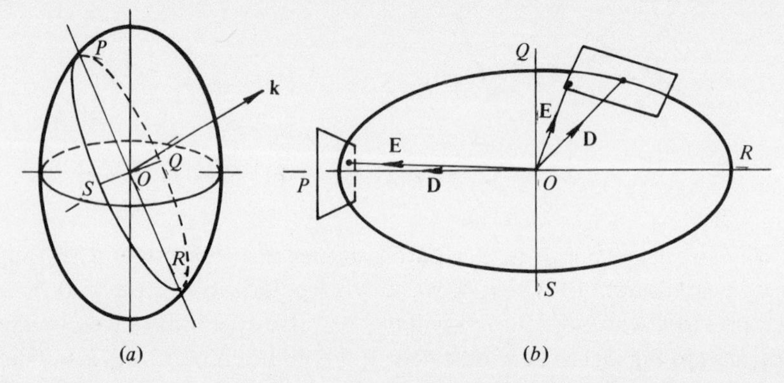

Fig. 5.12. (a) Section of the indicatrix normal to **k** and (b) the axes of the section.

There are two characteristic waves for each **k**-vector. Thus to determine the μ-surface we must draw the ellipse $PQRS$ corresponding to each direction of **k** and measure its major and minor axes; these give the radii of the two branches of the μ-surface along each vector **k**.

5.4.5 *The μ-surface.* A good idea of the form of the μ-surface results from considering its sections in the (x, y), (y, z) and (z, x) planes. We can standardize the indicatrix to have its major, intermediate and minor axes along the z-, y- and x-axes respectively.

We start with **k** along x and consider what happens as it rotates about z. When it is along OX the ellipse $PQRS$ has OZ and OY as major and minor axes, and therefore the two values of μ are μ_3 and μ_2 respectively. As **k** rotates about z, the section $PQRS$ still contains OZ as its major axis, but its minor axis shrinks from OY to OX, and μ from μ_2 to μ_1 (Fig. 5.13). The two refractive indices for such directions are thus μ_3 and a value between μ_2 and μ_1, corresponding to polarization along z and in the perpendicular plane. We can therefore draw the horizontal section of the

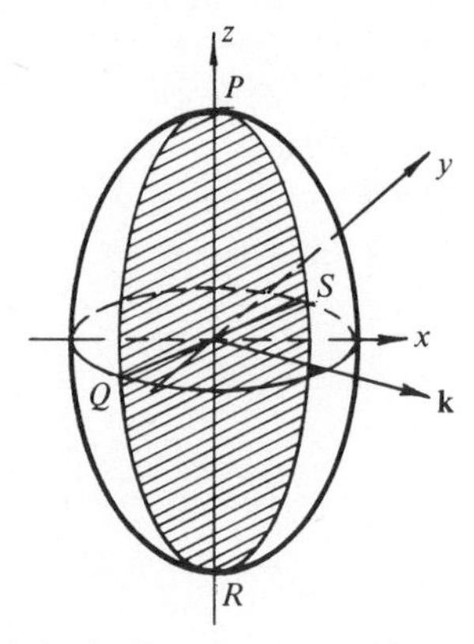

Fig. 5.13. Section of the indicatrix when **k** lies in the (x, y) plane.

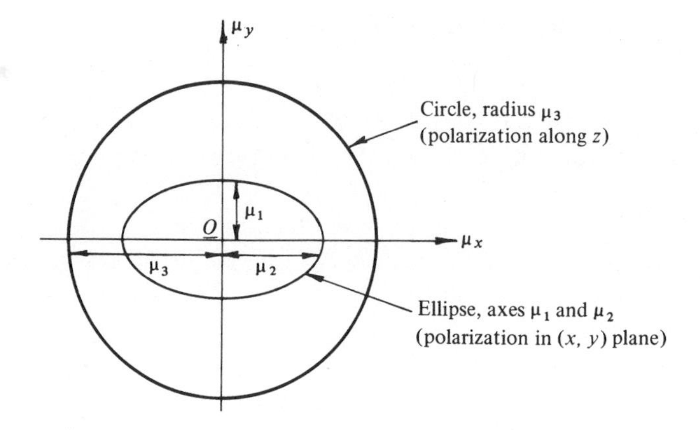

Fig. 5.14. Section of the μ-surface in the (x, y) plane.

μ-surface for directions of **k** in the (x, y) plane (Fig. 5.14). The symmetry of the section follows from the symmetry of the indicatrix. In the same way, by taking **k** from y to z by rotating about x we draw the section in the (y, z) plane (Fig. 5.15). The zx section is similarly constructed by taking **k** from z to x; the two curves cross at A. For **k** in the directions OA, then, the two polarizations have the same refractive index; these directions are called *optic axes*, and there are in general two of them, at equal angles from z in the (z, x) plane (Fig. 5.16). They are normal to the two circular diametric sections of the ellipsoid (Fig. 5.17). From this much information we can complete the surface by interpolation. The polarization of the two branches of it are not easily related to the axes except for the cases

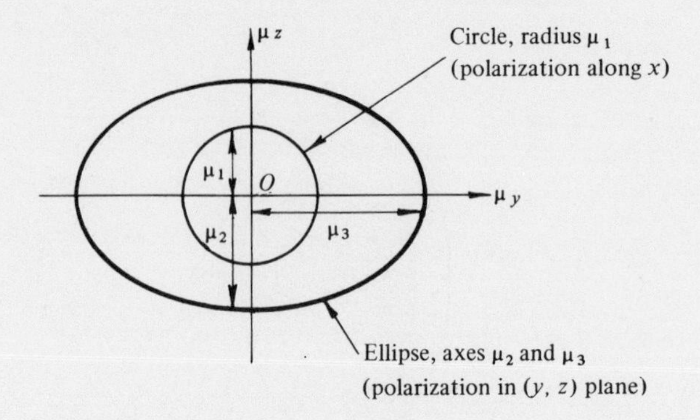

Fig. 5.15. Section of the μ-surface in the (y, z) plane.

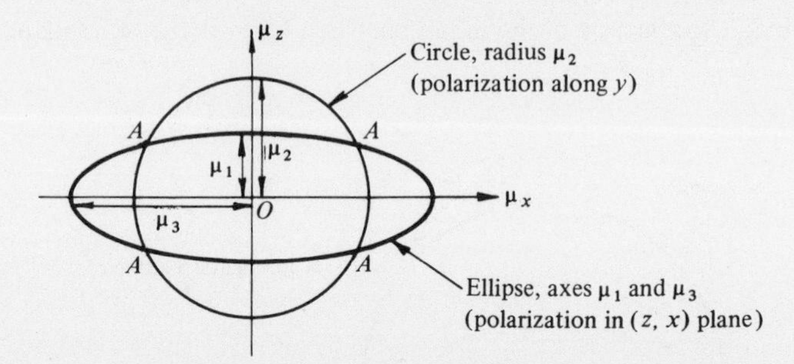

Fig. 5.16. Section of the μ-surface in the (z, x) plane.

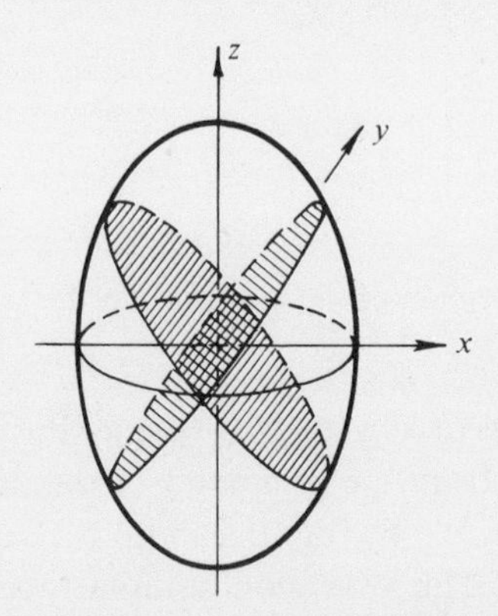

Fig. 5.17. Circular sections of the indicatrix.

above; for example, in Fig. 5.18 the major and minor axes of the intersection ellipse are shown for a general **k**. One octant of the interpolated surface is sketched in Fig. 5.19(a). It is clear that it consists of two surfaces, which touch. The two complete surfaces touch at four points related by symmetry. They define the two optic axes.

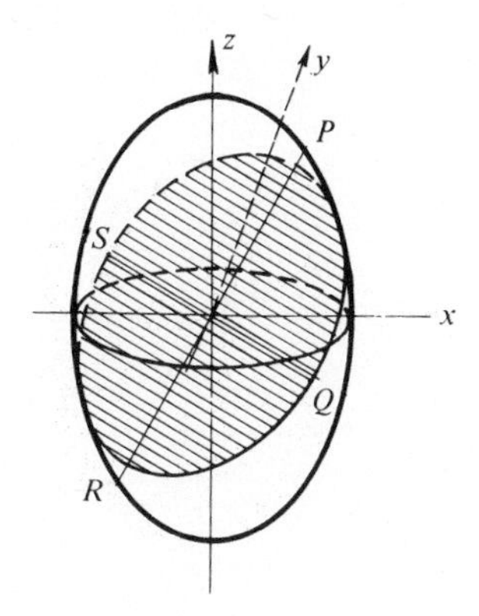

Fig. 5.18. For an arbitrary **k** the axes of the section-ellipse *PQRS* bear no simple relation to the principal axes.

5.4.6 *Ray vectors from the* μ*-surface.* Having constructed the μ-surface it is in principle a simple matter to deduce the direction of the normal **Π** at each point, which gives the ray-vector direction. Using the definition of the two types of rays:

ordinary, for which the ray vector **Π** is parallel to the wave-vector **k**,
extraordinary, for which **Π** is not parallel to **k**.

we can distinguish one general and two special cases (see Fig. 5.19(*b*)):

(a) for **k** in an arbitrary direction, $\mathbf{k}_1$, both surfaces give rise to extraordinary waves, as neither of the ray vectors $\mathbf{\Pi}_{i1}$ and $\mathbf{\Pi}_{o1}$ (from the inner and outer branches) is parallel to **k**.

(b) If **k** lies in a symmetry plane (x, y), (y, z) or (z, x), there is one ordinary ray corresponding to the section of the μ-surface which is a circle, and one extraordinary ray from the other branch. The ordinary ray always has a principal polarization: $\mathbf{k}_2$ is an example of this behaviour; $\mathbf{\Pi}_{o2}$ is parallel to **k**, and corresponds to polarization in the z-direction.

(c) If **k** lies in a principal direction $\mathbf{k}_4$ both rays are ordinary and the ray vectors $\mathbf{\Pi}_{i4}$ and $\mathbf{\Pi}_{o4}$ are both parallel to it.

Once the ray vectors have been determined it is easy to see the polarization direction of the extraordinary rays. We saw in § 5.4.4 that for a characteristic ray **k**, **D**, **E** and **Π** lay in the same plane normal to **H** and that **D**, **k** and **H** were mutually orthogonal. Since we can now construct **Π** for any given **k**-vector, **H** is determined as normal to the plane defined by **k** and **Π**, and hence **D** can be deduced. This is illustrated for $\mathbf{k}_5$ on Fig. 5.19(*b*).

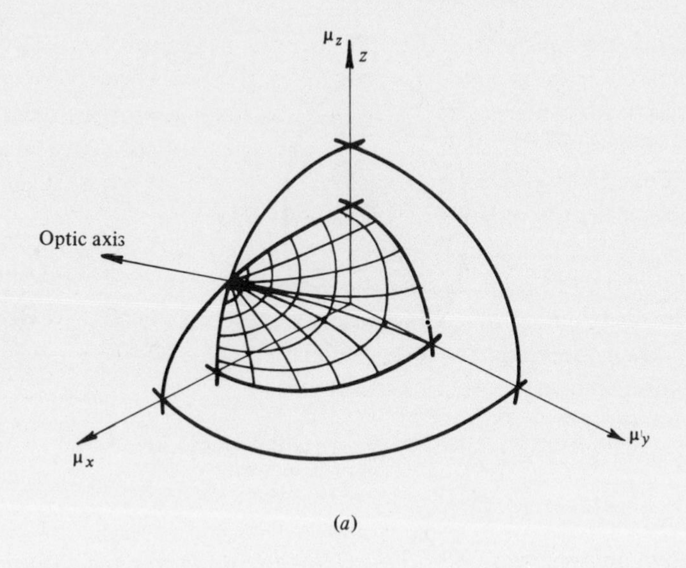

(a)

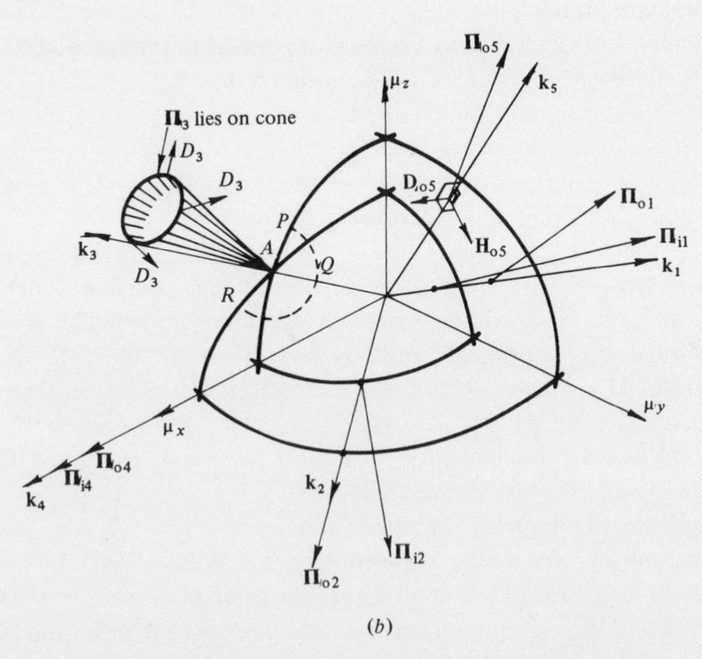

(b)

Fig. 5.19. (a) The μ-surface for a biaxial crystal; (b) the determination of propagation properties for various waves. The suffices i and o represent the inner and outer branches of the surface.

5.4.7 *Conical propagation.* A peculiar form of propagation occurs when the wave-vector $\mathbf{k}_3$ is along the optic axis. The ray vector is then not unique, as both branches of the μ-surface are dimpled at the point A. The possible direction of $\mathbf{\Pi}$ then lies on a cone, one edge of which is along $\mathbf{k} = \mathbf{k}_3$. This gives rise to a phenomenon known as external conical refraction; if an unpolarized light wave is incident normally on the surface of a crystal cut perpendicular to an optic axis, the observed ray directions inside the crystal lie on a cone, one edge of which is parallel to the optic axis. The degeneracy of characteristic rays is also evident from Fig. 5.12(b) when the ellipse becomes a circle. A second form of conical propagation, known as internal conical refraction, occurs as a result of all points on the circle PQR (Fig. 5.19(b)) having parallel ray vectors, although corresponding to wave-vectors lying on a cone.

5.5 Uniaxial crystals

The above analysis has been carried out for the most general type of crystal, which is known as a biaxial crystal since it has two optic axes. Uniaxial crystals, which have only one optic axis as a result of symmetry, frequently occur and have several important uses. They have the characteristic, already mentioned in § 5.4.1, that two of their principal dielectric

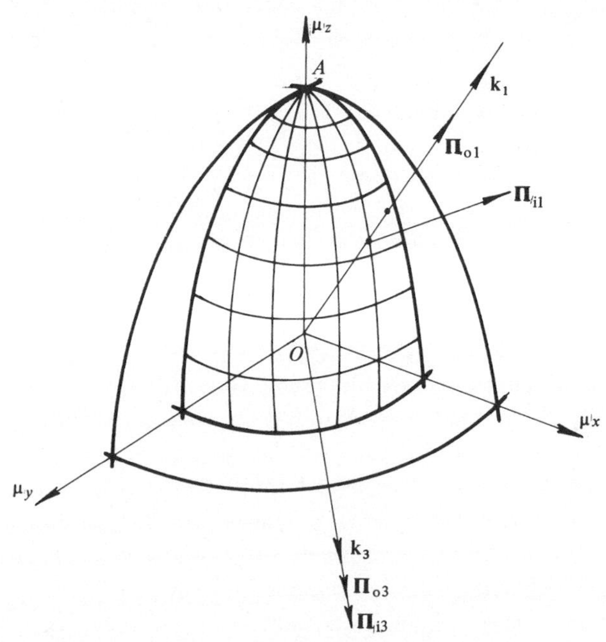

Fig. 5.20. The μ-surface for a uniaxial crystal.

constants are equal and their dielectric ellipsoids are spheroids. It follows that their optical indicatrices are also spheroids and therefore the one optic axis is the axis of unique dielectric constant. The μ-surface for such a crystal is considerably simpler than for a biaxial crystal, and it is easy to see that it breaks into two branches, one a sphere and the other a spheroid touching it along the optic axis (Fig. 5.20).

5.5.1 *Propagation in a uniaxial crystal.* By following through an analysis of the ray vector as a function of wave-vector, the reader can easily see that

(a) for a general $\mathbf{k}$ there is one ordinary and one extraordinary ray;

(b) for $\mathbf{k}_2$ along the optic axis OA there are two identical ordinary rays, with equal refractive indices, so that propagation is the same as in an isotropic medium;

(c) for $\mathbf{k}_3$ normal to the optic axis there are two ordinary rays with different refractive indices;

(d) conical propagation does not occur.

5.5.2 *The electro-optic effect.* The anisotropic behaviour of some materials can be modified by the application of an external electric field. There can be several results; for example an isotropic material might become uniaxial as a result, or a biaxial crystal might suffer a change in the directions of its optic axes. Basically the effects can be described as follows.

If a crystal has a centre of symmetry in its atomic arrangement, or the material is not crystalline at all – such as a liquid – the result of applying the external field E_0 usually depends on E_0^2. The relationship to an even power of E_0 results simply from the indistinguishability of positive and negative values of E_0 in such materials. There are many examples of the uses of these effects, one of the oldest being the Kerr cell. When an electric field is applied to glass or nitrobenzene (the former an amorphous solid; the latter a liquid) a quadratic electro-optic effect occurs and the medium behaves as a uniaxial crystal, with its optic axis along the applied field. As a result, a beam of light passing across the electric field and polarized at about 45° to it will become elliptically polarized to a degree depending on the thickness of liquid in the same way as it does in a quarter- or half-wave plate (§ 5.7.4); this is known as the *Kerr effect.* The *Kerr cell* is a shutter which consists of nitrobenzene between two crossed polaroids. By applying a suitable electric field the cell becomes transmitting. This effect is invaluable as a method of making a very fast optical shutter, since it involves no mechanical motion.

If the crystal has no centre of symmetry, the electro-optic effect can also depend on odd powers of E_0, resulting in the possibility of a linear relationship. In principle, an applied field E_0 in any direction can change the dielectric constant for wave field E in any other direction. The effect is usually related to the movement, within the crystal structure, of certain atoms or groups of atoms, and is often accompanied by a piezoelectric effect (change of dimensions induced by the external field E_0). The most well-known effect of this sort is known as the *Pockels effect* in which E_0 in the z-direction changes the dielectric constants for wave fields in the x- and y-directions by equal and opposite amounts. If the crystal were isotropic in zero field, the applied field would result in its becoming biaxial; but in most well-known examples the crystal is already uniaxial in zero field. Examples of crystals exhibiting this effect are KH_2PO_4, its deuterated equivalent, and ZnS. The formal description of the linear electro-optic effects involves a tensor relating the three components of the vector $\mathbf{E}_0$ with the six independent components ϵ_{ij} (5.38). This tensor r_{kl} (where k goes from 1 to 3 and l from 1 to 6) is discussed fully in more specialized texts (e.g. Yariv, 1975).

5.5.3 *The photoelastic effect.* Similarly to the electric field in the electro-optic case, strain in a transparent plastic material such as Perspex (polymethyl methacrylate) causes it to behave as a uniaxial crystal. The distribution of strains in a loaded beam, for example, can therefore be investigated in a plastic model by examining it between crossed polaroids (Fig. 5.21).

5.5.4 *Optical activity.* When a plane-polarized wave enters a quartz crystal along its optic axis, it is found that the plane of polarization rotates at a rate of about 22° per mm. Quartz is a uniaxial crystal, so that this behaviour is not consistent with the behaviour which we have described so far. The continuous rotation of the plane of polarization of a wave is known as *optical activity* and can occur in any material – crystalline or non-crystalline – having a helical structure. Quartz has a helical crystal structure, and may be right-handed or left-handed (Fig. 5.22). Sugar solutions are also well-known examples of non-crystalline optically active materials; dextrose solution rotates the polarization in a right-handed sense, whereas laevulose solution is the opposite.

The action of the oscillating magnetic wave-field on a helical molecule leads to oscillating electric fields in orthogonal directions, having phases differing by $\pi/2$. Two completely different molecular models illustrating

Fig. 5.21. An example of the photoelastic effect. A piece of strained Perspex is observed in monochromatic light between crossed polaroids. The incident light is polarized at 45° to the edge of the strip.

this result are described by Hartshorne & Stuart (1964). Such a result can be described by a dielectric tensor of the form:

$$\boldsymbol{\epsilon} = \begin{pmatrix} \epsilon_1 & \mathrm{i}a & 0 \\ -\mathrm{i}a & \epsilon_1 & 0 \\ 0 & 0 & \epsilon_3 \end{pmatrix}. \tag{5.52}$$

It is easy to show that this tensor is Hermitian ($\epsilon_{ij} = \epsilon_{ji}^*$) and has circularly polarized characteristic waves propagating along the z-direction (see Chapter 11 for a more detailed discussion of such phenomena). The right-handed and left-handed waves along z have refractive indices

$$\mu_{\mathrm{r,l}} = (\epsilon_1 \pm a)^{\frac{1}{2}} \tag{5.53}$$

and the slight difference between the two values explains the progressive rotation of the plane of polarization of the wave along the z-axis as follows.

When we add the two circularly polarized waves (§ 5.2.2)

$$\mathbf{E}_\mathrm{r} = (E_0, \mathrm{i}E_0) \exp\{\mathrm{i}(\omega t - \mu_\mathrm{r} kz)\}, \tag{5.54}$$

$$\mathbf{E}_\mathrm{l} = (E_0, -\mathrm{i}E_0) \exp\{\mathrm{i}(\omega t - \mu_\mathrm{l} kz)\}, \tag{5.55}$$

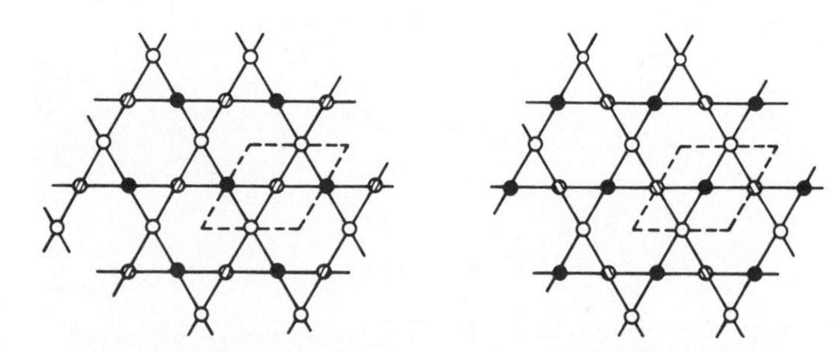

Fig. 5.22. Positions of the silicon atoms in right- and left-handed quartz (SiO_2) projected on a plane normal to the optic axis. The broken line outlines the unit cell, within which ● represents a Si atom at two-thirds of the cell dimensions normal to the paper, ▨ at one-third and ○ at 0. Oxygen atoms are omitted. The Si atoms are seen to lie on helices, the senses of the helices being opposite in the two stereo-isomers. This structure of quartz (β) is stable above 573 °C; below this temperature it is replaced by α-quartz which has a similar structure, the atoms being regularly displaced from the positions illustrated.

we have

$$\mathbf{E} = \mathbf{E}_r + \mathbf{E}_l = (E_0 \cos [\tfrac{1}{2}kz(\mu_r - \mu_l)], E_0 \sin [\tfrac{1}{2}kz(\mu_r - \mu_l)]) \times 2 \exp \{i(\omega t - k\bar{\mu} z)\}, \tag{5.56}$$

which represents a wave travelling with the mean refractive index $\bar{\mu}$ and whose plane of polarization is at angle $\frac{1}{2}kz(\mu_r - \mu_l)$ to the x-axis. This angle increases progressively with z.

One can follow through the geometrical construction of the μ-surface from the tensor (5.52) and show, by the methods described in more detail in Chapter 11, that the inner and outer surfaces characteristic of the uniaxial medium do not quite touch along the z-axis, remaining separated by the amount $\mu_r - \mu_l$ (Fig. 5.23). Certainly the effect is small; for quartz μ(ordinary) $= \bar{\mu} = 1.544$; μ(extraordinary) $= 1.553$; $\mu_r - \mu_l = 7 \times 10^{-5}$. However, since the effect is related to the internal helicity of the molecules, which is invariant whether the molecules are oriented or disoriented, it is the only anisotropic optical effect which can be found in disordered or amorphous materials – such as the sugar solutions already quoted. Optical activity can also be induced in materials by the application of a magnetic field, when it is called the *Faraday effect.* Notice the way in which birefringence (symmetric dielectric tensor) is related to an applied *electric* field through the electro-optic effect, whereas optical activity (antisymmetric dielectric tensor) is related to a *magnetic* field through the Faraday effect.

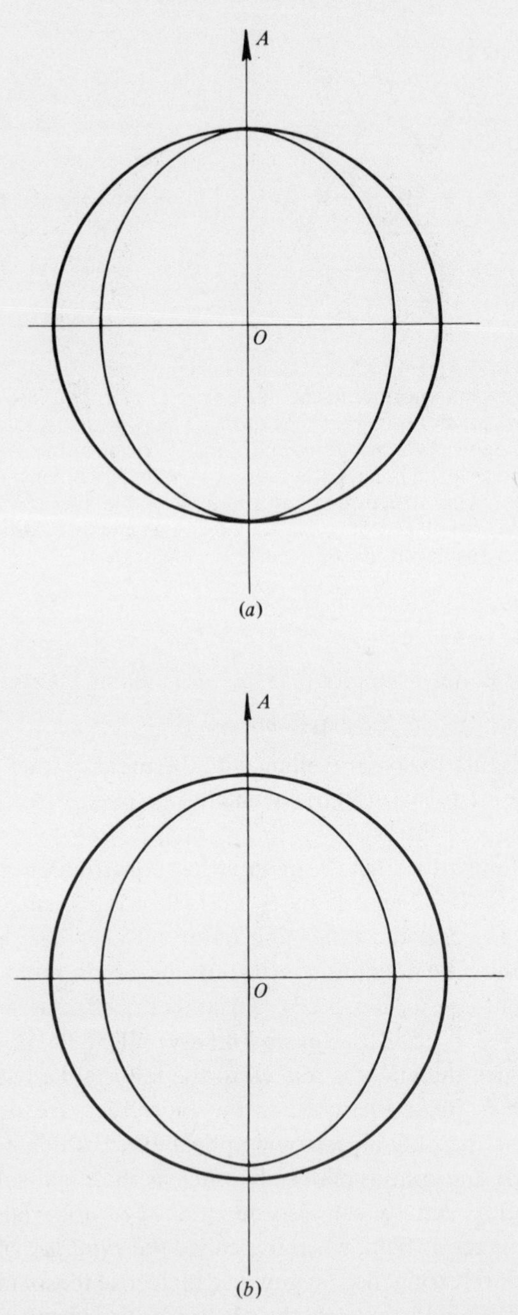

Fig. 5.23. (*a*) Section of the μ-surface in an inactive negative uniaxial crystal. *OA* is the optic axis. (*b*) Section of the μ-surface in an optically active negative uniaxial crystal, such as quartz.

5.6 Solving refraction problems in anisotropic materials

5.6.1 *The Poeverlein construction.* One of the great advantages of the 'μ-surface' method of representing the anisotropic propagation properties of crystals is the ease with which one can use it to solve refraction problems. We should remember that Snell's law, arising as it does from the continuity of E or H at a plane surface between two refractive media, is *always* true for the **k**-vector directions. Now suppose that we consider the refraction of light from, say, a homogeneous medium of index μ_1 into a crystal. Fig. 5.24 shows the sections of the μ-surface in the plane of

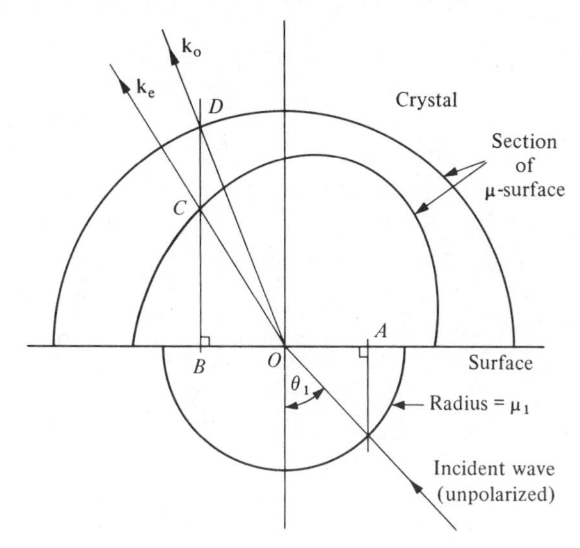

Fig. 5.24. Poeverlein's construction.

incidence, which contains the **k**-vectors of the incident and refracted waves, as well as the normal to the surface. The construction for tracing the **k**-vector from one medium to the other, due to Poeverlein, satisfies Snell's law, which requires that $\mu(\theta)\sin\theta$ be equal on the two sides of the interface.

Given the angle of incidence θ_1 in the homogeneous medium we construct $OA = OB = \mu_1 \sin\theta_1$. Then OC and OD represent the directions of the two characteristic **k**-vectors in the crystal. Their polarizations and ray-vectors are found in the usual manner. We shall illustrate the use of this method in the next section.

5.6.2 *Crystal polarizers.* Although polarized light can be produced from an unpolarized source either by Brewster-angle reflexion (§ 4.3.1) or by scattering (§ 10.2.2), by far the most important polarizers use birefringent materials. The first practical polarizer was the Nicol prism (1830) which was based on the natural calcite-crystal shape. Several variations followed as larger synthetic crystals and improved cutting methods became possible. These were all largely superseded by Polaroid sheet (§ 5.7.3). However, they have returned to the scene with the advent of high-power lasers since the absorption of the unwanted power by Polaroid sheet would cause its immediate destruction. In addition, plastic-based polarizers are unusable for optics at wavelengths where the plastic absorbs.

As examples of crystal polarizers we shall briefly discuss the Glan-air, Glan–Thompson and Nicol prisms, which are all based on a similar principle – the separation of two orthogonally polarized characteristic waves by means of critical reflexion (§ 4.3.2). They are all constructed from calcite ($CaCO_3$) which is uniaxial and has a particularly large difference between its principal refractive indices. They are: $\mu_o = \epsilon_1^{\frac{1}{2}} = 1.66$, $\mu_e = \epsilon_3^{\frac{1}{2}} = 1.49$. The suffices o and e represent ordinary and extraordinary, although the use of these terms is not exactly in keeping with our earlier definitions (§ 5.4.6).

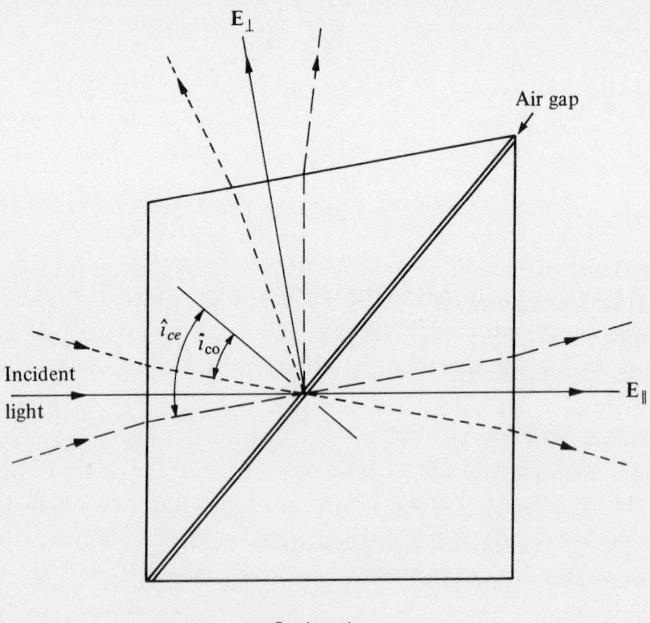

(*a*)

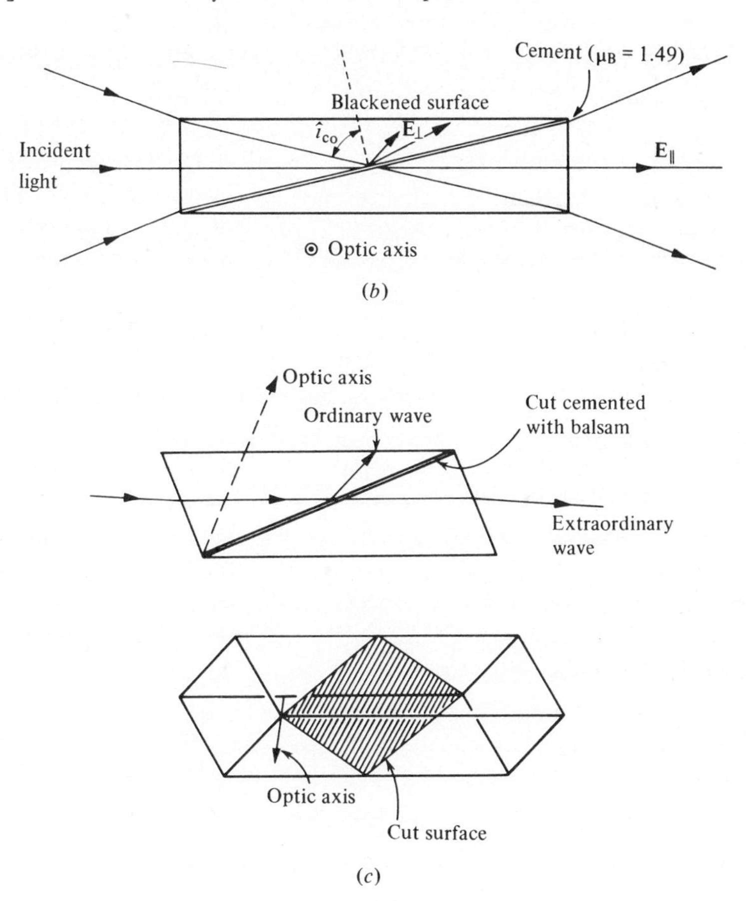

Fig. 5.25. Different types of polarizing prisms.

The discussion here is mainly intended to illustrate the use of the Poeverlein construction, and more detailed information can be found in the *Handbook of Optics* (Driscoll, 1978).

Consider first the Glan-air prism (Fig. 5.25(*a*)). The light is incident in a plane normal to the optic axis and therefore the component polarized with $\mathbf{E}_\perp$ normal to the paper is propagated with index μ_o and the other component with μ_e. In the angular range between $\hat{\imath}_{co} = \sin^{-1}(1/\mu_o)$ and $\hat{\imath}_{ce} = \sin^{-1}(1/\mu_e)$ the two polarizations are separated at the air gap, and only the second component is transmitted.

The Glan–Thompson prism has the two halves cemented together with an isotropic cement having refractive index μ_B between μ_o and μ_e (e.g. Canada balsam). The $\mathbf{E}_\parallel$ component is not critically reflected at any angle (Fig. 5.25(*b*)).

The Nicol prism (Fig. 5.25(*c*)) is similar to the Glan–Thompson in being cemented, but the optic axis now lies in the plane of incidence.

In Figs. 5.26 and 5.27 we see, side by side, the stages by which rays can be traced through the Glan-air and Nicol prisms. In the former, the optic axis is normal to the plane of incidence (the plane of the diagram) and the section of the μ-surface therefore consists of two concentric circles having radii μ_o and μ_e. In the Nicol prism, the optic axis lies in the plane of

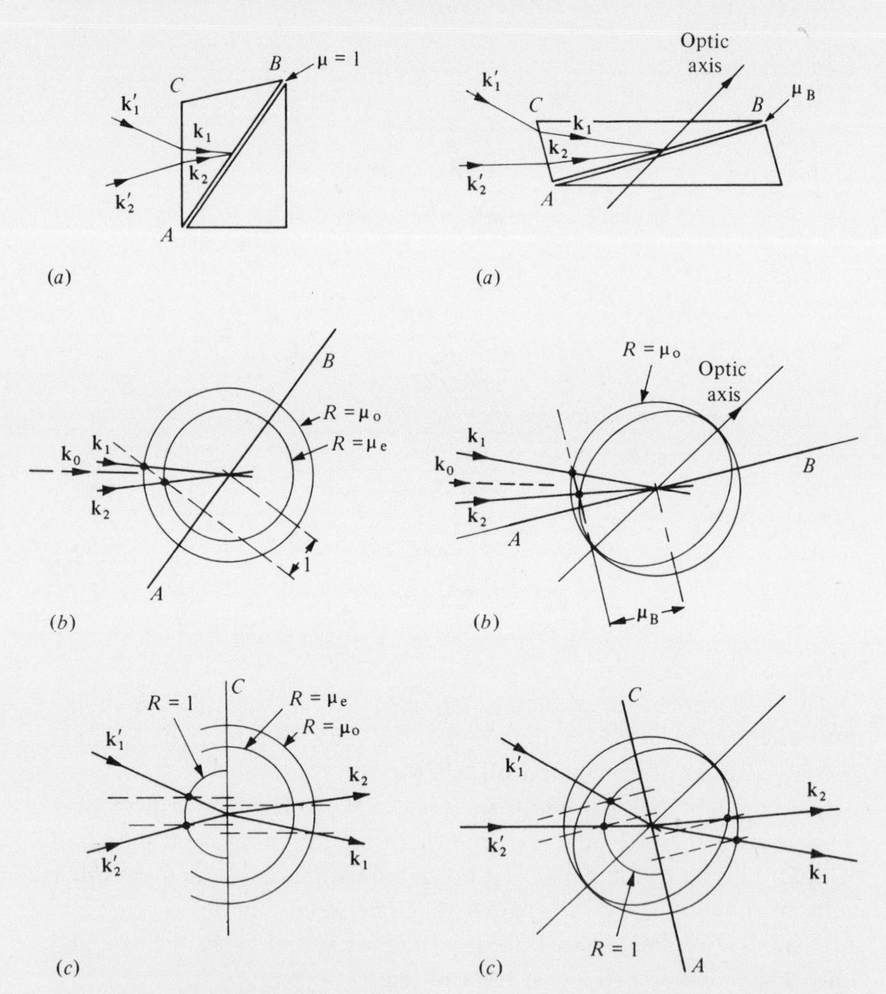

Fig. 5.26. Poeverlein's constructions for the Glan-air prism: (*a*) designation of vectors; (*b*) critical angles at the air gap; (*c*) refraction at the entrance surface.

Fig. 5.27. Poeverlein's constructions for the Nicol prism (cf. Fig. 5.26): (*a*) designation of vectors; (*b*) critical angles at the cement; (*c*) refraction at the entrance surface.

incidence and the section of the μ-surface consists of a circle enclosing an ellipse (cf. Fig. 5.23(*a*)).

The first diagrams (*a*) of Figs. 5.26 and 5.27 show the wave-vectors $\mathbf{k}_1$ and $\mathbf{k}_2$ which are incident on the hypotenuse *AB* at the critical angles for the two branches of the μ-surface. The corresponding wave-vectors $\mathbf{k}_1'$ and $\mathbf{k}_2'$ outside the prism represent the extremes of the angular field of view.

In (*b*) we see the Poeverlein construction used to deduce the directions of the vectors $\mathbf{k}_1$ and $\mathbf{k}_2$, remembering that they both refract to angles 90° in the gap medium (air or balsam). The geometrical construction for the Glan-air prism gives a trivial result; that for the Nicol represents the solution of a non-analytical equation which is difficult to solve otherwise.

In (*c*) we see a second Poeverlein construction used to deduce the wave-vectors $\mathbf{k}_1'$ and $\mathbf{k}_2'$ by refraction at the entrance surface *AC*. Again the construction for the Glan-air is trivial, while that for the Nicol is not.

Finally the median vector $\mathbf{k}_0$ in (*b*) can be drawn. In the Nicol prism this is arranged to be parallel to the long side *BC* in order to make best use of the size of the prism.

5.7 Some practical procedures in crystal optics

5.7.1 *General considerations.* Although, in this book, we are mainly concerned with general principles, it is probably worth while clothing the preceding part of this chapter with some practical illustrations. These have always been of importance in the study of solids, but have become particularly so since the discovery, by X-ray diffraction, of the ubiquity of the crystalline state. It is usually worth while beginning the study of a compound by an optical examination, even if only to discover fragments that are good single crystals.

We shall attempt to explain only the general principles of the various procedures to be described; for full details books such as those by Hartshorne & Stuart (1964) and by Bunn (1961) should be consulted. There are, however, some basic terms that are in common usage and a knowledge of these will be a help in obtaining a grasp of practical procedures.

5.7.2 *Nomenclature.* The term birefringence is used both as an alternative to double refraction (§ 1.2.2) and as a measure of the difference between the two refractive indices of a principal section of the indicatrix (§ 5.4.2) of a crystal. As we have seen in § 5.4.6, when a beam of light is passed through a biaxial crystal it is split into two extraordinary beams and when it is passed through a uniaxial crystal it is split into an ordinary beam (o) and an extraordinary beam (e).

Uniaxial crystals are said to be *positive* if $\mu_e > \mu_o$, and *negative* if $\mu_e < \mu_o$, where μ_e and μ_o are defined in § 5.6.2. When light propagates at an angle to the optic axis the ordinary wave has refractive index μ_o and the extraordinary wave has a value between μ_o and μ_e (Fig. 5.23(*a*)). The direction of the **E**-vector of the wave with the larger value of μ is called the *slow* direction, and that with the smaller μ the *fast* direction.

Some crystals have quite different absorptions for the ordinary and extraordinary beams. Tourmaline, for example, even in thin sections, absorbs the ordinary ray almost completely and so can act as a means of polarizing light; it is very inefficient, however, because the extraordinary ray is also greatly absorbed. A biaxial crystal showing this effect is said to be *pleochroic* and a uniaxial crystal is said to be *dichroic*.

5.7.3 *Polaroid.* A much simpler material – Polaroid – has displaced the crystal polarizers for many purposes. This makes use of pleochroism, but the use of single crystals is avoided. Instead, a stretched film of polyvinyl alcohol dyed with iodine is used; the oriented polymeric chains absorb light whose electric vector lies along their length.

Polaroid works well, although polarization is not as complete as that produced by the crystal polarizers; moreover, the transmitted beam is also appreciably absorbed. But it is *much* cheaper to produce, can be obtained in sheets of almost unlimited size, and is thin enough to be easily incorporated in scientific instruments.

5.7.4 *Some auxiliary devices.* Polarizers are usually used in pairs – a polarizer for producing plane-polarized light and an analyser for investigating the nature of such light after various adventures, to be discussed in the next section. Two polarizers set so that the vibrations that they transmit are perpendicular are said to be *crossed* and to produce *extinction.* The quality of Polaroid can be tested by producing extinction with two pieces; usually the residual light is coloured, sometimes a dark purple.

Plane-polarized light is the raw material for producing other forms of polarized light. The Fresnel rhomb for producing circularly polarized light has already been described (§ 5.2.4). A simpler method is to use what is known as a quarter-wave plate – a plate of anisotropic crystal, such as mica, of a thickness such that the ordinary and extraordinary beams are transmitted with a phase difference of $\pi/2$. If a plane-polarized beam is incident so that its vibration makes an arbitrary angle ϕ with one of the vibrations that is transmitted by the plate (Fig. 5.28(a)) it can easily be seen that the resultant of the ordinary and extraordinary beams is elliptically polarized, with an axial ratio equal to tan ϕ.

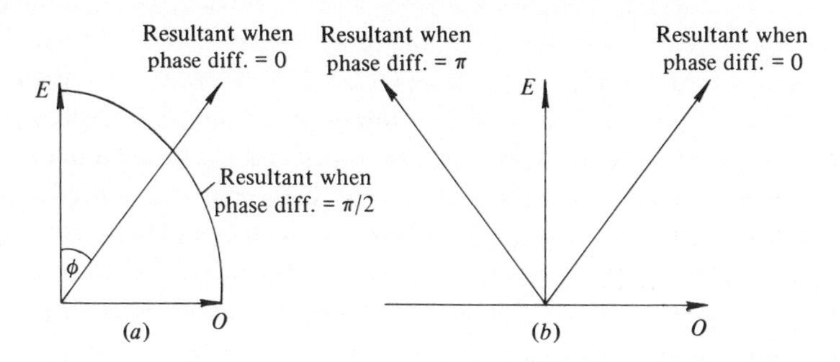

Fig. 5.28. Illustrating the action of (a) a quarter-wave plate and (b) a half-wave plate.

A half-wave plate – one that introduces a phase difference of π – is less useful; it would convert a plane-polarized beam at an angle ϕ (Fig. 5.28(b)) into one at an angle $-\phi$, and can therefore be regarded as turning the plane of polarization through 2ϕ. It will reverse the sense of rotation of incident circularly polarized light.

Quarter-wave and half-wave plates can be correct only for particular wavelengths, since refractive index is frequency-dependent and because λ varies by a factor of two over the visible spectrum. A quarter-wave plate of muscovite mica correct for $\lambda = 5900$ Å has a thickness of about 0.026 mm.

It is useful to be able to measure the phase difference between the ordinary and extraordinary beams by balancing it against a known phase difference. A device for producing such a phase difference to be used in this way is called a *compensator*. The simplest type is the quartz wedge, which has an angle of about 1°; the phase difference is linearly related to distance, the constant of proportionality being obtained from knowledge of the principal refractive indices.

The quartz wedge suffers from the disadvantage of having non-parallel faces, and because no part of it has zero phase difference. This is overcome in the Babinet compensator by using two quartz wedges, the slow direction in one being perpendicular to the slow direction in the other (Fig. 5.29). By moving the compensator, a given phase difference can be produced in any part of the field of view.

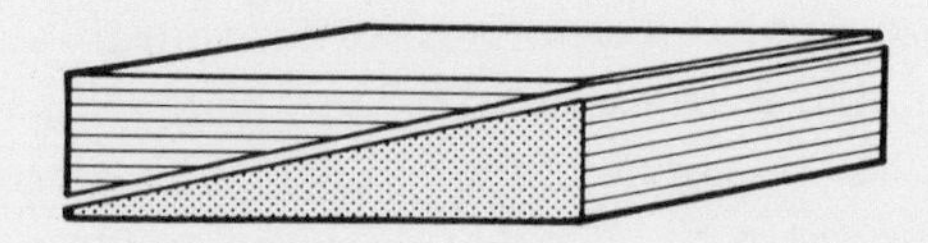

Fig. 5.29. Babinet's compensator. The parallel lines indicate the direction of the optic axes in the two wedges.

The most important instrument in this field is the polarizing microscope. Fundamentally it differs from the ordinary microscope in having a polarizer below and an analyser above the specimen stage; both these should be rotatable and it is convenient if the stage is rotatable also. Such a microscope should have both low-power and high-power condensers, corresponding to approximately parallel illumination and strongly convergent illumination respectively.

5.7.5 *Practical applications.* It is easy to test whether a beam of light is polarized by passing it through a sheet of Polaroid; if the intensity changes, then some polarization is present. But this is far from a complete answer; if the intensity does not vary, the beam may be unpolarized, or circularly polarized or a mixture; if the intensity changes somewhat it may be elliptically polarized or a mixture of elliptically polarized light (including plane-polarized) and unpolarized light; only if the intensity is completely extinguished can one make a definite deduction – that it is plane-polarized.

For a complete answer, auxiliary equipment is needed, and this can take the form of the quarter-wave plate; by trial one can find an orientation that will convert elliptically polarized light into plane-polarized. The reader may like to amuse himself by devising a logical and systematic procedure whereby the distinctions mentioned in the last paragraph can be made. To specify a beam completely one would require to know what proportion is polarized and, if it is elliptically polarized, what are the orientations and the ratio of the axes of the ellipse.

To measure refractive indices of isotropic crystals, the most sensitive method is that which makes use of the apparent disappearance of a body when it is immersed in a liquid of the same refractive index. Laboratories that have to carry out such measurements have available standard liquids, which can be mixed to give intermediate values. If anisotropic crystals are to be studied, the refractive index depends upon direction and upon the plane of polarization; different results will therefore be obtained for crystals lying in different orientations in the liquid. Bunn, in his book

Chemical Crystallography (1961), gives an account of the determination of the principal refractive indices in these circumstances.

Of course, if a good single crystal is obtainable, large enough to handle and to orientate accurately, more direct results can be obtained. To begin with, one can find, with the polarizing microscope (§ 5.7.4), whether the extinction directions are simply related to the external form of the crystal; if the extinction direction is, for example, parallel to an edge, the extinction is said to be *straight*, but if it is not, and has no obvious relation to any symmetry that appears to be present, it is said to be *oblique*. Such results can distinguish between the different symmetries of biaxial crystals.

The most important use of the polarizing microscope is to obtain evidence of the symmetry of a crystal from what is known as its *interference figure* – the pattern obtained by interference of the ordinary and extraordinary beams. At first sight, interference would seem to be impossible, since, although they are coherent, the ordinary and extraordinary beams are polarized at right angles; but if the two beams are passed through an analyser the components parallel to the vibration plane of the analyser *can* interfere.

If a high-power condenser is used to illuminate the crystal, convergent light is produced and quite beautiful interference patterns are obtained. These patterns represent a view of the crystal in terms of the direction of the light passing through it, and the pattern is in focus in the back focal plane of the objective where all rays parallel to a given direction intersect. An auxiliary lens – the Bertrand lens – enables the patterns to be seen on a larger scale through the eyepiece of the microscope.

It would take too long to describe the full implications of these patterns, but examples of figures for uniaxial and biaxial crystals are shown in Fig. 5.30. The uniaxial figure shows a dark cross which

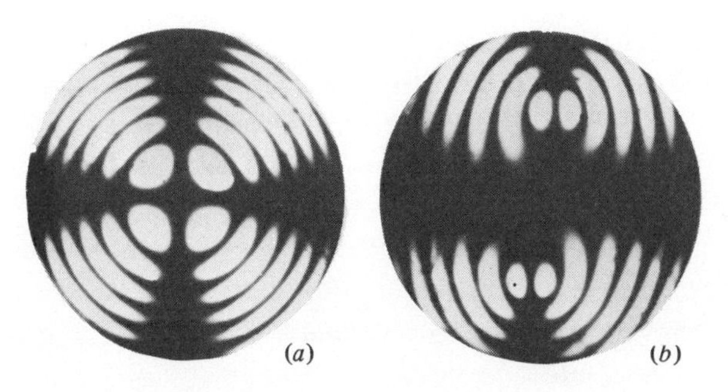

Fig. 5.30. Interference figures (*a*) in a uniaxial crystal; (*b*) in a biaxial crystal.

represents the two extinction directions of the polarizers used; the circular fringes are essentially similar to fringes of equal inclination (§ 7.1.6). The biaxial pattern is more complicated.

It must not be assumed that figures as clear as these can be obtained for *any* crystal; often absorption makes the figures dark and indistinct, and the smallness of the birefringence may make the distinction between uniaxial and biaxial figures difficult to observe. Nevertheless, the polarizing microscope is a weapon of considerable utility in the preliminary investigation of any reasonably transparent crystal.

6

Diffraction

6.1 Meaning of diffraction

As we saw in Chapter 1, the wave theory of light was not at first generally accepted because light did not appear to have any obviously wave-like properties; for example, it did not bend round obstacles as water waves are clearly seen to do. The reason why this difficulty no longer prevents our acceptance of the wave theory is that we are now aware of the relative scales of the two sorts of waves: water waves are coarse and we can see that they only bend round obstacles that have dimensions of the same order of magnitude as the wavelength; larger objects merely stop the waves in the sense that the waves bending round the edge produce negligible effects. But the wavelength of light is about 5×10^{-7} m and an object of about a hundred waves in size – sufficient to stop a light wave – is still very small by ordinary standards.

Nevertheless some bending of the light waves round the edges of obstacles does occur and can be observed over a range of conditions. For particles of the order of a few wavelengths in size, no special apparatus is needed; for example, the water droplets that condense on a car window are surprisingly uniform in size and show beautiful haloes round the street lights as the car passes by. For objects that are much larger, special apparatus is needed. The effects are called *diffraction* phenomena.

6.1.1 *Interference and diffraction.* There is another class of phenomena that is closely related to diffraction. For example, if we look at a distant source of light through our eyelashes we see odd patterns of light distribution. These are produced by the interaction of the waves passing through the spaces between the eyelashes; the effects are called *interference* phenomena.

Diffraction and interference are sometimes not clearly distinguishable, and different writers attach different meanings to the two words. We shall try to maintain the convention that interference involves the deliberate production of two or more separate beams and that diffraction occurs naturally when a single wave is limited in some way. We shall not always succeed in maintaining this convention because some names – the diffraction grating, for example – do not fit in with it and are too well established to change. But we shall try to preserve the distinction where we can. It is, in fact, identical with the distinction between the Fourier series (corresponding to interference) and the Fourier transform (diffraction), and it will be remembered from § 3.4.1 that the series can be deduced as a special case of the transform. In the same way all interference can be explained on the same basis as diffraction effects.

6.1.2 *Approaches to the theory of diffraction.* It is possible to produce a theory of diffraction on the basis of Huygens' principle (§ 2.7.1) of secondary emission, and by means of it to explain almost all observed diffraction phenomena. Before proceeding to do this, however, we ought to look at some of the assumptions implicit in such a theory. We can start by considering exactly how diffraction calculations should be carried out. We can write down Maxwell's equations and solve them subject to the boundary conditions imposed by the diffracting obstacle. For example, diffraction of an incident plane wave by a perfectly conducting sphere can be tackled by this method. But even this simple example is too complicated for general understanding, and a considerably more naive approach has been evolved to explain almost all observed diffraction effects. It makes the basic approximation that the amplitude of the electromagnetic wave can be adequately described by a scalar variable, and that effects arising from the polarization of waves can be neglected.

6.1.3 *The scalar-wave approximation.* In principle, a scalar-wave calculation should be carried out for each component of the vector wave, but in practice this is rarely necessary. On the other hand, we can see the type of conditions under which the direction of polarization might be important by considering how we would begin the problem of diffraction by a slit in a perfectly conducting sheet of metal. Considering each point on the plane of the sheet as a potential radiator, we see that

(a) points on the metal sheet will not radiate at all, because the field **E** must be zero in a perfect conductor;

(b) points well into the slit will radiate equally well in all polarizations, because the field can equally well be in any direction in free space;

(c) points close to the edge of the slit will radiate better when **E** is perpendicular to the edge of the slit than when **E** is parallel. This occurs because **E** (parallel) changes continuously from zero in the metal to a non-zero value in the slit, whereas **E** (perpendicular) is not continuous across a surface (§ 4.3).

The slit thus produces a diffraction pattern appropriate to a rather smaller width when the illumination is polarized parallel to its length. Because such differences are limited to a region within only about one wavelength of the edge of the obstacle, they become progressively more important at longer wavelengths. For example the efficiency of blazed reflexion gratings (§7.7.5) used in the infra-red is almost always polarization dependent, and closely-spaced wire grids are efficient polarizers at wavelengths above 2 μm. We conclude that the scalar-wave approximation is substantially correct only for visible light and shorter wavelengths.

The reader is therefore invited, for the time being, to forget that light consists of two oscillating vector fields, and imagine the vibration to be that of a single scalar variable ψ with angular frequency ω and wave-vector $\mathbf{k}$, which is of magnitude $2\pi/\lambda$ in the direction of travel of the wave. This direction is assumed also to be the ray direction; we shall not concern ourselves with anisotropic media here. Because ψ represents both amplitude and phase we shall consider it as a complex quantity. The time-dependent factor $\exp(\mathrm{i}\omega t)$ is of no importance in this chapter, since it is carried through all the calculations unchanged. It will therefore be omitted.

6.2 Fresnel and Fraunhofer diffraction

6.2.1 *Diffraction under arbitrary conditions.* Let us try intuitively to build a theory of diffraction based on the re-emission of scalar waves by points on a surface spanning the aperture. The reader who is not to be satisfied by this intuitive approach can be reassured that a more rigorous, but still scalar-wave, derivation of the same theory has been given by Kirchhoff and is discussed in § 6.5. But most of the parts of the integral formulation can be written down intuitively, and we shall first derive it in such a manner.

We shall consider the amplitude observed at a point P arising from light emitted from a point source Q and scattered by a plane obstacle R (Fig. 6.1). We shall suppose that if an element of area $\mathrm{d}s$ at S on R is disturbed by a wave ψ_1 this same point acts as a coherent secondary emitter of strength $f_S\psi_1\,\mathrm{d}s$, where f_S is called the *transmission function* of R at point S. In the simplest examples f_S is zero where the obstacle is opaque and unity where it is transparent, but it is easy to imagine intermediate cases. The coherence of the re-emission is important; the phase of the emitted

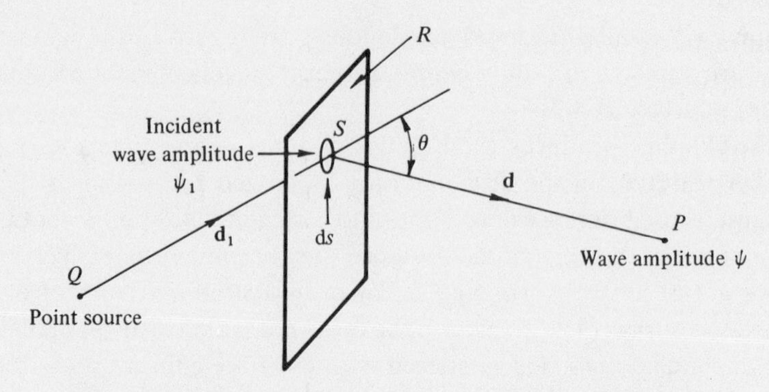

Fig. 6.1. Definition of quantities for the diffraction integral.

wave must be exactly related to that of the initiating disturbance ψ_1, otherwise the diffraction effects will change with time. The exact phase relation may vary over the screen, but at any one place it can be incorporated as a complex phase factor in f_S.

The scalar wave emitted from a point source Q of strength a_1 can be written as a spherical wave (§ 4.2.3),

$$\psi_1 = \frac{a_1}{d_1} \exp(-\mathrm{i}kd_1), \tag{6.1}$$

and consequently S acts as a secondary emitter of strength a,

$$a = f_S \psi_1 \, \mathrm{d}s, \tag{6.2}$$

so that the contribution of ψ received at P is

$$\begin{aligned} \mathrm{d}\psi &= f_S \psi_1 \frac{1}{d} \exp(-\mathrm{i}kd) \, \mathrm{d}s \\ &= f_S a_1 \frac{1}{dd_1} \exp\{-\mathrm{i}k(d+d_1)\} \, \mathrm{d}s. \end{aligned} \tag{6.3}$$

The total amplitude received at P is therefore the integral of this expression over the obstacle R:

$$\psi = \iint_R a_1 f_S \frac{1}{dd_1} \exp\{-\mathrm{i}k(d+d_1)\} \, \mathrm{d}s, \tag{6.4}$$

The quantities f_S, d and d_1 are all functions of the position S. It will be shown in § 6.5.2 that expression (6.2) should really contain an inclination factor too; i.e. the strength of a secondary emitter depends on the angle between the incident and scattered radiation θ in Fig. 6.1. We shall neglect this factor, except to consider its effect in general in § 6.4.4.

Diffraction calculations involve integrating the expression (6.4) under various conditions. We shall consider first of all some simplifying conditions which may help to make the principles clearer. First let us restrict our attention to a system illuminated by a plane-wave. We do this by taking the source Q to infinity and making it infinitely bright; we therefore make d_1 and a_1 go to infinity while maintaining their ratio constant:

$$\frac{a_1}{d_1} = A. \tag{6.5}$$

Now we shall only consider plane obstacles R normal to the incident illumination. The integral is then to be performed across the plane. If we denote position in the plane by a vector $\mathbf{r}$ from some origin O we write f_S as a function $f(\mathbf{r})$ and (6.4) becomes

$$\psi = A \exp(-\mathrm{i}kd_1) \iint_R f(\mathbf{r}) \frac{1}{d} \exp(-\mathrm{i}kd)\, \mathrm{d}^2\mathbf{r}, \tag{6.6}$$

The factor $\exp(-\mathrm{i}kd_1)$, being constant over the plane R as this is normal to the incident wave, will henceforth be absorbed into A. The intensity observed at P is

$$I = |\psi|^2 = \psi\psi^*. \tag{6.7}$$

6.2.2 *The distinction between Fraunhofer and Fresnel diffraction.* The classification of diffraction effects into Fresnel and Fraunhofer types depends on the way in which the phase kd changes as we cross the obstacle R. This depends on three factors: the distance d between the obstacle and the plane of observation, the size of the obstacle R (i.e. the extent of its transmitting region), and the wavelength $\lambda = 2\pi/k$. If kd varies linearly with the position of S in R, the diffraction is called Fraunhofer diffraction; if the variation has non-linear terms of size comparable with π, the diffraction is called Fresnel diffraction. We can translate this statement into quantitative terms if we define a circle of radius ρ which just includes all the transmitting regions of R (Fig. 6.2). Then at an arbitrary point P at distance $\mathbf{q}$ from the axis through the centre of this circle the phase kd of the wave from the point $\mathbf{r}$ in the circle is

$$kd = k(d_0{}^2 + |\mathbf{r} - \mathbf{q}|^2)^{\frac{1}{2}} \approx kd_0 + \tfrac{1}{2}kd_0{}^{-1}(q^2 - 2\mathbf{r}\cdot\mathbf{q} + r^2) + \ldots, \tag{6.8}$$

where we have assumed that r and q are small compared with d_0. This expression contains the constant term $k(d_0 + \frac{1}{2}q^2/d_0)$, the term $k\mathbf{r}\cdot\mathbf{q}/d_0$ which is linear in $\mathbf{r}$, and the quadratic term $\frac{1}{2}kr^2/d_0$. Now since the largest

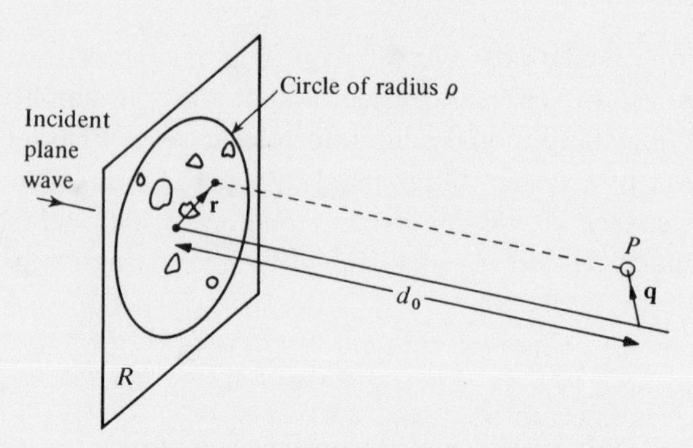

Fig. 6.2. Elements of a diffraction calculation.

value of r which contributes to the problem is ρ, we have a quadratic term of maximum size $\frac{1}{2}k\rho^2/d_0$. This means that Fresnel or Fraunhofer conditions are obtained depending on whether $\frac{1}{2}k\rho^2/d_0$ is greater or less than π. In terms of wavelength λ this gives us:

$$\text{Fresnel diffraction} \qquad \rho^2 > \lambda d_0,$$

$$\text{Fraunhofer diffraction} \qquad \rho^2 < \lambda d_0.$$

For example, if a hole of diameter 2 mm is illuminated by light of wavelength 5×10^{-4} mm, Fresnel diffraction patterns will be observed at distances d_0 less than 2 m, and Fraunhofer diffraction at greater distances. The calculation of the patterns will show that the transition from one type to the other is gradual.

To return to Fig. 6.1, it is easy to show that when the obstacle is illuminated by a point source at finite distance d_1, we need to replace d_0^{-1} by $(d_0^{-1} + d_1^{-1})$ in the foregoing discussion. When the object is one-dimensional, for example a slit or series of slits, it is common to replace the point source Q by a line or slit source. Each point of the line source produces a diffraction pattern from the obstacle, and provided these are identical and not displaced laterally they will lie on top of one another and produce an intensified version of the pattern from a point source. This requires the line source and slit obstacle to be accurately parallel. We shall deal in this chapter only with patterns produced by point sources; the experimental hazards of non-parallel slits must be left to fend for themselves.

6.2.3 *Experimental observation of diffraction patterns.* Given a point source of monochromatic light or a coherent wavefront from a laser, it is easy to observe diffraction patterns of both types. When using a point source, it is important to make sure that it is really small enough for the radiated wave to be a true spherical wave. (This amounts to saying that the radiated wave is coherent; see Chapter 8.) In other words, the spherical waves emitted by various points in the source, assuming it to have finite extent D, must coincide to an accuracy of about $\frac{1}{4}\lambda$ over the transmitting part of R, which is the circle of radius ρ. The requirement for this is easily seen to be

$$D\rho/d_1 < \tfrac{1}{4}\lambda. \tag{6.9}$$

For our 2 mm circular hole, at a distance $d = 1$ m the source must have dimensions $D < 0.1$ mm; at a distance of 1 km a 10 cm diameter street lamp will suffice.

To observe Fresnel patterns, it is only necessary to put a screen at the required distance d. To observe Fraunhofer patterns, using the point source at d_1 we need to put the observing screen at distance $d_0 = -d_1$ or in any plane optically conjugate to this. For example, we can look directly at the source, so that the retina of the eye is conjugate to the plane of the source, and by inserting the obstacle anywhere along the line of sight (usually close to the pupil) the Fraunhofer pattern can be observed. Defocusing the eye converts the pattern into a Fresnel pattern. For quantitative work one uses a point source with a lens giving a parallel beam, or else an expanded collimated laser beam, either of which is equivalent to infinite d_1. The Fraunhofer pattern is then observed on a screen at infinite d_0, or else in the focal plane of a converging lens, which plane is conjugate to the infinite d_0. Many of the photographs in this book were taken with an optical diffractometer, which is constructed on the above principle (Fig. 6.3).

6.3 Fresnel diffraction

This section and the next are to be devoted to a mathematical study of Fresnel diffraction in a few simple systems. Fraunhofer diffraction and its applications are a much more important subject, and are therefore given a chapter entirely to themselves (Chapter 7). The mathematics of Fresnel diffraction is more complicated, however, and as a result often receives much more attention than it deserves. The only real justification for this attention is that Fresnel diffraction affords an interesting example of the use of amplitude–phase diagrams, which form an important method of evaluating integrals of the form

$$I = \int f(x) \exp\{\mathrm{i}\phi(x)\}\, \mathrm{d}x; \tag{6.10}$$

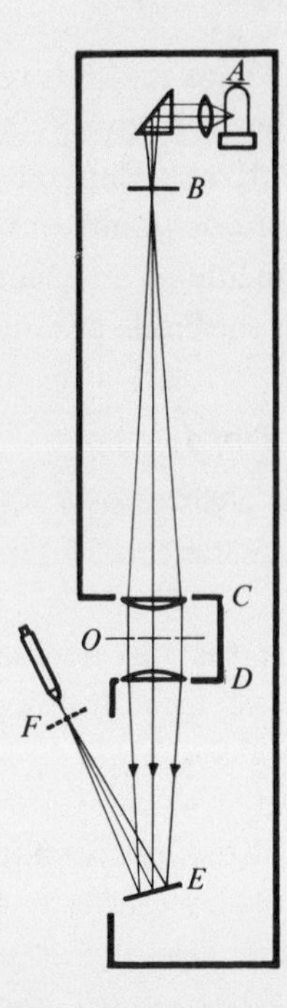

Fig. 6.3. Optical diffractometer. A is the light source; B is the pin-hole; C and D are the lenses; and E is an optically flat mirror. The diffraction pattern of an object at O is seen in the plane F.

this occurs frequently in many branches of physics. We do not, however, recommend any reader to spend a large amount of time simply in the mastering of Fresnel diffraction calculations once he has understood the principles.

It is often claimed that Fresnel diffraction patterns are also important as the basis of holography (§ 9.7.4). While this is undoubtedly true, it is rarely necessary to calculate the exact intensity distribution in a hologram. An isolated example of the need for such an exact calculation is the analysis of side-looking radar data, and their subsequent reconstruction into images. (See e.g. Leith, 1976.)

The basic integral to be evaluated is therefore equation (6.6)

$$\psi = \iint_R Af(\mathbf{r})\frac{1}{d}\exp(-\mathrm{i}kd)\,\mathrm{d}^2\mathbf{r}. \tag{6.11}$$

We shall expand d in the exponent by the binomial theorem, as in § 6.2.2, but we shall not do so in the reciprocal since it has relatively little effect there:

$$\psi = \iint_R Af(\mathbf{r})\frac{1}{d_0}\exp\left\{-\mathrm{i}k\left(d_0+\frac{1}{2}\frac{r^2}{d_0}+\ldots\right)\right\}\mathrm{d}^2\mathbf{r}$$

$$= \frac{A}{d_0}\exp(-\mathrm{i}kd_0)\iint_R f(\mathbf{r})\exp\left(-\mathrm{i}k\frac{r^2}{2d_0}\right)\mathrm{d}^2\mathbf{r}. \qquad (6.12)$$

It will be clear that (6.12) is the same type of integral as (6.10).

6.3.1 *Circular systems, where the integral can be evaluated exactly.* In a system with circular symmetry, the value of ψ on the axis of rotation can be evaluated exactly from (6.12). In such a system, $f(\mathbf{r})$ can be written as a function of the scalar r and therefore as a function of its square

$$s = r^2, \qquad (6.13)$$

$$f(r) = g(s). \qquad (6.14)$$

The element of area d^2r can also be written in terms of s

$$\mathrm{d}^2r = 2\pi r\,\mathrm{d}r = \pi\,\mathrm{d}s \qquad (6.15)$$

and therefore the integral (6.12) becomes (omitting the $\exp(-\mathrm{i}kd_0)$ which will be eliminated when we observe the intensity $\psi\psi^*$)

$$\psi = \frac{A\pi}{d_0}\int_R g(s)\exp\left(-\frac{\mathrm{i}ks}{2d_0}\right)\mathrm{d}s. \qquad (6.16)$$

This is a Fourier-transform integral (§ 3.4.1), in which ψ is a function of the variable $k/2d_0$, which is related to the position along the symmetry axis of the system. We cannot calculate the pattern off this axis analytically – although we can observe it quite easily (Fig. 1.2).

There are three systems of importance in this class:

(a) the circular hole of radius ρ, $g(s) = 1, s < \rho^2$,
$g(s) = 0, s > \rho^2$;

(b) the circular disc of radius ρ, $g(s) = 0, s < \rho^2$,
$g(s) = 1, s > \rho^2$;

(c) the zone plate, for which $g(s)$ is periodic.

6.3.2 *The circular hole.* The integral becomes, under the conditions (a) above,

$$\psi = \frac{A\pi}{d_0}\int_0^{\rho^2}\exp\left(-\frac{\mathrm{i}k}{2d_0}s\right)\mathrm{d}s \qquad (6.17)$$

$$= \frac{A\pi}{d_0}\frac{2d_0}{\mathrm{i}k}\left\{\exp\left(-\frac{\mathrm{i}k}{2d_0}\rho^2\right)-1\right\}. \qquad (6.18)$$

The observed intensity along the axis then becomes

$$\psi\psi^* = \frac{8A^2\pi^2}{k^2}\left(1 - \cos\frac{k\rho^2}{2d_0}\right). \tag{6.19}$$

As the point of observation moves along the axis, the intensity at the centre of the pattern alternates between zero and $16A^2\pi^2/k^2$ periodically with d_0^{-1}.

6.3.3 *The circular disc.* In a similar way, we must evaluate the integral for (b):

$$\psi = \frac{A\pi}{d_0}\int_{\rho^2}^{\infty} \exp\left(-\frac{ik}{2d_0}s\right) \mathrm{d}s \tag{6.20}$$

$$= \frac{A\pi}{d_0}\frac{2d_0}{ik}\left\{\exp(-i\infty) - \exp\left(-\frac{ik}{2d_0}\rho^2\right)\right\} \tag{6.21}$$

The exponential exp $(-i\infty)$ can safely be taken as zero. We have, it will be remembered, approximated the d^{-1} term in (6.11) by d_0^{-1}; this approximation will be invalid in the limit $s \to \infty$, and the fact that in this limit $d \to \infty$ and $d^{-1} \to 0$ makes it permissible to neglect the exp $(-i\infty)$ term. Thus the intensity

$$\psi\psi^* = 4\frac{A^2\pi^2}{k^2} \tag{6.22}$$

for all values of d_0. This surprising result, that there is always a bright spot at the centre of the diffraction pattern of a disc (Fig. 1.2), was the argument used finally to overthrow the opponents to the wave theory of light (§ 1.2.4). To this extent Fresnel diffraction has been of vital importance to the development of optics.

6.3.4 *The zone plate.* The zone plate is little more than an amusing physical toy to illustrate Fresnel diffraction, although its significance is now enhanced as providing an approach to the understanding of holograms (§ 9.7.3). It is usually made by photographing down a large drawing of alternate black and white rings of diameters such as to make $g(s)$ a periodic function (Fig. 6.4) with periodicity $2a_0^2$

$$\begin{aligned} g(s) &= 0 \quad \text{from} \quad s = 0 \text{ to } a_0^2 \\ &\qquad\qquad\qquad 2a_0^2 \text{ to } 3a_0^2 \\ &\qquad\qquad\qquad 4a_0^2 \text{ to } 5a_0^2, \text{ etc.,} \\ &= 1 \quad \text{from} \quad s = a_0^2 \text{ to } 2a_0^2 \\ &\qquad\qquad\qquad 3a_0^2 \text{ to } 4a_0^2 \\ &\qquad\qquad\qquad 5a_0^2 \text{ to } 6a_0^2, \text{ etc.} \end{aligned}$$

Fig. 6.4. Zone plate.

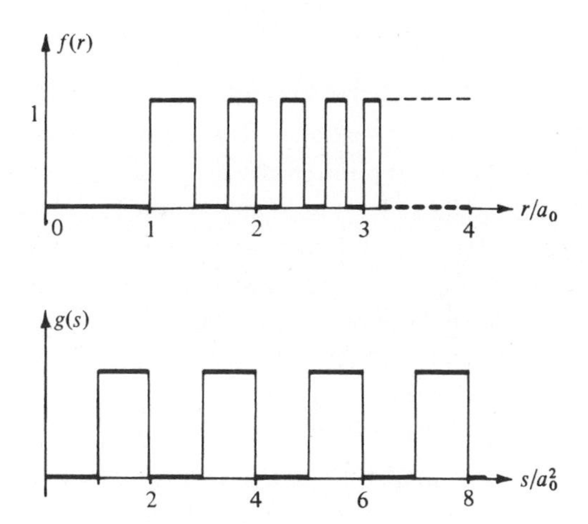

Fig. 6.5. Functions $f(r)$ and $g(s)$ $(s = r^2)$ for a zone plate.

If there is an infinite number of rings, the Fourier integral (6.16) for ψ will be zero except at discrete values of d_0 satisfying

$$\frac{k}{2d_0} = \frac{n2\pi}{2a_0^{\,2}} \qquad (n \text{ integral}). \tag{6.23}$$

The waveform of $g(s)$ dictates the amplitude of ψ at its various peaks. In the case of the square wave illustrated in Fig. 6.5 we have a Fourier series with amplitude $1/n$ for odd n and zero for even n (§ 3.3.2), so that ψ has values proportional to

$$\frac{A\pi}{d_0}\frac{1}{n} = \frac{A\pi}{d_0}\frac{2d_0\pi}{ka_0^{\,2}} = \frac{2A\pi^2}{ka_0^{\,2}}, \tag{6.24}$$

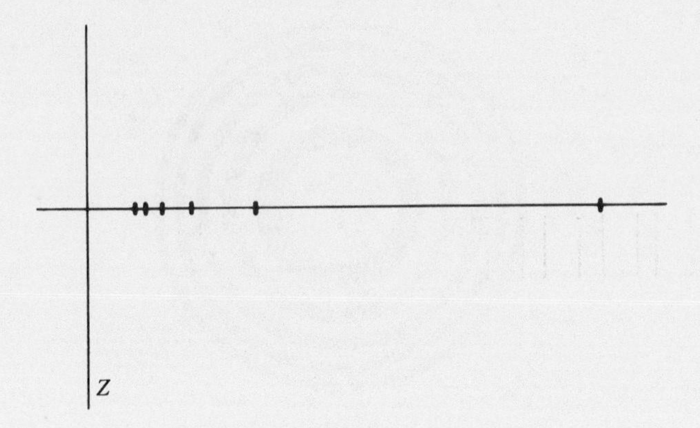

Fig. 6.6. Positions of the 'foci' of a zone plate, Z.

at values of $d_0 = ka_0^2/2\pi n$ for odd integers n (Fig. 6.6). Notice that the amplitude reduction of the orders of the Fourier series has been cancelled exactly by the $1/d_0$ before the integral.

The zone plate can be used in a similar way to a lens. If we concentrate on one particular order of diffraction, $n = 1$ say, we can see from § 6.2.2 that if illumination is provided by a point source at distance d_1 (Fig. 6.1) the position of the image moves out to satisfy

$$\frac{1}{d_0} + \frac{1}{d_1} = \frac{2\pi}{ka_0^2} \tag{6.25}$$

which is equivalent to a lens of focal length

$$f = \frac{ka_0^2}{2\pi} = \frac{a_0^2}{\lambda}. \tag{6.26}$$

It clearly suffers from serious chromatic aberration (see Problem 6.3).

6.4 Fresnel diffraction by linear systems

The Fresnel integral (6.12) must be evaluated graphically for systems without circular symmetry. We can write it in Cartesian coordinates, for example:

$$\psi = \frac{A}{d_0} \int\int_R f(x, y) \exp\left\{-\frac{ik}{2d_0}(x^2 + y^2)\right\} \mathrm{d}x\, \mathrm{d}y. \tag{6.27}$$

In this section we shall consider only systems in which $f(x, y)$ can be separated into the product of two functions

$$f(x, y) = g(x)h(y) \tag{6.28}$$

so that the integral is separable:

$$\psi = \frac{A}{d_0}\int_{-\infty}^{\infty} g(x)\exp\left(-\frac{ik}{2d_0}x^2\right)\mathrm{d}x\int_{-\infty}^{\infty} h(y)\exp\left(-\frac{ik}{2d_0}y^2\right)\mathrm{d}y. \tag{6.29}$$

The two parts can then be evaluated separately.

6.4.1 *The Fresnel integral.* Let us consider how to evaluate an integral of the type

$$\psi = \int_{x_1}^{x_2} f(x)\exp\{-\mathrm{i}\phi(x)\}\,\mathrm{d}x \tag{6.30}$$

by means of a graphical method. We shall eventually be led to the Cornu spiral, which is the particular method appropriate to Fresnel diffraction patterns. We represent each infinitesimal increment of ψ

$$\mathrm{d}\psi = f(x)\exp\{-\mathrm{i}\phi(x)\}\,\mathrm{d}x, \tag{6.31}$$

by a vector in the complex plane of length $f(x)\,\mathrm{d}x$ and at angle $\phi(x)$ to the real axis. The value of ψ is then the vector sum of the increments, which is the vector joining the x_1 and x_2 ends of the curve formed by all the increments head-to-tail (Fig. 6.7). This is called an *amplitude–phase diagram.*

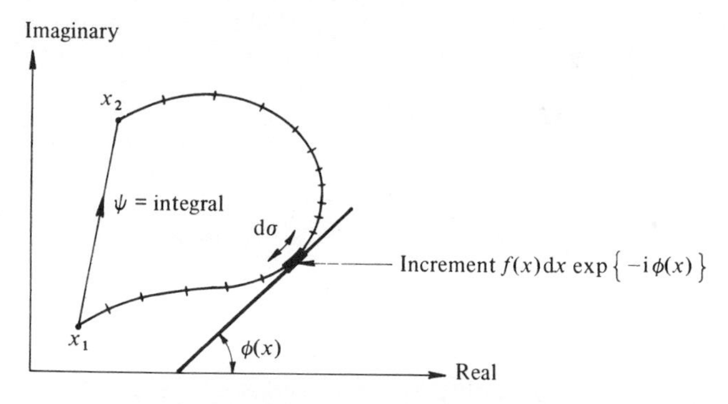

Fig. 6.7. Complex plane diagram of the integral $\int_{x_1}^{x_2} f(x)\exp\{-\mathrm{i}\phi(x)\}\,\mathrm{d}x$.

The problem is to calculate the form of the curve which is drawn schematically in Fig. 6.7. If we move a length $\delta\sigma$ along the curve, and the angle ϕ changes by $\delta\phi$ as a result, the radius of curvature of the curve is clearly

$$\rho = \frac{\delta\sigma}{\delta\phi} \rightarrow \frac{\mathrm{d}\sigma}{\mathrm{d}\phi} \quad \text{as} \quad \delta\sigma \quad \text{and} \quad \delta\phi \rightarrow 0 \tag{6.32}$$

and if we know ρ as a function of σ the complete curve can be drawn out. Now σ, being the length along the curve, is simply $\int_0^x f(x)\,\mathrm{d}x$ and therefore the functions necessary to plot the curve are

$$\sigma = \int_0^x f(x)\,\mathrm{d}x \quad \text{and} \quad \rho^{-1} = \frac{\mathrm{d}\phi}{\mathrm{d}\sigma} = \frac{1}{f(x)}\frac{\mathrm{d}\phi}{\mathrm{d}x}. \tag{6.33}$$

6.4.2 *Diffraction by a slit.* In the example of the Fresnel diffraction pattern we proceed as follows. Let us consider the problem of diffraction by a single slit, so that

$$g(x) = 1 \qquad x_1 < x < x_2,$$
$$g(x) = 0 \quad \text{otherwise}.$$

The integral then gives the amplitude and phase of the disturbance at P, which is opposite $x = 0$ (Fig. 6.8). To build up to the whole of the pattern

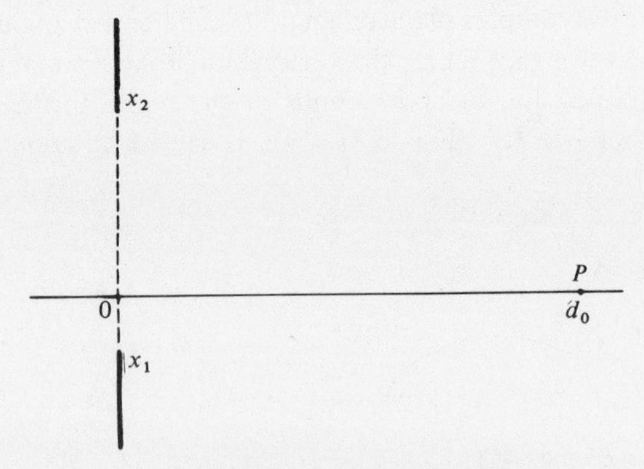

Fig. 6.8. Parameters of a slit.

we must repeat the calculation for P opposite various points on the slit by varying x_1 and x_2 so that $(x_1 - x_2)$ remains constant. The integral is therefore

$$\psi_P = \int_{x_1}^{x_2} \exp\left\{-\mathrm{i}\left(\frac{kx^2}{2d_0}\right)\right\}\mathrm{d}x. \tag{6.34}$$

The amplitude and phase are given by the vector between the points representing x_1 and x_2 on the curve defined by

$$\sigma = \int_0^x \mathrm{d}x = x, \tag{6.35}$$

$$\rho^{-1} = \frac{\mathrm{d}\phi}{\mathrm{d}x} = \frac{\mathrm{d}}{\mathrm{d}x}\frac{(kx^2)}{2d_0} = \frac{kx}{d_0} = \beta^2 x. \tag{6.36}$$

Thus x is the distance from the origin measured along the curve, and the curvature at that point is $\beta^2 x$, proportional to x. This curve is a function of β, which is inconvenient, but it can be written in terms of the dimensionless quantities;

$$x' = \beta x, \qquad \rho' = \beta \rho \tag{6.37}$$

(which simply involves increasing the scale of the drawing by a factor β); then we have:

$$\frac{1}{\rho'} = x'. \tag{6.38}$$

The curve in terms of ρ' and x' is thus independent of β. It is called the *Cornu spiral* and is illustrated in Fig. 6.9. To calculate the whole of the

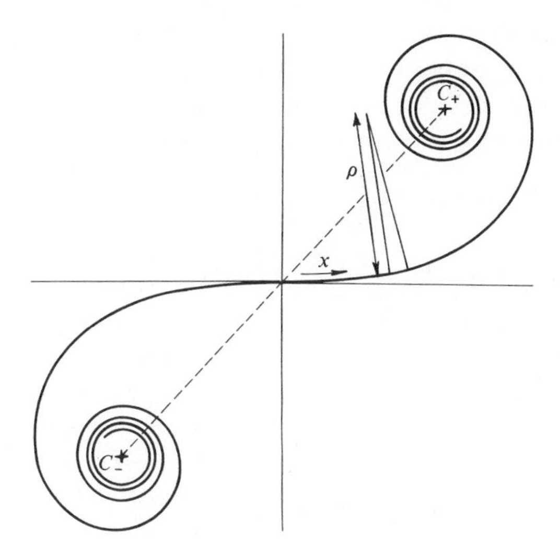

Fig. 6.9. The Cornu spiral.

diffraction pattern from such an aperture we take a series of values of x_1 and x_2, such that $(x_1 - x_2)$ is the width of the slit, and measure the vector length between the points on the spiral representing x_1 and x_2. This corresponds to the amplitude of the wave at P, which is opposite zero, or a distance x_1 from the x_1-end of the slit. To allow for the factor β, when using the dimensionless curve in Fig. 6.9, distances measured along the curve are βx rather than x. It will be seen, then, that diffraction patterns become rather complicated when both βx_1 and βx_2 are in the horns of the spiral – i.e. when βx is typically of the order of, or greater than, 10. As an example, we have calculated the diffraction pattern for a value of

$\Delta x' = 8.5$. Using light of wavelength 6×10^{-5} cm we then have

$$\Delta x' = \left(\frac{k}{d_0}\right)^{\frac{1}{2}} \Delta x = \left(\frac{2\pi}{\lambda d_0}\right)^{\frac{1}{2}} \Delta x = 8.5, \tag{6.39}$$

whence

$$\frac{(\Delta x)^2}{d_0} \approx 7 \times 10^{-4} \text{ cm.} \tag{6.40}$$

The same pattern is observed for all slit widths Δx and distances d_0, d_1 satisfying

$$\Delta x^2 (d_0^{-1} + d_1^{-1}) = 7 \times 10^{-4} \text{ cm.} \tag{6.41}$$

Fig. 6.10(*a*) shows the calculated intensity as a function of position, and Fig. 6.10(*b*) the diffraction pattern observed. The numerical calculations were made with the help of an accurate scale drawing of the Cornu spiral in the classic, *Physical Optics* (Wood, 1934).

We must not, however, neglect the y integral in (6.29). In this example of a long slit, $h(y)$ is always unity and the Cornu spiral gives the integral

$$\int_{-\infty}^{\infty} h(y) \exp\left(-\frac{\mathrm{i}k}{2d_0} y^2\right) \mathrm{d}y$$

$$= \int_{-\infty}^{\infty} \exp\left(-\frac{\mathrm{i}k}{2d_0} y^2\right) \mathrm{d}y = \text{vector } C_+C_-. \tag{6.42}$$

This vector is of constant fixed length, and is therefore rather irrelevant, but it does alter the phase of the whole pattern by $\pi/4$ since C_+C_- is at 45° to the axes of the diagram. The phase is, of course, unobservable, but as a mathematical curiosity it is interesting to notice that the diffraction pattern of an infinite aperture, for which the integrals in both x and y are like (6.42), is the incident wave shifted in phase by $\pi/4$ twice, or $\pi/2$. But since an infinite aperture should not affect the wave at all, it is hard to understand this phase shift. It is explained by the Kirchhoff diffraction integral in § 6.5.2.

6.4.3 *Diffraction by a single edge: the edge wave.* Some diffraction patterns have certain characteristics which can easily be recognized as geometrical properties of the Cornu spiral. The most characteristic property of the spiral is that for large values of x it becomes almost circular and converges very slowly towards its limits; over a considerable range of x it can be considered as a circle of radius x^{-1}.

The diffraction pattern of a straight edge, which can be considered as an aperture extending from a finite value of x to infinity, can be expressed in

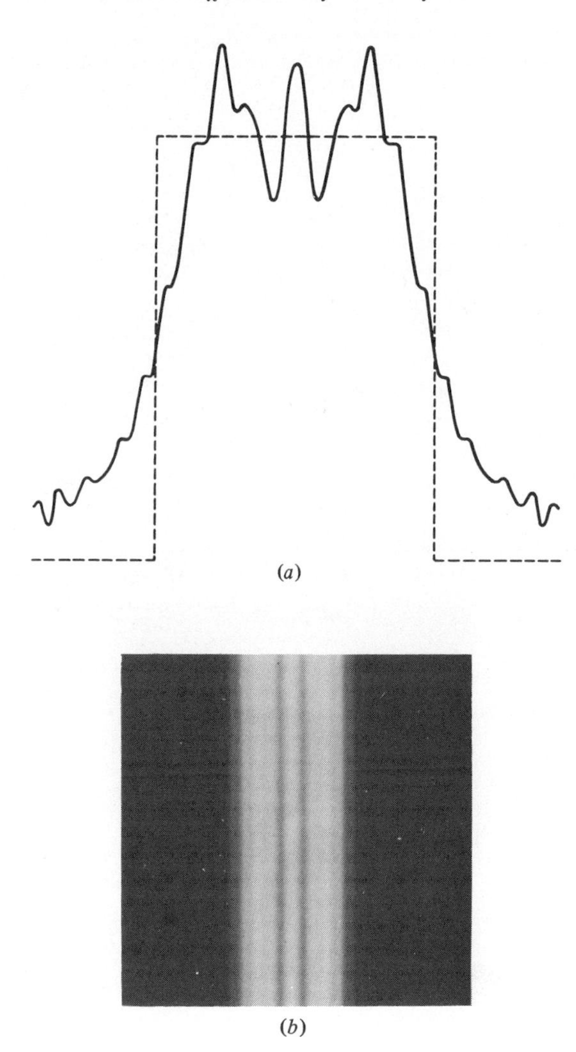

Fig. 6.10 (*a*) Amplitude of the Fresnel diffraction pattern calculated for a slit of width 0.9 mm observed with $d_0 = 20$ cm, $d_1 = 28$ cm and $\lambda = 6 \times 10^{-5}$ cm. The geometrical shadow is indicated by the broken lines. (*b*) Photograph of the diffraction pattern observed under the same conditions as (*a*).

terms of the Cornu spiral similarly to the previous example (Fig. 6.11). But the vector that joins the point βx to ∞ on the spiral can be investigated qualitatively with little difficulty.

When the point x representing the edge of the aperture is at a value so that $x = 0$ is in the geometrical shadow, the vector simply rotates about the centre of the horn, becoming slowly shorter as $\beta x \to \infty$. There are no

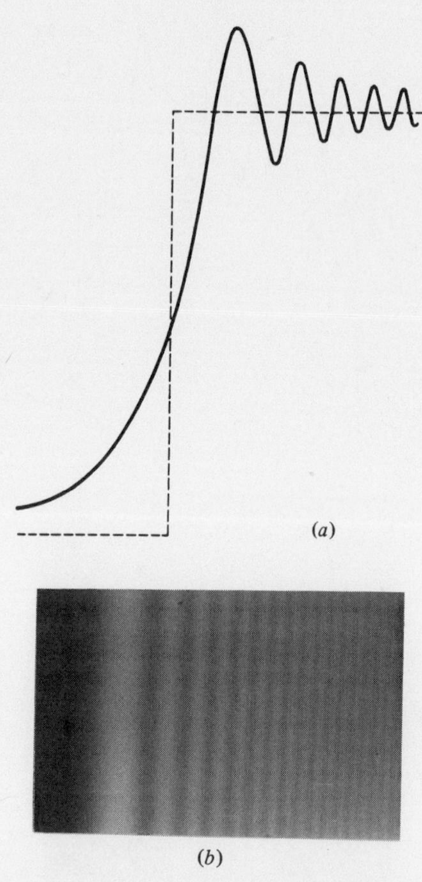

Fig. 6.11. (*a*) Amplitude of the Fresnel diffraction for a single straight edge. The geometrical shadow is indicated by the broken lines. (*b*) Photograph of the pattern from a single straight edge.

oscillations in its length so that there are no dark and light bands, and its phase changes steadily (Fig. 6.12):

$$\frac{\mathrm{d}\phi}{\mathrm{d}x} = \beta x. \tag{6.43}$$

This is identical with the wave coming from a line source, since the phase variation for such a wave is

$$\begin{aligned}\psi_P &= \exp\{-\mathrm{i}k(x^2 + d^2)^{\frac{1}{2}}\}\\ &\approx \exp\left\{-\mathrm{i}k\left(d + \frac{1}{2}\frac{x^2}{d}\right)\right\}\\ &= \exp\{-\mathrm{i}\phi(x)\},\end{aligned} \tag{6.44}$$

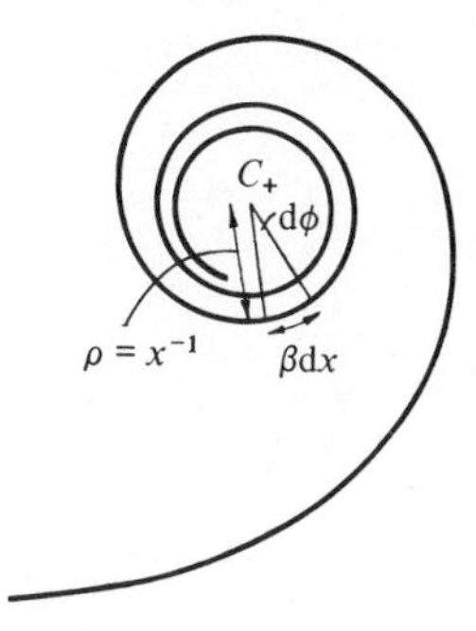

Fig. 6.12. Cornu-spiral construction for the edge wave observed in the shadow region.

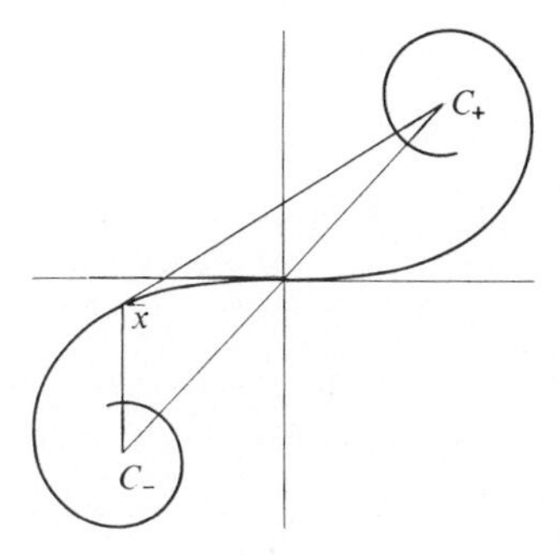

Fig. 6.13. Cornu-spiral construction for the edge wave observed in the bright region.

for which $\mathrm{d}\phi/\mathrm{d}x = kx/d = \beta x$ also. The amplitude is almost independent of x. As the phase and amplitude variations in the geometrical shadow are thus almost identical with those that would arise from a line source coincident with the edge, the appearance of the aperture is that of a line source, and this is known as the *edge wave*. Well into the bright part of the pattern a similar analysis can be applied, by representing the vector between $x = \infty$ and $x < 0$ by the sum of the two vectors C_+C_- and C_-x (Fig. 6.13). The appearance is thus of uniform illumination (C_+C_-) and a bright edge again. Interference between these two waves, plane and diverging, gives rise to a set of light and dark bands with a spacing which decreases as $x \to -\infty$. The edge wave is not observable in the region near $x = 0$.

6.4.4 *The effect of the approximation on the Cornu spiral.* Having derived the Cornu spiral as the amplitude–phase diagram for Fresnel-diffraction

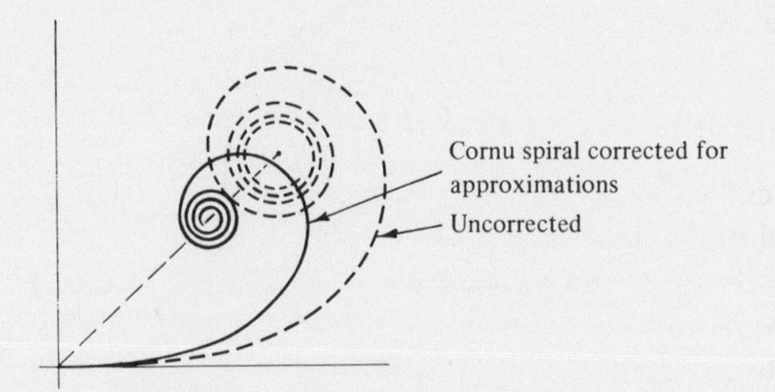

Fig. 6.14. Schematic corrections to the Cornu spiral.

calculations, we can see qualitatively what effect the various approximations might have.

The inclination factor and the inverse-square law can be included by a reduction of $g(x)$ below its value of unity at large x. This will make the spiral wind into its horns rather more quickly than in the uncorrected case (Fig. 6.14). Unfortunately these effects are not proportional to the dimensionless βx, and so the spiral would have to be recorrected for each calculation. The correction is of no practical importance.

The neglect of higher powers of x than the second in the binomial expansion of the exponent

$$\mathrm{i}k({d_0}^2+x^2)^{\frac{1}{2}} \approx \mathrm{i}k\left(d_0+\frac{1}{2}\frac{x^2}{d_0}+\ldots\right) \tag{6.45}$$

would have a similar effect on the spiral. Once again, experimental conditions are usually such that this effect is negligible also.

6.4.5 *Integrals performed by amplitude–phase diagrams.* The Cornu spiral is one example of an integral which can be carried out using an amplitude–phase diagram. The method is applicable to any integral of the form

$$\int_{x_1}^{x_2} f(x)\exp\{-\mathrm{i}\phi(x)\}\,\mathrm{d}x, \tag{6.46}$$

which includes the Fourier integral. Here the phase

$$\phi(x)=kx\sin\theta=ux. \tag{6.47}$$

The Fourier integral for a slit is therefore simple, since $f(x)$ is unity across the slit; thus

$$\sigma=\int_0^x f(x)\,\mathrm{d}x=x \tag{6.48}$$

and the curvature of the amplitude–phase diagram

$$\rho^{-1} = \frac{\mathrm{d}\phi}{\mathrm{d}s} = u, \tag{6.49}$$

which defines an arc of length x of a circle, the diameter of which is determined by the value of u or $k \sin \theta$.

The theory of integration by the method of stationary phase, or steepest descent, is another outcome of amplitude–phase diagrams. It concerns integrals of the type of (6.46) between infinite limits (for example, the vector C_+C_-) and shows that almost the entire integral comes from the contributions of the parts where the curvature

$$\frac{\mathrm{d}\phi}{\mathrm{d}s} = 0; \tag{6.50}$$

it is then possible to evaluate the integral approximately in terms of the derivatives of the function at such points. For the Cornu spiral this occurs only at the origin, but other functions may give a number of points of stationary phase. See, for example, Margenau & Murphy (1964).

6.5 The Kirchhoff diffraction integral

As an electromagnetic field everywhere in a bounded region of space can be uniquely determined by the boundary conditions around this region, it is of interest to see to what extent such an approach is consistent with the idea of re-radiation by points on a wavefront through the aperture. As a result of solving the scalar wave equation (from (6.1)):

$$\nabla^2\psi = -\frac{\omega^2}{c^2}\psi = -k^2\psi, \tag{6.51}$$

which refers to any component of the electric or magnetic wave-field, we shall see that the disturbance ψ_0 at a point inside the bounded region can be written in terms of the disturbances and its derivatives on the boundary of the region. In simple cases where these are determined by external waves originating from a point source the result is very similar to the one which we have already used intuitively.

6.5.1 *The exact mathematics for the diffraction integral.* In problems involving boundaries it is often convenient to study the properties of the differences between two solutions of an equation rather than of one solution alone, since the boundary conditions become simpler to handle.

The diffraction integral provides one such example, and we shall compare the required solution of (6.51) with a *trial solution*

$$\psi_1 = \frac{1}{r}\exp(-\mathrm{i}kr), \tag{6.52}$$

which is a spherical wave radiating from the origin $\mathbf{r} = 0$ (see § 4.2.3). This wave satisfies (6.51) except at $\mathbf{r} = 0$. This origin we shall define as the point of interest inside the bounded region, and ψ_0 is then the disturbance there:

$$\psi_0 = \psi(0). \tag{6.53}$$

The two wave-fields ψ (to be calculated) and ψ_1 (the convergent reference wave) satisfy the equation

$$\psi\nabla^2\psi_1 - \psi_1\nabla^2\psi = -\psi k^2\psi_1 + \psi_1 k^2\psi = 0 \tag{6.54}$$

at all points except $\mathbf{r} = 0$, because both ψ and ψ_1 are solutions of (6.51). This expression (6.54) can be integrated throughout the bounded region of space that we are considering, which is a volume V bounded by a surface S, and the integral changed by Green's theorem to a surface integral:

$$\iiint_V (\psi\nabla^2\psi_1 - \psi_1\nabla^2\psi)\,\mathrm{d}V = \iint_S (\psi\boldsymbol{\nabla}\psi_1 - \psi_1\boldsymbol{\nabla}\psi)\cdot\mathbf{n}\,\mathrm{d}s, \tag{6.55}$$

$\mathbf{n}$ being the outward normal to the surface S at each point. Because the integrand (6.54) is zero, the integrals (6.55) are also zero, provided that the region V does not include the origin $\mathbf{r} = 0$. The surface S is therefore chosen to be as illustrated in Fig. 6.15, consisting of an arbitrary outer surface S_1 and a small spherical surface S_0 of radius δr much less than one wavelength surrounding the origin. Volume V lies between the two surfaces, and $\mathbf{n}$ is the outward normal from V, which is therefore inward on S_0 and outward on S_1.

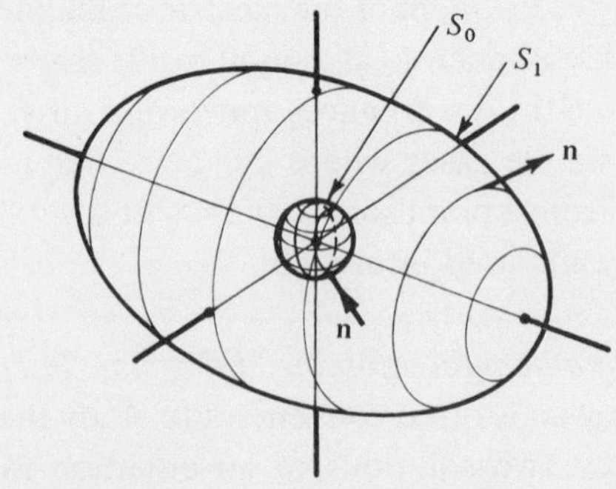

Fig. 6.15. The surface for integration. V lies between S_0 and S_1.

Over this two-sheet surface we thus have, for (6.55)

$$\iint_{S_1} (\psi\nabla\psi_1 - \psi_1\nabla\psi)\cdot\mathbf{n}\,\mathrm{d}s + \iint_{S_0} (\psi\nabla\psi_1 - \psi_1\nabla\psi)\cdot\mathbf{n}\,\mathrm{d}s = 0. \quad (6.56)$$

We can evaluate the functions of ψ_1:

$$\psi_1 = \frac{1}{r}\exp(-ikr) \quad (6.57)$$

$$\nabla\psi_1 = \frac{-\mathbf{r}}{r^2} ik\exp(-ikr) - \frac{\mathbf{r}}{r^3}\exp(-ikr) = \frac{-\mathbf{r}}{r^3}\exp(-ikr)(ikr+1) \quad (6.58)$$

and substitute in (6.56) to obtain

$$\iint_{S_1+S_0} \frac{\exp(-ikr)}{r^3}\{\psi(ikr+1)\mathbf{r} + r^2\nabla\psi\}\cdot\mathbf{n}\,\mathrm{d}s = 0. \quad (6.59)$$

The S_0 contribution can be evaluated, since over the small sphere of radius δr we can consider ψ to be constant (equal to ψ_0). Also since $\mathbf{n}$ is then the unit vector parallel to $-\mathbf{r}$, we can write $\mathbf{r}\cdot\mathbf{n}$ as $-r$ and substitute $r^2\,\mathrm{d}\Omega$ for $\mathrm{d}s$; thus

$$\iint_{S_0} \frac{\exp(-ikr)}{r^3}\{\psi_0(ikr+1)\mathbf{r} + r^2\nabla\psi(0)\}\cdot\mathbf{n}\,\mathrm{d}s$$

$$= \int_{S_0} -\exp(-ikr)\{\psi_0(ikr+1) - r\nabla\psi(0)\cdot\mathbf{n}\}\,\mathrm{d}\Omega, \quad (6.60)$$

evaluated at $r = \delta r$, $\mathrm{d}\Omega$ being the element of solid angle. In the limit as $\delta r \to 0$ there is only one term which does not approach zero and that is

$$\int_{S_0} -\exp(-ikr)\psi_0\,\mathrm{d}\Omega \to -4\pi\psi_0. \quad (6.61)$$

Equation (6.59) therefore gives

$$\iint_{S_1} \frac{\exp(-ikr)}{r^3}\{\psi(ikr+1)\mathbf{r} + r^2\nabla\psi\}\cdot\mathbf{n}\,\mathrm{d}s = 4\pi\psi_0. \quad (6.62)$$

This expression is the analytical result of the wave equation (6.51) and applies to any solution of ψ and to any surface S_1 which surrounds the origin. For a surface not enclosing the origin the introduction of S_0 was unnecessary and the right-hand side of (6.62) would be zero. To make calculation of ψ_0 possible in a practical case we shall now consider a particular system which gives rise to a ψ that can be calculated at all points on the surface S_0.

6.5.2 *Illumination by a point source.* If the disturbance on the surface originates from a point source, the value of ψ can be easily calculated. We

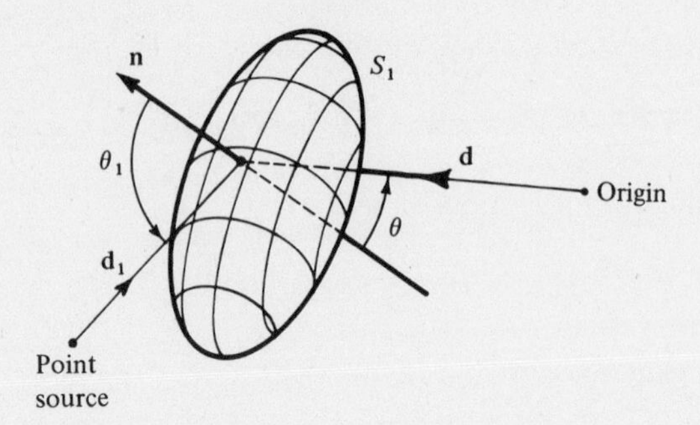

Fig. 6.16. Part of the surface S_1 showing normal and vectors.

consider a point on the surface S_1 to lie a distance $\mathbf{r}=\mathbf{d}$ from the origin, and $\mathbf{d}_1$ from the point source (Fig. 6.16 and also Fig. 6.1). The value of ψ is then

$$\psi = \frac{1}{d_1}\exp(-ikd_1) \tag{6.63}$$

and its gradient

$$\nabla\psi = \frac{-\mathbf{d}_1}{d_1^{\,3}}\exp(-ikd_1)(ikd_1+1), \tag{6.64}$$

which is the analogue of (6.58). Substituting these values into (6.62) gives

$$\iint_{S_1}\exp\{-ik(d+d_1)\}\left\{\frac{\mathbf{d}\cdot\mathbf{n}}{d_1d^3}(ikd+1)-\frac{\mathbf{d}_1\cdot\mathbf{n}}{d_1^{\,3}d}(ikd_1+1)\right\}\mathrm{d}s = 4\pi\psi_0. \tag{6.65}$$

When d and d_1 are both very much greater than the wavelength, which is not at all difficult to achieve, we can neglect 1 with respect to kd and therefore, substituting for the scalar products $\mathbf{d}\cdot\mathbf{n}$ and $\mathbf{d}_1\cdot\mathbf{n}$ in terms of the angles θ and θ_1 in Fig. 6.16, we have

$$4\pi\psi_0 = \iint_{S_1}\exp\{-ik(d+d_1)\}\frac{ik}{dd_1}(\cos\theta+\cos\theta_1)\,\mathrm{d}s. \tag{6.66}$$

This is the justification for the expression (6.4) which we have already used in our diffraction calculations. But it contains three extra pieces of information. The first is a definite form $(\cos\theta_1+\cos\theta)$ for the inclination factor; the second is a phase factor i; and the third a proportionality to k. The latter two, if introduced at the beginning, would have removed two slight inconsistencies that have been running through the calculations. The anomaly pointed out in § 6.4.2, in which diffraction by an infinite aperture appears to lead to a phase lagging $\pi/2$ behind the incident wave, is now resolved, since this factor of i produces a forward phase shift of $\pi/2$ to cancel it. The factor k has also appeared incongruously in the denominator of expressions such as (6.18) and (6.21); although it is not completely self-evident that the diffracted amplitudes should not depend on wavelength, it is reassuring to find the dependence to be a fiction. The inclination factor is more interesting, and is discussed in § 6.5.3 below.

We have, of course, omitted the variable transmission function f_s from this discussion. Returning it to its place, we write (6.66) as

$$\psi_0 = \iint f_s \frac{\mathrm{i}(\cos\theta + \cos\theta_1)}{2\lambda d d_1} \exp\{-\mathrm{i}k(d+d_1)\}\,\mathrm{d}s. \tag{6.67}$$

This should be compared with (6.4). For paraxial conditions the resemblance is further improved since $\cos\theta = \cos\theta_1 = 1$.

6.5.3 *The inclination factor.* The form of the inclination factor

$$F(\theta_1, \theta) = (\cos\theta_1 + \cos\theta) \tag{6.68}$$

is of considerable interest. If we assume that the surface S_1 is normal to the incoming radiation – i.e. that it lies in a wave-front – we have

$$F(\theta) = 1 + \cos\theta, \quad \text{since} \quad \theta_1 = 0. \tag{6.69}$$

This condition is the one which we have used in our discussion of Fresnel diffraction, where $d_1 \to \infty$ and S_1 is a plane. In Fig. 6.17 $F(\theta)$ is sketched as a polar plot of θ; it has value 2 in the forward direction, 1 at right angles, and zero backwards. It is therefore consistent with the simple Huygens' construction, which always neglects back-radiation (§ 2.7.1).

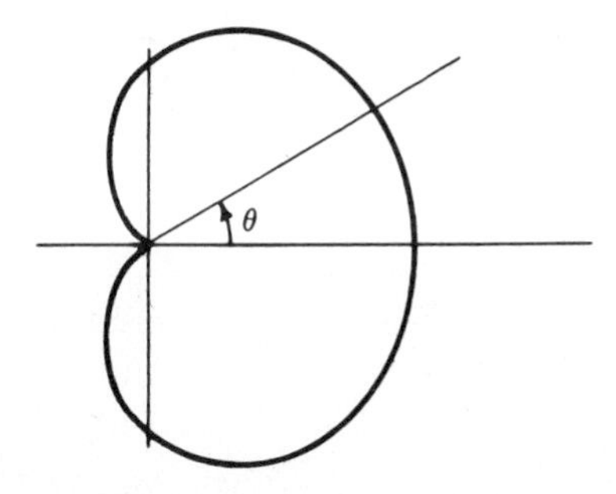

Fig. 6.17. The inclination factor $F(\theta)$ plotted in polar coordinates.

But why does the inclination factor depend on θ_1 and θ separately, so that $F(\theta_1, \theta)$ depends on the direction of the particular surface S that has been chosen? It would seem reasonable at first for it to depend on $(\theta - \theta_1)$, so that the direction of the surface would not matter, but on closer inspection we see that this is not so. The whole picture of *re-radiation* is a fiction of course; it is not as if we have a real screen of material at S which is scattering light uniformly in all directions. The important property of the integral is not that any one bit of the integrand

$$\exp\{-\mathrm{i}k(d+d_1)\}\frac{\mathrm{i}k}{dd_1}(\cos\theta_1+\cos\theta)$$

should be independent of S, but that the whole should be, and that has already been assured. Suppose we were to change the surface S_1 to another surface S_2, also satisfying the conditions that it surrounds the origin. Then the value of ψ_0 is given by

$$4\pi\psi_0=\iint_{S_2}=\iint_{S_1}+\left(\iint_{S_2}-\iint_{S_1}\right), \tag{6.70}$$

where the integrals are of the exact form of (6.62). The last term (bracketed) in (6.70) is the surface integral over a surface not enclosing the origin; its surface normals are outward over the part which is S_2 and inward over S_1, and thus the volume V lies between the two (Fig. 6.18).

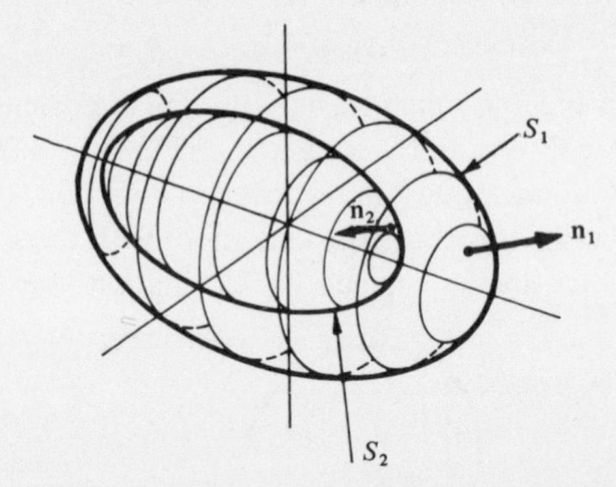

Fig. 6.18. Surface consisting of two neighbouring surfaces S_1 and S_2.

The integral over such a surface is zero, as explained in § 6.5.1. Therefore the value of ψ_0 calculated from any surface S_1 or S_2 is independent of the surface. The variation of inclination factor $F(\theta, \theta_1)$ with the surface is necessary to maintain the total integral invariant, despite the fact that the integrand itself changes. In practice, the surface is usually chosen to span the aperture; and if this lies in a single wavefront θ_1 is zero. Only the aperture itself is included in the integral, as the disturbance is zero everywhere else on the surface, and evaluation of (6.67) is not difficult. We are then faced with integrals exactly of the type we have already studied.

7

Fraunhofer diffraction and interference

7.1 Significance of Fraunhofer diffraction

The difference between Fresnel and Fraunhofer diffraction has been discussed in the last chapter. In § 6.2.2 we showed that Fraunhofer diffraction is characterized by a linear change of phase over the diffracting obstacle; in practice this linear change is achieved only in special circumstances, such as occur when the object is illuminated by a beam of parallel light. It is therefore necessary to use lenses, both for the production of the parallel beam and for observation of the resultant diffraction pattern (Appendix IV). The more usual statement is that Fraunhofer diffraction is the limit of Fresnel diffraction when the source and the observer are infinitely distant from the obstacle.

7.1.1 *Interference.* We stated in § 6.1 that we would try to draw the distinction that diffraction corresponds to the effects produced by limiting a beam of radiation, and interference to the effects produced when a finite number of beams overlap. Interference can take place under both Fresnel and Fraunhofer conditions, but all the important aspects of the phenomena are closely related to Fraunhofer conditions and we shall treat them solely in this way.

7.1.2 *Elementary treatment of interference.* If we have two point sources of radiation (Fig. 7.1) we can see that at the points indicated by the black spots, crests from one source arrive simultaneously with crests from the other, or troughs with troughs. The resultant wave-field therefore oscillates, and light is observed. But at the points indicated by open circles, troughs from one source arrive simultaneously with crests from the other and the two tend to cancel. What we see on a screen intercepting the waves is therefore a set of dark bands with illumination in between; these are called *interference fringes*, and their first production by Young (§ 1.2.4) was of importance in establishing the wave nature of light.

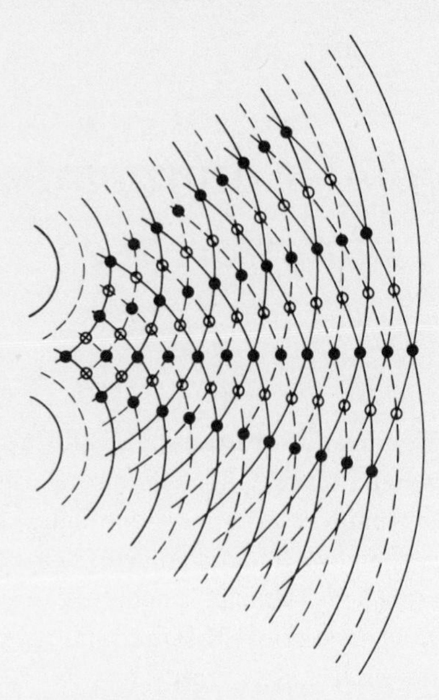

Fig. 7.1. Interference of two sets of spherical waves.

7.1.3 *Coherence.* It is a necessary condition for the production of interference fringes that the phase difference between the waves remain constant over a long period of time. If one wave alone changes in phase, the positions of the fringes on the observing screen – which may of course be the retina of the observer's eye – will change with time, and if the change is rapid, as it is if natural sources of light are used, then the effects average out and no interference fringes are seen.

The only way in which it is possible to create a set of light waves having fixed phase differences is to start with a single source and to divide the emitted light into a number of distinct waves which can then be allowed to interfere (Fig. 7.2). Such waves are then said to be *coherent* and will be discussed in considerable detail in the next chapter. For the purpose of the present chapter we shall simply assume that waves are either completely coherent or completely incoherent with respect to one another. No interference effects can be observed between waves which are mutually incoherent.

7.1.4 *Some simple devices for producing coherent sources.* Three general methods are available for producing two coherent beams: we can use the wavefront of a single wave (Young's fringes); we can use a source and its image; we can use two images of the same source. The last is the most

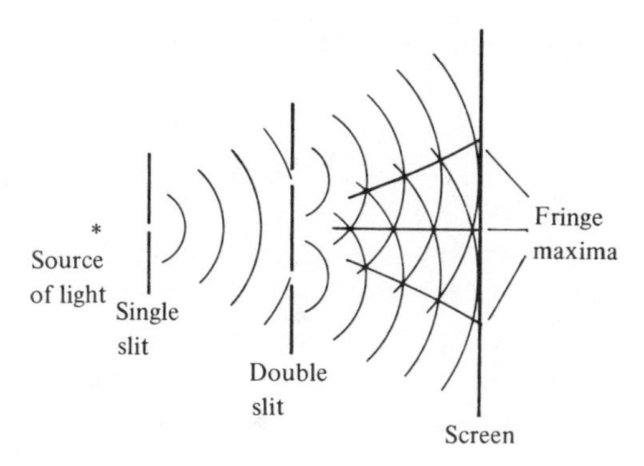

Fig. 7.2. Set-up for Young's fringes. Each of the pair of slits behaves as a separate source (cf. Fig. 7.1).

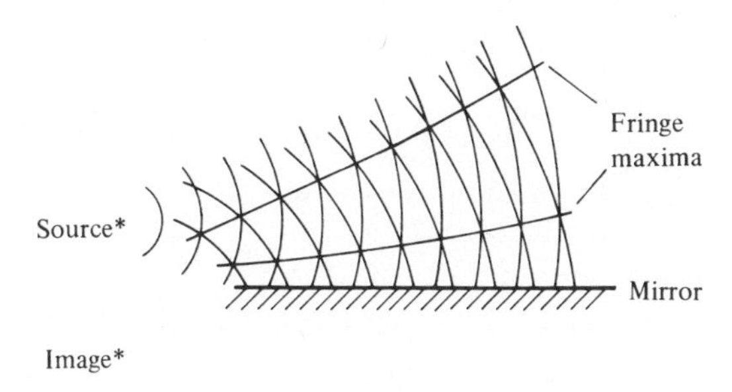

Fig. 7.3. Set-up for Lloyd's single-mirror fringes. The phase change on reflexion is simulated by making the radii of the circles representing the reflected wave interleave those representing the direct wave.

popular and several devices – the biprism, the double mirror, and the split lens, for example – are fully described in most textbooks on optics. These can all be used for measurements of wavelengths to an accuracy of a few per cent.

Interference between a source and its image is best illustrated by Lloyd's single mirror, with which interference occurs between the direct source and its image produced in a plane mirror (Fig. 7.3); if the source is nearly in the plane of the mirror the separation of the source and its image is quite small, and well-separated interference fringes can be produced. Obviously, however, the zero fringe – that which is equidistant from the two sources – cannot be produced by this method; but if one extrapolates back to the position where it should occur, one finds that there is a minimum of intensity there, not a maximum. There is therefore some asymmetry between the source and its image; this can be traced to the change of phase that occurs when light is reflected from a medium of higher refractive index (§ 4.3.5).

7.1.5 *Naturally occurring interference phenomena.* The coherence conditions described in § 7.1.3 sometimes arise naturally and produce interference patterns in quite simple conditions, as, for example, the colours produced by a thin film of oil on a wet road. Here, the interference takes place between the light reflected from the top and bottom of the film.

Such effects, however, are not interpretable in detail because they are too imprecise: the illumination is white, not monochromatic; the source (usually the sky) is large; and the film is non-uniform. For precise work all these variables must be controlled.

7.1.6 *Interference measurements.* We must first consider how the variables are related. Fig. 7.4 shows a ray incident upon the surface of a film of

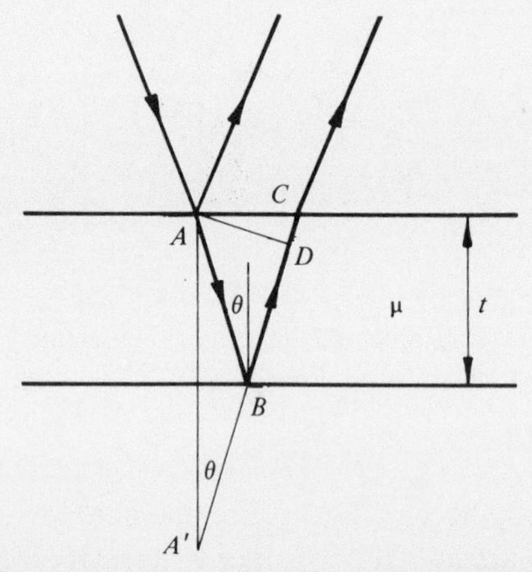

Fig. 7.4. Path differences for rays reflected from top and bottom of plane film.

refractive index μ; it divides into a reflected ray and a transmitted ray AB, which is incident at angle θ on the lower surface of the film and is then reflected along the path BC. We need to find the path difference between the directly reflected ray and that reflected at the lower surface.

Let A' be the mirror image of A in the lower surface, and let AD be the perpendicular from A onto $A'C$. Then the optical path difference required is $\mu(AB+BD)$, which is equal to $\mu A'D$ or $2\mu t \cos\theta$, where t is the thickness of the film. Thus for interference minima (not maxima, since the ray reflected at A suffers a phase change of π, as we noted in § 4.3.5)

$$2\mu t \cos\theta = m\lambda, \tag{7.1}$$

where m is an integer. This is the basic equation on which all thin-film interference theory is based.

If we illuminate a non-uniform film with parallel monochromatic light, a system of interference fringes is produced that follow the lines of constant t; such fringes are called *fringes of equal thickness* and are discussed further in § 11.2.2. If we illuminate a uniform film with an extended source of monochromatic light, fringes are seen that trace out lines of constant θ; these are called *fringes of equal inclination* and are used in spectroscopy, as we shall show in § 7.8.3. Finally, if we allow a beam of white light to fall at a fixed angle of incidence on a uniform film, the values of λ that satisfy equation (7.1) will be absent from the reflected light and so the film will be uniformly coloured; the colour can give a rough indication of the thickness of the film.

Variations on these three themes are possible. For example, if both t and λ vary we may write

$$\frac{t}{\lambda} = \frac{m}{2\mu \cos \theta}, \tag{7.2}$$

which is constant for a given m and a fixed value of θ. If the light is then dispersed, the spectrum will show fringes corresponding to different values of m; these fringes are called *fringes of equal chromatic order.*

7.1.7 *Localized fringes.* It may seem odd that fringes can be observed when extended sources of illumination are used. From our understanding of Young's fringes (§ 7.1.3) we should expect that space-coherent illumination would always be necessary. The answer is that if we require a fringe system throughout space, then space-coherence *is* necessary; but, even without space-coherent illumination we may find that there is some closely defined region in space in which the interference effects from the various time-coherent points all coincide. Fringes are then observed, and are said to be *localized.*

We can find the surface of localization for fringes of equal thickness (§ 7.1.6) by means of a simple construction (Fig. 7.5(a)). Let OP_1 and OP_2 represent the two reflecting plane surfaces. With centre O draw a circle – representing a cylinder – cutting the plane of illumination in the points A and B; the two chords A_1B_1 and A_2B_2 are the images of AB in OP_1 and OP_2 respectively. Therefore the points A_1A_2 are time-coherent and so are B_1 and B_2. Thus the fringe system produced by A_1 and A_2 has its zero order along a radius bisecting the chord A_1A_2 and that produced by B_1 and B_2 along the bisector of B_1B_2. These two fringes will intersect at O. This is the position where the fringes will be localized.

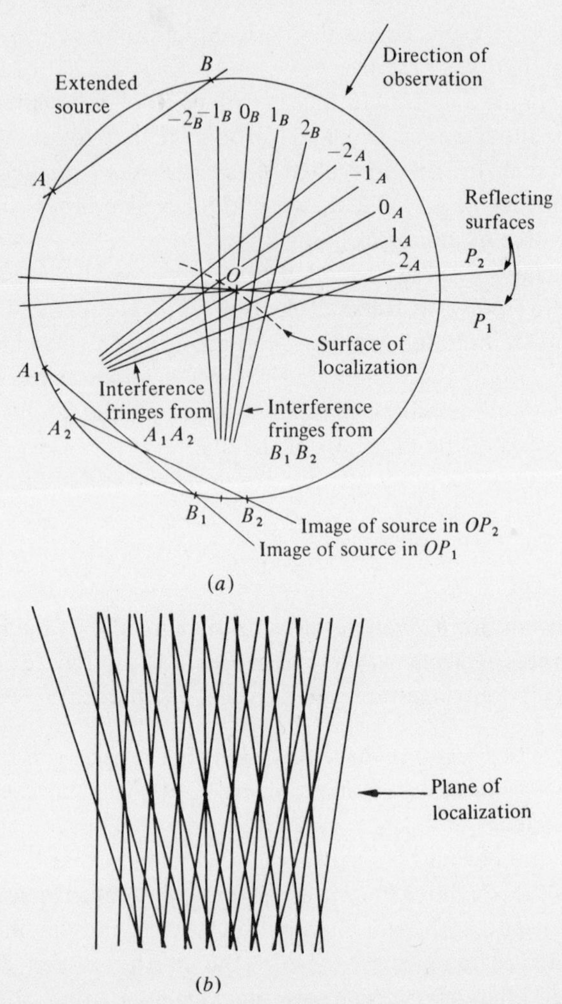

Fig. 7.5. (*a*) Construction for finding plane of localization for fringes from wedge-shaped film; (*b*) three sets of fringe maxima from three pairs of time-coherent point images, showing that only in one position is a recognizable fringe system produced.

As shown in Fig. 7.5(*b*), the orders other than the zero cannot be treated so simply, but we can see from this figure that the composite fringe system is recognizable only on one plane. The set of fringe maxima from A_1 and A_2 lie upon hyperbolae (cf. Fig. 7.1) that approximate to a set of straight lines diverging from a point midway between A_1 and A_2; the superposition of fringes from other coherent pairs of points leads to an unintelligible mixture except in the plane of localization.

If the illumination is roughly normal to the film, the fringes will be localized in the plane of the film. If the film surfaces are not both plane – as, for example, in the formation of Newton's rings – the construction does not apply; but we can assume that it applies approximately to successive portions of the film by taking tangents to the surfaces. For Newton's rings, this would lead to fringes localized in the plane surface of the film.

7.2 Fourier theory of Fraunhofer diffraction

The whole of Fraunhofer diffraction theory can be summed up in one statement: 'The Fraunhofer diffraction pattern of an object is the Fourier transform of that object.' This is true in the sense that the amplitude and the phase of the radiation at any point in the diffraction pattern are the amplitude and phase at the corresponding point in the Fourier transform.

We shall justify this statement and then proceed to use it in some examples; the reason for the great practical importance of Fraunhofer diffraction is contained in it, because the operation of deriving a Fourier transform is a mathematical operation and does not involve constants of the apparatus. A similar statement cannot be made about Fresnel diffraction.

Let us consider two points, A_1 and A_2 (Fig. 7.6), scattering radiation of wavelength λ. If the radiation is incident normally to the line separating the two points, which are at a distance x apart, the phase difference for waves diffracted at an angle θ is equal to $(2\pi x/\lambda)\sin\theta$, which we may write as $kx\sin\theta$. If we have a collection of points A, each scattering an amplitude $f(x)$, the complete wave diffracted in the direction θ is given by

$$\psi(k\sin\theta)=\int_{-\infty}^{\infty} f(x)\exp(-ikx\sin\theta)\,\mathrm{d}x. \qquad (7.3)$$

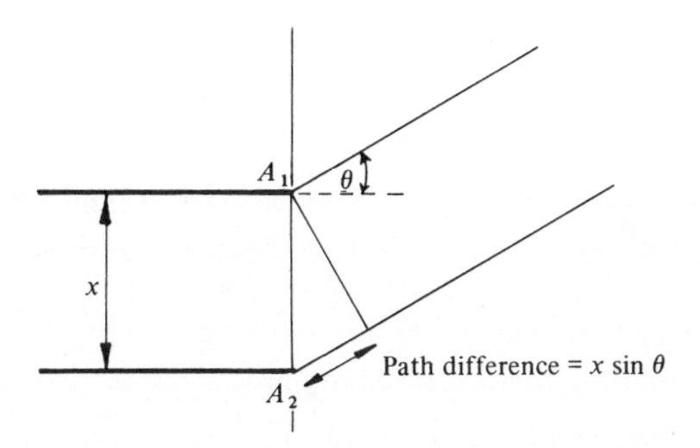

Fig. 7.6. Path difference for rays scattered at angle θ from two points A_1 and A_2.

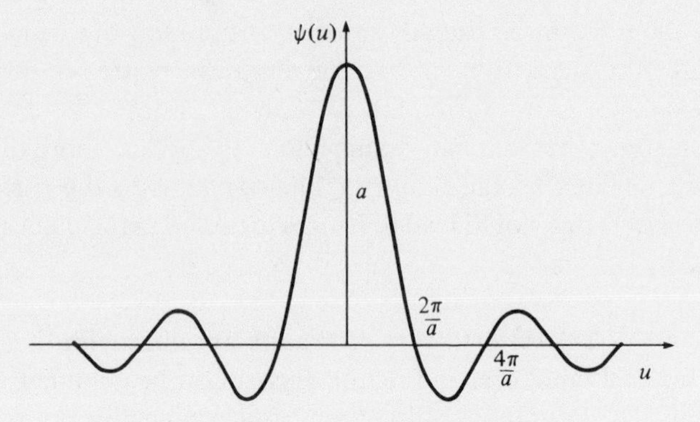

Fig. 7.7. Form of function a sinc $(\frac{1}{2}au)$.

Thus, ψ is the Fourier transform of $f(x)$ in terms of the variable $k \sin\theta$ (compare equation (3.28)), which we shall call u.

This treatment is one-dimensional. We shall consider later in this chapter transforms in two and three dimensions.

As examples of one-dimensional diffraction calculations we shall now discuss three simple diffracting obstacles which illustrate the way in which the prominence of diffraction fringes is related to sharp edges. They are: (a) a slit of width a; (b) a blurred slit with transmission function $f(x)$ represented by a triangular function; (c) a slit with a Gaussian transmission function. The last two examples would have to be produced in some way photographically.

7.2.1 *Diffraction by a sharp slit.* If the slit has width a, we have $f(x) = 1$ from $x = -a/2$ to $+a/2$. Outside the slit, $f(x) = 0$. For this function,

$$\psi(u) = \int_{-a/2}^{+a/2} \exp(-\mathrm{i}ux)\,\mathrm{d}x = a \operatorname{sinc}(au/2). \tag{7.4}$$

The intensity of the Fraunhofer diffraction pattern is

$$|\psi(u)|^2 = a^2 \operatorname{sinc}^2(au/2). \tag{7.5}$$

The form of $\psi(u)$ (Fig. 7.7) is important; it has a maximum of a at $u = 0$, and is zero at regular intervals where $au/2 = m\pi$, where m is a finite integer. The heights of the resulting maxima are approximately proportional to $1/(2m+1)$: this result arises if we assume that these maxima lie half-way between the zeros; their *exact* positions are not easy to find.

Fig. 7.8 shows the observed intensity, $|\psi(u)|^2$. The zeros in this function occur at angles given by

$$\tfrac{1}{2}ka \sin\theta = m\pi$$

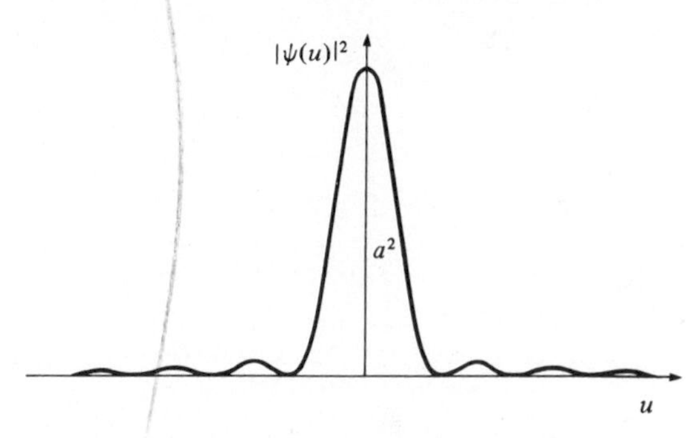

Fig. 7.8. Form of function $a^2 \operatorname{sinc}^2(\frac{1}{2}au)$.

or

$$\frac{\pi}{\lambda}\sin\theta = m\pi/a,$$

i.e.

$$\sin\theta = m\lambda/a. \tag{7.6}$$

7.2.2 *Diffraction by a blurred slit, represented by a triangular function.* The slit transmission function can be represented by

$$f(x) = 1 - |x|/a \quad \text{for } -a < x < a; \quad \text{zero otherwise.} \tag{7.7}$$

If we define the 'width' of the slit as the quotient of the area under the curve and its maximum height, i.e. $\int_{-\infty}^{\infty} f(x)\,\mathrm{d}x/f(0)$, this slit also has width a. Then

$$\begin{aligned}\psi(u) &= \frac{1}{a}\int_0^a (a-x)\exp(-\mathrm{i}ux)\,\mathrm{d}x \\ &\quad + \frac{1}{a}\int_{-a}^0 (a+x)\exp(-\mathrm{i}ux)\,\mathrm{d}x \\ &= a\operatorname{sinc}^2(au/2).\end{aligned} \tag{7.8}$$

The form of this expression is shown in Fig. 7.8. It is everywhere positive, reaching zero at values of u given by

$$au/2 = m\pi \tag{7.9}$$

or

$$a\sin\theta = m\lambda. \tag{7.10}$$

The angular positions of these zeros are thus exactly the same as for the uniform slit; since the average width of the wedge is the same as that of the slit, this result is not surprising. But the maxima of the sidebands

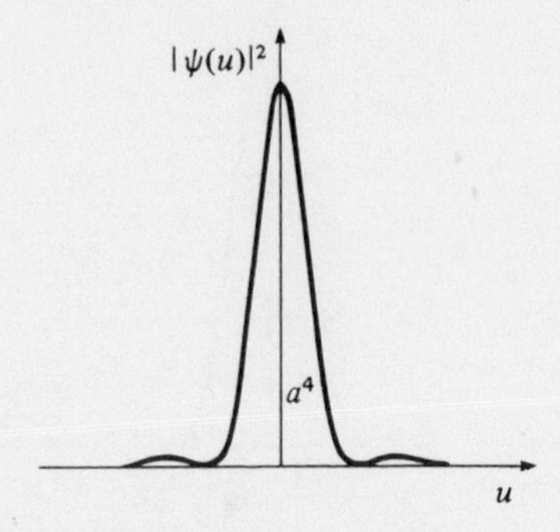

Fig. 7.9. Form of function $a^4 \operatorname{sinc}^4(\frac{1}{2}au)$, the intensity along the equatorial line of the diffraction pattern of a blurred slit.

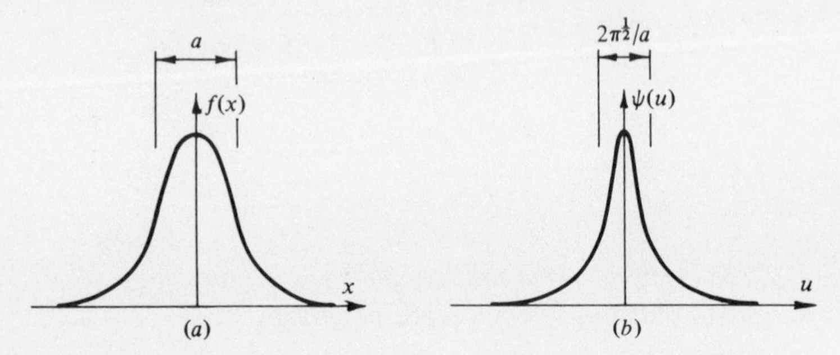

Fig. 7.10. (*a*) and (*b*) Gaussian function and its diffraction pattern.

produced (Fig. 7.9) are much less; their intensities are proportional to $(2m+1)^{-4}$.

It will also be seen that the *amplitude* function for the wedge-shaped slit is the same as the *intensity* function for the uniform slit.

7.2.3 *Diffraction by a slit with a Gaussian transmission profile.* A third slit of width a (as defined in § 7.2.2) and unit maximum transmission has the profile:

$$f(x) = \exp(-\pi x^2/a^2). \tag{7.11}$$

The diffraction pattern of this slit is (see § 3.4.2)

$$\psi(u) = a \exp(-a^2u^2/4\pi). \tag{7.12}$$

This is another Gaussian function (Fig. 7.10), of variance $2\pi^{\frac{1}{2}}/a$ and thus has no sidebands at all.

7.2.4 *Discussion of these results.* We have taken these three examples because they illustrate some important general principles. The two finite functions have Fourier transforms with sidebands; this result follows because, if we sum a Fourier series with a finite number of terms – the Fourier transform can be regarded as the limit of a Fourier series (§ 3.4.1) – there are bound to be ripples with periodicity of the same order of magnitude as that of the highest term introduced. If, however, we take an infinite function, such as a Gaussian, there is the possibility that the Fourier transform can be smoothed out by the inclusion of the higher-order terms.

Secondly, the widths of all the three central diffraction peaks is inversely proportional to the widths of the diffracting objects. And, finally, the scale of detail in the diffraction patterns of the two finite slits is also inversely proportional to the widths of the slits.

7.2.5 *Diffraction by an object with phase variation only.* There are many objects, particularly natural ones, that do not absorb light appreciably but change its phase on transmission. Any ordinary piece of window glass will do this; it is transparent, but its thickness is not uniform and light passing through different parts of it suffers a varying amount of phase retardation. If the refractive index of the glass is μ, the optical difference between two paths including different thicknesses t_1 and t_2 is

$$(\mu-1)(t_1-t_2) \tag{7.13}$$

and consequently the wavefront emerging from the glass sheet is no longer a plane as it was when incident (Fig. 7.11). Since waves of different

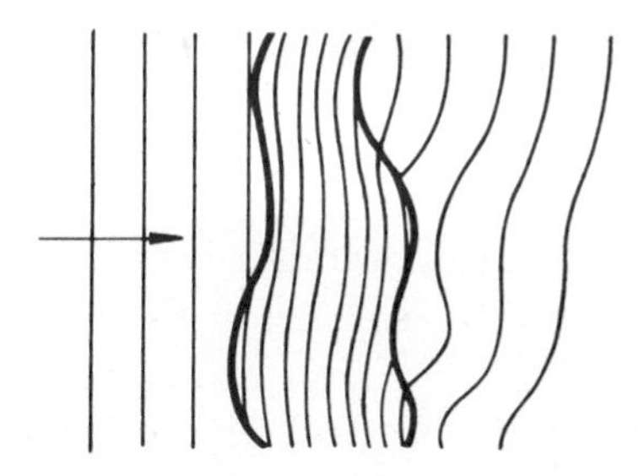

Fig. 7.11. Distortion of plane-wavefront by non-uniform glass plate.

phases but the same amplitudes are represented by complex amplitudes with the same modulus, such a state of affairs as this 'phase object' is represented by a complex transmission function $f(x)$ with constant modulus. We shall take as an example of this behaviour a thin prism, of

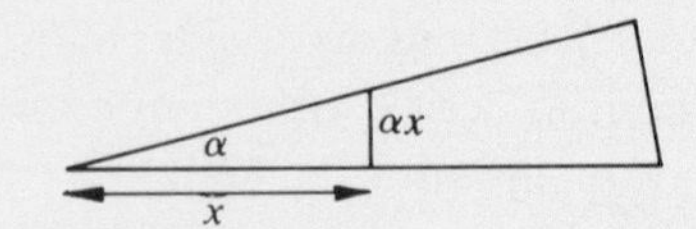

Fig. 7.12. Thickness of prism of angle α.

angle α and refractive index μ. The thickness of the prism at position x is αx (Fig. 7.12) and the phase lag introduced is thus

$$\begin{aligned} f(x) &= \exp\{\mathrm{i}k(\mu-1)t\} \\ &= \exp\{\mathrm{i}k(\mu-1)\alpha x\}. \end{aligned} \tag{7.14}$$

The prism is assumed to be infinite in extent along both x and y directions. The diffraction pattern corresponding to $f(x)$ is then

$$\begin{aligned} \psi(u) &= \int_{-\infty}^{\infty} \exp\{\mathrm{i}k(\mu-1)\alpha x\} \exp(-\mathrm{i}ux)\,\mathrm{d}x \\ &= \delta\{u-(\mu-1)k\alpha\}. \end{aligned} \tag{7.15}$$

The diffracted wave thus travels in the direction represented by

$$u = k(\mu-1)\alpha. \tag{7.16}$$

Substituting for u

$$\left.\begin{aligned} u &= k \sin\theta = k(\mu-1)\alpha. \\ \theta &= (\mu-1)\alpha \quad \text{for small } \theta. \end{aligned}\right\} \tag{7.17}$$

The light thus remains concentrated in a single direction, but is deviated from the incident by the same angle as deduced from geometrical optics.

7.3 Fraunhofer diffraction by a two-dimensional screen

Now that we have established the principles of Fraunhofer diffraction by a one-dimensional object we can easily extend them to two dimensions. It might seem logical to consider even a general three-dimensional object but it turns out that the third dimension introduces extra complications, and needs a rather different approach which is discussed in § 7.6.†

† The basic reason for this is that diffraction is a boundary-value problem, as was discussed in § 6.5. There, definition of the amplitude over a *two-dimensional* boundary surface allowed us to determine the diffracted wave at all points inside it. Thus *every* two-dimensional screen has a diffraction pattern. A three-dimensional object, however, over-defines the problem by stipulating amplitudes not only on a boundary surface but also within it. The problem may then not be self-consistent, and *may* not have a solution; there arises the possibility that a three-dimensional object will not have a diffraction pattern at all!

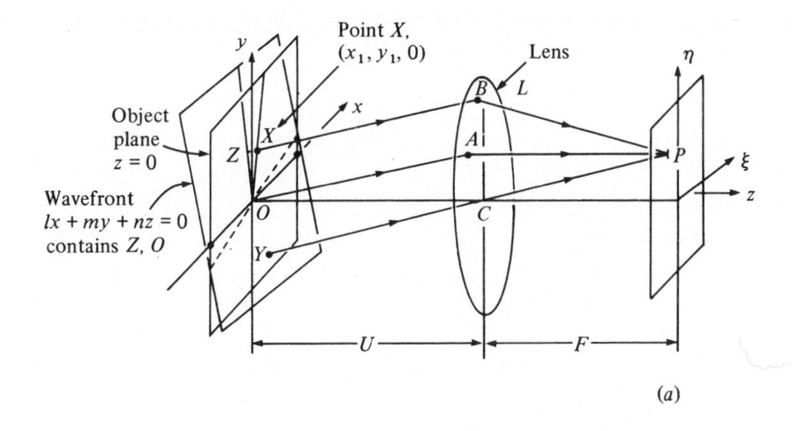

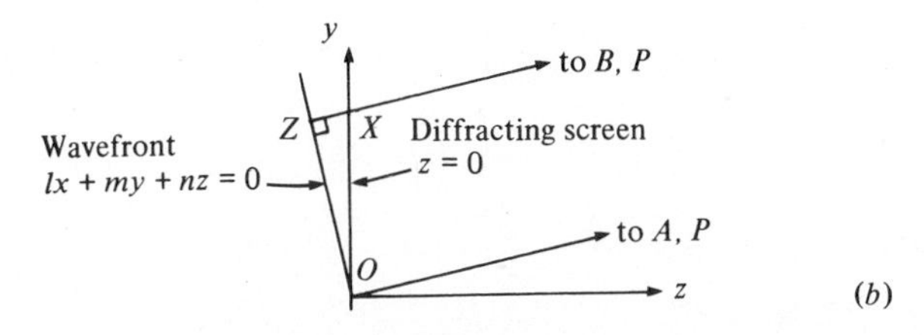

Fig. 7.13. (*a*) Illustrating Fraunhofer diffraction by a two-dimensional object; (*b*) detail of the region *OZX*.

7.3.1 *The two-dimensional Fourier transform.* Consider a plane-wave travelling along the z-axis (Fig. 7.13) and incident at $z = 0$ on a screen with amplitude transmission function $f(x, y)$. The diffracted light is collected by a lens L, of focal length F, in the plane $z = U$.

All light waves leaving the screen in a particular direction are focused by the lens to a point in the focal plane. In the figure XB, OA, YC are all parallel and are focused at P. The amplitude of the light at P is therefore the sum of the amplitudes at X, O, Y, etc., each with the appropriate phase retardation $\exp(\mathrm{i}k\overline{XBP})$, etc., where $\overline{XBP}$ indicates the optical path from X to P via B, including the path through the lens.

Now the amplitude at X, the general point (x_1, y_1) in the plane $z = 0$ is simply the amplitude of the incident wave, assumed unity, multiplied by the transmission function $f(x_1, y_1)$. To calculate the optical path $\overline{XBP}$ we remember that according to Fermat's principle (§ 2.7.2) the optical paths from the various points on a wavefront to its focus are all equal. Let the direction of XB, OA, ... be represented by direction cosines (l, m, n).

Then the wavefront normal to them, through O, which focuses at P, is the plane

$$lx + my + nz = 0 \tag{7.18}$$

and the optical paths from the wavefront to P, i.e. $\overline{OAP}$ and $\overline{ZBP} = ZX + \overline{XBP}$, are equal (Fig. 7.13($b$)). Now ZX is just the projection of OX onto the ray XB, and this can be expressed as the component of the vector $(x_1, y_1, 0)$ in the direction (l, m, n), namely:

$$ZX = lx_1 + my_1. \tag{7.19}$$

Thus:

$$\overline{XBP} = \overline{OAP} - lx_1 - my_1. \tag{7.20}$$

The amplitude at P is obtained by integrating $f(x_1, y_1) \exp \{ik\overline{XBP}\}$ over the screen:

$$\psi_p = \exp (ik\overline{OAP}) \iint f(x, y) \exp \{-ik(lx + my)\} \, dx \, dy \tag{7.21}$$

or

$$\psi(u, v) = \exp (ik\overline{OAP}) \iint f(x, y) \exp \{-i(ux + vy)\} \, dx \, dy, \tag{7.22}$$

where $u = lk$, $v = mk$. Notice in (7.22) that we have dropped the suffix 1 in the integrals as (x, y) can now be used for coordinates in the plane $z = 0$. The point P in the focal plane has coordinates (ξ, η) which can be related exactly to u and v only if the details of the lens are known. For Gaussian optics

$$\xi = F \tan (\sin^{-1} l) \approx uF/k, \tag{7.23}$$

$$\eta = F \tan (\sin^{-1} m) \approx vF/k. \tag{7.24}$$

It would be useful if the linear approximation could be preserved out to larger angles, and lenses with this property have been designed. In general, however, one has to work at small angles for the (ξ, η): (u, v) relationship to be linear.

When we observe a diffraction pattern, or photograph it, we measure the intensity $|\psi(u, v)|^2$, and the exact value of $\overline{OAP}$ is irrelevant. Although there are some experiments in which the phase of the diffraction pattern is important, and which we shall consider in the next section, it is usual to ignore the phase factor and write for (7.22):

$$\psi(u, v) = \iint f(x, y) \exp \{-i(ux + vy)\} \, dx \, dy. \tag{7.25}$$

7.3.2 *The phase of the Fraunhofer diffraction pattern.* The intensity of the diffraction pattern is independent of the exact position of the screen relative to the lens, in that the distance OC only affects the phase factor $\exp(ik\overline{OAP})$. For some purposes, it is necessary to know the phase of the diffraction pattern also, for example if the diffracted wave is to be allowed to interfere with another coherent light wave as in some forms of pattern recognition (§ 9.8).

Now the factor $\exp\{ik\overline{OAP}\}$ is quite independent of $f(x, y)$, since it is determined by the geometry of the optical system. It is very easy to calculate it if we allow $f(x, y)$ to be the pinhole $\delta(x)\delta(y)$. Then the diffraction pattern is, from (7.22),

$$\begin{aligned}\psi(u, v) &= \exp(ik\overline{OAP}) \int \delta(x) \exp(-iux)\, dx \int \delta(y) \exp(-iuy)\, dy \\ &= \exp(ik\overline{OAP}). \qquad (7.26)\end{aligned}$$

However, we know that the action of the lens is to produce an image of the pinhole; if OC is the object distance U, the image is at the conjugate distance V given by $V^{-1} = F^{-1} - U^{-1}$. So the wave on the right-hand side of the lens L is just a spherical wave converging on the plane $z = U + V$. In the plane F:

$$\psi(\xi, \eta) = \exp[-ik\{(V-F)^2 + \xi^2 + \eta^2\}^{\frac{1}{2}} + ik\overline{OI}], \qquad (7.27)$$

where $\overline{OI}$ is the optical path from O to the image of the pinhole. Thus, for any particular object position, $\psi(\xi, \eta) = \exp(ik\overline{OAP})$ can be evaluated. Of particular interest is the case where the screen is in the front focal plane ($U = F$). Then $\overline{OI} \to \infty$ and it is easy to see that the diffracted wave is a plane-wave:

$$\psi(\xi, \eta) = \exp(ik\overline{OAP}) = \text{constant}. \qquad (7.28)$$

Therefore, when the object is situated in the front focal plane, the Fraunhofer diffraction pattern represents the true complex Fourier transform of $f(x, y)$. For all other object positions the intensity of the diffraction pattern is that of the Fourier transform, but the phase is not.

7.3.3 *Fraunhofer diffraction in obliquely incident light.* If the plane-wave illuminating the mask in Fig. 7.13 does not travel along the z-axis, the foregoing treatment (§ 7.3.1) can be adjusted in a rather trivial manner. Specifically, when the incident wave-vector has direction cosines (l_o, m_o, n_o) the phase of the wave reaching the point (x, y) on the mask is advanced by $k(l_o x + m_o y)$ with respect to that at the origin. Thus the retardation of the component from (x, y) is $k\{(l-l_o)x+(m-m_o)y\}$. The integral (7.21) is now:

$$\psi_p = \exp(\mathrm{i}k\overline{OAP}) \iint f(x, y) \times \exp[-\mathrm{i}k\{(l-l_o)x+(m-m_o)y\}]\,\mathrm{d}x\,\mathrm{d}y. \tag{7.29}$$

This can still be written in the form (7.22)

$$\psi(u, v) = \exp(\mathrm{i}k\overline{OAP}) \iint f(x, y) \exp\{-\mathrm{i}(ux+vy)\}\,\mathrm{d}x\,\mathrm{d}y, \tag{7.30}$$

provided that u and v are redefined:

$$u = k(l-l_o), \qquad v = k(m-m_o). \tag{7.31}$$

These definitions revert to lk and mk when l_o, m_o are zero, i.e. for normal incidence. Now remembering the definition of (l, m, n) as cosines of the angles $(\theta_x, \theta_y, \theta_z)$ between the wave-vector and the (x, y, z)-axes, we see that the definition of u, for example, is

$$u = k(\cos\theta_x - \cos\theta_{xo}). \tag{7.32}$$

A given feature of the diffraction pattern, corresponding to a certain value of u, therefore appears at an angle of deviation $(\theta_x - \theta_{xo})$ with respect to the incident wave-vector. Putting u in (7.32) equal to a constant, one can easily show (by expanding the difference between the two cosines) that the angle of deviation is minimized when

$$\theta_x = \pi - \theta_{xo}, \tag{7.33}$$

i.e. when the diffracting screen bisects the obtuse angle (Fig. 7.14) between incident and diffracted radiation. This condition of minimum

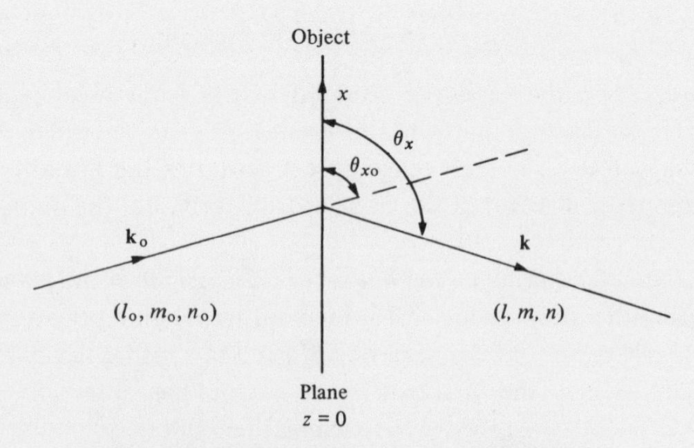

Fig. 7.14. The angle of minimum deviation.

deviation can be met for only one component u at a time, and is often used for diffraction gratings (§ 7.7) and holograms (§ 9.7), where the diffraction order or the reference-beam direction define the value of u very closely.

7.3.4 *Diffraction pattern of a rectangular hole.* If we consider a rectangular hole whose sides are parallel to the x- and y-axes, the two variables have independent limits and the integral (7.25) can be written as a product:

$$\psi(u, v) = \int_{-a/2}^{a/2} \exp(-iux)\, dx \int_{-b/2}^{b/2} \exp(-ivy)\, dy, \qquad (7.34)$$

the function $f(x, y)$ being constant (we may take it as unity) over the area of the aperture, of sides a and b. We take the origin to be at the centre of the aperture so that it represents an even function and thus has a real transform.

Thus

$$\psi(u, v) = ab \operatorname{sinc}(\tfrac{1}{2}ua) \operatorname{sinc}(\tfrac{1}{2}vb), \qquad (7.35)$$

each factor being similar to that derived for a uniform slit (equation (7.4)).

The diffraction pattern has zeros at values of u and v given by ua and $vb = 2m\pi$, except for $m = 0$, for which the value of ψ is unity. Thus the zeros lie on lines parallel to the edges of the slit, given by the equations

$$ua = m_1\lambda \quad \text{and} \quad vb = m_2\lambda. \qquad (7.36)$$

Thus the centre peak, for example, is bounded by lines given by $m_1 = \pm 1$ and $m_2 = \pm 1$, which form a rectangle whose dimensions are inversely proportional to those of the diffracting aperture (Fig. 7.15).

7.3.5 *Diffraction pattern of a circular hole.* For a circular hole the integral (7.25) is more difficult to evaluate since the limits are not now independent. It is best to use polar coordinates (Fig. 7.16) for points in the aperture and in the diffraction pattern. If (ρ, θ) are the polar coordinates in the aperture

$$x = \rho \cos\theta \quad \text{and} \quad y = \rho \sin\theta, \qquad (7.37)$$

and if (ζ, ϕ) are the polar coordinates in the diffraction pattern

$$\zeta \cos\phi \equiv u, \qquad \zeta \sin\phi \equiv v. \qquad (7.38)$$

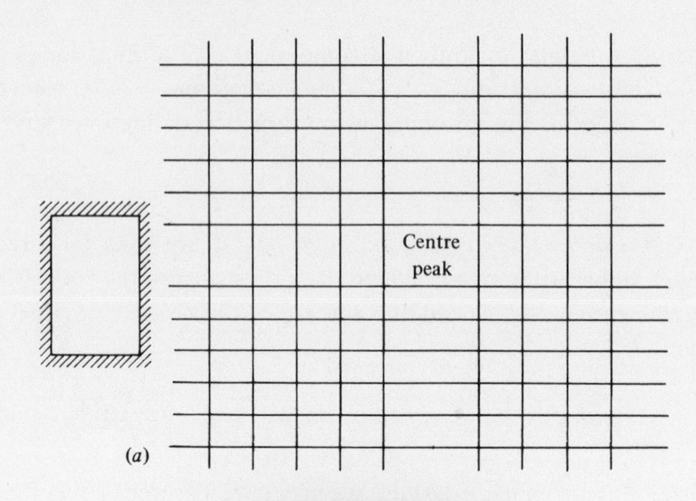

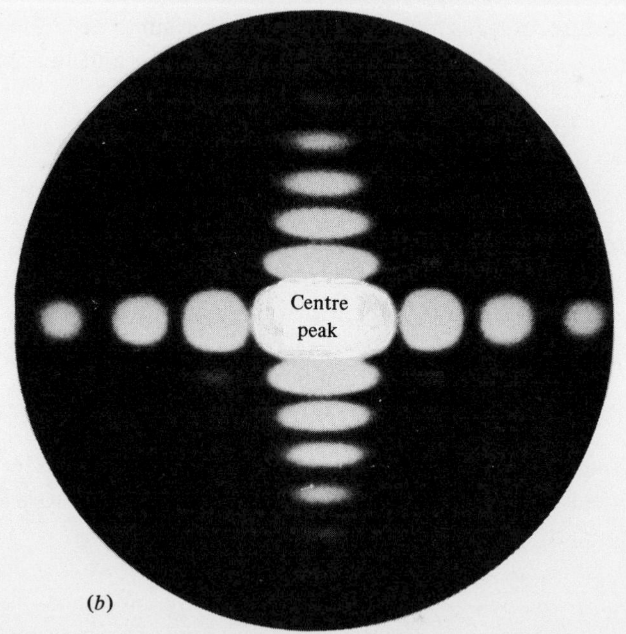

Fig. 7.15 (*a*). Lines of zero intensity in diffraction pattern of aperture shown at left of diagram. (*b*): the observed diffraction pattern.

Thus equation (7.25) becomes

$$\psi(u, v) = \int_0^a \int_0^{2\pi} \exp\{-\mathrm{i}(\rho\zeta \cos\phi \cos\theta + \rho\zeta \sin\phi \sin\theta)\}\rho \, \mathrm{d}\rho \, \mathrm{d}\theta$$

$$= \int_0^a \int_0^{2\pi} \exp\{-i\rho\zeta \cos(\theta - \phi)\}\rho \, \mathrm{d}\rho \, \mathrm{d}\theta. \qquad (7.39)$$

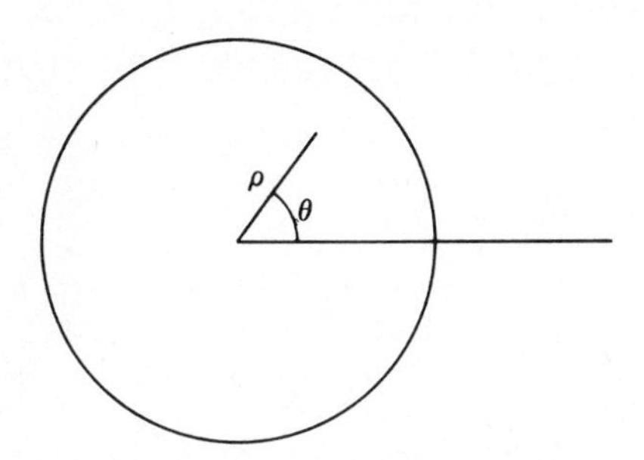

Fig. 7.16. Polar coordinates of a point in a circular aperture.

To evaluate this expression we can make use of the properties of Bessel functions. A Bessel function $J_n(z)$ of order n can be defined as

$$J_n(z)=\frac{\mathrm{i}^{-n}}{2\pi}\int_0^{2\pi}\exp(\mathrm{i}z\cos x')\exp(\mathrm{i}nx')\,\mathrm{d}x' \tag{7.40}$$

(where x' is a dummy variable), whence

$$J_0(z)=\frac{1}{2\pi}\int_0^{2\pi}\exp(\mathrm{i}z\cos x')\,\mathrm{d}x'. \tag{7.41}$$

If we put

$$z=-\rho\zeta \quad \text{and} \quad x'=\theta-\phi,$$

then

$$\psi(\zeta,\phi)=2\pi\int_0^a J_0(\rho\zeta)\rho\,\mathrm{d}\rho. \tag{7.42}$$

Now from a recurrence relation for Bessel functions,

$$xJ_1(x)=\int_0^x x''J_0(x'')\,\mathrm{d}x''; \tag{7.43}$$

if we put

$$x''=\zeta\rho \quad \text{and} \quad x=\zeta a$$

$$\begin{aligned}\zeta aJ_1(\zeta a)&=\int_0^{\zeta a}\zeta\rho J_0(\zeta\rho)\,\mathrm{d}\zeta\rho\\&=\zeta^2\int_0^a J_0(\zeta\rho)\rho\,\mathrm{d}\rho\\&=\frac{\zeta^2}{2\pi}\psi(\zeta,\phi).\end{aligned} \tag{7.44}$$

Thus

$$\begin{aligned}\psi(\zeta,\phi)&=\frac{2\pi aJ_1(\zeta a)}{\zeta}\\&=\pi a^2\left\{\frac{2J_1(\zeta a)}{\zeta a}\right\}.\end{aligned} \tag{7.45}$$

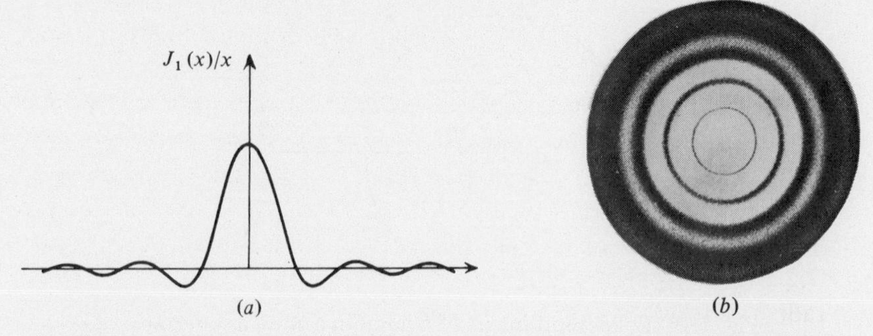

Fig. 7.17. (*a*) Form of function $J_1(x)/x$, the diametral amplitude distribution in the diffraction pattern of a circular aperture; (*b*) Fraunhofer diffraction pattern of circular hole.

The pattern has circular symmetry; $\psi(\zeta, \phi)$, as one would expect, has no dependence on ϕ.

The form of the function $2J_1(x)/x$ is interesting. $J_1(x)$ is zero at $x = 0$, but – like sinc x – the function has a finite value of unity there. It then decreases to zero, becomes negative and continues to oscillate with a gradually decreasing period that tends to a constant (Fig. 7.17(*a*)).

The diffraction pattern is shown in Fig. 7.17(*b*). The central peak is known as the Airy disc, and it extends to the first zero, which occurs at $x = 3.832$, or at angle $\zeta/k = 0.61\lambda/a$. We see, therefore, as we expect, that the radius of the Airy disc is inversely proportional to the radius of the hole.

It should also be noted from equations (7.35) and (7.45) that the amplitude – not the intensity – at the centre of the diffraction pattern is proportional to the *area* of the hole. This result makes sense when we realize that the linear dimensions of the diffraction pattern are inversely proportional to those of the hole, and thus that the total intensity in the diffraction pattern is proportional to the area of the hole, as it should be. As the hole varies in size, the intensity at the centre varies proportionately to the square of the area.

7.3.6 *A crude derivation of the size of the Airy disc.* The derivation of the form of the diffraction pattern in terms of a Bessel function does not really throw any light upon the physics of the problem. If it had not been that Bessel functions appear in other physical problems, the function $J_1(x)$ would not have been worth tabulating and we should be no nearer an acceptable solution when the equation (7.45) had been derived. It is, however, possible to see a rough solution in terms of the concepts discussed in §§ 7.2.1–7.2.4. There we derived the diffraction patterns of a sharp slit, a blurred slit and a Gaussian function.

We saw that the diffraction patterns of the first two had sidebands, and that that of the last had not; we therefore deduced that the sidebands were the result of the sudden termination of the diffracting object. We also saw that the positions of the minima of the diffraction pattern of the sharp slit and the blurred slit were identical when their widths, as defined in § 7.2.2, were equal. Let us extend this idea to a circular hole. If a hole has radius r, the side of a rectangle of equal area, of which the other side is $2r$, must be $\pi r/2$ (Fig. 7.18). The position of the first minimum in the

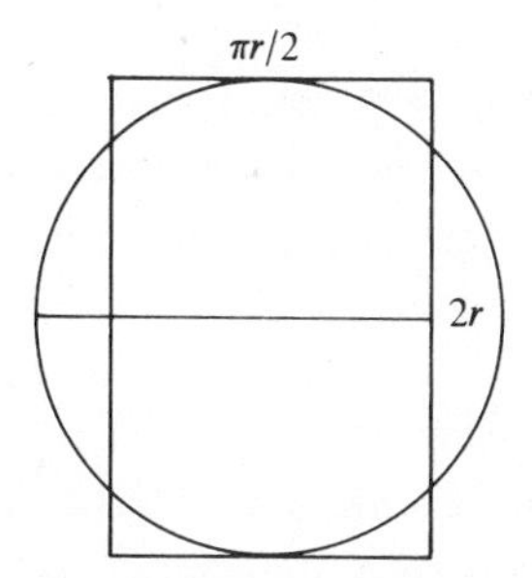

Fig. 7.18. Rectangle and circle of equal area.

diffraction pattern of this slit is, as we can see from equation (7.6), given by

$$\sin\theta = 2\lambda/\pi r, \tag{7.46}$$

which is a factor of $4/\pi$ greater than is given by a slit of width $2r$. This factor is 1.27 – a fair approximation to the correct value 1.22.

This idea is included here because, although it cannot be quantitatively justified, it gives some impression of the way a crude approximation can sometimes give a reasonably good value for a physical quantity much more quickly than elaborate mathematics.

7.3.7 *Addition of diffraction patterns.* The additive property of transforms sometimes enables the diffraction patterns of relatively complicated objects to be derived, if their shapes can be expressed as a sum of simpler ones. The transforms of the separate components of the object must be expressed with respect to the same origin, and the real and imaginary parts of the complete transform are then obtained by adding the real parts and the imaginary parts, respectively.

The process is particularly simple if the separate components are centro-symmetric about the same point; then the separate transforms merely add algebraically. For example, it is possible to derive the diffraction pattern of three slits by adding the transform of the two outer ones to that of the inner one; the diffraction pattern of four slits can be obtained by regarding them as two pairs, one with three times the spacing of the other. Some examples of this sort are included in the problems at the end of the book.

An opaque obstacle can be regarded as giving a negative transform. For example, a thick rectangular frame may be regarded as the difference between the outer rectangle and the inner one. (Note in working out such an example that the height of the central peak of the transform of a rectangle is proportional to the area, as shown in § 7.3.4.) The diffraction pattern of an annular ring is given by the difference between the diffraction patterns of the outer and inner circles; the result has practical implications that will be discussed later in § 9.6.6.

7.3.8 *Complementary screens; Babinet's theorem.* An important theorem in optics is concerned with the interference patterns of two complementary screens. This theorem was first stated by Babinet and has some practical applications.

Two screens are said to be complementary if they each consist of openings in opaque material, the openings in one corresponding exactly to the opaque parts of the other (Fig. 7.19). Babinet's theorem says that the interference patterns of two such screens are exactly the same except for a small region near the centre. For example, the pattern of a set of opaque discs should be the same as that of a set of equally-sized holes similarly arranged. The theorem is illustrated by the diffraction patterns shown in Fig. 7.20.

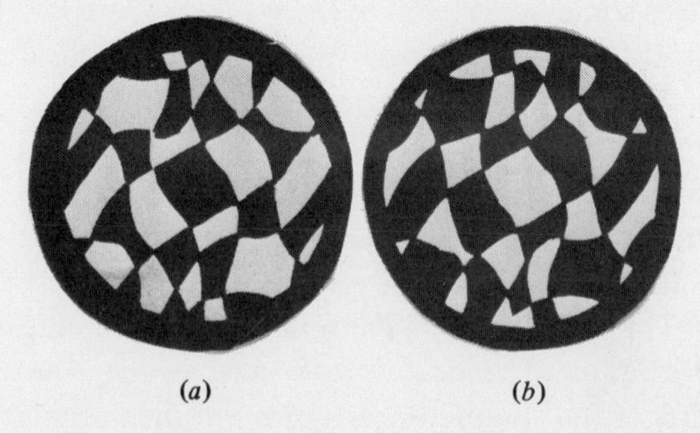

Fig. 7.19. (*a*) and (*b*) complementary screens.

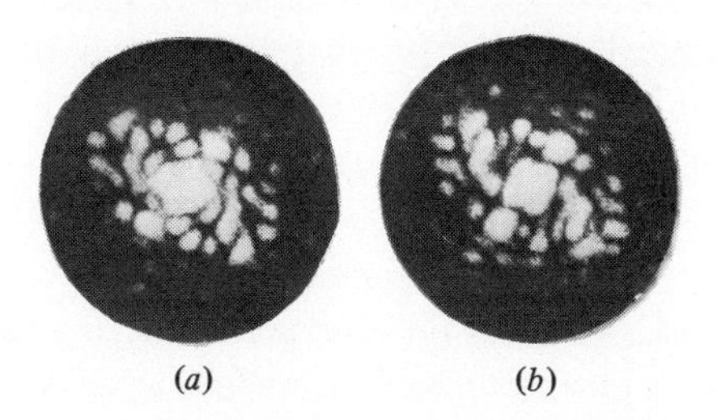

Fig. 7.20. (*a*) and (*b*) diffraction patterns of Fig. 7.19.

The theorem can be proved on general grounds. Suppose that the amplitudes of the diffraction patterns of two complementary screens are ψ_1 and ψ_2. Now, the diffraction function for a combination of apertures can be obtained by adding the separate (complex) functions. If we add ψ_1 and ψ_2 we should obtain the diffraction function for the complete opening. If this is circular, the sum of ψ_1 and ψ_2 is thus the Airy-disc pattern, which is confined to a small region round the centre; the rest is blank. Therefore the sum of ψ_1 and ψ_2 must be zero everywhere except over the region of the Airy-disc pattern. The moduli of ψ_1 and ψ_2 must be equal, their phases differing by π. The intensity functions are therefore the same.

This reasoning, however, is deceptively simple; experiments show that Babinet's theorem is not even approximately obeyed for simple screens, and certain extra conditions are necessary to produce results such as those shown in Fig. 7.20.

Screens can be complementary only over a limited area, which is most simply taken to be circular. Babinet's theorem assumes that we can ignore all but the centre part of the diffraction pattern of this circle – i.e. that the transform tends to a δ-function. This is not so; as we can see from Fig. 7.17(*a*), the non-origin troughs and peaks of the function $2J_1(x)/x$ are not negligible: they have heights of -0.13, 0.06, -0.04, ... For Babinet's theorem to be true it is therefore necessary that the two screens should have transforms that are much stronger than the function $2J_1(x)/x$ over a large range. Two conditions are necessary:

(a) Both screens should have a large transmitting region, so that their diffraction patterns are strong; since the two screens are complementary, this means that each must be about 50 per cent transparent.

(b) They must contain fine detail, so that they have high-order Fourier coefficients greater than the corresponding values of $2J_1(x)/x$. The finer the detail, the greater the area over which Babinet's theorem is acceptably true.

The screens shown in Fig. 7.19 satisfy the first condition well enough, but barely satisfy the second; this is the reason why only the outer parts of the two diffraction patterns appear similar to each other.

7.4 Interference

We have so far considered only the effect of modifying a single wavefront; we now wish to consider the effects occurring when two or more wavefronts interact. These effects are called interference. We shall concern ourselves only with wavefronts from identical objects; any other cases are extremely difficult to deal with theoretically and are of little practical importance.

For identical apertures we can make use of the principle of convolution (§ 3.6.2). For example, two similar parallel apertures can be considered as the convolution of one aperture with a pair of δ-functions representing the separation. The interference pattern is therefore the product of the diffraction pattern of one aperture and that of the pair of δ-functions (§ 3.6.4). We can therefore always divide an interference problem into two parts – the derivation of the Fourier transform of the single aperture and that of the set of δ-functions.

The transform of the single aperture is called the *diffraction function* and that of the set of δ-functions is called the *interference function*; the complete diffraction pattern is the product of the two (Fig. 7.21).

7.4.1 *Interference pattern of two pinholes.* We can regard a pair of pinholes, with separation a, as the result of convoluting a single pinhole

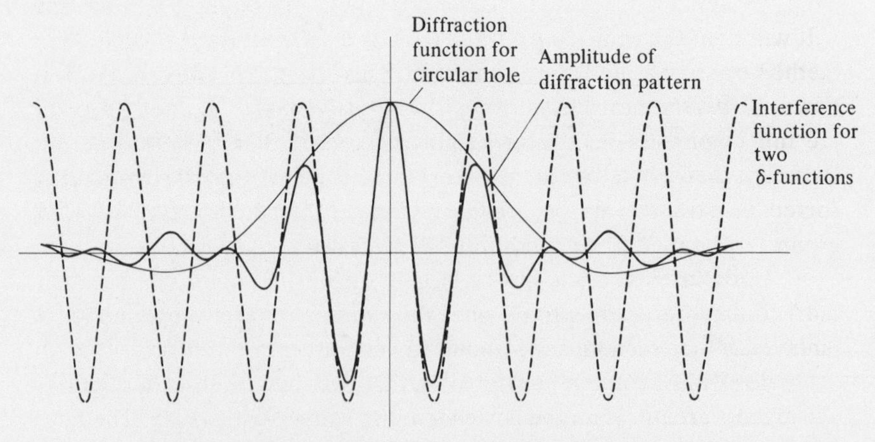

Fig. 7.21. Product of diffraction function and interference function gives amplitude of complete diffraction pattern of two circular holes.

with a pair of δ-functions. Now we have seen in § 3.4.5 that the transform of two δ-functions is given by

$$\psi(u) = 2\cos(ua/2). \tag{7.47}$$

Thus the diffraction pattern of the two pinholes is the diffraction pattern of one of them multiplied by a cosinusoidal function, varying in a direction parallel to the separation a. A practical illustration for holes of finite size is shown in Fig. 7.22.

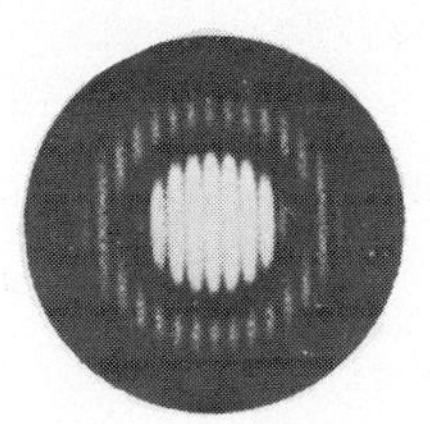

Fig. 7.22. Fraunhofer diffraction pattern of two circular holes.

The zeros of the function occur at values of θ given by

$$ua/2 = (m + \tfrac{1}{2})\pi, \tag{7.48}$$

where m is an integer and $u = k\sin\theta = 2\pi\sin\theta/\lambda$.

Thus

$$\frac{2\pi\sin\theta}{\lambda}\frac{a}{2} = (m + \tfrac{1}{2})\pi,$$

or

$$a\sin\theta = (m + \tfrac{1}{2})\lambda. \tag{7.49}$$

It will be realized that what we have achieved is a rather round-about method of deriving an expression for Young's fringes (§ 1.2.4). There are, however, several reasons for using this approach: first, we have derived the full expression for the profile of the fringes, not just the spacing; secondly, we have demonstrated that the convolution method gives the correct result for a simple example; and, thirdly, we have prepared the ground for more complicated systems, such as that which follows.

7.4.2 *Interference pattern of a regular series of pinholes.* An array of pinholes can be regarded as the convolution of a set of δ-functions with one pinhole. In § 3.4.5 we showed that, if the δ-functions form a regular one-dimensional lattice with spacing d, the transform is

$$\psi(u) = \sum_{n=0}^{N-1} \exp(-\mathrm{i}und), \tag{7.50}$$

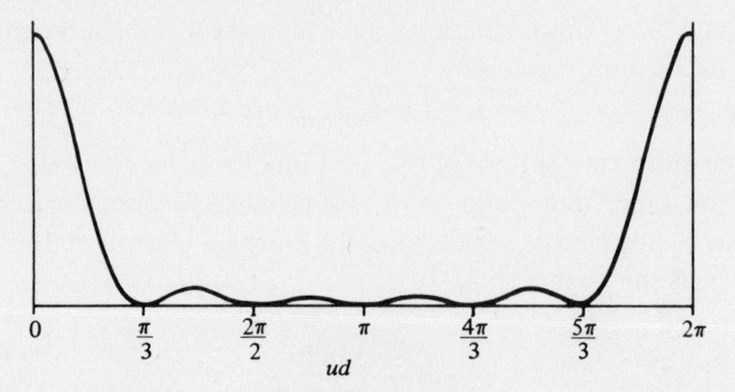

Fig. 7.23. Form of function $\sin^2(uNd/2)/\sin^2(ud/2)$ for $N=6$.

where N is the number of pinholes. When $N\to\infty$, the sum of the series is given by (3.41), namely:

$$\psi(u)=\sum_{m=-\infty}^{\infty}\delta(u-2\pi m/d). \tag{7.51}$$

The index m is called the *order of diffraction*. When N is finite, the sum of the geometrical series is

$$\psi(u)=\frac{1-\exp(-\mathrm{i}uNd)}{1-\exp(-\mathrm{i}ud)}. \tag{7.52}$$

The intensity is given by the product of this expression and its complex conjugate, which is easily shown to be

$$I=\frac{\sin^2 uNd/2)}{\sin^2(ud/2)}. \tag{7.53}$$

This expression, which is plotted in Fig. 7.23 for $N=6$, has some interesting properties. It is zero whenever the numerator is zero except when the denominator is also zero; then it is N^2. As the number of pinholes increases, the number of zeros increases and the pattern becomes more detailed. The peaks of intensity N^2 – called the principal maxima – become outstanding compared to the smaller subsidiary maxima, of which there are $N-2$ between the principal maxima. In fact, these maxima approximate to the δ-functions of (7.51), namely $\delta(u-2m\pi/d)$.

The conditions for the production of principal maxima are that $ud/2=m\pi$. Since $u=2\pi\sin\theta/\lambda$, we have

$$d\sin\theta=m\lambda, \tag{7.54}$$

the well-known equation for the diffraction grating. We shall return to this later, when we discuss diffraction gratings (§ 7.7).

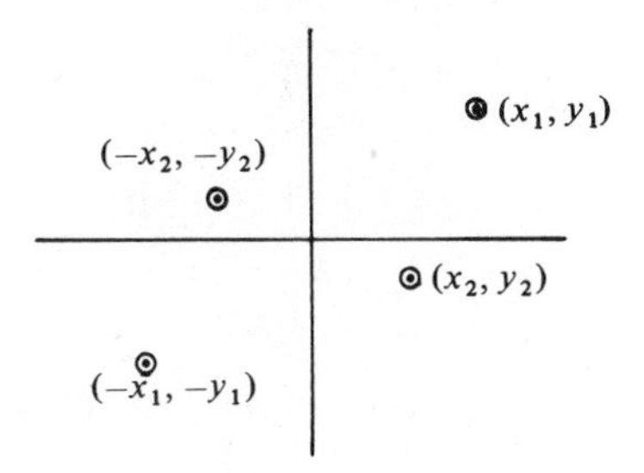

Fig. 7.24. Two pairs of pinholes.

7.4.3 *Interference pattern of a lattice of pinholes.* We can now extend our results to a regular array of pinholes, which we may call a two-dimensional lattice. We can approach this through a set of four pinholes, at positions $\pm(x_1, y_1)$, $\pm(x_2, y_2)$ (Fig. 7.24). We have to evaluate the expression

$$\begin{aligned}\psi(u, v) &= \sum \exp\{-i(ux+vy)\}\\ &= 2[\cos(ux_1+vy_1)+\cos(ux_2+vy_2)\}\\ &= 4\cos\left(u\frac{x_1+x_2}{2}+v\frac{y_1+y_2}{2}\right)\cos\left(u\frac{x_1-x_2}{2}+v\frac{y_1-y_2}{2}\right).\end{aligned} \tag{7.55}$$

As in § 7.41, we see that this function has maxima at values of u and v given by the equations

$$\left.\begin{aligned}u\frac{x_1+x_2}{2}+v\frac{y_1+y_2}{2} &= h\pi\\ \text{and}\qquad u\frac{x_1-x_2}{2}+v\frac{y_1-y_2}{2} &= k\pi,\end{aligned}\right\} \tag{7.56}$$

where h and k are integers. The interference pattern is therefore the product of two sets of linear fringes, each set being perpendicular to the separation of the pairs of holes (Fig. 7.22). Such fringes are called *crossed fringes* and are shown in Fig. 7.25.

By reasoning analogous to that of § 7.4.2, and which we do not need to reproduce here, we can see that as the lattice of pinholes, with these four points providing the unit cell, increases in extent, the conditions for constructive interference become more precisely defined, and in the limit the interference pattern becomes a collection of points, also arranged on a lattice (Fig. 7.26). This is called the *reciprocal lattice* of the original lattice, because u and v are reciprocally related to the separations of the pairs of holes in Fig. 7.24.

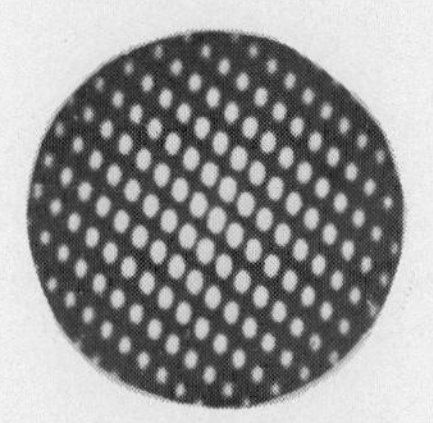

Fig. 7.25. Diffraction pattern of four holes at corners of a parallelogram, showing crossed fringes.

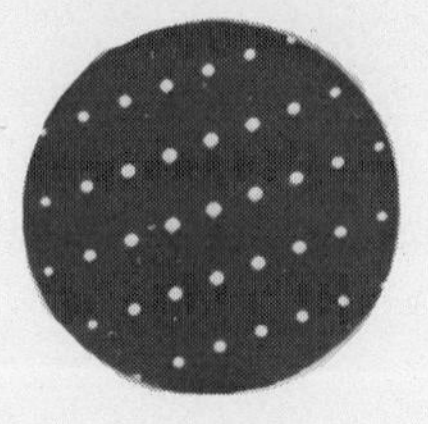

Fig. 7.26. Diffraction pattern (reciprocal lattice) of a lattice of holes.

Fig. 7.27. Pair of parallel apertures.

This concept of the reciprocal lattice becomes more important in connexion with three-dimensional interference which we shall discuss in § 7.6.

7.4.4 *Interference pattern of two parallel apertures of arbitrary shape.* We can regard a pair of similar parallel apertures (Fig. 7.27) as the convolution of a single aperture with two δ-functions. The interference pattern is therefore the product of the diffraction pattern of a single aperture and the interference function, which is a set of sinusoidal fringes. This is illustrated in Fig. 7.28, and is one of the most important results in diffraction theory.

7.4.5 *Interference pattern of a periodic array of parallel apertures.* If we have an extended lattice of similar apertures (Fig. 7.29), we may consider

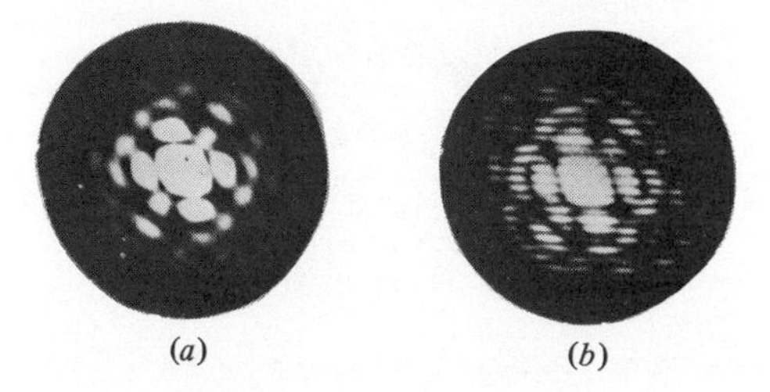

Fig. 7.28. (*a*) Diffraction pattern of one of the apertures in Fig. 7.27. (*b*) Complete diffraction pattern of the mask in Fig. 7.27.

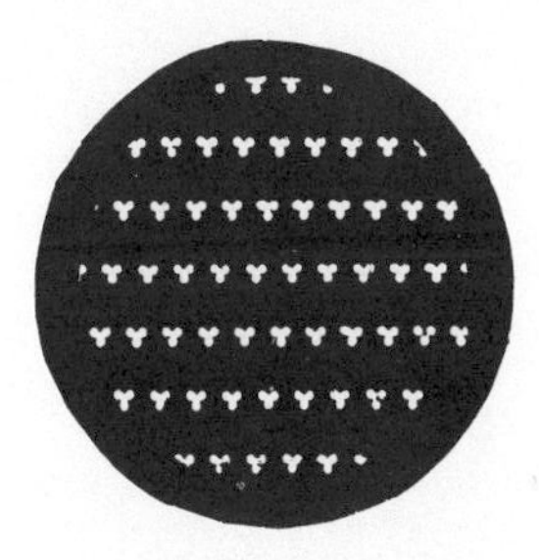

Fig. 7.29. Lattice of parallel apertures.

it as the convolution of a single aperture with the lattice having translations a and b. Then the diffraction pattern (Fig. 7.30) is the product of the diffraction patterns of the single aperture and that of the lattice. In other words, the reciprocal-lattice pattern is multiplied by the diffraction pattern of the unit.

If the unit is a circular hole, fairly small with respect to the lattice spacings, then the influence of the diffraction pattern is easily seen (Fig. 7.26); if the pattern is more complicated (Fig. 7.31), such as a set of holes representing a chemical molecule, the result is less clear (Fig. 7.32) but the correspondence is quite definite.

We may look upon the diffraction pattern in another way. A single unit of Fig. 7.31 gives a particular diffraction pattern (Fig. 7.33); the effect of putting the units on a lattice is, apart from making the pattern stronger, to make the diffraction pattern observable only at the reciprocal-lattice points. This process is called *sampling*; it is important in dealing with diffraction by crystals, and has many applications in communication theory.

If we regard the set of apertures as a two-dimensional diffraction grating, the reciprocal lattice represents its set of orders. Each reciprocal-lattice point is an order of diffraction (§ 7.4.2), specified now by *two*

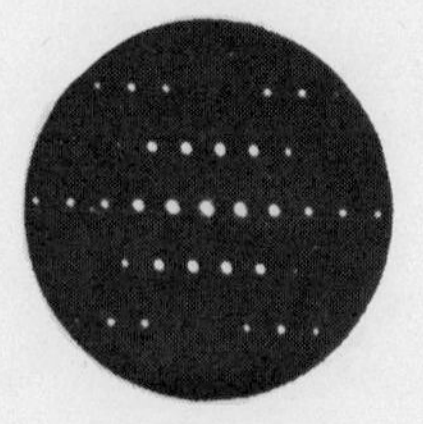

Fig. 7.30. Diffraction pattern of Fig. 7.29.

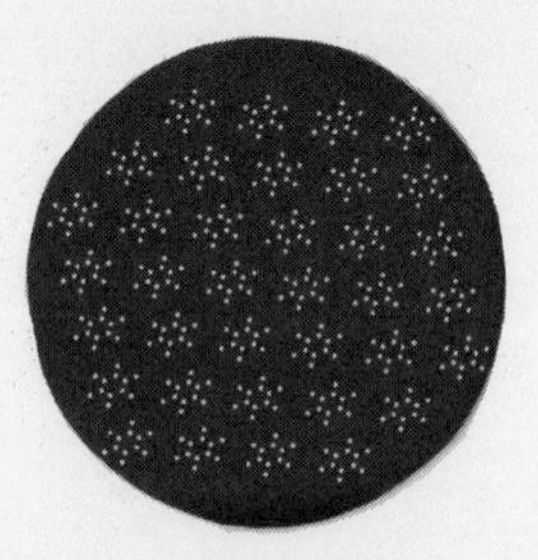

Fig. 7.31. Set of holes representing a lattice of chemical molecules.

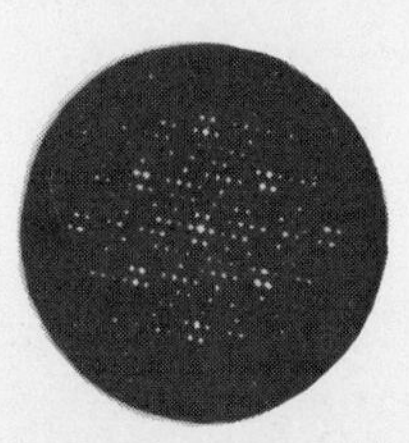

Fig. 7.32. Diffraction pattern of Fig. 7.31.

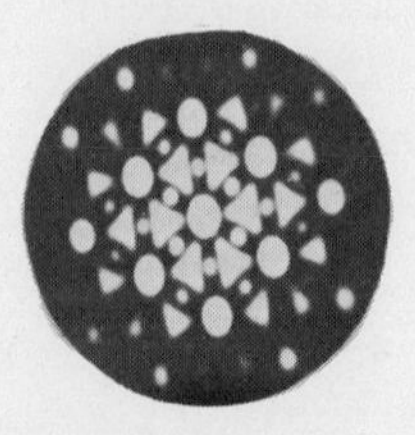

Fig. 7.33. Diffraction pattern of a unit of Fig. 7.31.

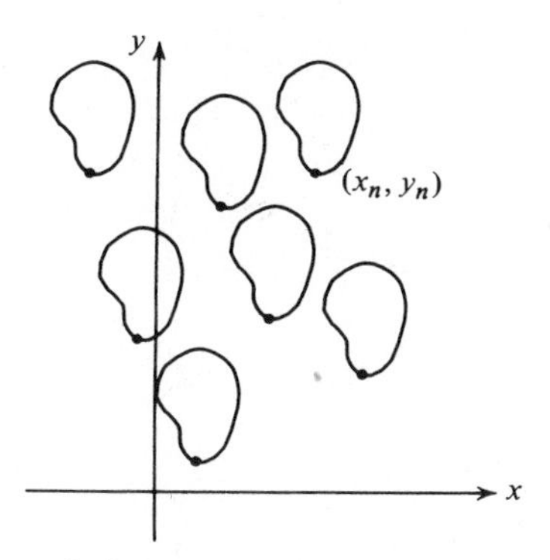

Fig. 7.34. Random set of similar apertures, showing the origin (x_n, y_n) of an individual aperture.

integers, h and k, instead of one. In three dimensions (§ 7.6) we shall see that *three* integers are needed.

7.4.6 *Diffraction by a random collection of parallel apertures.* Suppose that the diffracting object consists of a collection of parallel apertures arranged randomly. We can regard the collection as the convolution of the single aperture with a set of δ-functions representing the aperture positions.

We therefore need to determine the diffraction pattern of a set of N randomly arranged δ-functions. This problem is expressed mathematically as

$$\begin{aligned}\psi(u, v) &= \sum_{n=1}^{N} \delta(x - x_n)\delta(y - y_n) \exp\{-i(ux + vy)\}\\ &= \sum \exp\{-i(ux_n + vy_n)\},\end{aligned} \tag{7.57}$$

where the nth aperture has random origin at (x_n, y_n) (Fig. 7.34). This sum cannot be evaluated in general. But the *intensity* of the transform

$$I(u, v) = |\psi(u, v)|^2 \tag{7.58}$$

can be evaluated by writing the square of the sum (7.57) as a double sum:

$$\begin{aligned}|\psi(u, v)|^2 &= \left| \sum_{n=1}^{N} \exp\{-i(ux_n + vy_n)\} \right|^2\\ &= \sum_{n=1}^{N} \sum_{m=1}^{N} \exp[-i\{u(x_n - x_m) + v(y_n - y_m)\}].\end{aligned} \tag{7.59}$$

Now since x_n and x_m are random variables, $(x_n - x_m)$ is also random and so the various terms usually make randomly positive and negative contributions to the sum. There are two exceptions to this statement. Firstly, the terms $n = m$ in the double sum all contribute a value $e^{i0} = 1$, and there are N of them, so that the expected value of the double sum (7.59) is N. Secondly, when $u = v = 0$, *all* the terms in the sum contribute 1, and the value of (7.59) is N^2, so that we can write the statistical expectation:

$$I(u, v) = N + N^2 \bar{\delta}(u, v), \tag{7.60}$$

where $\bar{\delta}(u, v)$ has the value of unity at $(u, v) = (0, 0)$ and is zero elsewhere. The function (7.60) represents a bright spot of intensity N^2 at the origin and a uniform background of intensity N.

Of course a truly random distribution does not exist and the above description must really be modified. If the N points are all within a finite region (say a square of side D) the terms in the double sum (7.59) will all have positive values even if u and v deviate from zero by as much as $\pi/2D$. So the spot at the origin has a finite size, of this order of magnitude. In addition, the randomness of the distribution is usually restricted to avoid the overlapping of neighbouring apertures. This can be shown to result in a weak structure appearing in the background term; the subject can be treated fully by the self-correlation method described in § 11.4.1.

Returning to the diffraction pattern of the random array of apertures, we now recall that the object was expressed as the convolution of a single aperture with the random array of δ-functions. Its diffraction pattern is then the product of the diffraction pattern of a single object and the function (7.60). At all points except the origin and its immediate vicinity the result is an intensity just N times the intensity of the single aperture's diffraction pattern. Only at the origin itself there appears a bright spot, with intensity N^2 times that of the zero order of the single aperture diffraction pattern. The result is illustrated by Fig. 7.35; in Fig. 7.35(c) the spot at the origin has been emphasized by under-exposing the central region of the photograph. If the number of apertures becomes *very* large, the bright spot is the only observable feature.

7.4.7 *Effects of coherence.* It is important to remember that these results are true only if the apertures are coherently illuminated; if the illumination is incoherent no interference effects will arise. In practice it is difficult to attain either complete coherence or complete incoherence, and the results will depend upon the degree of coherence (§ 8.6.6) that exists in the incident beam. An important intermediate case exists when the mean

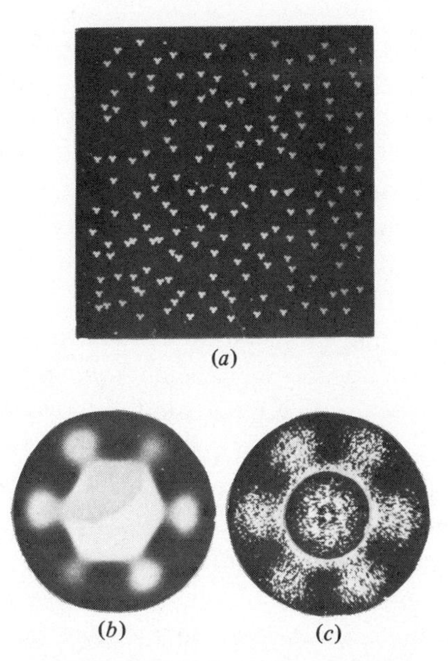

Fig. 7.35. (*a*) Mask of random parallel apertures; (*b*) diffraction pattern of one unit of Fig. 7.35(*a*); (*c*) complete diffraction pattern of Fig. 7.35(*a*); the centre inset is an under-exposed part of the diffraction pattern showing the strong spot at the centre.

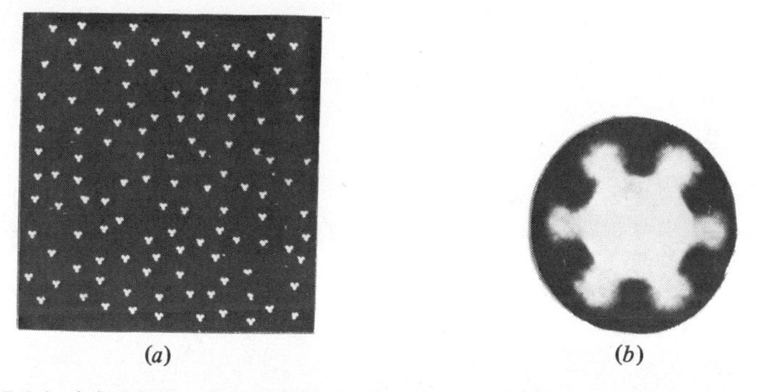

Fig. 7.36. (*a*) Mask of parallel apertures at roughly equal separations; (*b*) diffraction pattern of Fig. 7.36(*a*) with coherence only over individual apertures.

separation of the apertures – either randomly or regularly arranged – is great enough with respect to the size of the apertures for the illumination to be reasonably coherent over each aperture, but incoherent from one aperture to another. Then the complete diffraction pattern is simply the diffraction pattern of the single aperture multiplied in the intensity by the number of apertures. This is illustrated in Fig. 7.36.

7.5 Multiple-beam interference

Two-beam interference provided both an initial verification of the wave theory of light and a method of measuring wavelengths with an accuracy of a few per cent. Because, as we showed in § 7.4.1, the intensity distribution in two-beam interference is sinusoidal, the positions of the maxima and minima cannot be located with any great accuracy. Multiple-beam interference overcomes this difficulty; the conditions for reinforcement of several beams are much more precise than those for the reinforcement of two, and very sharp maxima can be obtained.

The first discovery of this sort arose when Fraunhofer was investigating diffraction effects with his newly invented spectroscope; he wanted to increase the intensity of the pattern that he was observing and decided to use a row of apertures instead of one. To his surprise, he found that he obtained a completely different diffraction pattern consisting of isolated lines. He had invented the diffraction grating.

Other devices for producing multiple beams are now also in use. Chief amongst these are the accurately-made highly-reflecting surfaces which reflect waves backwards and forwards a large number of times. Such instruments can attain a great precision which makes optics the most highly accurate branch of physics.

7.5.1 *Multiple reflexions.* We have seen in § 7.1.5 that interference fringes can be obtained with light reflected from the top and bottom of a thin film but obviously there are in practice more than two beams interfering; some light will continue to be reflected within the film as shown in Fig. 7.37. But the two-beam approximation is adequate because after the first reflexion there is a rapid reduction in the amplitude of successive reflexions; therefore the observed fringes are approximately sinusoidal. How can we increase the sharpness?

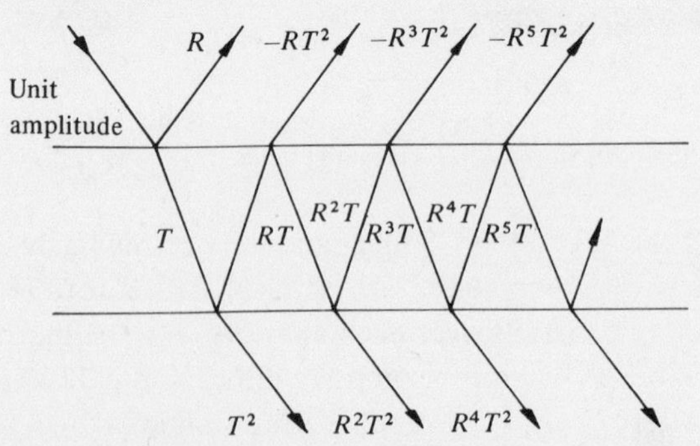

Fig. 7.37. Multiple internal reflexions in a glass plate.

Clearly, we must reduce the rate of fall-off of amplitude; we can, if the film is solid, deposit a highly-reflecting metal such as silver on the surface. The more highly-reflecting the deposit, the sharper the fringes, but the weaker the light that will go to form them. Therefore we must try to see theoretically what practical conditions are to be aimed at.

Fig. 7.37 shows a film at the surfaces of which a fraction R of the incident amplitude is reflected and fraction T transmitted. By reflexion and by transmission we have a succession of waves whose phases change by constant increments in a similar way to the operation of the diffraction grating (§ 7.4.2). The difference now, however, is that the amplitudes of the waves also vary. We can see, with reference to Fig. 7.37 that the series which we have to sum has reflected terms R, $-RT^2$, $-R^3T^2, \ldots$, and transmitted terms $T^2, R^2T^2, R^4T^2, \ldots$. The minus sign in the first set expresses the change of phase that takes place at one of the reflexions (§ 4.3.5).

Let us consider the transmitted light. By comparison with § 7.4.2, we see that the series that we have to sum is

$$A(k) = T^2 \sum_{p=0}^{\infty} R^{2p} \exp(-\mathrm{i}kps) = T^2 \sum_{p=0}^{\infty} R^{2p} \exp(-\mathrm{i}pu), \quad (7.61)$$

where $u = ks$. This function can be evaluated by two methods:

(a) as a Fourier series with coefficients $a_p = R^{2p}$;

(b) as a geometric series with factor $R^2 \exp(-\mathrm{i}u)$.

First we treat (7.61) as a Fourier series. It represents (§ 3.2.3) a periodic function with period $\Delta u = 2\pi$. The function within each period is the Fourier transform of the coefficients R^{2p}, where p is considered as a continuous variable. Writing

$$R^{2p} = \exp(2p \ln R), \quad (7.62)$$

we see that we need the transform of

$$f(p) = \exp(-\alpha p)\ (p \geqslant 0); \qquad f(p) = 0\ (p < 0) \quad (7.63)$$

which is

$$a(u) = \int_0^{\infty} \exp\{-(\alpha + \mathrm{i}u)p\}\, \mathrm{d}u = (\alpha + \mathrm{i}u)^{-1}, \quad (7.64)$$

where u is the spatial frequency variable conjugate to p. Substituting $2 \ln R$ for α we have the required result:

$$a(u) = (2 \ln R + \mathrm{i}u)^{-1}; \qquad |a(u)|^2 = [4(\ln R)^2 + u^2]^{-1}. \quad (7.65)$$

The function $a(u)$ is called a 'Lorentzian function'. Its intensity $|a(u)|^2$ is illustrated in Fig. 7.38 (broken line); superficially it looks like a Gaussian

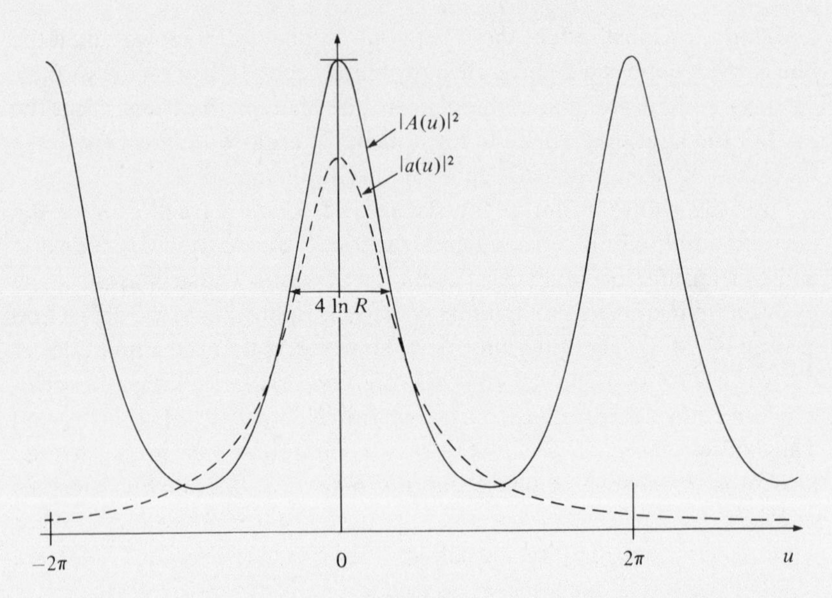

Fig. 7.38. The Lorenzian function (broken line); periodically repeated Lorentzian (full line).

curve, but it decays more slowly in the wings. Now, the Fourier series (7.61) represents $a(u)$ repeated periodically at intervals $\Delta u = 2\pi$; this function has intensity $|A(u)|^2$ and is shown in Fig. 7.38 (full line).

We can define a measure of the width of a peak by finding the points on its two sides where the intensity has fallen to half the maximum value. The separation between these points is called the 'half-peak width' (§ 3.4.6). For the Lorentzian (7.65) we find, writing ks again for u, that half the maximum occurs when

$$u = ks = \pm 2|\ln R| \tag{7.66}$$

so the half-peak width is given by:

$$(ks)_{\mathrm{H}} = 4|\ln R|. \tag{7.67}$$

Either k or s can be the variable in this expression. In general, R can be made very close to unity, in which case the overlap of neighbouring peaks is negligible, and to a very good approximation, the half-peak width is

$$(ks)_{\mathrm{H}} \approx 4(1-R). \tag{7.68}$$

In terms of wavelength λ, remembering that k_{H} is the *difference* between the values of k on the two sides of the peak, we use

$$\delta k/k = -\delta\lambda/\lambda; \qquad \delta\lambda = \delta k \lambda^2/2\pi;$$

and find

$$(\lambda s)_{\mathrm{H}} = 2\lambda^2(1-R)/\pi. \tag{7.69}$$

Secondly, we can evaluate (7.61) as a geometric series. This is the conventional method of attacking the problem. We write

$$A(k) = T^2 \sum_{p=0}^{\infty} \{R^2 \exp(-\mathrm{i}ks)\}^p, \tag{7.70}$$

which is a geometrical series having sum

$$A(k) = T^2/\{1 - R^2 \exp(-\mathrm{i}ks)\}. \tag{7.71}$$

The intensity

$$I(k) = |A(k)|^2 = T^4/(1 + R^4 - 2R^2 \cos ks). \tag{7.72}$$

This expression has a maximum value of $(TT')^2/(1-R^2)^2$ when s is an integral number of wavelengths, since $k = 2\pi/\lambda$; it has a minimum value of $(TT')^2/(1+R^2)^2$ when s is an odd number of half-wavelengths. Thus we have that

$$\frac{I_{\max}}{I_{\min}} = \left(\frac{1+R^2}{1-R^2}\right)^2, \tag{7.73}$$

which shows that the contrast in the fringes increases with R, as we should expect. Fig. 7.39 shows the form of $I(k)$ in detail.

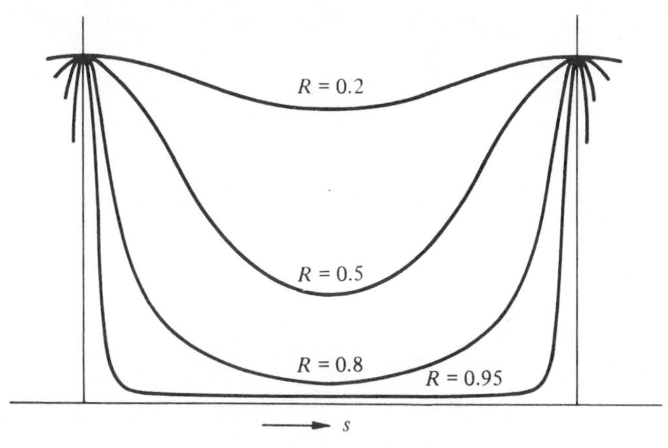

Fig. 7.39. Form of curves of $I/I_{\max}$ for different values of R.

Calculation of the half-peak width of this function is not difficult. For $I(k)$ of (7.72) we have

$$\frac{I}{I_{\max}} = \frac{1 + R^4 - 2R^2}{1 + R^4 - 2R^2 \cos ks} \tag{7.74}$$

Table 7.1. *Values of half-peak width $(ks)_H$ for different values of R*

R	0.5	0.6	0.7	0.8	0.9	0.95	0.98
$(ks)_H$	0.54	0.36	0.24	0.15	0.07	0.03	0.01

and thus the value of ks for which this quantity is one-half is given by

$$\cos ks = -(1+R^4-4R^2)/2R^2 = \cos \tfrac{1}{2}(ks)_H. \qquad (7.75)$$

Table 7.1 gives the values of $(ks)_H$ corresponding to different values of R. For comparison it should be noted that the corresponding value for Young's fringes is 0.50, so that until we attain a value of R greater than 0.6 no improvement in sharpness is obtained. (For values of R less than about 0.5, I does not reach values less than $\tfrac{1}{2}I_{max}$.) When R is very close to unity, (7.75) can be expanded in terms of $\Delta = (1-R)$ as:

$$\cos ks \approx 1-2\Delta^2, \qquad (7.76)$$

whence

$$(ks)_H = 4\Delta = 4(1-R). \qquad (7.77)$$

This limiting value agrees exactly with the value calculated previously (equation (7.68)).

An interesting paradox is associated with the maximum value of $I(k)$. This, according to (7.72) is $T^4/(1-R^2)^2$. Now if the coatings are non-absorbing, we can see that the fractions of transmitted and reflected intensity must add up to one, i.e. $T^2+R^2=1$. Thus, for non-absorbing coatings, all the incident light is transmitted at the peak of the interference fringes *even if T is very close to zero*. The reason, of course, is the constructive interference of a large number of weak transmitted waves.

Interference effects of this sort can be observed as either k or s changes. In § 7.1.6 we saw that

$$s = 2\mu t \cos \theta, \qquad (7.78)$$

indicating that interference peaks are observed whenever

$$2n\pi = ks = 2k\mu t \cos \theta,$$

or

$$n\lambda = 2\mu t \cos \theta. \qquad (7.79)$$

7.5.2 *Multiple reflexions in an amplifying medium.* A subject which has become of great importance with the advent of the laser is the effect of an amplifying medium on the behaviour of a multiple reflexion interferometer, since this is the basis of laser resonators.

If the amplitude amplification in one round trip of the light wave is G, we modify (7.61) to:

$$A(k)=T^2 \sum_{p=0}^{\infty} (R^2G)^p \exp(-ikps). \tag{7.80}$$

The result obtained is similar to (7.72) provided that R^2G remains less than unity. If G is large enough, the value becomes unity and the sum is:

$$\begin{aligned} A(k) &= T^2 \sum_{p=0}^{\infty} \exp(-ikps) \\ &= T^2 \sum_{q=-\infty}^{\infty} \delta(k-2\pi q/s). \end{aligned} \tag{7.81}$$

The spectrum is a series of ideally sharp lines; for this function $k_H=0$. This is the basic reason that laser lines are so sharp. Now if one asks what happens if GR^2 becomes greater than unity, one is asking a question which mathematics cannot answer. In practice, in a continuous laser the amplification factor G eventually settles down, at high enough intensity, to a value which is equal to R^{-2}, so that stability is achieved at that intensity. In a pulsed laser the amplification is large as the pulse starts, and gradually gets smaller as the population inversion is wiped out. So that the number of terms in the series for which GR^2 is greater than unity is finite; towards the end of the pulse G falls below unity and the series terminates.

One would expect from the above arguments that a continuous laser would emit a number of perfectly sharp lines separated by $\delta k = 2\pi/s$ (from (7.81)) indicating a wavelength separation $\delta\lambda = \lambda^2/s$. As will be discussed in § 8.7.4, the lines are not ideally sharp for practical reasons, and the number of lines emitted is rather small (sometimes only one) centred around the wavelength for which the gain G is maximum. A pulsed laser does not produce ideally sharp lines; if the wave train continues only for a time T, the linewidth must be greater than λ^2/cT.

7.5.3 *The confocal resonator.* A multiple-reflexion interferometer which has become very important as a laser resonator consists of a pair of spherical mirrors between which the light bounces backwards and forwards. Its advantage over the pair of plane mirrors discussed in §§ 7.5.1 and 7.5.2 is that the light does not easily escape from the interferometer even if the mirrors are not aligned properly, and even when the separation between them is large compared with their physical size. We shall not discuss this interferometer in detail, but just use it to illustrate the idea of transverse mode patterns.

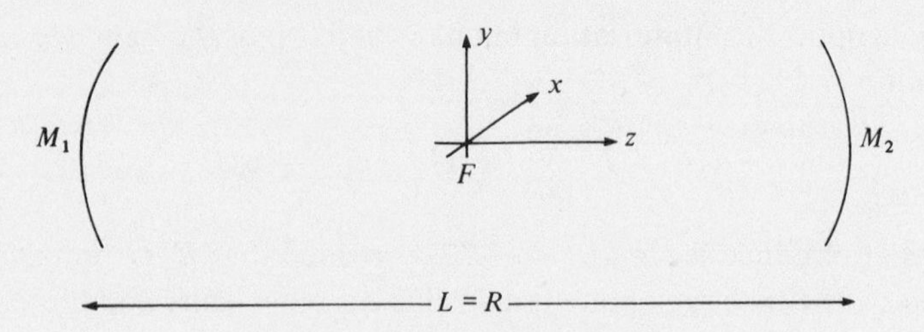

Fig. 7.40. Confocal resonator.

Suppose that the two mirrors have equal radii of curvature R and are separated by distance L along the line containing the two centres of curvature. For quite a large range of L/R (actually $0 < L/R < 1$) it can be shown that a light ray entering one mirror from the other will never escape from the cavity between them.† Notice in passing that $L/R = 0$, corresponding to a pair of plane mirrors, is on the extreme margin of this range; unless the plane mirrors are exactly parallel, any ray will eventually walk out of the space between the two. This defect was not critically important in the pre-laser days, when the light was attenuated at every reflexion and in any case decayed to negligible intensity after several reflexions. The possibility of amplifying media has changed this, and for lasers it is necessary to design resonators from which escape by geometrical optics is impossible.

Just to illustrate the idea of a mode pattern, we shall consider the case $L = R$, for which the centre of each mirror lies on the other one. Then the two foci (at distance $F = R/2$) coincide, and the interferometer is called a 'confocal resonator'. Consider the amplitude $f(x, y)$ of light travelling to the right in the common focal plane (Fig. 7.40). Since this is the light amplitude in the focal plane, the amplitude at the other focal plane of the mirror M_2 (which is coincident with it) must be the Fourier transform $a(u, v)$ of $f(x, y)$ (§ 7.3.2), but the light is travelling to the left. However, this system is symmetrical about the focal plane and so the direction of travel of the waves is unimportant, and a stable mode of operation is seen when the Fraunhofer diffraction pattern is identical, in amplitude and phase at each point in the plane (x, y), with the original function $f(x, y)$.

Very few functions have this property: that the function and its Fourier transform are identical in form. Two such functions have already been

† This proof is given in almost every book on lasers (e.g. Siegman, 1971). Actually the two mirrors may have different radii, but we shall assume them to be equal in this discussion. See also Fig. 8.17.

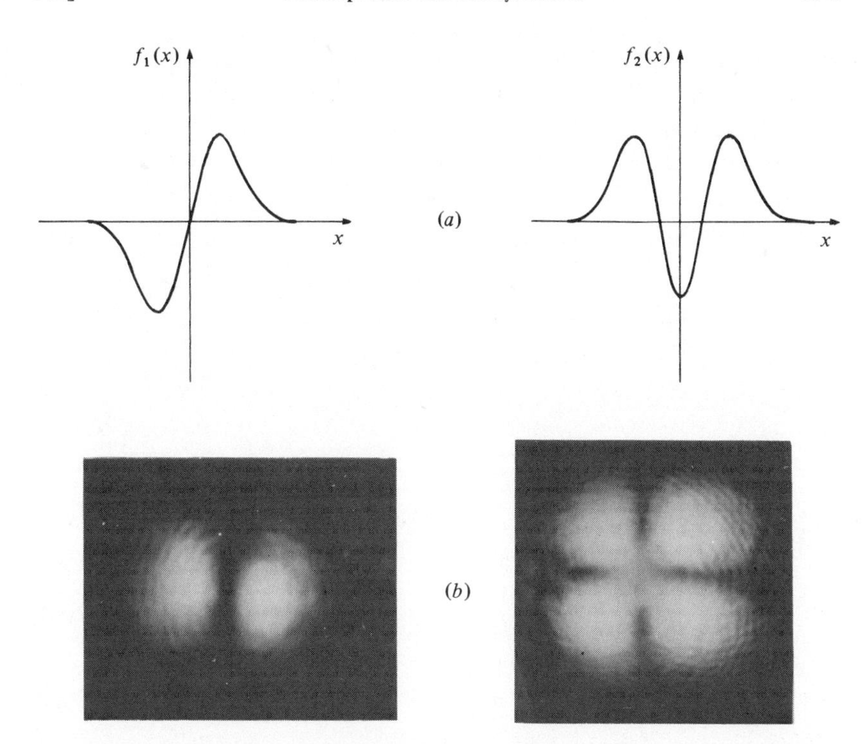

Fig. 7.41. (*a*) Examples of Gauss–Hermite functions; (*b*) photographs of laser modes.

introduced: the infinite periodic array of δ-functions, and the Gaussian function. The former is of infinite extent and is not useful in this particular problem. The latter, the Gaussian function, is in fact the first of a set of functions all of which have the required properties of transforming into themselves and also being finite in extent. The set of functions is the set of Gauss–Hermite polynomials which should be familiar to any student of quantum mechanics as the radial wave functions of a harmonic oscillator. They are expressed as

$$f_n(x) = H_n(x/\sigma) \exp(-x^2/2\sigma^2), \tag{7.82}$$

where the functions H_n obey the recurrence relation:

$$2xH_n = H_{n+1} + 2nH_{n-1}; \qquad H_0 = 1, \tag{7.83}$$

which gives $H_1 = 2x$, $H_2 = 4x^2 - 1$, etc. In Fig. 7.41(*a*) we have drawn two examples of these functions.

In two dimensions any product of the form

$$f_{lm}(x, y) = H_l(x/\sigma)H_m(y/\sigma) \exp\{-(x^2 + y^2)/2\sigma^2\} \tag{7.84}$$

satisfies our requirements. They can be seen as the intensity distribution across the output beam of a slightly misaligned continuous-wave laser; two examples are shown in Fig. 7.41(b). The various functions are known as 'transverse modes' and are referred to by the number pair (l, m).

The requirement of identity of scale allows us to calculate the focal size at the common focus F. For the mode (0, 0) which is a Gaussian spot,

$$f(x, y) = \exp\{-(x^2+y^2)/2\sigma^2\}, \tag{7.85}$$

the diffraction pattern $a(u, v)$ is the transform of (7.85):

$$a(u, v) = \exp\{-(u^2+v^2)\sigma^2/2\}. \tag{7.86}$$

The physical size of the pattern is given (for small angles) by (7.23)–(7.24):

$$\xi = uF/k, \qquad \eta = vF/k \tag{7.87}$$

and we required ξ and η to be identical coordinates with x and y. Thus for $a(\xi, \eta)$ and $f(x, y)$ to be the same function:

$$\exp\{-(x^2+y^2)/2\sigma^2\} = \exp\{-k^2(x^2+y^2)\sigma^2/2F^2\} \tag{7.88}$$

giving

$$k^2\sigma^2/F^2 = 1/\sigma^2, \qquad \sigma^2 = F/k. \tag{7.89}$$

Thus the radius of the spot, of order $\sigma/\sqrt{2}$, is $(F/2k)^{\frac{1}{2}}$ or $(F\lambda/4\pi)^{\frac{1}{2}}$. This of course, is the spot at the centre of the laser tube. When the beam reaches the mirrors, it will have the form of the Fresnel diffraction pattern of this spot at distance F, which is easily seen to have about twice the above radius.

7.5.4 *Multilayers.* Another way of producing multiple reflexions is to superimpose regular layers of materials of different refractive index. Some light will be reflected at each discontinuity.

The theory is basically the same as for multiple reflexions from two surfaces, but becomes much more involved as the various permutations of reflexion and transmission by the numerous interfaces are taken into account. There are some very elegant theoretical treatments for periodic multilayers, but we shall not go into the subject here. A very important practical application arises in the design of multilayer dielectric mirrors and interference filters, and some practical details will be given in § 7.9.2.

7.6 Three-dimensional interference

It might be thought that there is no need to extend the subject to three dimensions, since diffraction and interference patterns can exist only on

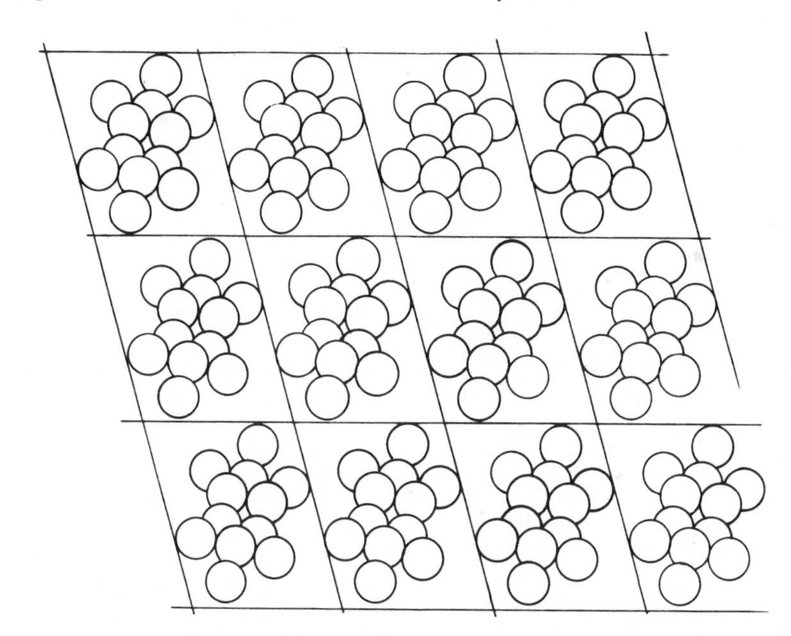

Fig. 7.42. Two-dimensional representation of a crystal structure.

two-dimensional screens. The need arose, however, when the diffraction of X-rays by crystal was discovered; crystals are three-dimensional diffraction gratings, and if we wish to investigate crystal structures we must make use of three-dimensional diffraction.

A crystal is a collection of atoms. For simplicity let us consider a crystal composed of identical atoms only. From the point of view of X-ray diffraction, since X-rays are scattered only by electrons, a crystal can be considered as a set of atomic positions – δ-functions – convoluted by the electron density function for one atom. The atomic positions repeat on a lattice; i.e. a small group of atoms – the unit cell – is repeated regularly in three dimensions. Therefore we can regard the crystal as composed of the single group convoluted with the lattice positions. These ideas are illustrated in two dimensions in Fig. 7.42. (It should be noted that the lattice is not synonymous with the structure; it is merely a name for the framework upon which the structure is built.) This would lead to an infinite crystal. We limit it therefore by multiplying the convolution by the shape function – the external boundary.

From the convolution theorem, therefore, we see that the diffraction pattern is the product of the transform of the atom, the transform of the set of δ-functions representing the atomic positions in the unit cell, and

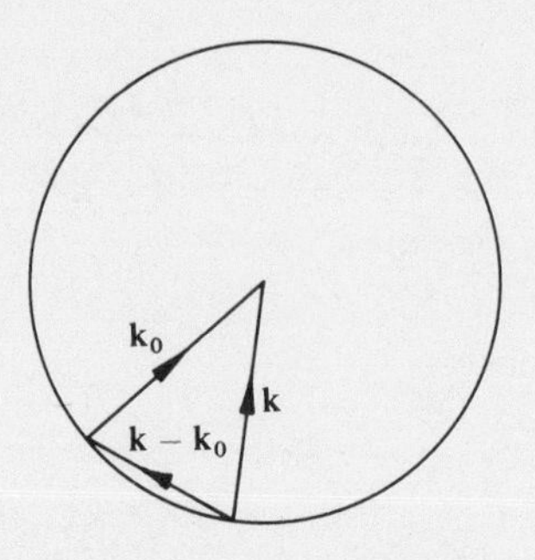

Fig. 7.43. Sphere of observation.

the transform of the crystal lattice; this product must then be convoluted with the transform of the shape function of the crystal.

This is a *complete* outline of the theory of X-ray diffraction. All that remains is to fill in the details. Unfortunately, this would require several textbooks since each aspect is complicated. We shall deal fully, therefore, only with the third aspect – the transform of the lattice – since this is directly related to the previous parts of this chapter.

7.6.1 *Diffraction by a three-dimensional grating.* We are concerned with the diffraction pattern produced by a three-dimensional lattice of δ-functions. Suppose we have an incident wave with wave-vector $\mathbf{k}_0$, and that it is diffracted to a direction with vector $\mathbf{k}$. In order to conserve energy, the incident and diffracted waves must have the same frequency

$$\omega = ck \tag{7.90}$$

and therefore the moduli of $\mathbf{k}$ and $\mathbf{k}_0$ must be equal;

$$|\mathbf{k}| = |\mathbf{k}_0|. \tag{7.91}$$

Alternatively, we can say that the waves must have the same time-variation, $\exp(\mathrm{i}\omega t)$, since this must pass unchanged through the calculation of diffraction by a *stationary* lattice. (Diffraction by a moving lattice is different and is dealt with in § 11.4.4.) The condition (7.91) can be represented geometrically by saying that $\mathbf{k}_0$ and $\mathbf{k}$ must be radius-vectors of the same sphere, which is called the *reflecting sphere* or *sphere of observation* (Fig. 7.43). An order of diffraction produced in this way is called a Bragg reflexion, after W. L. Bragg who, in 1912, introduced the idea of reflexion of X-rays by lattice planes.

Let us proceed with the calculation of the amplitude of the wave diffracted in the direction of $\mathbf{k}$. The δ-function of the lattice point $\mathbf{r}'$ acts as

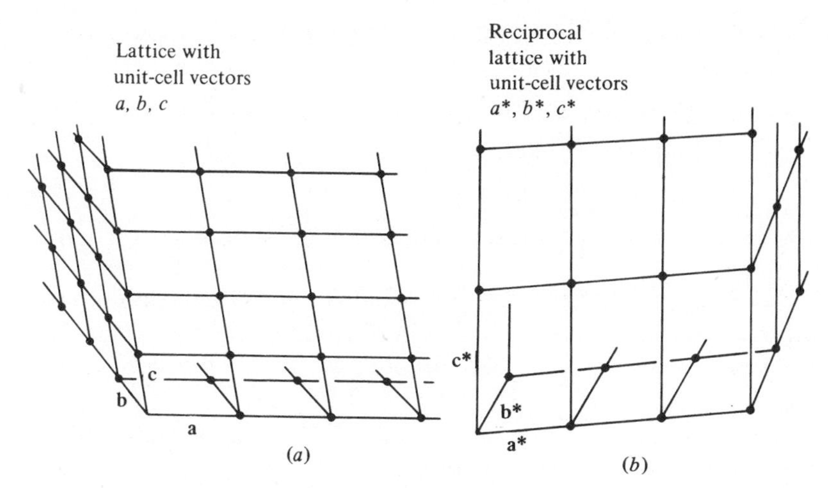

Fig. 7.44. Reciprocal lattice with unit-cell vectors.

a secondary source of strength equal to the value of the incident wave at the point,

$$\exp\{-i(\mathbf{k}_0 \cdot \mathbf{r}')\}, \tag{7.92}$$

and this scatters a wave in the direction **k** which can be written as

$$\psi(\mathbf{k}) = \exp[-i\{(\mathbf{k} \cdot \mathbf{r}) + \alpha\}], \tag{7.93}$$

where α is, as yet, an arbitrary phase. But this wave originates from $\mathbf{r}'$ as (7.55), so that

$$\exp\{-i(\mathbf{k} \cdot \mathbf{r}' + \alpha)\} = \exp\{-i(\mathbf{k}_0 \cdot \mathbf{r}')\}. \tag{7.94}$$

Thus

$$\alpha = (\mathbf{k}_0 - \mathbf{k}) \cdot \mathbf{r}'$$

and

$$\psi(\mathbf{k}) = \exp[-i\{(\mathbf{k} \cdot \mathbf{r}) + (\mathbf{k}_o - \mathbf{k}) \cdot \mathbf{r}'\}]. \tag{7.95}$$

The total diffracted beam with wave-vector **k** is therefore given by integrating (7.95) over all space $\mathbf{r}'$, which for the lattice of δ-functions with unit-cell vectors **a**, **b**, **c** (Fig. 7.44(*a*))

$$f(r) = \delta(\mathbf{r} - l\mathbf{a} - m\mathbf{b} - n\mathbf{c}) \qquad (l, m, n \text{ integers}) \tag{7.96}$$

reduces to the summation

$$\psi(\mathbf{k}) = \exp(-i\mathbf{k} \cdot \mathbf{r}) \sum_{l,m,n=-\infty}^{\infty} \exp[-i\{(\mathbf{k} - \mathbf{k}_0) \cdot (l\mathbf{a} + m\mathbf{b} + n\mathbf{c})\}]. \tag{7.97}$$

The summation is clearly zero unless the phases of all the terms are multiples of 2π:

$$(\mathbf{k}-\mathbf{k}_0)\cdot(l\mathbf{a}+m\mathbf{b}+n\mathbf{c})=2\pi s \qquad (s \text{ is an integer}). \qquad (7.98)$$

One trivial solution to this equation is

$$\mathbf{k}-\mathbf{k}_0=0 \qquad (s=0), \qquad (7.99)$$

which also satisfies (7.91). But there is also a host of other solutions.

7.6.2 *Reciprocal lattice.* These other solutions to (7.98) can be derived by means of a reciprocal lattice (§ 7.4.3). The vectors between points in the reciprocal lattice are then solutions of (7.98), and in order to represent observable diffracted beams must also satisfy (7.91).

We define three vectors $\mathbf{a}^*$, $\mathbf{b}^*$ and $\mathbf{c}^*$ in terms of the real-lattice constants by means of the equations

$$\left.\begin{aligned} \mathbf{a}^* &= V^{-1}\mathbf{b}\times\mathbf{c}, \\ \mathbf{b}^* &= V^{-1}\mathbf{c}\times\mathbf{a}, \\ \mathbf{c}^* &= V^{-1}\mathbf{a}\times\mathbf{b}, \end{aligned}\right\} \qquad (7.100)$$

where V is the volume of the unit cell in real space:

$$V=\mathbf{a}\cdot\mathbf{b}\times\mathbf{c}.$$

It now follows that if $(\mathbf{k}-\mathbf{k}_0)/2\pi$ can be written as the sum of integral multiples of $\mathbf{a}^*$, $\mathbf{b}^*$ and $\mathbf{c}^*$:

$$(\mathbf{k}-\mathbf{k}_0)/2\pi = l^*\mathbf{a}^*+m^*\mathbf{b}^*+n^*\mathbf{c}^* \qquad (l^*, m^*, n^* \text{ are integers}), \quad (7.101)$$

the left-hand side of (7.98) becomes, on substituting (7.101) for $(\mathbf{k}-\mathbf{k}_0)$,

$$2\pi(l^*\mathbf{a}^*+m^*\mathbf{b}^*+n^*\mathbf{c}^*)\cdot(l\mathbf{a}+m\mathbf{b}+n\mathbf{c})=2\pi(ll^*+mm^*+nn^*),$$

which can always be written $2\pi s$ since l, m, n, l^*, m^*, n^* have been defined as integers. Clearly, this simple result follows because, from the definition (7.100), $\mathbf{a}^*$, for example, is normal to $\mathbf{b}$ and $\mathbf{c}$ and therefore $\mathbf{a}^*\cdot\mathbf{b}$ and $\mathbf{a}^*\cdot\mathbf{c}$ are zero.

Now, if $(\mathbf{k}-\mathbf{k}_0)$ can be written in the form (7.101), it is clearly a vector of a lattice with unit-cell vectors $\mathbf{a}^*$, $\mathbf{b}^*$, $\mathbf{c}^*$; this is the reciprocal lattice (Fig. 7.44(*b*)). Its name arises because it has dimensions reciprocally related to those of the real lattice (§ 7.4.3); if we reduce the scale of any of $\mathbf{a}$, $\mathbf{b}$, $\mathbf{c}$, the corresponding reciprocal vector increases.

The observed diffraction pattern consists of those beams which satisfy both (7.91) and (7.101). The two conditions are represented geometrically by

(a) the observation sphere,

(b) the reciprocal lattice,

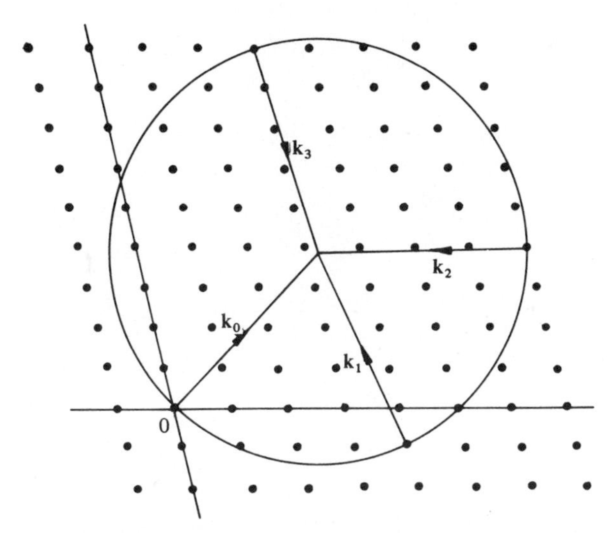

Fig. 7.45. Two-dimensional representation of intersection of sphere of observation with reciprocal lattice, showing directions of incident beam $\mathbf{k}_0$ and of three possible diffracted beams $\mathbf{k}_1$, $\mathbf{k}_2$ and $\mathbf{k}_3$.

One therefore draws the observation sphere and the reciprocal lattice superimposed and looks for intersections (Fig. 7.45). The observation sphere passes through the origin of reciprocal space (because $\mathbf{k}-\mathbf{k}_0=0$ is a point on it) and its centre is defined by the direction of the vector $\mathbf{k}_0$. Mathematically, the exact intersection of a sphere and a set of discrete points is negligibly probable; but because neither an exactly parallel beam nor a purely monochromatic source of X-rays exists, diffraction by a crystal does in fact occur. (One important point is the trivial solution (7.99) which ensures that at least one 'diffracted' beam (the undeflected one) exists to carry away the incident energy.) By controlling $\mathbf{k}_0$ – the direction of the incident beam – and moving the crystal and recording screen in appropriate ways, it is possible to produce a section of the reciprocal lattice with, say, one of the indices l, m, n constant. Such a photograph is shown in Fig. 7.46.

7.6.3 *Diffraction by a complete crystal.* We can see from Fig. 7.46 that the patterns cover a circular patch; the reason for this is that the atomic scattering factor, or the transform of the electron density, falls off with increasing angle of diffraction, producing negligible intensities at a fairly definite angle. The product of the transform and the reciprocal lattice therefore covers a spherical volume and Fig. 7.46 shows a circular section.

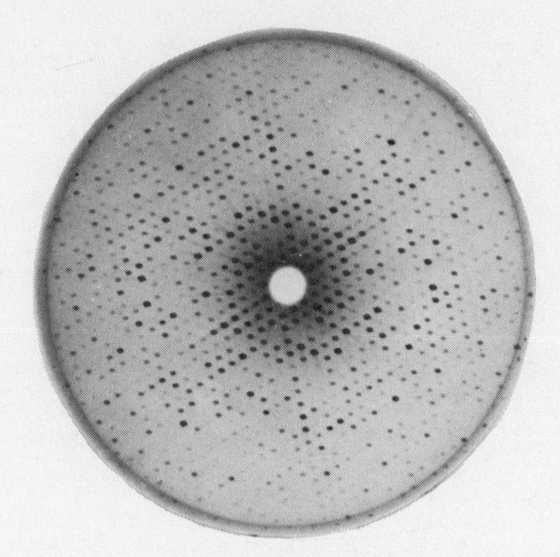

Fig. 7.46. Precession photograph of haemoglobin. (By courtesy of Dr M. F. Perutz.)

We can see also that the intensities within the observed reciprocal lattice vary in unpredictable ways; some orders of diffraction are strong and some weak. This variation is a result of multiplying by the transform of the atomic positions.

Finally, the spots in Fig. 7.46 are of finite size. This is caused by the geometry of the apparatus – finite size of X-ray focus, angular divergence of beam and so on. Even if these factors could be allowed for, however, the spots would still have a finite size; because of the size of the crystal the complete diffraction pattern convolutes the shape transform of the crystal, and thus each diffraction spot would have a shape corresponding to this transform. This statement is exactly equivalent to the result that each of the principal maxima from a diffraction grating has the same shape. This effect cannot be observed directly in practice, because a crystal small enough to scatter X-rays with appreciable broadening of the orders of diffraction would not scatter enough radiation to be observed. But the broadening can be observed from an agglomeration of small crystals.

7.7 Diffraction gratings

A diffraction grating is a periodic one-dimensional array of similar apertures, usually narrow slits or mirrors. We have seen in § 7.4 that the diffraction pattern of such an array is a periodic series of δ-functions, whose strengths are determined by the exact shape and dimensions of the apertures. The positions of the δ-functions are determined only by the period of the array. A given δ-function, called an 'order of diffraction' appears at a specified value of $u = u_m = 2\pi m/d$ (from (7.51)), so its angle θ depends on the wavelength $\lambda = 2\pi/k$. This relationship makes diffraction gratings important tools for spectroscopy.

7.7.1 *Production of diffraction gratings.* Although this book is concerned primarily with general principles, methods of production of diffraction gratings must be briefly described because some of the succeeding theory depends upon acquaintance with them.

The first serious gratings were made by scratching lines on glass or metal with a fine diamond. Rowland used an accurate screw to translate the diamond through a small distance after a short line of two or three centimetres had been ruled. Obviously much is implied in this sentence: the diamond and the flat upon which the grating is to be ruled must be carefully chosen; the screw and flat must be accurately adjusted relative to each other; the diamond point must not change during the ruling operation; and the temperature of the whole apparatus must be kept constant so that no irregular expansions occur. Thus machines – which are called ruling engines – for ruling gratings are extremely complicated and costly.

Recent years have seen a vast change in the methods of production of gratings. First, gratings have been ruled by cutting a very fine screw thread on a cylinder. The cylinder is then coated with a plastic which forms a surface-replica that can be removed after setting. In this way surprisingly good gratings can be produced very cheaply.

But more important is the development of holographic diffraction gratings. As will be pointed out in § 9.7.3, holograms are essentially *complicated* diffraction gratings. They can also be made to be *simple* diffraction gratings. The development of high-resolution emulsions, particularly photo-resists for the microelectronic industry, has made it possible to photograph a very fine interference pattern between plane-waves, which produces, in a single exposure, a grating with many thousands of lines. For example if two plane parallel coherent laser beams with wavelengths of say 0.5 μm interfere at an angle of $2\alpha = 60°$, the interference pattern has Young's fringes with a spacing $\lambda/\sin\alpha = 1$ μm. Because the laser lines are very sharp the number of fringes is enormous, and a grating many centimetres long can be produced in a single exposure. This technique completely avoids the problem of errors in line position, which is very troublesome in ruled gratings.

Another advantage of the holographic grating is that the line-spacing can be arranged to be non-uniform in a planned way so as to correct for known aberrations in the associated optics or to reduce the number of accessory optical elements required. For example, a self-focusing grating can be produced by using as the source of the grating the interference pattern between a spherical wave and a plane-wave.

Most diffraction gratings are reflexion gratings, either being ruled on an optically flat reflecting surface, or being produced holographically on such a surface by etching through the developed photoresist. Gratings can also be produced on cylindrical and spherical surfaces in order to add a further dimension to the possible correction of aberrations.

7.7.2 *Resolving power.* One of the important functions of a diffraction grating is the measurement of the wavelengths of spectral lines; since we know the spacing of the grating we can use equation (7.54) to measure wavelengths absolutely. The first question we must ask about a grating is 'What is the smallest separation between two wavelengths that will result in two separate peaks in the spectrum?' In other words, what is the limit of resolution?

This problem can be considered in terms of (7.53), which gives the shapes of the principal and secondary maxima. In § 7.4.2 we showed that there were $N-1$ zero values of the intensity lying between the principal maxima produced by a grating containing N lines. If two different wavelengths are present in the radiation falling on a grating, the intensity functions will add together; we need to find the conditions that decide whether single or double principal maxima are produced. We therefore have to consider in more detail the exact shape of the interference function.

First of all there are the heights of the secondary maxima. It is difficult to find the values precisely but a close approximation can be made by assuming that they lie exactly half-way between the minima. From equation (7.53),

$$I=\frac{\sin^2(uNd/2)}{\sin^2(ud/2)}, \tag{7.102}$$

we can see that the maxima lie close to positions given by

$$\frac{ud}{2}=\left(m+\frac{2p+1}{2N}\right)\pi, \tag{7.103}$$

where m denotes the principal maxima and p the secondary maxima. If we substitute these values in (7.53), we find that

$$I=\frac{\sin^2\{(2p+1)/2\}\pi}{\sin^2\{(2p+1)/2N\}\pi}$$

$$=\frac{1}{\sin^2\{(2p+1)/2N\}\pi}. \tag{7.104}$$

This result indicates a series of gradually decreasing maxima as p increases from unity. If p is small compared with N, the first subsidiary maxima have heights, relative to the principal maxima, of

$$\left(\frac{2}{3\pi}\right)^2, \quad \left(\frac{2}{5\pi}\right)^2, \quad \left(\frac{2}{7\pi}\right)^2, \ldots,$$

which are equal to 0.045, 0.016, 0.008, ... These values, particularly the first, are by no means negligible.

We can plot a graph (Fig. 7.47) showing these maxima, and we can find the effect of adding to it another graph on a slightly different scale. If we adopt the Rayleigh criterion for resolution, that the maximum for one curve should lie on the first minimum of the other, the resultant shows a definite dip (Fig. 7.47). If we reduce the separation to 0.9 of that given by Rayleigh (Fig. 7.48(*a*)), the dip is still present but at 0.8 (Fig. 7.48(*b*)) it has almost disappeared. Since resolution is never as clear-cut as the precise equation would seem to indicate, the Rayleigh criterion is seen to be a reasonable one to use.

If Fig. 7.47 represents the first order, we can see that the limit of resolution, represented by $\delta\lambda/\lambda$, is equal to $1/N$, since $1/N$ is the fractional displacement of the first minimum from the principal maximum. For the order n, the fractional displacement is $1/nN$. We therefore arrive at the result that the resolving power – the reciprocal of the limit of resolution – of a grating is equal to nN.

This result shows that the resolving power obtainable does not depend solely upon the number of lines; if a coarse grating is made, a higher order can be used and the resolving power may be as good as that of a finer grating. If L is the total length of the grating the spacing is L/N, and the resolving power is equal to

$$Nn=L\sin\theta/\lambda. \tag{7.105}$$

Thus for a given angle θ, the resolving power depends upon the length L, and not upon the spacing d. Gratings should therefore be made as long as possible. The *reductio ad absurdum* argument that we might as well use

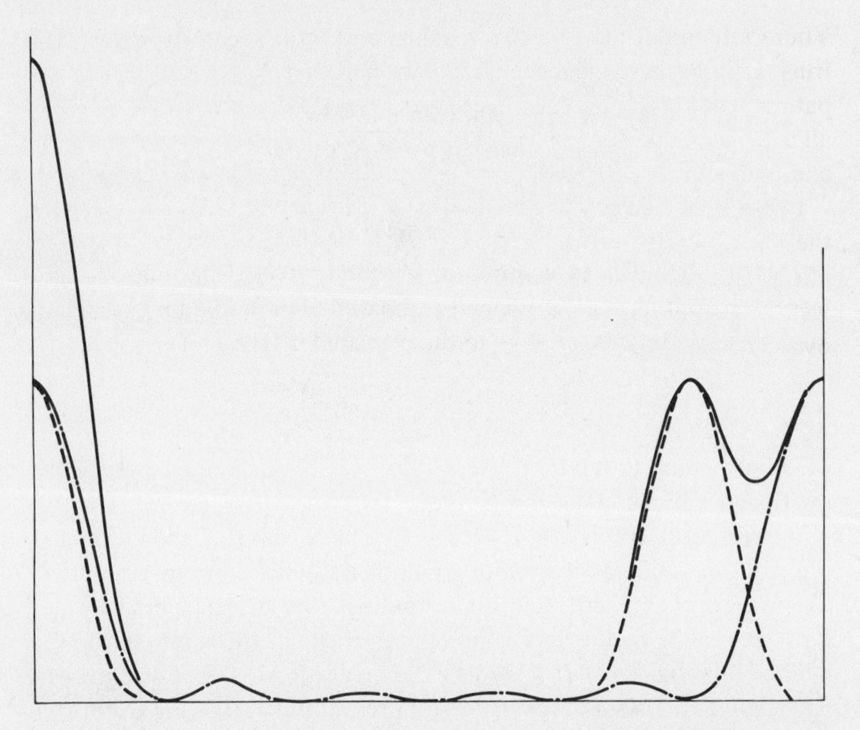

Fig. 7.47. The addition of two diffraction-grating functions (Fig. 7.23) for different wavelengths, showing resolution of the two wavelengths according to the Rayleigh criterion.

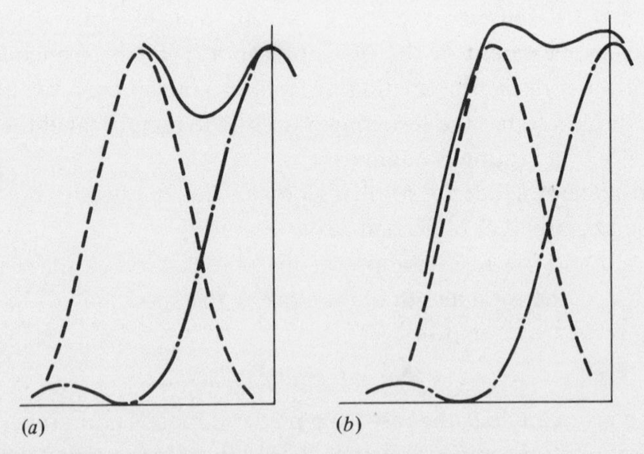

Fig. 7.48. Addition of principal maxima of diffraction grating functions for (*a*) 0.9 and (*b*) 0.8 of the Rayleigh criterion.

Young's fringes from two slits a distance L apart is not unsound; such fringes would be so fine that they *would* give high resolution, but the pattern would be unintelligible! (Nevertheless, this is the principle behind all two-beam interferometry – including the Michelson interferometer – and the patterns *can* be interpreted by means of a computer.)

In practice, it is rare for a grating to have a resolving power equal to the theoretical value; errors in ruling are inevitably present and they can affect the performance considerably. With a very good grating and well-corrected (Appendix II) optical components, resolving powers of over half the theoretical value can be obtained, but this is unusual.

7.7.3 *Effects of periodic errors.* There is one type of error that does not affect the resolving power but is nevertheless undesirable for other reasons; this is a periodic error in line position. It can arise from a poor screw or by a badly designed coupling between the screw and the table carrying the grating (§ 7.7.1), and has the effect of enhancing some of the secondary maxima.

One possible treatment is to consider that an error in line position is repeated every qth line; the true spacing is qd, and therefore q times as many orders will be produced. Most of them will be very weak, but some may be strong enough to be appreciable compared with the main orders. More generally, we may deduce the interference pattern of a grating in which the positions of the lines vary sinusoidally; this is the method that we shall use here.

Let us suppose that the position of the pth line is

$$x_p = pd + \epsilon \sin 2\pi p/q,$$

where ϵ is small compared with d. Then the series that we have to sum is:

$$\begin{aligned}\sum_p \exp(-\mathrm{i}ux_p) &= \sum_p \exp\{-\mathrm{i}u(pd + \epsilon \sin 2\pi p/q)\} \\ &= \sum_p \{\exp(-\mathrm{i}upd)\exp(-\mathrm{i}u\epsilon \sin 2\pi p/q)\}. \quad (7.106)\end{aligned}$$

If ϵ is small compared with d, this summation is equal to

$$\sum_p \{\exp(-\mathrm{i}upd)\cdot(1 - \mathrm{i}u\epsilon \sin 2\pi p/q)\}. \quad (7.107)$$

The first term represents the ordinary diffraction-grating summation (7.50); the second term represents the effect of the sinusoidal error. Thus

(7.107) can be written as

$$\sum_p \exp(-iupd)+\sum_p \exp(-iupd)\left[-\frac{u\epsilon}{2}\left\{\exp\left(\frac{2\pi ip}{q}\right)-\exp\left(-\frac{2\pi ip}{q}\right)\right\}\right]$$

$$=\sum_p \exp(-iupd)-\frac{u\epsilon}{2}\sum_p \exp\left\{-ip\left(ud+\frac{2\pi}{q}\right)\right\}$$

$$+\frac{u\epsilon}{2}\sum_p \exp\left\{-ip\left(ud-\frac{2\pi}{q}\right)\right\}$$

$$=\sum_m \delta\left(u-\frac{2\pi m}{d}\right)-\frac{u\epsilon}{2}\sum_m \delta\left\{u-\frac{2\pi}{d}\left(m+\frac{1}{q}\right)\right\}$$

$$+\frac{u\epsilon}{2}\sum_m \delta\left\{u-\frac{2\pi}{d}\left(m-\frac{1}{q}\right)\right\}, \tag{7.108}$$

where m is the order of diffraction defined by (7.51).

The summations show that, in addition to the principal maxima, there are also maxima at angles given by the two quantities $m+1/q$ and $m-1/q$. That is, each order m is flanked by two lines (satellites) at a separation of $1/q$th of the orders in reciprocal space (Fig. 7.49(a)). The intensities of these lines are proportional to ϵ^2, and to $\sin^2\theta$; thus they should increase rapidly with order. They should not exist around the zero order.

If ϵ is not small compared with d, we should have to include further terms in expression (7.108); we should find that further satellites at $2/q$, $3/q, \ldots$ would appear. But since we are concerned here only with basic principles, it is not worth while taking the theory further.

Another type of periodic error can occur if the scratches on the grating are made in the correct places but have periodically varying intensity. The amplitudes of the terms in (7.50) are then no longer constant, but have to be replaced by such an expression as $A+B\sin 2\pi p/q$. The manipulation

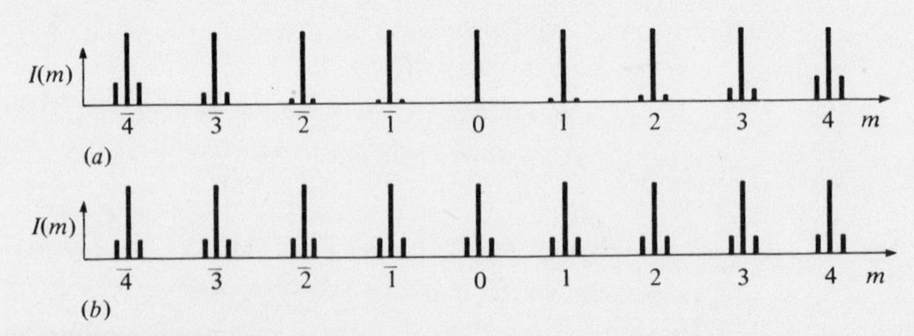

Fig. 7.49. Representations of sidebands from periodically deformed gratings. (a) Result of periodic error in line position; (b) result of periodic error in line intensity.

of the summation will be left to the reader, who should find that satellites again arise, but they do not now depend upon sin θ. Thus all the orders, including the zero order, are flanked by satellites (Fig. 7.49(*b*)).

Further complications can arise if both the effects co-exist; then the satellites become of unequal intensity on each side of the order that they accompany. Effects of this sort might seem to be of purely academic interest, but they can in fact arise in alloys, where heavier and larger atoms may upset the regularity of a matrix of lighter atoms. The effect may not be periodic, but the analysis of the periodic problem provides a useful start for the more general one.

Random errors in gratings also occur. We shall not treat them here, but simply point out that the treatment and results are similar to those for thermal disorder in crystals, discussed in § 11.4.3.

7.7.4 *Diffraction efficiency*. The discussion of gratings has so far concentrated on the interference function, the Fourier transform of the set of δ-functions representing the positions of the individual lines. This transform has now to be multiplied by the diffraction function, which is the transform of the individual apertures.

Let us first consider a transmission grating for which the lines are slits, each with width b (which must be less than their separation d). The diffraction function is then the transform of such a slit (equation (7.4)),

$$\psi(u) = b \operatorname{sinc}(bu/2). \tag{7.109}$$

At the order m, $u_m = 2\pi m/d$. For the first order, as b is varied, the maximum value of $\psi(u_1)$ is easily shown to occur when $b = d/2$; thus the optimum slit width is half the spacing. But even with this value the efficiency of the grating is dismally small. The light power P_m reaching the various orders is proportional to the values of $|\psi(u_m)|^2$ namely,

$$P_0 \propto d^2/4, \qquad P_1 \propto d^2/\pi^2, \qquad P_2 = 0, \text{ etc.} \qquad (\text{for } b = d/2). \tag{7.110}$$

The constant of proportionality can be deduced by putting $b = d$, whereupon the grating becomes completely transparent. Then all the incident light falls in the zero order and we have powers

$$P_0' \propto d^2, \qquad P_i' = 0 \quad (i > 0) \quad (\text{for } b = d). \tag{7.111}$$

The *diffraction efficiency* is defined as the fraction of the incident light diffracted into the strongest non-zero order. Maximized by the choice of $b = d/2$, it still only reaches $P_1/P_0' = \pi^{-2}$, about 10 per cent. The figure can be improved slightly, in theory, by choosing a slit with a sinusoidal profile, but the difficulty of producing such slits prohibits their use in practice.

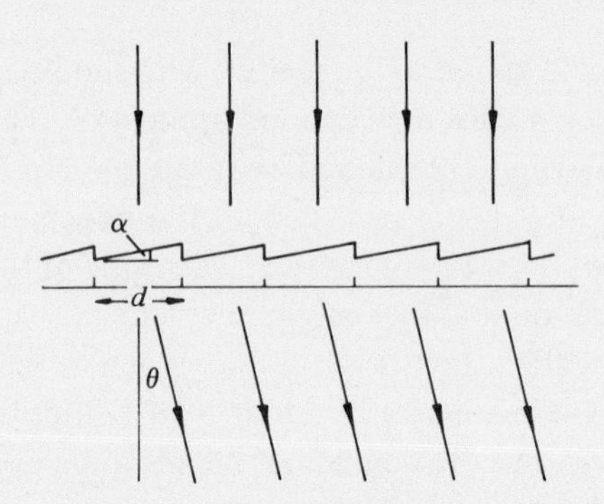

Fig. 7.50. Blazed transmission grating. The value of θ must satisfy the two equations $n\lambda = d \sin \theta$ and $\theta = (\mu - 1)\alpha$.

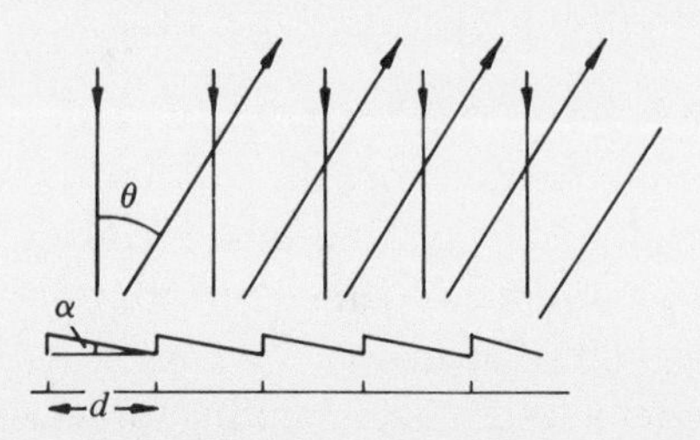

Fig. 7.51. Blazed reflexion grating. The value of θ must satisfy the two equations $n\lambda = d \sin \theta$ and $\theta = 2\alpha$.

7.7.5 *Blazed gratings.* The discussion in the previous section shows us how inefficient an amplitude transmission grating must necessarily be. To improve the situation we must allow the 'slit' to include phase changes too, and Lord Rayleigh put forward what he thought was a hypothetical idea, of combining the effects of refraction or reflexion with interference to make a grating in which most of the intensity is concentrated in one particular order. The principle is illustrated by Figs. 7.50 and 7.51. Each element in the transmission grating shown in Fig. 7.50 is made in the form of a prism, of which the angle is such that the deviation produced is equal to the angle of one of the orders of diffraction; correspondingly, in Fig. 7.51 a reflexion grating is shown in which each element is a small mirror.

Such gratings can now be made and are used very considerably. Instead of using any available sharp diamond edge for ruling a grating, a special edge is selected that can make optically-flat cuts at any desired angle. Gratings so made are called 'blazed gratings'. It should be noted that a diffraction grating can be blazed only for one particular order and one wavelength, and it can therefore be used only for a restricted wavelength region. The angle α (Figs. 7.50 and 7.51) fixes the angle between the zero order and the direction of maximum intensity, and ideal blazing is obtained only at the wavelength for which this is equal to the angle of diffraction.

It is instructive to consider blazing of gratings from the Fourier-transform point of view. Suppose that we make a grating of slits whose width is equal to the spacing of the grating; obviously, we have merely made a large rectangular aperture, whose diffraction pattern is confined to small angles – the zero order of diffraction. This result arises because the other orders of diffraction must all fall at zero positions on the transform of the single slit (equation (7.111)). But if we now introduce a phase variation (compare § 7.2.5) along the width of each slit – either by reflexion or transmission – then the transform will be displaced. Blazing consists in displacing the transform so that its peak coincides with a particular order, usually the first (Fig. 7.52).

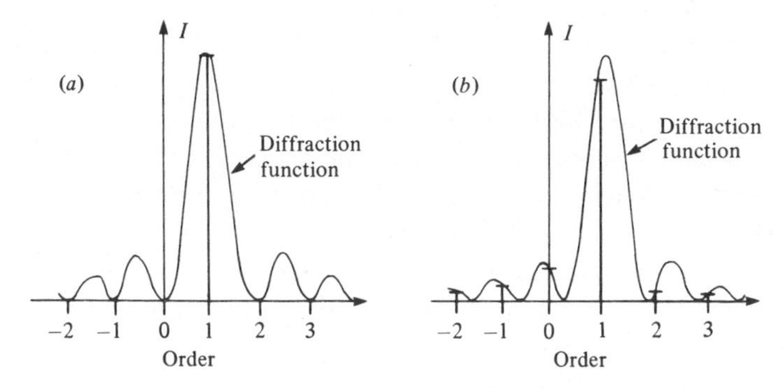

Fig. 7.52. Diffracted intensity in the orders of a blazed grating; (*a*) at the wavelength for which the blazing was designed (all energy goes, theoretically, into the +1 order); (*b*) at a slightly different wavelength (the +1 order predominates, but other orders appear weakly).

The matching cannot be exact because the width of each element cannot be exactly equal to the spacing. Moreover, the shift of the transform depends upon the wavelength of the radiation, as we saw in the last paragraph.

7.7.6 *Echelon gratings.* We have seen in § 7.7.2 that the resolving power does not depend directly upon the spacing of a grating or upon the number of lines; it depends on their product – the length L of the grating. It should therefore be possible to obtain a high resolving power by a grating with relatively few elements of large spacing. At the same time we

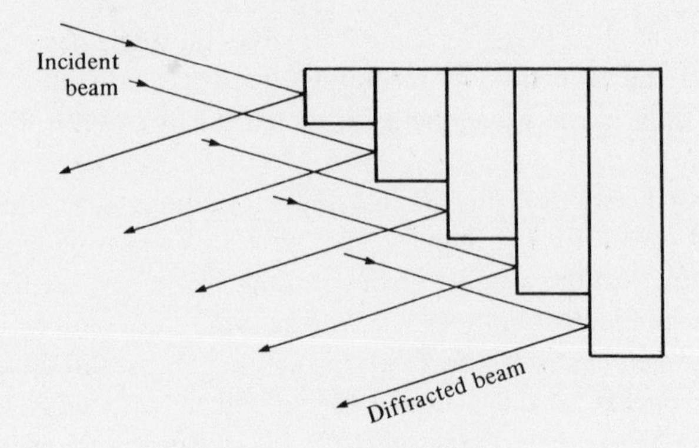

Fig. 7.53. Echelon grating.

can, in effect, make use of the 'blazing' technique described in the last section.

Michelson first made such gratings, and now they are in common use. A typical construction is shown in Fig. 7.53; this is a reflexion grating made from precisely equal flat plates, accurately stepped back in sequence. A typical construction would have a step depth of the order of 1 mm and 'treads' somewhat less. Such a grating can give orders of diffraction up to about 2000. Resolving powers of the order of 10^5 are attainable. Echelon gratings can also be used in transmission.

Echelon gratings are useful in investigating fine structures, but not as primary means of investigating spectra; the various orders of diffraction are very close together, and consequently the patterns obtained are rather complicated. In the visible region, of course, the colours of the lines establish their identities, but in the infra-red, where echelon gratings are of most use, the patterns have to be photographed, and the spectral lines are not easily recognized.

7.8 Interferometers

The phenomenon of interference cleared up a fundamental problem in optics, but at the same time, as is usual in physics, it opened up completely new fields. For example, the experiment of Young's fringes (§ 7.1) enables us to measure distances of the order of 10^{-5} cm. If such a simple piece of apparatus has such capabilities, what could we not do with properly-designed equipment? Multiple-beam interference (§ 7.5) can radically improve the accuracy obtainable, to the order of 10^{-7} cm, which is of molecular dimensions. Diffraction gratings can have resolving power of the order of 10^6 (§ 7.7.2); thus wavelengths differing by 10^{-10} cm can be separated, and measured perhaps to 10^{-11} cm. Thus the simple experiments opened up a world of accuracy, and of measurements of small dimensions, that was unthinkable before the nature of light was understood.

These great steps forward could not, however, be accomplished without careful work and minute attention to experimental detail, resulting in instruments known generally as interferometers. In the next few sections we shall discuss the basic principles of these instruments, giving only the essential minimum of practical detail. Their uses, and the influence that they have had on the rest of physics, will be discussed in Chapter 11.

7.8.1 *Rayleigh refractometer.* Since the Young's slit experiment produces interference fringes from two separate beams, it should be possible to vary the conditions in one of the beams and to observe the change in the fringe pattern. Lord Rayleigh, for example, showed that it was possible in this way to compare refractive indices that are very close to unity – for example, the refractive indices of gases.

Since the two beams must be separated over appreciable distances, it is first of all necessary to have a rather large slit separation – of the order of 1 cm; also, the beams must be parallel and so a lens (Fig. 7.54) has to be

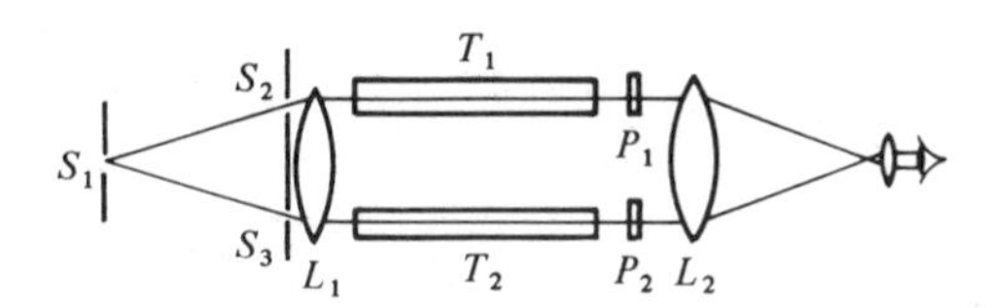

Fig. 7.54. Rayleigh refractometer. S_2 and S_3 are slits illuminated by light from the slit S_1, L_1 is the collimating lens, T_1 and T_2 are the two tubes, P_1 and P_2 are compensating plates which can be tilted in order to adjust the position of the fringes, L_2 brings the two beams to a focus. (From Thewlis, 1958.)

used to give parallel light from the source. The two beams then pass through separate tubes and are caused to interfere in the focal plane of a second lens.

With a large separation of the slits, the fringes are very fine; if the maxima are separated by 5×10^{-5} radians, with a second lens of focal length 20 cm the fringes are only 10^{-3} cm apart. Their measurement therefore requires a high-power microscope, with correspondingly small intensity. Morever, the centre fringe is not easily identified, and the use of

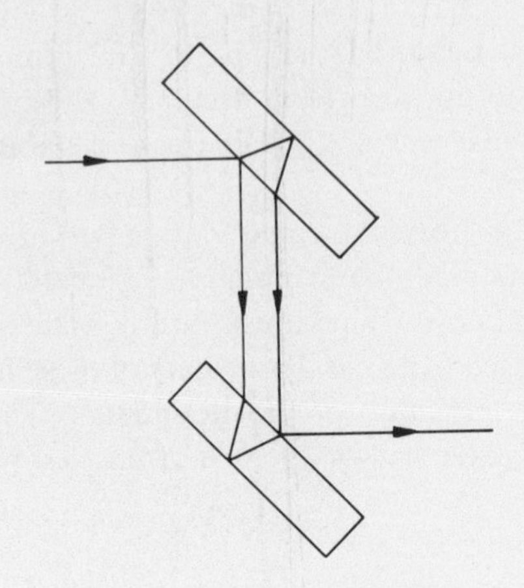

Fig. 7.55. Jamin interferometer.

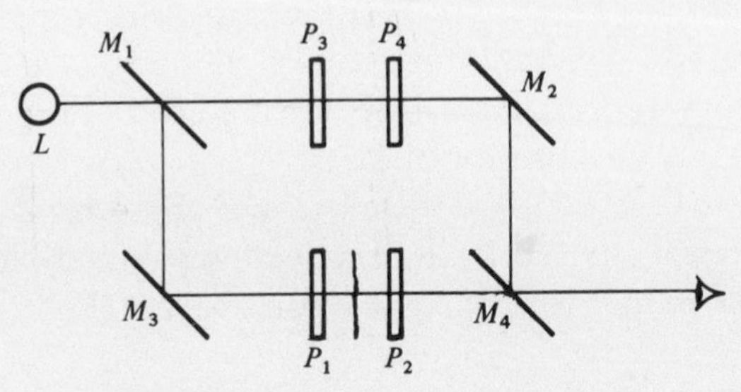

Fig. 7.56. Mach–Zehnder interferometer. L is the source, M_1, M_2, M_3, M_4 are the mirrors, P_1, P_2, P_3, P_4 are plates enclosing spaces in which the optical paths are to be compared. (From Thewlis, 1958.)

white light as a preliminary is helpful; the centre fringe is white and the others are coloured.

With this instrument Rayleigh was able to show that the refractive index of a gas varied linearly with pressure, and measured refractive indices of gases to about 1 in 10^6. Variations on the apparatus have been made by Jamin, who produced two beams by internal and external reflexion from a glass plate (Fig. 7.55), and by Mach and Zehnder, who used separate reflexions from mirrors (Fig. 7.56). In some ways, the last interferometer is more akin to the Michelson interferometer, which will be described in the next section.

7.8.2 *Michelson interferometer.* The Michelson interferometer differs from the others so far described in that it produces beams that are not only widely separated but are also directed into directions at right angles.

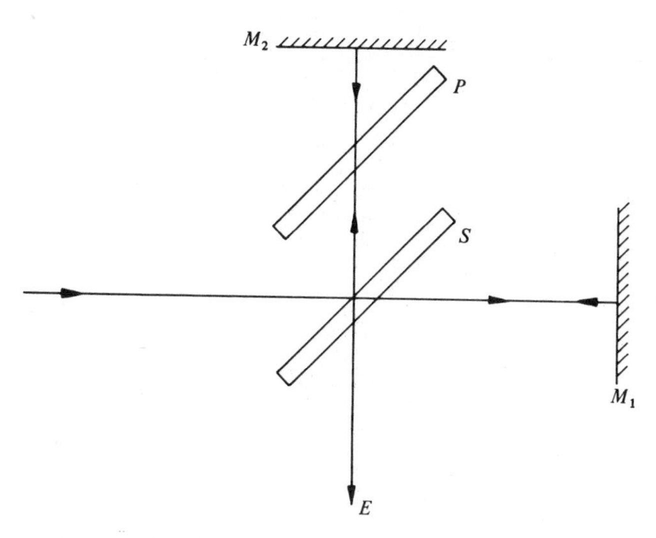

Fig. 7.57. Michelson interferometer.

These two features make it a very versatile instrument and, with its modifications, it is the best known of all the interferometers. It should not be confused with the Michelson stellar interferometer, which is described in § 8.6.6.

The principle is illustrated in essence in Fig. 7.57. Light enters from the left and is partly reflected and partly transmitted by the semi-silvered mirror S; the two beams are reflected from the mirrors M_1 and M_2, and the resultant interference fringes are observed by the eye at E. Since the rays reflected from M_1 have to pass through three thicknesses of the mirror S, whereas those reflected from M_2 have to pass through only one, an extra plate at P is inserted to give equality between the two paths. This plate is needed because glass is dispersive and so only by having the same amount of the same glass in both beams can the optical paths be made equal at *all* wavelengths. This compensating plate must be of the same thickness as S and placed at the same angle.

With the Michelson interferometer many different sorts of fringes can be obtained – straight, curved, or completely circular, in monochromatic or white light. These can all be understood in terms of a single theory if we regard the problem as a three-dimensional one, the different sorts of fringes resulting from looking at the same pattern from different directions.

The source must be a broad one, but we can simplify the understanding of the interferometer by considering one 'ray' at a time, coming from one

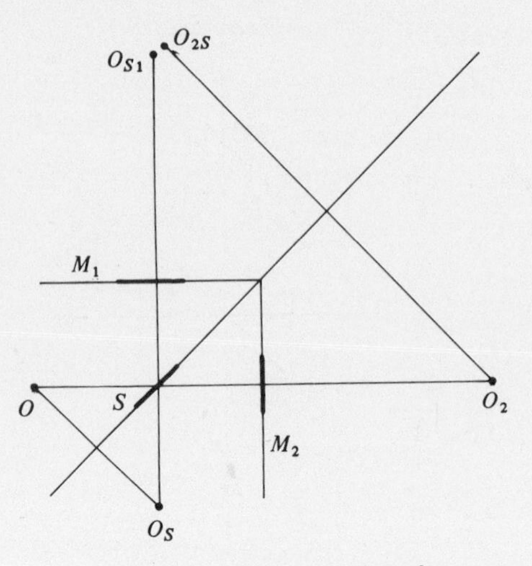

Fig. 7.58. Principle of Michelson interferometer.

point on the source. If we ignore the finite thickness of the components, we see from Fig. 7.58 that O has an image in S at O_S and an image at O_2 in M_2; O_S has an image O_{S1} in M_1, and O_2 has an image O_{2S} in S. The images O_{S1} and O_{2S} are the two virtual sources that give rise to interference.

It can easily be seen that O_{S1} and O_{2S} can be brought as closely together as we require. Small adjustments in M_1 and M_2 can change their relative positions, and it is even possible to have one behind the other. The different sorts of fringes arise from the relative positions of O_{S1} and O_{2S}, and the scales of the fringes depend upon their separation.

All these facts can be understood by considering the Fourier transform (§ 3.4.1) of the two points; this is a set of planar sinusoidal fringes, represented with artificial limits (they would extend to infinity for purely monochromatic radiation if the two sources were precise points) in Fig. 7.59. The different fringes observed are different aspects of this Fourier transform.

To understand this statement we make use of the concepts introduced in § 7.6.1 – particularly the sphere of observation. Now, however, we are dealing with coherent *sources* and not scatterers so that their phase difference is always zero and does not depend upon an incident beam $\mathbf{k}_0$. The only vector that is involved is $\mathbf{k}$, and this, being of constant length $2\pi/\lambda$, describes a sphere with the origin of reciprocal space at its centre. The sphere of observation therefore penetrates the Fourier transform, with its centre on a maximum (Fig. 7.59).

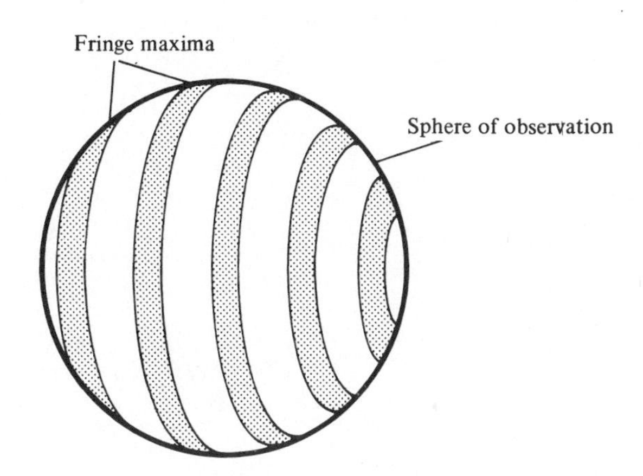

Fig. 7.59. Representation of Fourier transform of two points – plane sinusoidal fringes – cutting sphere of observation.

As the points O_{S1} and O_{2S} become closer, the scale of the fringes becomes larger, and as the disposition of the points changes the transform rotates into different orientations. Fig. 7.60 shows how different types of fringes arise. If white light is used, the sphere of observation must be considered as having a finite thickness; as we can see from Fig. 7.60(*d*), the centre fringe only is sharp and the others are coloured and soon merge together. Such coloured fringes are the best way of identifying the zero-order interference.

So far we have considered only point O on the source, and it might be thought that we have shown that fringes are produced from this one point. In fact, what we have shown is that different intensities are produced in different directions of $\mathbf{k}$; and each point on the observer's retina, or on a photographic plate, corresponds to such a direction. But the ray from O to the eye or camera lens also selects a well-defined direction, and so what one sees is an image of the source with each point giving its own contribution to the interference pattern. The fringes are therefore localized in space (§ 7.1.7) in the image of the source produced by reflexion in S and M_1 or M_2 and S.

7.8.3 *Fabry–Pérot interferometer.* In complete contrast to the Michelson interferometer is the Fabry–Pérot plate or etalon which, instead of versatility, aims only at high resolution. It achieves this by using multiple reflexion (§ 7.5.1). The interferometer consists simply of two optically worked, precisely parallel, glass plates (Fig. 7.61).

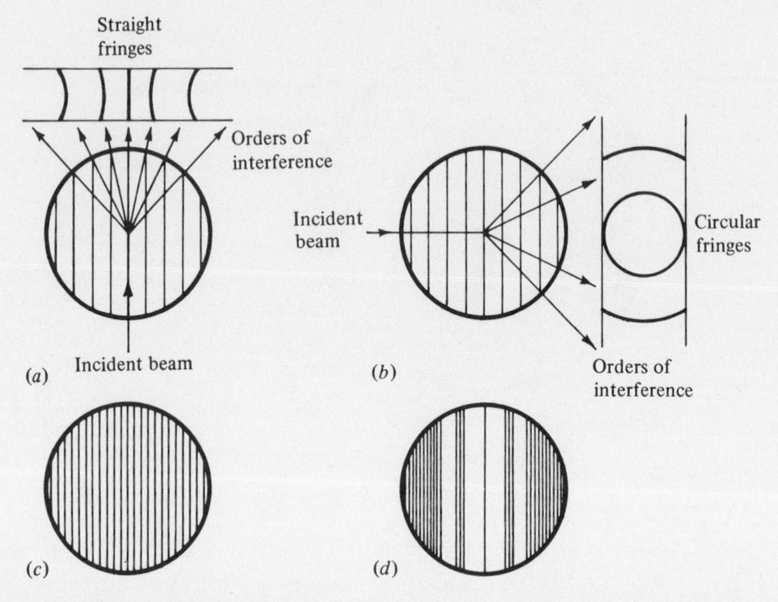

Fig. 7.60. Different types of fringes from the Michelson interferometer: (*a*) and (*b*) show how straight and circular fringes are produced, and obviously intermediate types of fringes are possible; (*c*) shows how the fringes become finer as O_{S1} and O_{2S} move further apart; (*d*) shows how broadened fringes are produced if a range of wavelengths (e.g. white light) is used.

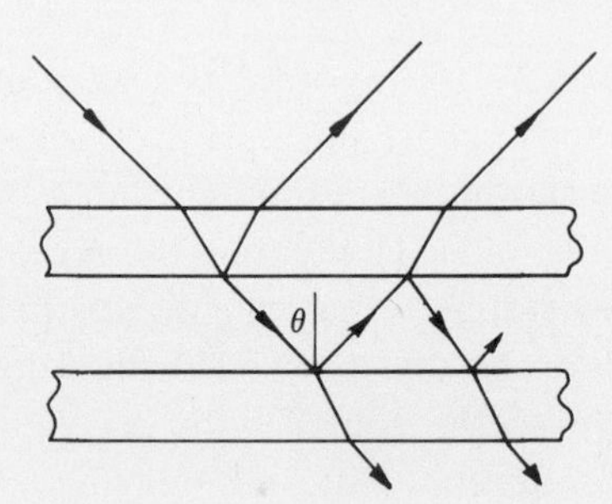

Fig. 7.61. Fabry–Pérot etalon.

We have seen (§ 7.1.6) that the condition for production of an interference maximum is that

$$2t \cos \theta = m\lambda, \tag{7.112}$$

where θ is the angle of inclination of the beam within the air space between the plates. If the plates are silvered the directions θ become very well defined and a high resolution results.

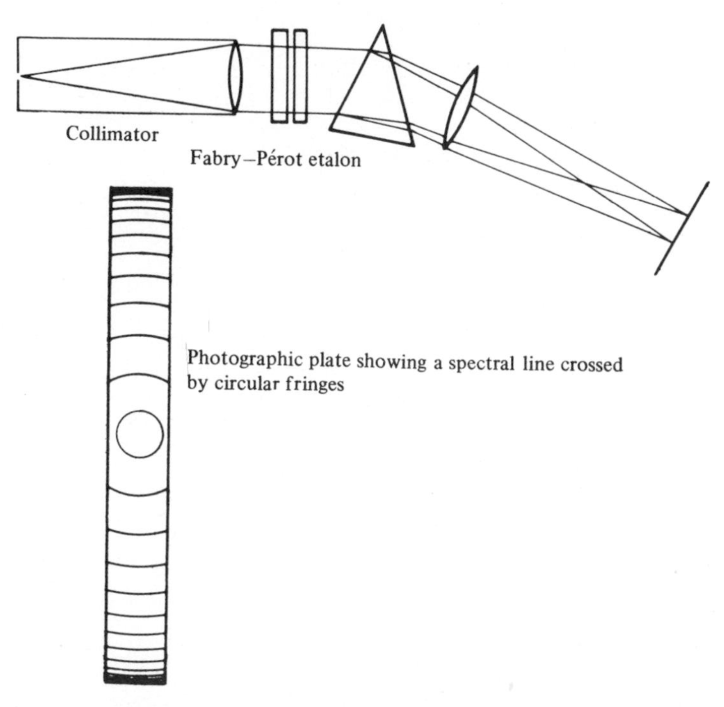

Fig. 7.62. Use of Fabry–Pérot etalon for high-resolution spectroscopy.

One of the simplest ways of using the Fabry–Pérot etalon is to insert it in front of the dispersing system (grating or prisms) on a spectrometer (Fig. 7.62). The spectral lines are then crossed by circular fringes that satisfy equation (7.112) and thus any fine structure becomes evident. If only one wavelength is present, only one set of fringes is observed, but if there are two or more wavelengths – too close together of course to be resolved by the dispersing system – more complicated groupings are observed, as shown diagrammatically in Fig. 7.63.

7.9 Some practical applications

Interferometry is one of the most useful branches of optics, in both pure and applied physics. Some applications, such as the determination of the refractive indices of gases by the Rayleigh refractometer, have already been mentioned (§ 7.8.1); others, of fundamental importance to the development of physics – varying from the standardization of the metre to applications in radio-astronomy – will be dealt with in Chapter 11. Here we shall deal only with a few practical matters that cannot strictly be

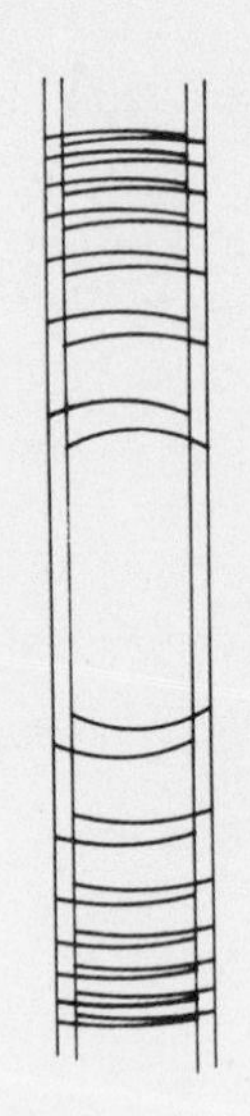

Fig. 7.63. Diagrammatic representation of pattern produced by Fabry–Pérot etalon for a close doublet. The straight lines represent the images of the edges of the slit of the spectrometer.

classed as fundamental but that nevertheless deserve some mention in a textbook on optics.

7.9.1 *Non-reflecting films.* When light passes from one medium to another, a certain amount of light is reflected, depending upon the refractive indices of the two media. For certain purposes – notably in image formation, where the reflected light reduces the intensity of the image and also may produce unwanted light elsewhere – it would be helpful if the reflexion could be eliminated. This can be done fairly well by depositing upon the surface a thin layer with appropriate refractive index and of a thickness such that it produces reflected beams from its two surfaces that are roughly equal in intensity and have a phase difference of π.

There is obviously no general solution that is appropriate to all wavelengths and all angles of incidence. We therefore concentrate upon the middle of the visible spectrum – say, 5500 Å – and normal incidence. To produce beams of equal intensity, equation (4.38a) shows that the refractive index of the film must be $(\mu_1\mu_2)^{\frac{1}{2}}$, where μ_1 and μ_2 are refractive indices of the two media. For a glass lens in air we therefore require a medium of refractive index of about 1.25. Its optical thickness must be $\lambda/4$ – about 1600 Å – and its actual thickness about 1300 Å.

It is impossible to find a material that has the required general properties – hardness, stability, adhesion to glass – with such a low refractive index, and materials such as magnesium fluoride ($\mu = 1.38$) are the best that can be obtained.

The process of depositing such a film is called 'blooming'; it is an example of impedance-matching (§ 4.2.5). It is carried out by evaporating the material *in vacuo* onto the glass surface, continuing until a characteristic purple coloration is seen; this colour results from the combination of blue and red resulting from the elimination of the middle of the spectrum.

7.9.2 *Multilayer dielectric mirrors.* In the previous section (§ 7.9.1) we described the use of a single dielectric layer to decrease the reflexion coefficient of a glass surface. The reflexion coefficient was reduced because we were able to break the reflected wave into two parts of approximately equal amplitude having a phase difference of π which interfered destructively.

The same principle can be applied to more complicated systems. An important example is the production of highly reflecting surfaces by the deposition of a series of layers of dielectric material. The result can be achieved once again by using layers of optical thickness $\mu d = \lambda/4$, but one now requires *constructive* interference between the various reflected layers. This is achieved when the layers have alternately high and low refractive index. The amplitude reflexion coefficient $(\mu_1 - \mu_2)/(\mu_1 + \mu_2)$ is then fairly large (but never more than about 0.25), and changes sign at alternate interfaces. The latter fact, together with the extra path length of $\frac{1}{2}\lambda$ for each successive wave, ensures constructive interference between all the reflected waves. The complete theory, taking into account multiple reflexions, is more complicated than this, but leads us to the same conclusion. The reflectivity of such a system is high only at wavelengths around four times the optical thickness of the individual layers, so that it is a wavelength-selective mirror, but its amplitude reflexion coefficient approaches unity approximately as

$$R \sim \frac{\mu_H^{2m} - \mu_L^{2m}}{\mu_H^{2m} + \mu_L^{2m}} \approx 1 - 2\left(\frac{\mu_L}{\mu_H}\right)^{2m}, \tag{7.113}$$

where μ_L and μ_H are the refractive indices of the two types of layer, and m is the number of pairs. Notice the formal similarity between (7.113) and (4.38a) for a single interface. For example, using ZnS ($\mu_H = 2.32$) and

MgF_2 ($\mu_L = 1.38$) in five pairs of layers, the amplitude and intensity reflexion coefficients R and R^2 are respectively

$$R = 99.35 \text{ per cent and } R^2 = 98.7 \text{ per cent.}$$

The latter value should be compared with the intensity reflexion coefficient for silver and aluminium layers evaporated on glass which reach 92–95 per cent at visible wavelengths. Multilayer mirrors are particularly used for laser resonators (§ 7.5.3), where any losses have to be offset by increased amplification in the lasing material, and for high-power optical systems where reflexion losses cause heating of the mirrors. In addition, more complicated arrangement of layers of different thicknesses and refractive indices allows great variety in the reflectivity as a function of wavelength, and almost any reasonable function is achievable in practice.

7.9.3 *Interference filters.* A filter is a system which transmits certain frequencies or wavelengths, and rejects others. Although many types of optical filter are available as coloured glass or other materials, their properties are mainly determined by the chance properties of these materials, and cannot usually be designed to achieve special characteristics. The use of multilayer dielectric systems also allows us to design filters with almost any required character. We shall only mention here one important class of filter: a narrow-band filter which transmits a certain wavelength λ and the region surrounding it ($\lambda \pm \delta\lambda$) to the exclusion of all other wavelengths. The use of multilayers for this purpose is based on the Fabry–Pérot interferometer (§§ 7.5.1 and 7.8.3). We saw there that the intensity transmitted by such an interferometer is generally almost zero, but peaks to a high value whenever the thickness of the plate is an integral number of half-wavelengths (Fig. 7.39). Moreover the half-widths of the transmission peak gets smaller as the reflexion coefficient approaches unity, and the maximum intensity transmission coefficient is $T^4(1-R^2)^2$. The use of dielectric multilayer mirrors as the reflecting elements in a Fabry–Pérot interferometer makes both narrow linewidth and high peak transmission possible. First, reflexion coefficients very close to unity can be achieved (much closer than with metal films), and secondly there is no absorption in a system constructed entirely of transparent materials, so that $T^2 = 1 - R^2$ and the peak transmission is unity. Both statements in the previous sentence are limited practically by manufacturing tolerance.

As an example, we can work out approximately the properties of a filter of this type. The reflectors are made with layers of thickness $\lambda/4$; the

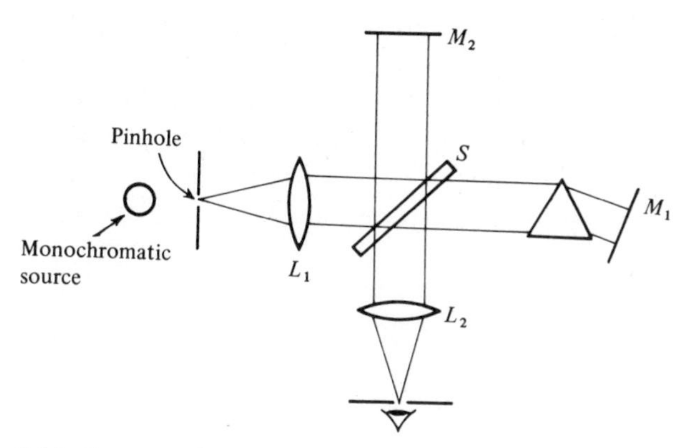

Fig. 7.64. Twyman–Green interferometer. (From Thewlis, 1958.)

spacer has thickness $\lambda/2$ or any multiple of this. If there is to be only one pass band in the visible spectrum, the spacer must have thickness $\lambda/2$. We denote quarter-wave layers of the high and low refractive index by H and L and the half-wave spacer by $2H$ and have a filter described conventionally by

$$(HL)^m 2H(LH)^m,$$

where each reflector has m pairs of layers. The half-width of the pass band of this filter will be, from (7.69)

$$\delta\lambda = -\lambda^2 \ln R/\pi s \approx 2\lambda^2(1-R)\pi s. \qquad (7.114)$$

In this equation s is the path difference between successive reflected beams, here $\lambda/2$, whence:

$$\delta\lambda/\lambda = \lambda(\mu_L/\mu_H)^{2m}/\pi. \qquad (7.115)$$

Substituting the numbers used previously for μ_L, μ_H and m we have

$$\delta\lambda/\lambda = 1/450; \text{ for } \lambda = 5000 \text{ Å}, \quad \delta\lambda = 11 \text{ Å}.$$

This is, of course, a calculated specification and will not be achieved exactly in practice, because of inaccuracies in the exact thicknesses of the twenty-one thin films involved. It is, however, typical of what can be achieved with interference filters.

7.9.4 *Twyman–Green interferometer.* The Michelson interferometer can be adapted for use as a testing device for optical components. Two lenses L_1 and L_2 (Fig. 7.64) are inserted in the position shown so that the incident light is parallel and the emergent light is observed in only one direction. Under these conditions, as we have seen in § 7.8.2, only a uniform field will be seen in the ordinary Michelson interferometer. If an optical component such as a prism is introduced, the field will remain uniform if the prism is perfect, but otherwise defects will show up as variations of light intensity. Usually, however, the instrument is adjusted to give a moderate number of fringes and the defects of the prism show up as a bending of the fringes.

8

Coherence

8.1 Introduction

The coherence of a wave describes the accuracy with which it can be represented by a pure sine wave. So far we have discussed optical effects in terms of waves whose wave-vector **k** and frequency ω can be exactly defined; in this chapter we intend to investigate the way in which uncertainties and small fluctuations in **k** and ω can affect the observations in optical experiments. Waves which appear to be pure sine waves if they are observed only in a limited space or for a limited period of time are called *partially coherent waves*, and we shall devote the first part of this chapter to developing measures of the deviations of such impure waves from their pure counterparts. These measures of the coherence properties of the waves are functions of both time and space; we shall therefore define two distinctly different criteria of coherence. The first criterion expresses the correlation to be expected between a wave now and a certain time later; the other between here and a certain distance away. Theoretically these two criteria are related, but it is very convenient to consider them independently in order to get a clear understanding of their meaning. Fig. 8.1 illustrates, in a very primitive manner, one wave which is partially coherent in time (it appears to be a perfect sine wave only when observed for a limited time) and a second wave which is partially coherent in space (it appears to be a sinusoidal plane-wave only if observed over a limited region of its wavefront).

The understanding of the coherence properties of light has had numerous practical consequences. Amongst these are Michelson's and Brown and Twiss's methods of measuring the angular diameters of stars, and the technique of Fourier-transform spectroscopy. But the most outstanding influence on the concept of coherence has been made by the laser. These subjects form the basis of the second half of the chapter.

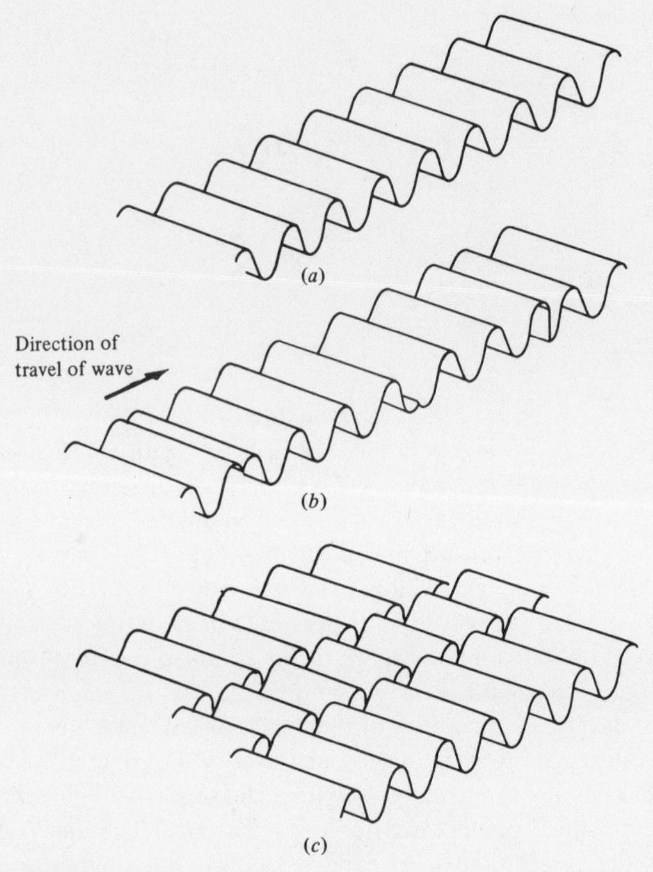

Fig. 8.1. Partially coherent waves: (*a*) perfectly coherent wave; (*b*) wave with space-coherence only; (*c*) wave with time-coherence only.

8.2 Temporal coherence

Let us try to define clearly what we know about a real light wave, from a classical monochromatic light source (not a laser). We know that the light we see at any moment is emitted by a number of atoms, each making a transition between the same pair of energy levels, but that the emission from any one atom is no way related to that from any other atom. In fact a careful spectroscopic analysis shows us that the light is not really monochromatic in the strict sense of the word; it contains components of various wavelengths within a certain range, called the 'linewidth'. Generally one uses the term 'quasi-monochromatic' for such radiation. The physical reasons for the linewidth are discussed in more detail in

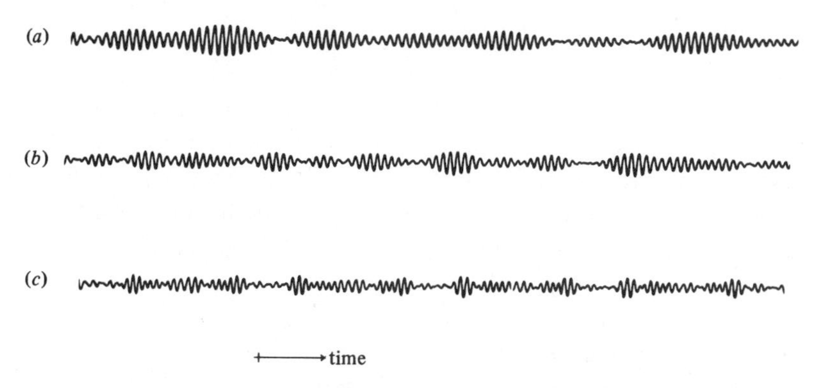

Fig. 8.2. Three impure sine waves, showing the amplitude fluctuations resulting from a spread in component frequencies. Each is generated from five components randomly distributed in the range $\pm\frac{1}{2}\epsilon$ about ω_0 where the values of ϵ/ω_0 are as follows: (*a*) 0.08, leading to groups containing about $1/0.08 = 12$ waves; (*b*) 0.16, leading to groups containing about 6 waves; (*c*) 0.25, leading to groups containing about 4 waves. As can be seen, the definition of a group containing so few waves is very hazy.

§ 8.3, but just to fix our ideas we might notice that at a finite temperature all the atoms in the emitting material (usually a gas) are moving randomly in various directions, and so the emission from each atom is Doppler-shifted by a different amount.

If we now ask exactly what the light wave looks like, we can answer the question by performing a Fourier synthesis based on the remarks in the previous paragraph. We take a number of sine waves, having frequencies randomly chosen within a specified range – the linewidth of the radiation – and add them together. We have done this in Fig. 8.2, where examples of continuous waves have been generated each from five sine waves with frequencies randomly chosen within a specified interval. What we see is a complicated beat phenomenon; the amplitude of the wave is not a constant, but fluctuates in a rather haphazard fashion. The average length of a beat is related to the range of frequencies involved.

The wave trains in Fig. 8.2 can also be looked at in a different way. We can consider each beat as an independent wave-group, and the complete wave train as the emission of a series of such wave-groups at random intervals. This description turns out to be convenient for some purposes, and we shall show in § 8.2.2 that the spectrum for such a process is precisely what we have described above as the spectrum of a quasi-monochromatic light source – it consists of a line of finite width and there is no phase relationship between the components at the various frequencies. The description must therefore be mathematically correct.

However, it does have a danger; the individual wave-groups must not be interpreted as photons. Apart from the fact that the model has been created by completely classical thinking, and therefore cannot produce a quantized particle, we must point out that the rate of occurrence of the wave-group is determined entirely by the spread of frequencies. Now if the wave-groups were photons, their average rate of occurrence would depend on the intensity of the wave, and not the linewidth. So despite the attractiveness of such an idea, it is not consistent with the facts.

8.2.1 *The amplitude and phase of quasi-monochromatic light.* Let us try to develop the ideas of this section a little further. We assume that the light beam is represented by the superposition of N waves of equal amplitude a and random phase ϕ_n having frequencies ω_n randomly chosen within the range $\omega_0 \pm \epsilon/2$. The amplitude and intensity of the combined wave are:

$$A(t) = a \sum_{n=1}^{N} \exp\{\mathrm{i}(\omega_n t + \phi_n)\}, \tag{8.1}$$

$$I(t) = |A(t)|^2 = a^2 \left| \sum_{n=1}^{N} \exp\{\mathrm{i}(\omega_n t + \phi_n)\} \right|^2, \tag{8.2}$$

which can be written as a double sum (cf. § 7.4.6)

$$I(t) = a^2 \sum_n \sum_m \exp[\mathrm{i}\{(\omega_n - \omega_m)t + \phi_n - \phi_m\}]. \tag{8.3}$$

What is noticeable about the waves in Fig. 8.2 is that their amplitude and phase fluctuate. This effect should be brought out by a comparison of long-term and short-term averages of (8.3) and (8.1); the average over a period T_0, long compared with the duration of a beat, should iron out the fluctuations, whereas that over a period T_1, short compared with the beat, should not. We can evaluate such averages:

$$[\overline{I(t)}]_T = T^{-1} a^2 \int_0^T \sum_n \sum_m \exp[\mathrm{i}\{(\omega_n - \omega_m)t + \phi_n - \phi_m\}]\,\mathrm{d}t. \tag{8.4}$$

When the integration time T is T_0, the term $(\omega_n - \omega_m)t$ can have any value up to ϵT_0, which according to the definition of T_0 can be as large as we like. Thus there is a tendency to cancellation of the oscillating parts of the integral, all of them going through many cycles at different rates during the long period T_0. Only the terms for which $n = m$, all of which have value $\mathrm{e}^{-\mathrm{i}0} = 1$, make a consistent positive contribution to the integral and thus

$$[\overline{I(t)}]_{T_0} = a^2 T_0^{-1} \int_0^{T_0} \sum_n 1\,\mathrm{d}t = a^2 N. \tag{8.5}$$

This equation simply tells us that, because the waves are uncorrelated, the intensity of their sum is equal to the sum of their intensities.

Now let us calculate the short-term average. For this, we write equation (8.4) again, with $T = T_1$. The situation in the integral will be different from that in the long-term average provided that T_1 is short enough for the maximum phase, $|(\omega_n - \omega_m)T_1|$, to be less than $\pi/2$, in which case none of the terms in the sum can be detected as oscillatory during the period T_1. This occurs if $\epsilon T_1 < \pi/2$. The sum can then be simplified by assuming all terms $\exp\{i(\omega_n - \omega_m)t\}$ to be approximately unity. Then

$$\begin{aligned}[\overline{I(t)}]_{T_1} &= a^2 T_1^{-1} \int_0^{T_1} \sum_n \sum_{m=1}^{N} \exp\{i(\phi_n - \phi_m)\}\,dt \\ &= a^2 N + 2a^2 \sum_n \sum_{m<n} \cos(\phi_n - \phi_m), \qquad (8.6)\end{aligned}$$

where in the last line we have separated the terms involving $n = m$ from the rest. We cannot predict the exact value of the second term. We just know that we will generally have a non-zero value, and that this value does not change during the interval T_1. Equation (8.6) therefore tells us that the intensity remains approximately constant during the interval T_1, but has a value differing from the long-term mean a^2N. It is interesting to estimate the size of the fluctuating second term of (8.6). Its *square* has expected value

$$4a^4 \cdot \tfrac{1}{2}N^2 \cdot \overline{\cos^2(\phi_n - \phi_m)} = a^4 N^2, \qquad (8.7)$$

and so the term fluctuates with root-mean-square amplitude a^2N. The fluctuations are therefore comparable with the mean intensity, also a^2N, even as $N \to \infty$. So we deduce that the short-term mean fluctuates macroscopically about the long-term mean a^2N. The critical time $\pi/2\epsilon$ which divides the long term from the short term is called the *coherence time* of the wave, τ_c, and is typically of the order of 10^{-9} s for classical light sources.

During the short period T_1 we can also show that the wave behaves very much like a pure sine wave. We write the amplitude $A(t)$ (8.1) in the form:

$$A(t) = a \exp(i\omega_0 t) \sum_n \exp\{i(\omega_n - \omega_0)t + i\phi_n\}. \qquad (8.8)$$

During the short term T_1, we have seen that $|(\omega_n - \omega_0)t| < |(\omega_n - \omega_0)T| < \epsilon T_1/2 < \pi/4$. Therefore the error introduced by writing $\exp\{i(\omega_0 - \omega_n)t\} \approx 1$ is always small. From (8.8),

$$\begin{aligned}[A(t)]_{T_1} &\approx a T_1^{-1} \exp(i\omega_0 t) \int_0^{T_1} \sum_n \exp(i\phi_n)\,dt \\ &= a \exp(i\omega_0 t) \sum_n \exp(i\phi_n). \qquad (8.9)\end{aligned}$$

The sum $\Sigma \exp(\mathrm{i}\phi_n)$ will in general have a non-zero value $\alpha_1 \exp(\mathrm{i}\Phi_1)$, where Φ_1 is quite indeterminate. We conclude that during the period T_1 we can represent the wave by a harmonic wave $\exp\{\mathrm{i}(\omega_o t+\Phi_1)\}$, whose phase Φ_1 is constant but unknown. If we were to repeat the measurement starting at time T_2 we should find (see (7.7)) the same result with a new phase Φ_2 defined by:

$$\alpha_2 \exp(\mathrm{i}\Phi_2)=\sum_n \exp\{\mathrm{i}(\phi_n+\omega_n T_2)\}, \tag{8.10}$$

which in general is quite unpredictably different from Φ_1 in view of ω_n's being randomly chosen. There is no correlation between the phases Φ_1 and Φ_2 measured in separate intervals of length T_1.

We conclude that measurements made during a short term T_1 indicate a simple harmonic wave of constant intensity. Observation over a long term T_2 shows the intensity to fluctuate and there to be no correlation between the phases measured during intervals separated by much more than T_1.

8.2.2 *The spectrum of a random series of wave-groups.* We have seen that the beat patterns in Fig. 8.2 can alternatively be described as a random succession of wave-groups. Such a series does indeed have the same spectral characteristics. Consider for example the Gaussian wave-group of § 3.6.6.

$$f(t)=A\exp(\mathrm{i}\omega_0 t)\exp(-t^2/2\sigma^2), \tag{8.11}$$

whose Fourier transform is (from (3.71))

$$a(\omega)=2\pi A(2\pi\sigma^2)^{\frac{1}{2}}\exp\{-(\omega-\omega_0)^2\sigma^2/2\}. \tag{8.12}$$

A random series of such groups is:

$$f_r(t)=\sum_{n=1}^{N} f(t-t_n), \tag{8.13}$$

where t_n is the random centre point of the nth wave-group. Now the transform of (8.13) is:

$$a_r(\omega)=a(\omega)\sum_n \exp(-\mathrm{i}\omega t_n)=a(\omega)\beta\exp\{-\mathrm{i}\psi(\omega)\}. \tag{8.14}$$

By equating the square moduli of the expressions in (8.14) and taking the time-average it is easy to show that $\beta^2=N$.

The factor $\beta\exp\{-\mathrm{i}\psi(\omega)\}$ has an indeterminate phase, $\psi(\omega)$, different for each value of ω, (cf. (8.10)). So we conclude that the spectrum is like that of the single wave-group (8.11), but has random phase. The spectral intensity $|a_r(\omega)|^2$, however, is well defined, having the value

$$|a_r(\omega)|^2 = \beta^2 |a(\omega)|^2 = N|a(\omega)|^2. \tag{8.15}$$

The series of wave-groups reproduces exactly the known spectrum and is therefore a good physical representation of the wave.

8.2.3 *White light.* One limiting case is of particular interest; if we make the wave-group shorter, so that it eventually becomes a δ-function pulse, the function $a(\omega)$ becomes a constant, independent of ω. Thus a series of δ-pulses emitted at random times transforms to a set of all frequencies occurring with random phases. This is white light; it contains all frequencies in equal amounts, and they are completely uncorrelated in phase.

8.2.4 *The temporal coherence function.* We have described in § 8.2.1 how we can usefully think of quasi-monochromatic light as a continuous wave with a random phase modulation. In addition, the amplitude fluctuates in a similar manner to the phase. To describe mathematically the 'randomness' of the wave we define a correlation coefficient $\gamma(\tau)$ between the complex wave amplitudes at t and $t+\tau$.

Following the usual definition of a correlation coefficient in statistical work we write

$$\begin{aligned}\gamma(\tau) &= \overline{f(t)f^*(t+\tau)}/(\overline{|f(t)|^2}\,\overline{|f(t+\tau)|^2})^{\frac{1}{2}} \\ &= \overline{f(t)f^*(t+\tau)}/\overline{|f(t)|^2},\end{aligned} \tag{8.16}$$

where the mean values (indicated by a bar) are evaluated by integrating for a time long compared with all fluctuations of the light beam. Notice that the term $\overline{|f(t)|^2}$ is just the mean intensity of the beam.

Now for a pure sine wave we have

$$f(t) = A\exp(\mathrm{i}\omega_0 t), \tag{8.17}$$

whence

$$\gamma(\tau) = \exp(-\mathrm{i}\omega_0\tau). \tag{8.18}$$

It is therefore usual to refer to $|\gamma(\tau)|$, whose departure from unity represents the departure of the waveform from a pure sinusoid, as the *temporal coherence function*, since the phase factor $\exp(\mathrm{i}\omega_0\tau)$ conveys no useful information about the statistical properties of the wave.

For a quasi-monochromatic source, $|\gamma(\tau)|$ has a typical form illustrated in Fig. 8.3. When $\tau = 0$, by definition $\gamma(\tau) = 1$. As τ increases, $|\gamma(\tau)|$ falls monotonically to zero on a scale determined by τ_c. This form is not typical for laser light, which will be discussed in § 8.7.3.

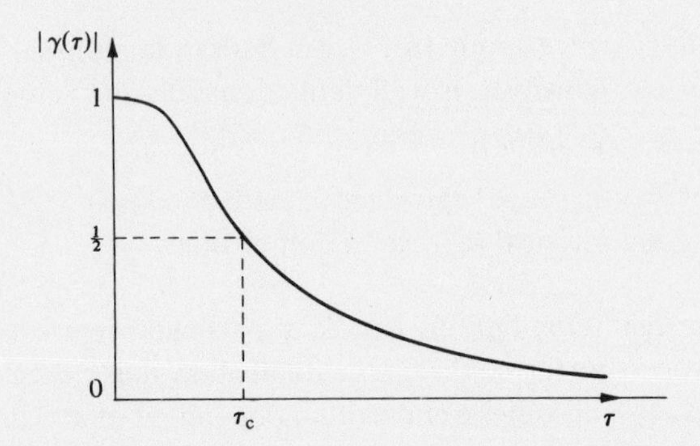

Fig. 8.3. Coherence function for a typical quasi-monochromatic source.

8.2.5 *The Wiener–Khinchin theorem.* The Fourier transform of $\gamma(\tau)$ (considering τ as the time-variable) is easily shown (§ 3.6.7) to be the spectral intensity $|a(\omega)|^2 = I(\omega)$ of the light, for unit intensity $|f(t)|^2$. This describes the spectrum one would see with a spectrometer on analysing the wave. Since the phase of the spectrum $a(\omega)$ is random, the spectral intensity $I(\omega)$ contains all useful information about the statistical properties of the light, and so does its transform $\gamma(\tau)$. The Fourier relationship between correlation function and spectral intensity is known as the *Wiener–Khinchin theorem.*

8.2.6 *Visibility of interference fringes, and its relationship to* γ. Suppose that we carry out an interference experiment in which fringes are formed between two samples from a light source, which are taken at times separated by an interval τ. This can be done, for example, by bringing the light from the source to two slits by different routes, one of which takes a time τ longer than the other. An important application of this idea will be discussed in § 8.5. In general, interference fringes can be formed between the waves radiated by the two slits. The *visibility* of the fringes can be defined as:

$$V = \frac{I_{\text{max}} - I_{\text{min}}}{I_{\text{max}} + I_{\text{min}}}. \tag{8.19}$$

If the fringes are clear, then $I_{\text{min}} \ll I_{\text{max}}$ and $V \approx 1$; if the fringes have poor contrast, then $I_{\text{min}} \approx I_{\text{max}}$ and $V \to 0$. When the two interfering waves have the same intensity, we shall now show that $V = |\gamma|$.

Consider the conceptual experiment in Fig. 8.4. The mechanism for illuminating the two slits is irrelevant to this discussion. The coherence

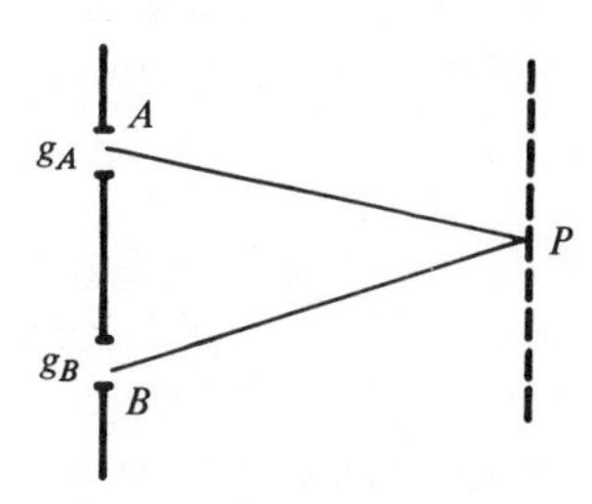

Fig. 8.4. Conceptual interference experiment.

function between the two waves, of complex amplitudes g_A and g_B radiated by the two slits is γ_{AB}, defined by (8.16):

$$\gamma_{AB} = \overline{g_A g_B{}^*}/(\overline{|g_A|^2}\,\overline{|g_B|^2})^{\frac{1}{2}}. \tag{8.20}$$

The amplitude at P is

$$f_p = \frac{1}{L}\{g_A \exp(-ikAP) + g_B \exp(-ikBP)\}$$

$$= \frac{\exp(-ikAP)}{L}\{g_A + g_B \exp(-i\phi)\}, \tag{8.21}$$

where

$$\phi = k(BP - AP). \tag{8.22}$$

The time-average of the intensity at P is

$$\bar{I} = \overline{|f_p|^2} = L^{-2}\{\overline{|g_A|^2} + \overline{|g_B|^2} + \overline{g_A g_B{}^*} \exp(i\phi) + \overline{g_A{}^* g_B} \exp(-i\phi)\}$$

$$= L^{-2}\{\overline{I_A} + \overline{I_B} + 2|\overline{g_A g_B{}^*}| \cos(\phi + \delta)\}, \tag{8.23}$$

where

$$\tan \delta = \operatorname{Im}(\overline{g_A g_B{}^*})/\operatorname{Re}(\overline{g_A g_B{}^*}). \tag{8.24}$$

This can be rewritten in terms of intensities $\bar{I} = \overline{|g|^2}$:

$$\bar{I} = L^{-2}\{\bar{I}_A + \bar{I}_B + 2(\bar{I}_A \bar{I}_B)^{\frac{1}{2}}|\gamma_{AB}| \cos(\phi + \delta)\}. \tag{8.25}$$

Now the maximum and minimum intensities correspond to $\cos(\phi + \delta) = \pm 1$, whence the visibility

$$V = 2|\gamma_{AB}|(\bar{I}_A \bar{I}_B)^{\frac{1}{2}}/(\bar{I}_A + \bar{I}_B). \tag{8.26}$$

When $\bar{I}_A = \bar{I}_B$, the visibility (8.19) is:

$$V = |\gamma_{AB}|. \tag{8.27}$$

It is only fair to point out from the historical point of view that the visibility was originally chosen by Zernike, the father of coherence studies, as a measure of coherence; afterwards it was found that the complex γ could conveniently replace V in the mathematical development of the subject.

8.3 Physical origins of linewidths

So far we have introduced the width of a spectral line, or the finiteness of a wave train, simply as a parameter to be reckoned with; now we shall enquire briefly into the physical causes of line-broadening. A spectral line has its origin in a quantum transition in which an atom changes its stationary state from level A to level B, of energies E_A and E_B; a wave of frequency $\omega_0 = (E_A - E_B)/\hbar$ is emitted at the same time.

8.3.1 *Doppler broadening.* Let us consider radiation from an isolated atom in a gas at temperature T. If the atom, mass m, has velocity v_x along the line of sight while the transition is taking place, the spectral line will appear shifted by the Doppler effect. The distribution of velocities along a particular axis (x) in a perfect gas is Gaussian

$$f(v_x)\,\mathrm{d}v_x = C \exp\left(\frac{-mv_x^2}{2k_\mathrm{B}T}\right)\mathrm{d}v_x \tag{8.28}$$

and the Doppler shift in the observed frequency is

$$\omega - \omega_0 = \omega_0 v_x/c \tag{8.29}$$

so that

$$a(\omega) = C \exp\left\{\frac{-m(\omega-\omega_0)^2c^2}{2\omega_0^2k_\mathrm{B}T}\right\}. \tag{8.30}$$

This effect has broadened an ideally sharp spectral line into a line with a Gaussian profile, with half-width

$$\sigma = \omega_0(k_\mathrm{B}T/mc^2)^{\frac{1}{2}}. \tag{8.31}$$

In terms of wavelength, rather than frequency, we find the half-width to be $\lambda_0(k_\mathrm{B}T/mc^2)^{\frac{1}{2}}$.

As an example we can take the krypton line for which $\lambda = 5.6\times10^{-5}$ cm, $m = 84\times1.7\times10^{-24}$ g, $T = 80$ K, giving a half-width of 1.6×10^{-11} cm ≈ 0.002 Å. This agrees reasonably with the observed half-width of 0.003 Å.

8.3.2 *Collision broadening.* Considering an isolated atom does not give us the whole story. There will always be collisions between the various atoms in a real gas. On the average, a particular atom will expect to be free for a time

$$\tau_1 = (4NvA)^{-1} \tag{8.32}$$

between collisions, where N is the number of molecules per unit volume and A their collision cross-section. Now both N and v depend on temperature T and pressure P, and according to kinetic theory for a perfect gas (see e.g. Jeans, 1960)

$$Nv = P(3/mk_B T)^{\frac{1}{2}}. \tag{8.33}$$

So

$$\tau_1 = bT^{\frac{1}{2}}/P, \tag{8.34}$$

where b is a constant.

Now consider what happens if an emitting atom suffers a collision. We may suppose that the shock of the collision will at the very least destroy phase correlation between the emitted waves before and after the collision. The emission from all the atoms in the gas will therefore appear like a series of uncorrelated bursts of radiation each of average duration τ_1. The actual durations have a Poisson distribution of mean value τ_1; then the probability of there being a burst of length between τ and $\tau+\delta\tau$ is $p(\tau)$, where

$$p(\tau) = \tau_1^{-1} \exp(-\tau/\tau_1). \tag{8.35}$$

So the emitted wave consists of wave trains with frequency ω_0 and random phase, starting at random moments and having durations statistically distributed according to (8.35).

Following § 8.2.2, we deduce that the spectral intensity of such an emitted wave is the mean spectral intensity of the individual wave trains, the spectral phases being random. The spectrum is thus

$$I(\omega) = \int_0^\infty \left[\int_{-\tau/2}^{\tau/2} \exp(\mathrm{i}\omega_o t) \exp(-\mathrm{i}\omega t)\, \mathrm{d}t\right]^2 p(\tau)\, \mathrm{d}\tau. \tag{8.36}$$

In (8.36) the inner integral (in square brackets) is the Fourier transform of a sine wave lasting for a duration τ; the outer integral is the statistical average. The inner integral is well-known to us as $\tau \operatorname{sinc}\{(\omega-\omega_0)\tau/2\}$ (§ 3.4.2). Thus (8.36) becomes

$$I(\omega) = \int \frac{4}{\tau_1(\omega-\omega_0)^2} \exp(-\tau/\tau_1) \sin^2\{(\omega-\omega_0)\tau/2\}\, \mathrm{d}\tau \tag{8.37}$$

$$= 2\tau_1^{\,2}/\{1+(\omega-\omega_0)^2\tau_1^{\,2}\}. \tag{8.38}$$

The value of τ_1 can now be substituted from (8.34).

The function (8.38) is known as a *Lorenzian function* with half-width τ_1^{-1} and has already been introduced in § 7.5.1 as the line-shape of Fabry–Pérot fringes. If we draw (8.38) as a function of ω, we see that the

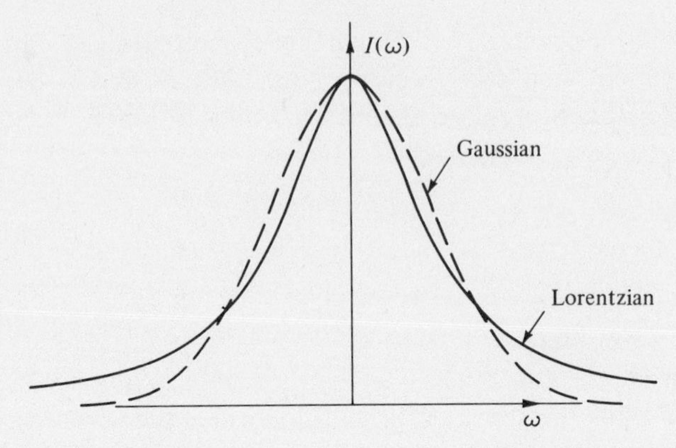

Fig. 8.5. Comparison between Lorentzian and Gaussian functions.

function is superficially similar to the Gaussian but has a much slower decay in its wings (Fig. 8.5).

In practice, temperature and pressure cause both Doppler and collision broadening in various degrees, and observed spectral lines are rarely exactly Gaussian or exactly Lorenzian. The Doppler contribution can be calculated exactly from the physical conditions; careful measurement of the line-shape (usually by Fourier transform spectroscopy – § 8.5) then allows the collision cross-section A in (8.32) to be estimated.

In air at s.t.p., for example, the free time τ_1 is of the order of 5×10^{-9} s, leading to a collision linewidth $\delta\lambda = 0.01$ Å.

8.4 Beats between light waves

When two waves, of the same amplitude A and frequencies ω_1 and ω_2 strike a detector simultaneously, their combined amplitude and intensity are:

$$f(t) = A\{\exp(\mathrm{i}\omega_1 t) + \exp(\mathrm{i}\omega_2 t)\}$$

$$= A\exp(\mathrm{i}\bar{\omega}t)\cos\{\tfrac{1}{2}(\omega_1-\omega_2)t\}, \tag{8.39}$$

$$I(t) = A^2\cos^2\{\tfrac{1}{2}(\omega_1-\omega_2)t\} = \tfrac{1}{2}A^2\{1+\cos(\omega_1-\omega_2)t\}. \tag{8.40}$$

The intensity (8.40) oscillates with the difference frequency $(\omega_1-\omega_2)$. This beat phenomenon is very common in acoustics; the question here is whether it can be observed with light waves. The major problem is one of the accuracy of the frequencies. Although the basic frequencies ω_1 and ω_2 may be defined to better than one part in 10^5, the uncertainties will be additive in the beat frequency, and if the proportional difference between ω_1 and ω_2 is only 10^{-4} the beat frequency will have an uncertainty of at least one part in 10. It can only just be defined as having a recognizable frequency at this level of uncertainty. It is therefore necessary to look for beats between light waves with greater frequency difference than this limit.

Looked at from the wave-group point of view (§ 8.2.2) the problem is a little clearer. We are presented with two wave-groups, each consisting of some 10^5 waves and therefore lasting only for a finite time. If we want to observe beats between them we must have a beat frequency which goes through a reasonable number of cycles during the coherence time, the lifetime of the wave-group. If we make this 'reasonable number' ten we arrive at exactly the same criterion as before. Every pair of wave-groups will, of course, interfere, and will produce beats of random phase, so the total beat disturbance has the same coherence time as the original wave, but of course a much lower frequency, and so the relative uncertainty in the beat frequency is much larger.

8.4.1 *Beats between discharge-tube sources.* Beats between conventional light sources have been observed, with great difficulty, by Forrester, Gudmundsen & Johnson (1955). The two frequencies used were Zeeman components of a single spectral line, and the compound disturbance was picked up by a photocell and analysed in the expected frequency range of the beats. Sufficient magnetic field was applied to give a Zeeman splitting an order greater than the width of the spectral line to satisfy the uncertainty criterion, but even so the signal-to-noise ratio was very poor. This was improved sufficiently to make the observations possible by switching the magnetic field on and off at a low frequency and looking for beats appearing and disappearing at the same frequency.

8.5 Fourier transform spectroscopy

8.5.1 *Spectral analysis using the Michelson interferometer.* In the Michelson interferometer (§ 7.8.2) one observes the interference pattern between two light waves which, after originating at the same source, have travelled along different paths on their way to the detector (Fig. 8.6). If the two paths have lengths differing by d, it is clear that the waves interfering at the detector at a given instant originated at the source at times separated by $\tau = d/c$. The coherence function, measured as the visibility of the interference pattern, is (§ 8.2.4)

$$\gamma(\tau) = \overline{f(t)f^*(t+\tau)}/\overline{|f(t)|^2}. \qquad (8.41)$$

By the Wiener–Khinchin theorem, the Fourier transform of $\gamma(\tau)$ is the spectral intensity $I(\omega)$.

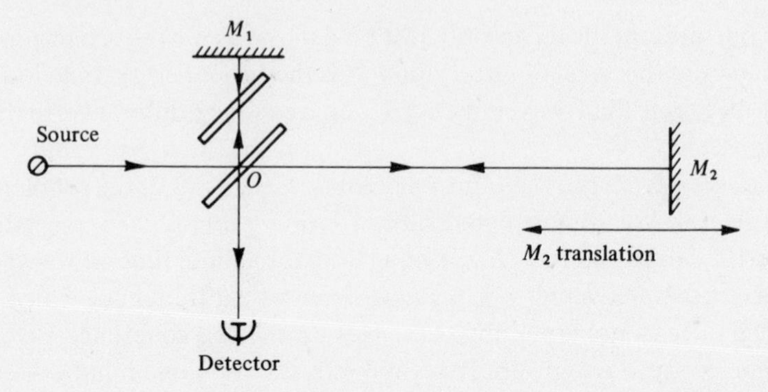

Fig. 8.6. Michelson interferometer. The path difference is $d = 2(OM_2 - OM_1)$.

This method of spectral analysis, which was originally invented by Michelson and is described in detail in *Studies in Optics* (Michelson, 1962), was at first not very popular because of the necessity for a Fourier transform in order to convert the observations into a conventional spectrum. The advent of electronic computers has now changed the situation. Because of the basic simplicity of construction of a Michelson interferometer, and its ease of application to wavelengths other than the visible (in particular infra-red), the method of 'two-beam interference' has become of considerable practical importance in modern physics and chemistry. It is now called 'Fourier Transform Spectroscopy' and can already claim books (e.g. Bell, 1972), review articles and conferences entirely devoted to its technical aspects and applications.

The above argument bears closer inspection, because visibility only measures the absolute value $|\gamma(\tau)|$. Let us write the source disturbance $f(t)$ as a Fourier transform:

$$f(t) = \int_{-\infty}^{\infty} a(\omega) \exp(-\mathrm{i}\omega t)\, \mathrm{d}\omega. \tag{8.42}$$

The amplitude at the detector, $s(t)$ is proportional to $f(t) + f(t+\tau)$:

$$\begin{aligned} s(t) &= \int_{-\infty}^{\infty} a(\omega)[\exp(-\mathrm{i}\omega t) + \exp\{-\mathrm{i}\omega(t+\tau)\}]\, \mathrm{d}\omega \\ &= 2\int_{-\infty}^{\infty} a(\omega) \exp\{-\mathrm{i}\omega(t+\tfrac{1}{2}\tau)\} \cos \tfrac{1}{2}\omega\tau\, \mathrm{d}\omega. \end{aligned} \tag{8.43}$$

The resultant mean intensity $\bar{I} = \overline{|s(t)|^2}$ can then be calculated by a double integral technique which is the continuous equivalent of the double sum technique in (8.4) and (8.5). We write, from (8.43)

$$I = |s(t)|^2 = 4\left|\int_{-\infty}^{\infty} a(\omega)\exp\{-\mathrm{i}\omega(t+\tfrac{1}{2}\tau)\}\cos(\tfrac{1}{2}\omega\tau)\,\mathrm{d}\omega\right|^2$$

$$= 4\iint_{-\infty}^{\infty} a(\omega)a^*(\omega')\exp\{-\mathrm{i}(\omega-\omega')(t+\tfrac{1}{2}\tau)\}$$

$$\times\cos(\tfrac{1}{2}\omega\tau)\cos(\tfrac{1}{2}\omega'\tau)\,\mathrm{d}\omega\,\mathrm{d}\omega', \tag{8.44}$$

where ω' has been introduced as a dummy variable. The mean value $\bar{I}$ of this double integral comes from averaging it over a long time, t. Because of the term $\exp\{-\mathrm{i}(\omega-\omega')(t+\tfrac{1}{2}\tau)\}$ in the integrand, all terms average to zero except those for which $\omega = \omega'$. This gives us $\bar{I}$ a function of τ only:

$$\bar{I}(\tau) = 4\int_{-\infty}^{\infty} |a(\omega)|^2\cos^2(\tfrac{1}{2}\omega\tau)\,\mathrm{d}\omega$$

$$= 4\int_0^{\infty} I(\omega)(1+\cos\omega\tau)\,\mathrm{d}\omega. \tag{8.45}$$

By writing the integral (8.45) from 0 to ∞, we have acknowledged the fact that positive and negative frequencies are physically identical, namely $I(\omega) = I(-\omega)$. The relationship can then be written as the real Fourier transform:

$$\bar{I}(\tau) - \bar{I}_0 = 4\int_0^{\infty} I(\omega)\cos\omega\tau\,\mathrm{d}\omega, \tag{8.46}$$

where I_0 is a constant background level to the whole interferogram. This is particularly evident in the example in Fig. 8.9,

$$\bar{I}_0 = 4\int_0^{\infty} I(\omega)\,\mathrm{d}\omega. \tag{8.47}$$

The spectrum $I(\omega)$ is then calculated by the inverse transform to (8.46):

$$I(\omega) = \frac{2}{\pi}\int_0^{\infty}\{\bar{I}(\tau) - \bar{I}_0\}\cos\omega\tau\,\mathrm{d}\omega. \tag{8.48}$$

Of course, in practice τ is limited to some maximum value τ_0 by the physical size of the instrument. One therefore has

$$I(\omega) = \frac{2}{\pi}\int_0^{\tau_0}\{\bar{I}(\tau) - \bar{I}_0\}\cos\omega\tau\,\mathrm{d}\tau \tag{8.49}$$

as the best approximation to the spectrum. Furthermore, the sharp cut-off at τ_0 introduces 'false detail' into the spectrum, in the same way as will be discussed in § 9.4.5, and the technique of apodization (§ 9.6.6) is often used mathematically to improve the line-shape obtained.

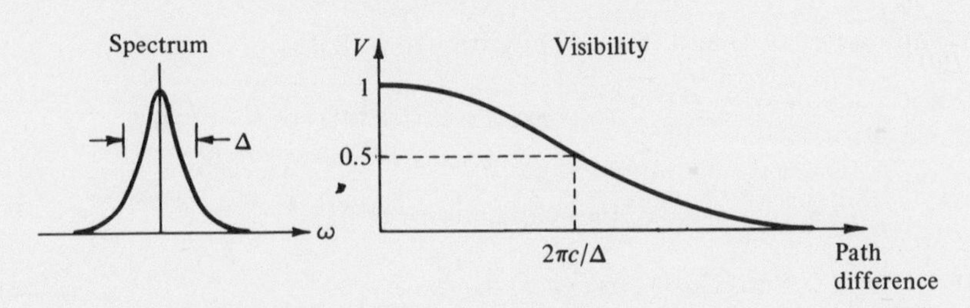

Fig. 8.7. Visibility of the fringes in a Michelson interferometer when illumination is by a single wide spectral line.

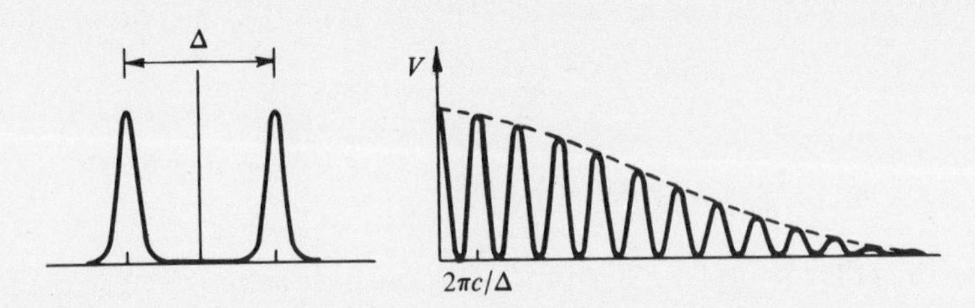

Fig. 8.8. As Fig. 8.7, illumination by a doublet.

In Figs. 8.7–8.9 we show three examples of Fourier spectroscopy. The first two illustrate Michelson's original approach, via the visibility function; the third is an example from a modern commercial Fourier spectrometer.

(a) Suppose we consider a single spectral line, of frequency ω_0 and half-width Δ. Then the fringes have visibility which falls to zero as d is increased; the visibility becomes 0.5 at a path difference $d = 2\pi c/\Delta$ (Fig. 8.7).

(b) A spectral doublet at $\omega = \omega_0 \pm \frac{1}{2}\Delta$ gives visibility which oscillates as the path difference is increased. From unit visibility when the path difference is zero, it falls to zero when $d = \pi c/\Delta$ and returns almost to unit visibility, if the lines are fairly sharp, whenever d is a multiple of $2\pi c/\Delta$ (Fig. 8.8).

(c) An example of the interferogram recorded for an infra-red source is shown in Fig. 8.9 together with its transform. Notice that the interferogram is measured with d varying from $-D$ to $+D$, although in principle 0 to $+D$ would be sufficient, since the spectral intensity is a real function of ω and therefore the interferogram must be an even function. However, the extra data allow the removal of some instrumental artifacts.

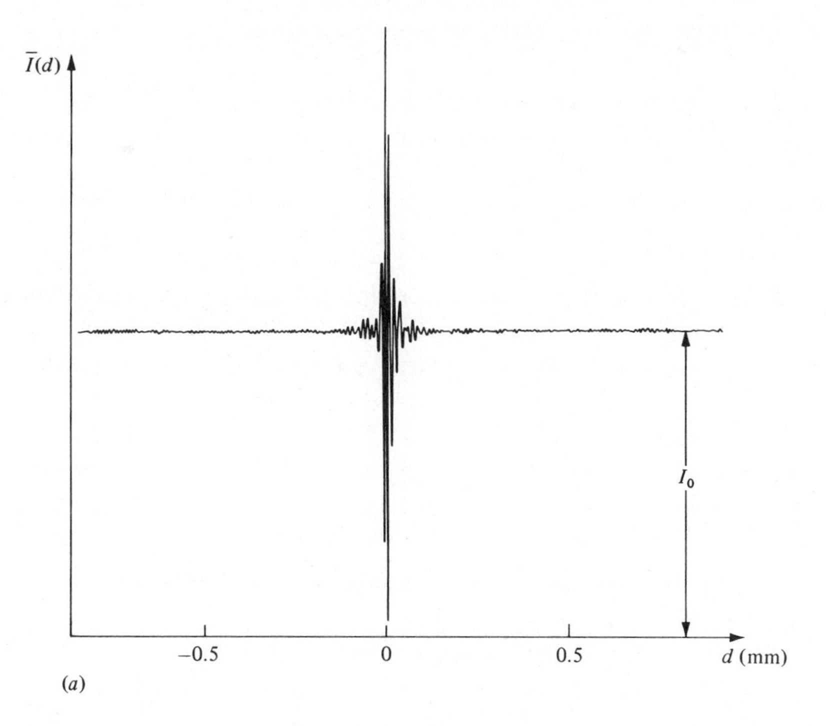

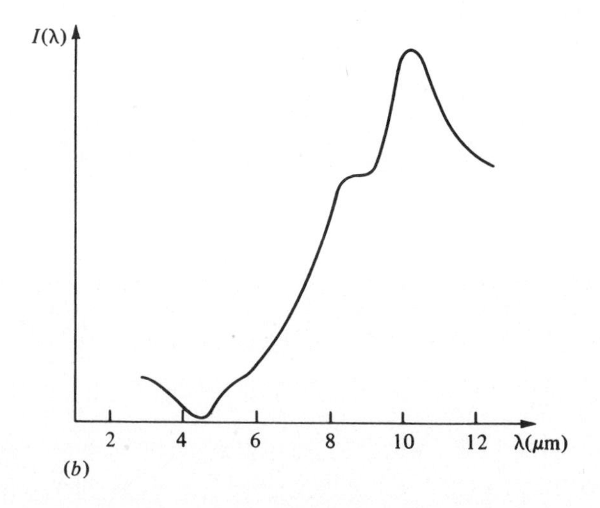

Fig. 8.9. Interferogram from a broad-band infra-red source, and the calculated spectrum.

8.5.2 *Resolution limits.* The resolution limit of Fourier spectroscopy can be calculated from the above examples. When d has a maximum value of D, from example (b) we see that we could just resolve two lines separated by

$$\Delta_{\min} = \pi c/D \tag{8.50}$$

or

$$(\delta\lambda/\lambda)_{\min} = \lambda/2D. \tag{8.51}$$

This is identical with the resolution attainable with a *perfect* diffraction grating of the same length D used with incident light at grazing angle ($\theta_0 = -\pi/2$) and observing in the highest order possible ($\theta = \pi/2$). Since these conditions cannot be achieved in practice, one can say that the Fourier spectrometer has a better resolution than a diffraction grating of the same dimensions. Another way of deriving the resolution limit considers the instrument as a diffraction grating with $N = 2$ lines only, separated by distance D. The order of diffraction n is the path difference between the two beams in units of λ, i.e. D/λ. Thus, using the formula from § 7.7.2

$$(\delta\lambda/\lambda)_{\min} = 1/Nn = \lambda/2D. \tag{8.52}$$

8.6 Spatial coherence

The concept of temporal coherence was introduced as an attempt to give a quantitative answer to the following question. At a certain instant of time we measure the phase of a propagating light wave at a certain point. If the wave were a perfect sinusoidal plane-wave, $A \exp(\mathrm{i}\omega_0 t)$, we should then know the phase at any time in the future. But in a *real* situation, for how long after that instant will an estimate made in the above way be reliable? The gradual disappearance of our knowledge of the phase was seen to result from uncertainty of the exact value of ω_0, and was related to the finite width of the spectral line representing the wave.

The second coherence concept, that of *spatial coherence*, is concerned with the phase relationship *at a given instant* between various points in a plane normal to the direction of propagation. If the wave were a perfect plane-wave, this plane would be a wavefront, and definition of the phase at one point P in it would immediately define the phase at every other point – even if the wave were not at all monochromatic. In practice, we can ask the question: if we know the value of the phase at P, how far away from P can we go and still make a correct estimate of the phase?

In a similar way that we found temporal incoherence to be related to uncertainty in the frequency ω_0 of the wave (and hence in the *magnitude* of the wave-vector, $|\mathbf{k}|$), we shall see the spatial incoherence to be related to uncertainty in the *direction* of the wave-vector $\mathbf{k}$. And uncertainty in the direction of $\mathbf{k}$ arises when the source of the light is not a point source, but is extended.

8.6.1 *A qualitative investigation of spatial coherence.* In § 8.2.6 we saw that the result of limited temporal coherence is the inability to create an interference pattern between two samples of the same wave if the times of the samples were separated by more than the coherence-time τ_c. Spatial coherence can be approached qualitatively in the same way.

Suppose that an incoherent, quasi-monochromatic source, of linear dimensions a, is used to illuminate the pair of pinholes used in a Young's interference experiment. We shall assume the existence of a fringe pattern on a screen to indicate coherence between the wave amplitudes at the two pinholes (Fig. 8.10). The source is distant $L \gg a$ from the pinholes, whose separation is x, and the screen is a further distance H along the axis.

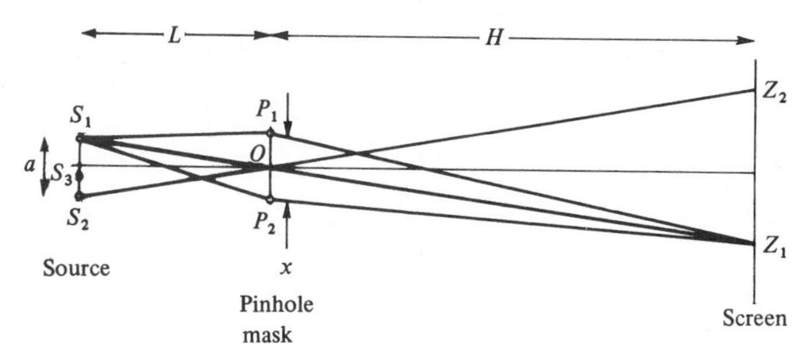

Fig. 8.10. Spatial coherence.

Consider the point S_1 at one end of the source. This point alone illuminates the pinholes coherently and therefore produces a fringe pattern on the screen. The zero order of the fringe pattern appears at Z_1, the point corresponding to zero difference between the optical paths $S_1P_1Z_1$ and $S_1P_2Z_1$. By symmetry, Z_1 lies on the line joining S_1 to O, the point half-way between the two pinholes. The period of the interference fringes is given by $H\lambda/x$. Now consider the other end of the source, S_2. This gives a fringe pattern with the same period, centred on the point Z_2, on the line S_2O. The two sets of fringes do not overlap, and since S_1 and S_2 are uncorrelated, tend to cancel one-another out. If Z_1Z_2 is equal to half the fringe spacing, the fringe patterns from S_1 and S_2 will be precisely in antiphase, and the fringe pattern will disappear. This implies that there will be no spatial coherence when

$$\tfrac{1}{2}H\lambda/x = Z_1Z_2 = aH/L, \qquad (8.53)$$

assuming all the angles to be small. The criterion for disappearance of the fringes is thus:

$$x = L\lambda/2a. \qquad (8.54)$$

The result can be stated as follows. Because of the size of the source, a, or more usefully its *angular* size, $\theta = a/L$, coherence between neighbouring points on the screen only occurs if the distance between the points is less than

$$x_c = \lambda/2\theta. \qquad (8.55)$$

This maximum distance x_c is called the *coherence distance* in the plane of the pinholes.

We have neglected, in this discussion, the effect of all points such as S_3 in-between S_1 and S_2, and thereby introduced an error of about 2. This will be corrected in § 8.6.3 by a more complete analysis.

When the argument is extended to two dimensions, it turns out that a source of limited angular dimensions defines a two-dimensional region within which both pinholes must be situated in order to be coherently illuminated. This region is called the *coherence area* and is generally inversely related in dimensions to those of the source.

The relationship between the coherence area, or strictly the coherence function, and the source dimensions will be shown to be that between Fourier transforms, at least when the source has a small angular diameter α. This relationship can be very useful in practice, and is the basis of the technique of 'aperture synthesis' which will be discussed briefly in § 11.6.7.

8.6.2 *The spatial coherence function.* In a manner which is formally analogous to our earlier definition of a temporal coherence function we define the correlation coefficient between the wave amplitudes f_P and f_Q at P and Q separated by the vector $\mathbf{r}$ in the observation plane. This is:

$$\gamma(\mathbf{r}) = \overline{f_P f_Q{}^*}/(\overline{|f_P|^2}\,\overline{|f_Q|^2})^{\frac{1}{2}}. \qquad (8.56)$$

Once again the averages are taken during periods long compared with typical fluctuations in the light waves, and the two averages $\overline{|f_P|^2}$ and $\overline{|f_Q|^2}$ are the mean light intensities $\bar{I}_P$ and $\bar{I}_Q$ at the two points. We call $\gamma(\mathbf{r})$ the *spatial coherence function* and assume its value to be dependent mainly on the value of $\mathbf{r}$ and not on the exact choice of the point P.

For advanced work one combines (8.16) and (8.56) to form a mutual coherence function of both time and space

$$\gamma(\mathbf{r}, \tau) = \overline{f_P(t) f_Q(t+\tau) \exp(\mathrm{i}\omega_0\tau)}/(\bar{I}_P \bar{I}_Q)^{\frac{1}{2}}. \tag{8.57}$$

We shall not need to use this complete form here.

8.6.3 *The van Cittert–Zernike theorem.* The spatial coherence function $\gamma(\mathbf{r})$ is related by a Fourier transform to the intensity distribution $I(\theta, \phi)$ in the source. This relationship was proved independently by van Cittert and Zernike and is the basis of a number of methods (for example, the Michelson stellar interferometer (§ 8.6.6) and the Mills cross (§ 11.6.6)) for measuring sizes of remote stellar objects. To derive the relationship in one dimension, we consider a large and distant incoherent source (Fig. 8.11) illuminating the observation plane. Distances in the observation

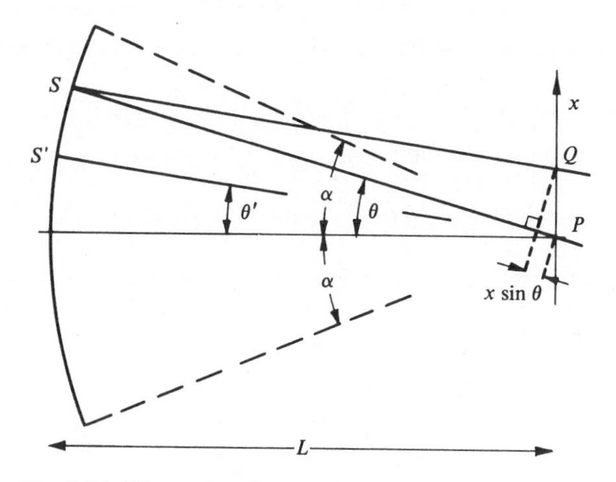

Fig. 8.11. Illustrating the van Cittert–Zernike theorem.

plane will be assumed small compared with the source and its distance L, but the angular diameter of the source, 2α, is not so limited. It is convenient to assume the source to lie on a sphere centred on P, so that all points on it are equidistant from P. The amplitude at point S on the source is $g(\theta)$, and the amplitude received at P $(x=0)$ is then

$$f(0) = \frac{1}{L}\int g(\theta) \exp(-\mathrm{i}kSP)\, \mathrm{d}\theta. \tag{8.58}$$

At Q we have

$$f(x) = \frac{1}{L}\int g(\theta) \exp\{-\mathrm{i}k(SP - x\sin\theta)\}\, \mathrm{d}\theta. \tag{8.59}$$

The product $c(x)=f(0)f^*(x)$ is then given by

$$c(x)=\frac{1}{L^2}\int g(\theta)\exp(-\mathrm{i}kSP)\,\mathrm{d}\theta\int g^*(\theta)\exp\{\mathrm{i}k(SP-x\sin\theta)\}\,\mathrm{d}\theta$$

$$=\frac{1}{L^2}\iint g(\theta)g^*(\theta')\exp\{-\mathrm{i}k(SP-S'P)\}\exp(-\mathrm{i}kx\sin\theta')\,\mathrm{d}\theta\,\mathrm{d}\theta', \tag{8.60}$$

where the product of two integrals has been converted to a double integral by the use of a second θ variable, namely θ' (cf. (8.44)).

Now the coherence function $\gamma(x)$ can be written from its definition (8.56) in terms of the time-average of $c(x)$, i.e.

$$\gamma(x)=\overline{c(x)}/\overline{c(0)}, \tag{8.61}$$

since the intensities at both P and Q, $\overline{|f(0)|^2}$ and $\overline{|f(x)|^2}$ respectively, can both be considered to be equal to $\overline{c(0)}$. The time-average $\overline{c(x)}$ of (8.60) involves the quantity $\overline{g(\theta)g^*(\theta')}$. When the source is incoherent, $g(\theta)$ and $g(\theta')$ are uncorrelated and so this average is zero unless $\theta=\theta'$, when it equals $\overline{|g(\theta)|^2}=\overline{I(\theta)}$, the source intensity at the point S. Introducing this into the double integral we have

$$\overline{c(x)}=\frac{1}{L^2}\int\overline{I(\theta)}\exp(-\mathrm{i}kx\sin\theta)\,\mathrm{d}\theta. \tag{8.62}$$

This integral is reminiscent of Fraunhofer diffraction (§ 7.2); in particular, when the source has small angular dimensions, α, we approximate $\sin\theta$ by θ and see that the value of $\gamma(x)$ is

$$\gamma(x)=\overline{c(x)}/\overline{c(0)}$$

$$=G(kx)=\frac{\int\overline{I(\theta)}\exp(-\mathrm{i}kx\theta)\,\mathrm{d}\theta}{\int\overline{I(\theta)}\,\mathrm{d}\theta}; \tag{8.63}$$

$G(kx)$ is the Fourier transform of $\overline{I(\theta)}$ normalized to unity at $x=0$. This theorem can easily be extended to the two dimensions of the source $I(\theta,\phi)$.

Consider a circular star as an example. The intensity is unity within a circle of angular radius α and zero outside it. The correlation function is therefore the Fourier transform which is (§ 7.3.5)

$$\gamma(x)=\frac{2J_1(k\alpha x)}{k\alpha x}, \tag{8.64}$$

which has its first zero at a separation

$$x=\frac{1.2\pi}{k\alpha}=\frac{0.61\lambda}{\alpha}. \tag{8.65}$$

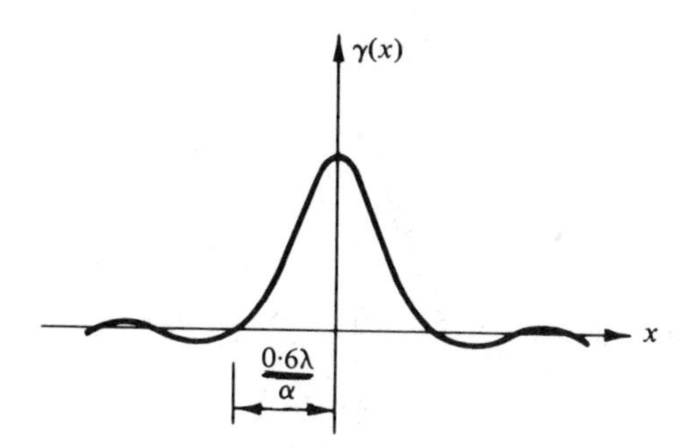

Fig. 8.12. Correlation function between P and Q as a function of their separation x, when illumination is by a circular source of angular radius α.

For a star of angular diameter 0.7 s, or about 0.5×10^{-6} radians, coherence exists over a circle of radius about 50 cm (Fig. 8.12). Admittedly, after the first zero there are further regions of correlation, both negative and positive, but these result from the sharp cut-off at the edges of the star and are rarely observed except in academic experiments.

8.6.4 *Partial coherence and the superposition of plane-waves.* In many problems of imaging in partially coherent light it is convenient to consider the light as a superposition of uncorrelated plane-waves in various directions. This attitude follows directly from the van Cittert–Zernike theorem, the origin of the uncorrelated plane-waves being the extended source. If the source has a large angular size we use the result (8.63) without approximation to calculate $\gamma(x)$. The result is then

$$\gamma(x)=\frac{\int\{\overline{I(\theta)}/\cos\theta\}\exp(-ikx\sin\theta)\,\mathrm{d}(\sin\theta)}{\int\{\overline{I(\theta)}/\cos\theta\}\,\mathrm{d}(\sin\theta)},\tag{8.66}$$

which is again a Fourier transform relation. What is of particular interest in microscopy is to relate the result to the cone angle, α, or numerical aperture, $\sin\alpha$, of the illumination (§ 9.5.1). With a Lambertian circular source of such a cone angle we find (cf. (8.45))

$$\gamma(x)=J_1(kx\sin\alpha)/kx\sin\alpha,\tag{8.67}$$

which falls to zero when $x=0.61\lambda/\sin\alpha$. In particular, uniform illumination from an infinitely extended source ($\alpha=\pi/2$) gives a coherence diameter $x=0.61\lambda$.

8.6.5 *A laboratory demonstration of spatial coherence.* The way in which the visibility (§ 8.2.6) of the interference pattern in a Young's slit experiment varies with the coherence between the sources can be

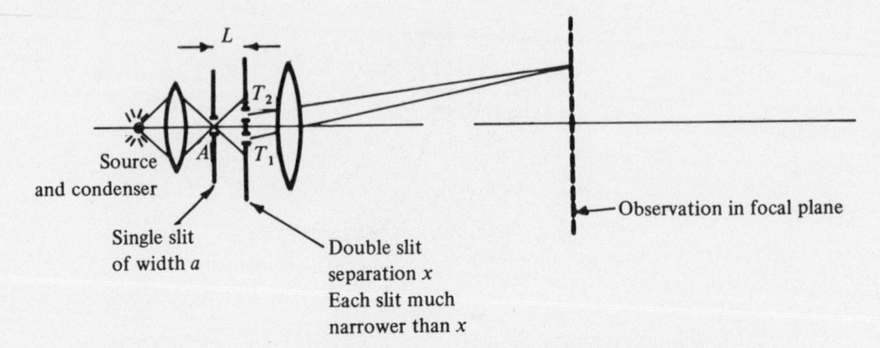

Fig. 8.13. Young's fringes experiment to illustrate partial coherence.

investigated in the simple experiment illustrated in Fig. 8.13, which is an experimental interpretation of Fig. 8.10. The set-up is one-dimensional, and the form of the correlation function between pairs of points in the plane T resulting from a wide slit A is the familiar sinc β (§ 7.2.1), where

$$\beta = \frac{kax}{L} = \frac{2\pi ax}{\lambda L}, \tag{8.68}$$

this function being the Fourier transform in (8.63). As a is increased from zero, the correlation between slits T_1 and T_2 begins at unity, becomes zero when

$$a = \frac{\lambda L}{2x} \tag{8.69}$$

and has the usual series of maxima and minima as a is increased beyond this value. The interference patterns observed for various values of a are illustrated in Fig. 8.14, in which the visibility clearly follows the same pattern. Notice that the effect of negative values of γ is to shift the pattern by half a fringe.

This experiment is the basis of two fundamentally important interferometers which are used to determine the angular diameters, $\alpha = a/L$, of inaccessible sources such as stars. As in the Young's slit experiment, the *Michelson stellar interferometer*† measures the coherence between the illuminations at T_1 and T_2 by observing the visibility of the fringes formed by their interference; the *Brown–Twiss interferometer* carries out the correlation by electronic means.

† Because of the presence of atmospheric turbulence, which is discussed more fully in § 11.5.2, no astronomical telescope can achieve even approximately the Rayleigh limit of resolution (§ 9.4.1). The Michelson stellar interferometer was the first instrument to allow this limit to be reached, and even exceeded.

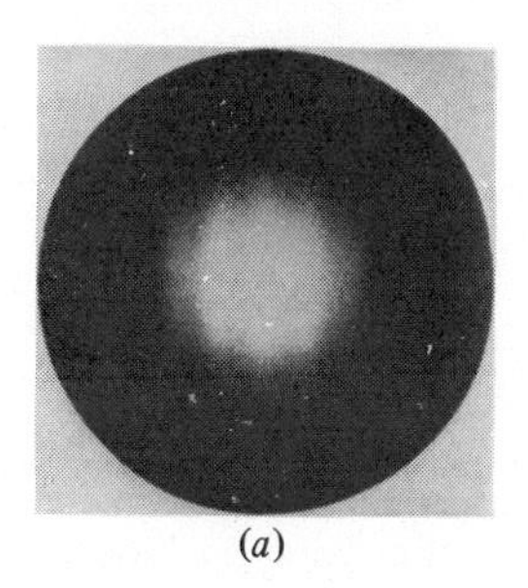

(a)

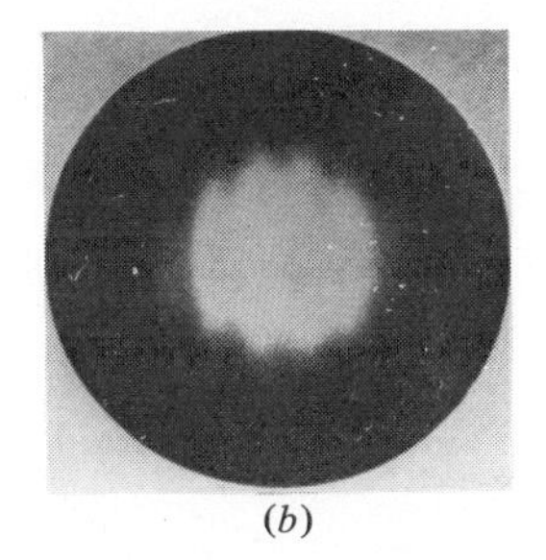

(b)

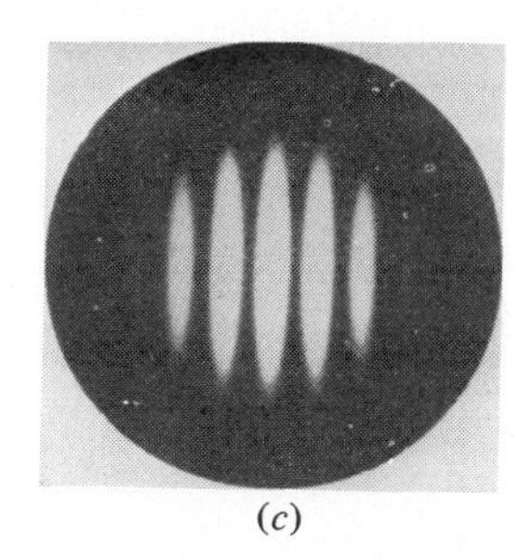

(c)

Fig. 8.14. Young's fringes with different degrees of spatial coherence. (a) $\gamma = +0.70$; (b) $\gamma = -0.132$; (c) $\gamma = +0.062$. Note particularly the minimum at the middle of (b), produced by the negative coherence. (From Thompson, 1958.)

8.6.6 *The Michelson stellar interferometer.* Michelson used the principle of the above experiment to design a stellar interferometer which enabled dimensional measurement of stars too small to be resolved by an astronomical telescope. We saw in the experiment (§ 8.6.5) that the visibility of the fringes, measured as a function of the slit separation x, was directly related to the size of the source; one can alternatively state that the fringe visibility measures directly $\gamma(x)$ and this function can be Fourier transformed to yield the stellar intensity distribution (§ 8.6.3). There is a minor catch in this statement: the fringe visibility measures $|\gamma(x)|$ and the phase is not known, and so the Fourier transform cannot be performed completely. However, it is usually sufficient to assume that a star has a centre of symmetry which makes the problem soluble; this point is discussed more fully in another connexion in § 9.3.4.

Michelson constructed a series of stellar interferometers, the most ambitious of which was built around the 100 inch Mount Wilson telescope. In principle one puts a screen over the objective of the telescope and makes two holes in it in such a way that their separation is variable. For a point source and round holes in the screen the interference pattern

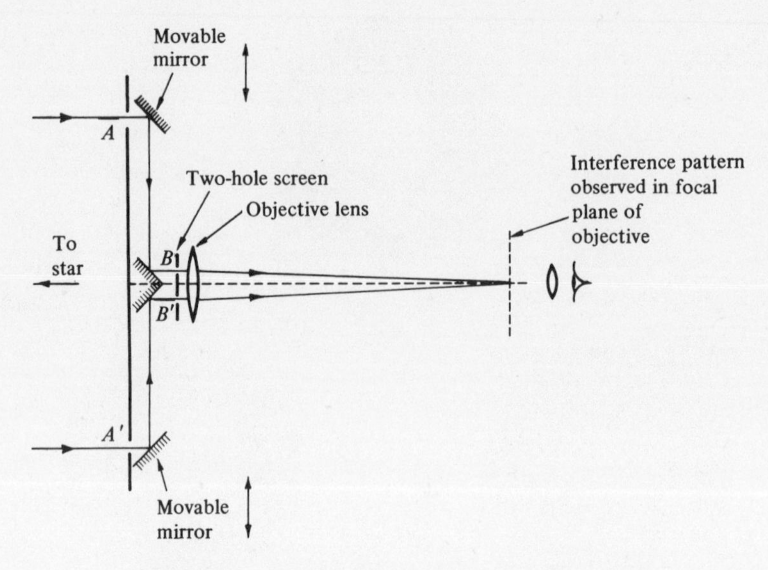

Fig. 8.15. The Michelson stellar interferometer.

observed in the telescope will look like Fig. 7.22: this is the diffraction pattern of two holes (producing a circular ring pattern crossed by fringes representing the *two* in the problem). This should be compared to the *Airy disc* normally observed, which is the circular ring pattern corresponding to the finite aperture of the objective (Fig. 7.17(*b*)).

When a source of finite diameter replaces the point source the continued existence of the fringes depends upon the coherence between the illumination of the two circular holes. The circular pattern is produced by each hole separately, but the straight fringes are joint property of the two. If the separation of the holes is increased, the fringes will become less and less clear, disappearing completely when the separation is $1.22\lambda/\alpha$; this property can be used to measure α.

If the angular diameter α of the star is very small, the separation of the holes may need to be very large before disappearance of the fringes is observed. This has two disadvantages. First, the fringes become extremely closely-spaced when the holes are well apart, and secondly, a telescope of very large aperture is needed, although the greater part of the expensive lens is never used. These difficulties were overcome in an ingenious manner by a mirror system illustrated in Fig. 8.15. The coherence measured is clearly that between the light at A and A', which can conveniently be mounted on racks to alter their separation. The interference pattern observed is that arising from the two apertures B and

B', and can therefore be made conveniently large by making the holes B and B' small and close together. Moreover, the dimensions of the pattern do not change as A and A' are separated. As an example, a star of diameter 0.01 arcsec would require separation

$$AA' = 1.2\lambda/(6\times 10^{-8}) \approx 2\times 10^{7}\lambda;$$

for $\lambda = 5\times 10^{-5}$ cm this makes $AA' = 10$ m. Unfortunately, the small intensity of starlight limited the use of this instrument to stars of exceptional brilliance; the original experiments on it were carried out on the star Betelgeuse in the Orion constellation.

It should be clear from the foregoing discussion that the diameter of the telescope objective itself is quite irrelevant to the results of the experiment. Michelson used the 100 inch telescope only because its construction was sufficiently rigid to bear the weight of the extra mechanical structures needed (see Michelson, 1962).

8.6.7 *Correlation of intensity fluctuations: the Brown–Twiss stellar interferometer.* Brown and Twiss showed that the coherence coefficient between two light waves can also be evaluated by measuring electronically the correlation between the fluctuations in their intensities. Referring back to § 8.2.1, and in particular to Fig. 8.2, we recall that the fluctuations in phase resulting from a coherence time τ_c give rise to fluctuations in intensity with the same coherence time. If the two partially-coherent waves which interfere to form fringes in the Michelson stellar interferometer are examined in detail we find that the intensity fluctuations of the waves are correlated only under the same conditions as give rise to correlation between the individual wave disturbances. The intensity fluctuations occur with a characteristic frequency τ_c^{-1} and if the light is filtered through a narrow band interference filter this frequency can be reduced to about 100 MHz, which is within the capability of electronic instrumentation. (It should be pointed out that the operation of the Michelson stellar interferometer, in contrast to the Brown–Twiss instrument, does not involve the existence of a coherence *time* at all.)

The stellar interferometer consists of two photomultipliers backed by reflectors which are focused on the star in question and separated by the variable distance x. The outputs from the photomultipliers are then correlated by electronic means, and the correlation coefficient evaluated as a function of x.

Brown and Twiss's experiments and the associated theory are described in a number of papers published during 1956 and 1957. The main experimental results appear in *Nature* (Brown & Twiss, 1956). Some further aspects of intensity fluctuations will be discussed in § 8.8.

8.7 The laser as a source of coherent light

Since the operation of the first ruby laser by Maiman in 1960, the subject of physical optics has undergone a renaissance both in understanding and in techniques. In particular, the subject of coherence, which was discussed classically along the lines of §§ 8.1–8.4 in the pre-laser days, has been reconsidered in every detail as a result of interest in the new sources. In the first edition of this book we felt it necessary to include a fairly detailed discussion of the physics of lasers; during the intervening ten years so many excellent monographs on lasers (e.g. Siegman, 1971) have been published that it now seems hardly necessary to include such a discussion in a book on the properties of light waves. In this section we shall try to understand laser light and where it is different from the radiation from classical sources, leaving a detailed description of the lasers themselves to other works.

8.7.1 *The basic elements of a laser.* Any laser contains three basic elements. There is a lasing material having a number of energy levels, between a certain pair of which a light-emitting transition is possible. There is a pump mechanism, by which atoms are excited to occupy the upper level of the pair we have just mentioned. As a result of the pumping the occupancy of the upper level becomes greater than that of the lower level; this situation – called population inversion – does not correspond to thermal equilibrium at any positive temperature. Finally there is an optical resonator, in its simplest form a Fabry–Pérot interferometer, which traps most of the emitted light in an optical path which returns many times backwards and forwards through the laser material, and thus increases the electromagnetic energy density within it. The fraction of light which escapes from the resonator is the useful laser light output.

Just in order to make this description more realistic, we shall describe the elements of the most well-known laser, the helium–neon laser (although this is not actually the simplest of lasers to begin with!). The lasing material is a gaseous mixture of helium and neon (ratio 6 : 1 at a pressure of about 1 mm Hg). The level scheme of these gases includes the levels indicated in Fig. 8.16, of which L and M are the lasing pair of levels in the neon gas. The pumping mechanism consists of an electrical discharge through the gas, and excites helium atoms from the ground state C to the excited level D. As a result of collisions between excited

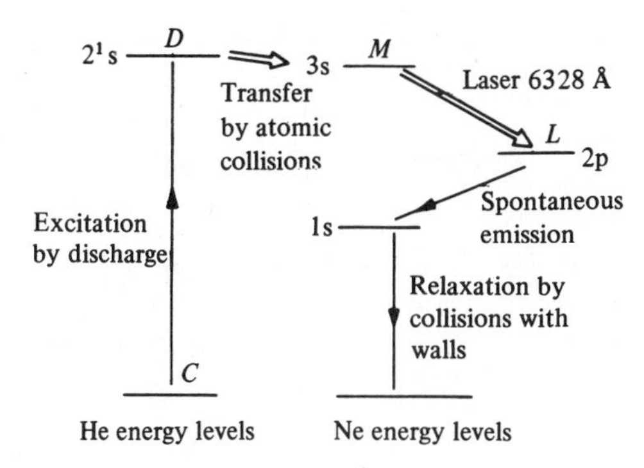

Fig. 8.16. Level scheme for 6328 Å emission in an He–Ne laser.

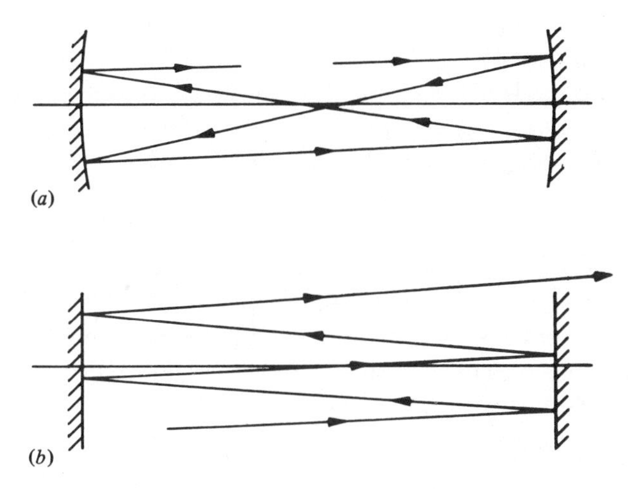

Fig. 8.17. Off-axis rays (*a*) in a confocal resonator (*b*) between two parallel mirrors.

helium atoms and unexcited neon atoms, a proportion of the neon atoms are excited to level M (with consequent de-excitation of the colliding helium atom). Now L is not the ground state, and is virtually unpopulated at the working temperature. So that population inversion has now been created between the levels M and L; i.e. there are more atoms in the upper level M than there are in the lower level L.

We shall leave the system in this metastable state for a moment to describe the optical resonator. This usually consists of two similar spherical mirrors separated by their radius of curvature. One can see fairly easily (Fig. 8.17(*a*)) that without any very critical requirement for alignment of the mirrors, a light ray reflected from one of the mirrors will

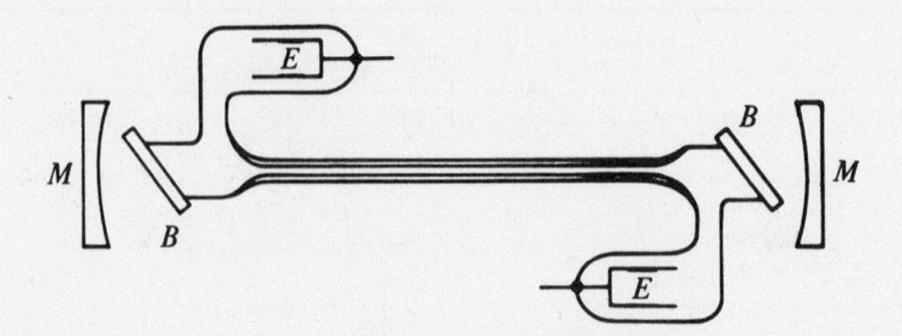

Fig. 8.18. Diagram of an He–Ne laser. E: electrodes to excite discharge in gas; B: Brewster angle windows (no reflexion losses for one polarization); M: confocal resonator mirrors.

be focused onto the other and will continue backwards and forwards between them. Contrast this with a pair of plane parallel mirrors separated by a distance (the length of the laser) large compared to their dimensions (Fig. 8.17(*b*)). There, only if the mirrors are exactly parallel to one another, and the initial ray is exactly normal to them, will the ray survive many reflexions to and fro without walking out of the intervening region. A complete laser is illustrated in Fig. 8.18.

8.7.2 *Spontaneous and stimulated emission.* We now return to the lasing material which is waiting with the populations of levels L and M inverted. Naively we expect that the situation should be unstable; the system will emit radiation spontaneously and revert to thermal equilibrium. Such a process does occur, but we should enquire a little more deeply and ask how.

Levels L and M are eigenstates of the neon atom. By this we mean that they are exact solutions of the Schroedinger equation in the potential field of the atom and correspond to stable states which therefore have infinite lifetime. A transition from one level to another occurs only as a result of some external perturbation, which changes the potential field of the atom slightly and results in the original 'stable' state becoming a slightly inexact solution of the Schroedinger equation in the new field. The solution then changes with time and a transition occurs; the atom emits radiation.

The perturbation which is relevant in this problem is the electromagnetic field of the light wave, and in a laser we introduce the optical resonator so as to increase this field to a high level, and thus to stimulate many transitions from level M to L. This is called 'stimulated emission'. In fact, as we have already pointed out, all emission must really be stimulated by some perturbation, but in the absence of the laser resonator the only light-field present is due to the zero-point field of the electromagnetic wave, and this should be orders of magnitude smaller than the field magnified by the resonator. The zero-point field is responsible for what is called 'spontaneous emission' which is the source of most transitions in ordinary discharge sources.

The likelihood of stimulated emission depends on a number of factors, but can be *compared* with spontaneous emission by a simple argument due to Einstein. This argument puts its finger on several points of importance in any discussion of lasers.

In thermal equilibrium at temperature T the numbers of atoms in the levels L and M (assumed to be non-degenerate) are related by the Boltzmann relation:

$$\frac{n_M}{n_L} = \exp\{-(E_M - E_L)/k_B T\} = \exp(-\hbar\omega/k_B T). \tag{8.70}$$

Now transitions between the two levels can be either spontaneous or stimulated. We consider the presence of stimulating radiation of intensity $I(\omega)$ at frequency ω. This is the only radiation which can interact with the L–M transition. The spontaneous transitions are – by definition – independent of $I(\omega)$, whereas the rate of stimulated transition will be (to a first approximation) proportional to $I(\omega)$. Thus we can write:

(a) rate of transition from L to M, which must be entirely stimulated because each transition involves *absorption* of energy from the stimulating source, is $BI(\omega)n_L$;

(b) rate of transition from M to L which includes stimulated and spontaneous transitions is $BI(\omega)n_M + An_M$.

The constants B and A are called the 'Einstein coefficients' and are functions of the wavelength of the light; $BI(\omega)/A$ is the ratio of the probability of stimulated to spontaneous transition in the presence of the light intensity $I(\omega)$. The same factor B is assumed to apply in each direction.

In the equilibrium state, the rates in the two directions must be equal. Now in the equilibrium state we know the relative values of n_L and n_M; we also know the equilibrium value of $I(\omega)$ since the equilibrium state includes black-body radiation at the temperature T:

$$I(\omega) = \frac{\hbar\omega^3}{4\pi^2 c^2}\left\{\exp\left(\frac{\hbar\omega}{k_B T}\right) - 1\right\}^{-1}. \tag{8.71}$$

Thus, equating the two directions in thermal equilibrium:

$$Bn_L I(\omega) = Bn_M I(\omega) + An_M, \tag{8.72}$$

$$\frac{n_M}{n_L} = \exp(-\hbar\omega/k_B T) = \frac{BI(\omega)}{BI(\omega) + A}. \tag{8.73}$$

This can be rewritten

$$\frac{BI(\omega)}{A} = \{\exp(\hbar\omega/k_{\mathrm{B}}T) - 1\}^{-1}. \tag{8.74}$$

And substituting for $I(\omega)$ from (8.71) gives

$$\frac{B}{A} = 4\pi^2c^2/\hbar\omega^3. \tag{8.75}$$

This argument shows that spontaneous emission A becomes more and more predominant as the frequency ω increases. The dependence of the ratio is on the cube of ω; one can therefore understand why the maser (microwave amplification by the stimulated emission of radiation) involving the lowest frequencies was the first to be invented, and the laser (light amplification . . .) followed, despite considerable odds in the ω^3 factor, and it is proving difficult to extend the laser region much into the near ultra-violet despite very considerable effort. The prospects of an X-ray laser seem remote.

Actually this discussion should be accompanied by some numbers to illustrate the considerable achievement of the laser. As a figure of merit we can calculate from (8.75) the intensity of radiation necessary to equate stimulated and spontaneous emission. This is a qualitative threshold for laser action. We find at $\omega = 10^{11}\ \mathrm{s}^{-1}$ (wavelength about 2 cm), $I(\omega) = 3\times10^{-20}\ \mathrm{J\ m^{-3}}$. This corresponds to about 3×10^4 photons per cubic metre – a very, very dilute energy density. It appears that spontaneous emission at microwave frequencies is a rare and virtually negligible phenomenon. At optical frequencies the picture is rather different. When $\omega = 3\times10^{15}\ \mathrm{s}^{-1}$ corresponding to green light (5000 Å) we find the threshold to be $I(\omega) = 7\times10^{-7}\ \mathrm{J\ m^{-3}}$. This is equivalent to a photon density of 2×10^{13} photons per cubic metre, or a unidirectional light flux of $7\times10^{-7}c = 20\ \mathrm{W\ m^{-2}}$ – quite a considerable energy density. Now, taking into account the fact that in a successful laser stimulated emission must dominate considerably over spontaneous emission we begin to see why lasers are necessarily intense light sources. Of course, we have calculated the energy density *within* the optical resonator which exists for the express purpose of increasing the light intensity inside the laser medium in order to overcome this threshold, and the useful light emitted from the laser tube may be considerably weaker. In addition, the only part of the spontaneous emission which degrades the stimulated output is that which travels along the axis of the laser tube; spontaneous emission is uniformly distributed in all directions, and only a small fraction of its intensity lies within the limited solid angle of the resonator.

8.7.3 *The coherence properties of stimulated emission.* One of the most unusual features of laser radiation is its natural collimation. The radiation emitted by a laser is usually parallel as far as the theoretical limit for a beam of finite diameter according to diffraction theory. This fact alone shows that the emitted wave must be spatially coherent across its entire wavefront. The coherence properties of laser radiation arise as a result of the mechanism of stimulated emission.

When the stimulating radiation strikes an excited atom it causes the atom to radiate in phase with it. Otherwise there would be destructive interference between the stimulating and radiated waves. Now the situation when a stimulating wavefront passes through an excited laser medium is just like a 'real-life Huygens' principle' situation; every point on the wavefront re-radiates in phase with the stimulating wavefront, and the re-radiated waves reinforce one another only in the forward direction (§ 2.7.1). In any other direction they cancel out. This means that whatever the direction of the stimulating wave, the stimulated wave reinforces and continues it. Thus the wave is spatially coherent. In addition, the stimulating and stimulated waves have the same phase; this results in time-coherence to a much higher degree than in any other light source.

8.7.4 *Practical limits to the coherence of laser sources.* The above remarks are usually limited by practical considerations. The spatial coherence of the laser beam is limited by inhomogeneities in the laser medium or the resonator mirrors, through which the beam passes many times. In practice this results in a rather mottled wavefront leaving the laser. It is usual, for careful work, to filter out the inhomogeneities by focusing the beam to a point which is matched to a tiny pinhole. By choosing the size of the pinhole correctly, all the mottled components (Fourier components with non-zero spatial frequencies) are blocked and a uniform, homogeneous, spatially coherent beam emerges (Fig. 8.19). The time-coherence of the beam is limited in a different way. Ideally the linewidth of a Fabry–Pérot interferometer which includes an amplifying medium should be zero (§ 7.5.2) implying infinite coherence time. In practice we have to allow for mechanical instabilities and temperature changes which result in fluctuations of the length of the cavity and thus of the resonance wavelength. In addition, any amount of spontaneous emission which is present will add new uncorrelated wave-packets and reduce the coherence time. Another effect which is sometimes disturbing occurs when the laser operates at several frequencies at once. Each one corresponds to a maximum of the Fabry–Pérot spectrum (§ 7.5.1). One sees immediately that the emission

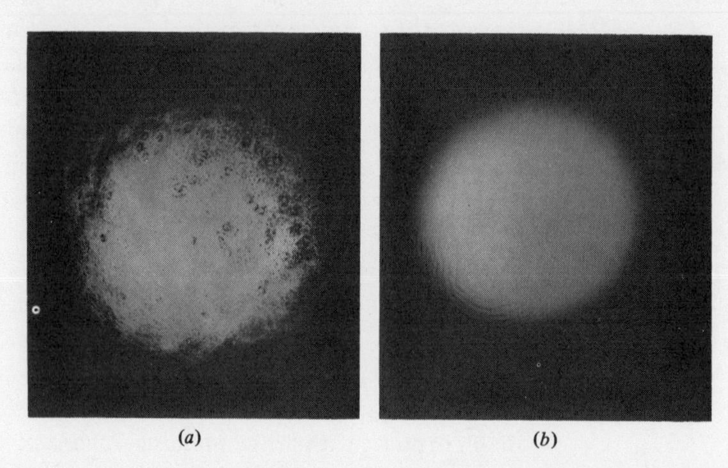

Fig. 8.19. Spatial inhomogeneities in a laser beam (*a*) as it leaves the laser, (*b*) after filtering through a pinhole filter.

spectrum now consists of a number of equally spaced line components, and so the temporal coherence function is periodic (the transform of the spectral intensity). Now the various peaks in the Fabry–Pérot function satisfy

$$n\lambda = 2L; \quad \omega = 2\pi c/\lambda = \pi cn/L, \tag{8.76}$$

where L is the cavity length and n is any integer. The periodicity in $\gamma(\tau)$ is thus $2\pi/$(periodicity in ω) which is just $2L/c$. The coherence length of the emitted wave, $l_c = c\tau_c$ is L – the length of the laser. But although the coherence function has fallen to zero at distance L, the coherence has revived by a distance $2L$, and continues periodically. Much more complicated coherence functions can be found in lasers which also allow transverse modes (§ 7.5.3).

8.8 Fluctuations in light beams: photon statistics

The experiments described in §§ 8.4.1 and 8.5.1 initiated a wave of interest in the fluctuations of partially coherent light beams during the late 1950s. The invention of the laser intensified this interest, because its output constituted a completely new type of light beam, and various treatments of the subject according to both classical and quantum theory have since been published. In this section we shall present a very simplified treatment of the classical theory of light fluctuations; what is interesting about it is that it produces results which can be directly related to the Bose characteristics of the photon, despite the fact that quantum theory has not been introduced.

Before studying the light beam itself, we shall ask what exactly one measures in an experiment to detect fluctuation. The answer is, of course, an electric current; we actually observe the emission of photoelectrons from the cathode of a photomultiplier. Any treatment of the subject must take this into account. Very briefly, the emission process can be described as follows. An electromagnetic field of frequency ω is incident on the cathode. Electrons in the cathode, in a stationary quantum state of energy E_1, feel the electromagnetic field as a perturbation at frequency ω and can then be excited to a new state of energy $E_2 = E_1 + \hbar\omega$. The probability of excitation within a certain period is proportional to the mean light intensity $\overline{I(t)}$ during that period. If the upper level is an unbound electron state, the electron is ejected from the cathode and is observable. We shall assume this to be the case. During the excitation process the electron absorbs energy $\hbar\omega$ from the light beam, but the exact moment of electron emission is a random process for which we can only state the probability of its occurring during a given period.

Now experimentally we observe the ejection of photoelectrons from the cathode. The stream of photoelectrons fluctuates for two reasons. The first is that the ejection process is random, with a mean rate proportional to the intensity $I(t)$. The second is that $I(t)$ itself fluctuates, in a manner that we have already illustrated in § 8.2. Now the first of these fluctuations can be treated easily. Assuming there to be no correlation between the emission of the electrons, this process is described exactly by Poisson statistics. The probability of n electrons being emitted during a period T is (see any text on statistics)

$$p(n) = \mathrm{e}^{-\bar{n}}\bar{n}^n/n!, \tag{8.77}$$

where $\bar{n}$ is the mean number emitted in that period:

$$\bar{n} = T\overline{I(t)}\eta. \tag{8.78}$$

The factor η depends on the efficiency of the photocathode, and is assumed to be constant. From this distribution one can easily derive the size of the fluctuations, which is described by the variance

$$\overline{\Delta n^2} = \overline{n^2} - \bar{n}^2 \tag{8.79}$$

and for the Poisson distribution (8.77) is easily shown to be

$$\overline{\Delta n_1^2} = \bar{n}. \tag{8.80}$$

The second independent source of fluctuation gives an additional variance $\overline{\Delta n_2^2}$, so that the observed variance is

$$\overline{\Delta n^2} = \overline{\Delta n_1^2} + \overline{\Delta n_2^2}. \tag{8.81}$$

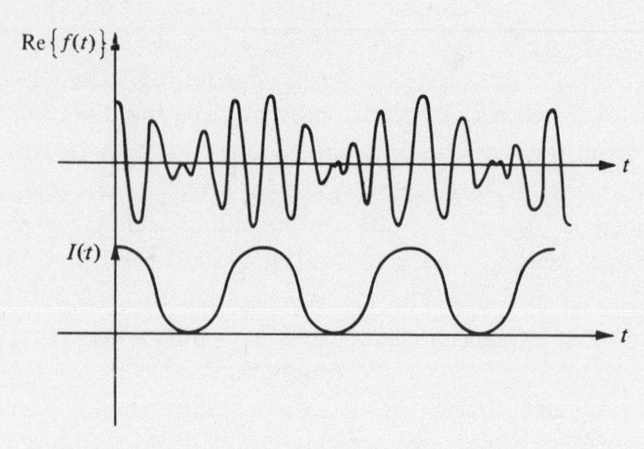

Fig. 8.20. Amplitude and short-term average intensity.

Remembering that the source of fluctuation is now that of $I(t)$ we write

$$\bar{n} = \eta T\overline{I(t)}, \tag{8.82}$$

$$\overline{n^2} = \eta^2 T^2 \overline{I(t)^2}. \tag{8.83}$$

We can use a very simple model to illustrate the type of results obtained. In § 8.2.1 we have described a partially coherent light wave by the Fourier synthesis of several continuous waves in a small frequency interval. If we had used only two waves of frequencies $\omega_0 \pm \frac{1}{2}\epsilon$ we should have created beats of the form

$$f(t) = 2E_0 \exp(\mathrm{i}\omega_0 t) \cos(\tfrac{1}{2}\epsilon t). \tag{8.84}$$

Now, $I(t)$ is the short-term average of the intensity $|f(t)|^2$ taken over a few cycles of ω_0 and (Fig. 8.20)

$$I(t) = |f(t)|^2 = 4E_0{}^2 \cos^2(\tfrac{1}{2}\epsilon t). \tag{8.85}$$

We can evaluate $\overline{I(t)}$ and $\overline{I(t)^2}$ for this very primitive beat model. Using (8.85) we take averages over many beat periods ($T \gg \pi/\epsilon$):

$$\overline{I(t)} = 4E_0{}^2 \overline{\cos^2(\tfrac{1}{2}\epsilon t)} = 2E_0{}^2, \tag{8.86}$$

$$\overline{I(t)^2} = 16E_0{}^4 \overline{\cos^4(\tfrac{1}{2}\epsilon t)} = 6E_0{}^4. \tag{8.87}$$

Substituting in (8.82) and (8.83), we have

$$\begin{aligned}\overline{\Delta n_2{}^2} &= \overline{n^2} - \bar{n}^2 = \eta^2 T^2 (6E_0{}^4 - 4E_0{}^4) \\ &= \tfrac{1}{2}\eta^2 T^2 \overline{I(t)}^2 = \tfrac{1}{2}\bar{n}^2.\end{aligned} \tag{8.88}$$

Adding together the two results (8.80) and (8.88) we find the variance in the total number of photoelectron counts:

$$\overline{\Delta n^2} = \bar{n} + \tfrac{1}{2}\bar{n}^2. \tag{8.89}$$

The first term in (8.89) represents the 'natural' fluctuations in the number n resulting from the random emission process; the second term $\frac{1}{2}\bar{n}^2$ represents 'excess fluctuations' resulting from the structure of the light wave. We must emphasize that in the above treatment we selected a very particular example of a fluctuating light wave, but nevertheless got an answer which is very close to that obtained by a full statistical treatment (Mandel, 1958). In fact the correct result is that $\overline{\Delta n^2}$ is a function of the measuring time T, varying from $\bar{n}+\bar{n}^2$ when $T \ll \tau_c$ to $\bar{n}+(2\tau_c/T)\bar{n}^2$ when $T \gg \tau_c$. However, the full calculation is too long to be reproduced here.

What is certainly intriguing about the above result is that it reproduces the variance calculated for a gas of Bose–Einstein photons. When $T \ll \tau_c$ the Heisenberg uncertainty principle does not allow us to distinguish the wave from a pure sine wave and therefore all photons associated with it must be considered as being in the same quantum state. For such an assembly the variance in the number of photons, considered as Bose–Einstein particles, is

$$\overline{\Delta n^2} = \bar{n} + \bar{n}^2 \tag{8.90}$$

(see e.g. Landau & Lifshitz, 1963). The resemblance is remarkable.

The above arguments show that when a cathode is illuminated by a continuous partially coherent wave, the emission of photoelectrons is not a random process governed by Poisson statistics. The photoemissions are therefore correlated. Another way of looking at the result is to calculate the auto-correlation function for the intensity $I(t)$, i.e.

$$c(\tau) = \overline{I(t)I(t+\tau)}. \tag{8.91}$$

When $\tau \gg \tau_c$, $I(t)$ and $I(t+\tau)$ are uncorrelated – this is the meaning of the coherence time τ_c – and so

$$\lim_{\tau\to\infty} c(\tau) = \overline{I(t)}^2, \tag{8.92}$$

When $\tau \ll \tau_c$, $I(t)$ and $I(t+\tau)$ are equal and

$$\lim_{\tau\to 0} c(\tau) = \overline{I(t)^2}. \tag{8.93}$$

The transition from one limit to the other occurs around τ_c. As a result, since $\overline{I(t)^2}$ must always be greater than $\overline{I(t)}^2$, the excess correlation appears as a peak in $c(\tau)$ near zero τ (Fig. 8.21). This phenomenon is

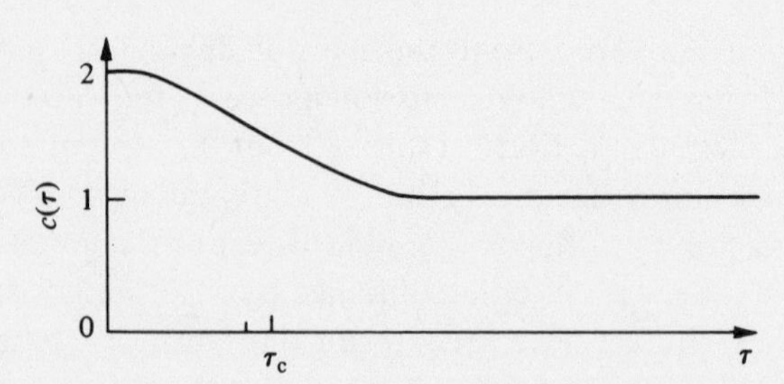

Fig. 8.21. Excess correlation in photoelectron counts when $\tau \lesssim \tau_c$.

called 'photon bunching'; if one photoelectron is observed, the probability of observing another during a time-interval $\delta t \ll \tau_c$ within the period τ_c following it is greater than the probability in the same interval δt later on. Substituting from (8.86) and (8.87) would suggest the ratio between the two probabilities to be 3 : 2; in fact the full calculation shows it to be just 2.

9

Optical instruments and image formation

9.1 Some aspects of the geometrical theory of instruments

As we have seen in Chapter 1, one of the first applications of the newly discovered physics of the seventeenth century was the use of lenses for the production of magnified images of objects. At first single lenses were used, but soon the advantages of lens combinations were discovered and the microscope and telescope were invented.

In the following sections we shall deal briefly with the principles of these instruments; since they are dealt with thoroughly in elementary textbooks we shall give only enough detail for the more important practical and theoretical aspects to be properly appreciated.

9.1.1 *Magnifying glass.* The magnifying glass – or its more elaborate form, the eyepiece – is an important component of most optical instruments and its function should be clearly understood. Its main purpose is to allow us to bring an object well within the near point by throwing an image to a distance at which the eye can produce a sharply focused image. This is illustrated in Fig. 9.1 in terms of paraxial rays, i.e. rays that are near to, and nearly parallel to, the axis.

The magnifying power is a more useful quantity than the magnification; it can be defined in two equivalent ways:

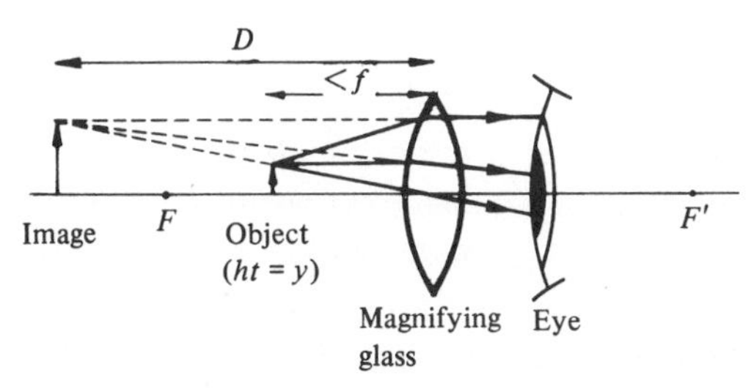

Fig. 9.1. Ray diagram of magnifying glass. In practice, $f \ll D$ and the object distance would be nearly equal to f.

(a) the ratio of the angle subtended at the eye by the image formed by the instrument to the angle subtended at the eye by the object at the near point;

(b) the ratio of the linear dimensions of the retinal image produced with the lens to those of the largest clear retinal image produced without the lens. Since the latter image is produced when the object is at the near point – a distance D, say, from the eye – the magnifying power can easily be seen (Fig. 9.1) to be approximately

$$\frac{y}{f}\Big/\frac{y}{D}=\frac{D}{f}.$$

The single lens is now used only for informal purposes. For precise work eyepieces are generally used and the general principles upon which these are based are described in Appendix III.

9.1.2 *Microscope*. The essential principle of the microscope is that a lens (called the objective) of high power is used to form a highly magnified real image of the object; this object must therefore be placed just outside the focus of the lens: an eyepiece is then used to examine this image. A ray diagram of the formation of the image is shown in Fig. 9.2. It will be seen that the final image is virtual and inverted.

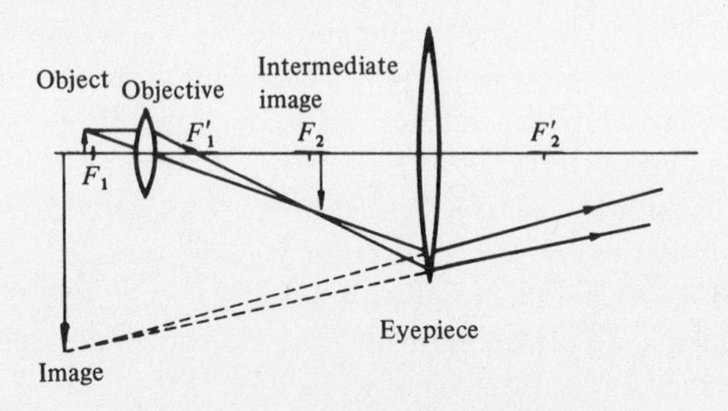

Fig. 9.2. Ray diagram of microscope.

9.1.3 *Telescope*. The principle of the telescope is similar to that of the microscope, except that, of course, the object is not accessible and must therefore be considered as being infinitely distant from the lens, which is again called the objective. The intermediate image is thus formed in the second focal plane of the objective and is very small. This image is then viewed by means of an eyepiece and is again virtual and inverted.

The geometrical theory of image formation in the microscope and telescope is straightforward and will not be dealt with here.

9.2 Practical considerations

9.2.1 *Entrance and exit pupils.* The elementary theory outlined in the previous sections says nothing about the sizes of the lenses, although in practice their dimensions are very important. It is wasteful to use large lenses if the outer parts do not contribute to the production of the image; it is wrong to use a lens that is so small that it reduces the efficiency of the others in the combination. The systematic study of these considerations is called the theory of stops, and a brief résumé of some of its aspects will be given here.

Three considerations are of importance. First, for efficiency we should try to see that each lens plays its full share in producing the image; secondly, the brightness of the image should be as high as possible and of reasonable uniformity; and thirdly, the image should not, over the field of view, have any appreciable distortion or loss of sharpness. None of these requirements can be *exactly* realized; we therefore design optical equipment around the most expensive item – the objective of a microscope or of a telescope – and make the other components match it as well as possible.

A stop – despite its name – is an opening in a coaxial system of lenses, a coaxial system being one for which the centres of curvature of all the refracting surfaces lie on a straight line – the axis. A stop may be an actual hole, such as the iris diaphragm of a camera, or it may be a lens itself. For a particular object point on the axis *one* stop must limit the cone of light forming the image (Fig. 9.3); this is called the *aperture stop.* The aperture stop is not necessarily the same component for different positions of the object, but since most optical instruments are used without much drastic variation, the aperture stop does not usually vary.

The image of the aperture stop in the lenses preceding it is called the *entrance pupil.* In the telescope or microscope this should be the objective itself, and in the camera it is the image of the iris diaphragm in the lens or in that part of the lens system that precedes it. The image of the aperture

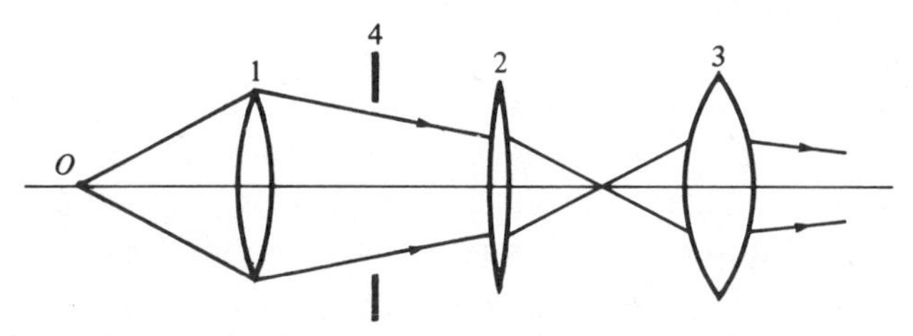

Fig. 9.3. System of three lenses with an object at O; lens 1 is the aperture stop. If the stop 4 is reduced in size, it could replace 1 as the aperture stop.

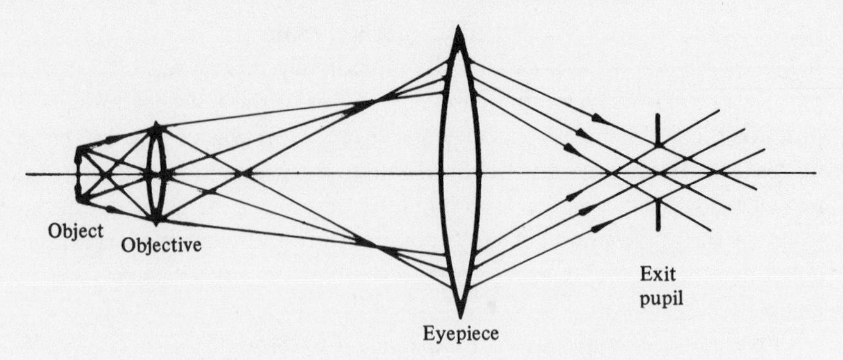

Fig. 9.4. Ray diagram for microscope showing position of exit pupil.

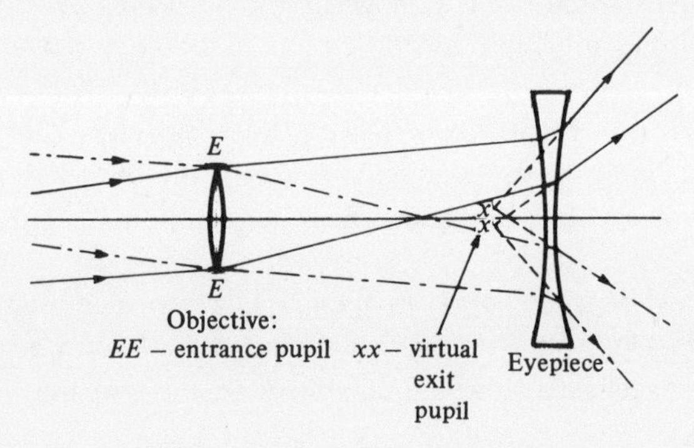

Fig. 9.5. Galilean telescope showing position of exit pupil.

stop in the lenses that follow it is called the *exit pupil*. In the telescope or microscope this is the image of the objective in the eyepiece, and, as we can see from Fig. 9.4, it is the plane where there is the greatest density of light. It is thus the place where the entrance pupil of the eye should be located. An optical system should therefore be designed so that its exit pupil is accessible; some instruments, such as the Galilean telescope (Fig. 9.5), do not satisfy this condition.

It must be remembered that the eye also forms part of any optical system that is used for direct viewing, and it is no use designing a system that is wasteful in under-using either the abilities of the eye or those of the instrument. The condition to be satisfied is that the exit pupil of the instrument should match the entrance pupil of the eye in both size and position but, as we have seen, this is not always possible.

9.2.2 *Definitions of aperture*. The cone of light that goes to form the centre of the image decides the brightness of the image. More precisely, the brightness depends upon the angle subtended by the entrance pupil at the centre of the object; for a camera lens this angle is determined by the *f*-number – the ratio of the focal length to the diameter of the entrance pupil. For the telescope, since all objects are effectively at infinity, the diameter of the objective – which is also the aperture stop and the entrance pupil – is usually given. For microscopes we use the quantity $\sin\theta$, where θ is the semi-angle of the cone subtended by the objective – which is also the aperture stop and the entrance pupil – at the centre of the object; this is called the *numerical aperture* (NA). If the medium in which the object is placed has a refractive index μ, the NA is $\mu \sin\theta$.

This section is not intended to be more than introductory. For a more detailed and quantitative discussion of these definitions, a textbook on the more technical aspects of optics should be consulted (e.g. Smith, 1966).

9.2.3 *Entrance window*. The entrance and exit pupils decide only what happens at the *centre* of the image; this is of course the most important part, but the image is always of finite size and off-axis parts should also be acceptable. For this requirement, other considerations enter. For example, we can see from Fig. 9.6 that, although the first lens of the system is

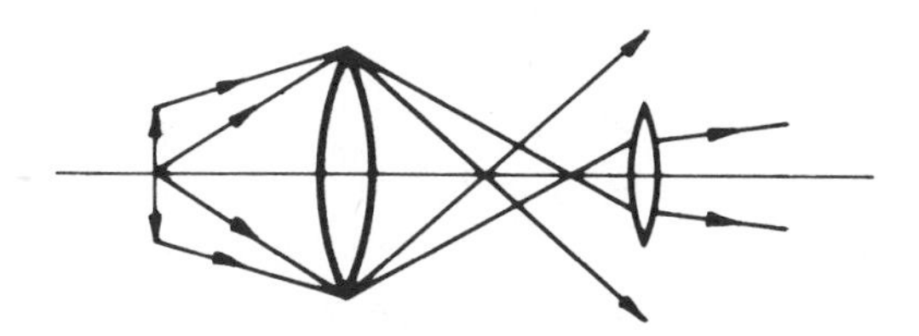

Fig. 9.6. Two-lens system showing how rays from periphery of object may not enter the second lens.

obviously the aperture stop, some of the off-axis rays passing through it will not pass through the second lens. In other words, the second lens decides *how much* of the object is seen. The stop that decides the visible area of the image is called the *field stop*.

The field stop is almost always an actual aperture inserted in the system. In its simplest form, in the camera, it is simply the edge of the plate or film holder, but in the microscope a special stop is inserted so that its image is in focus together with that of the object. This merely satisfies the aesthetic requirements that the image field shall have a sharp edge.

We can therefore see that the image of the field stop is important. The image in the lenses preceding it is called the *entrance window*, since this defines the area of the object that we are looking at; and the image in the lenses following it is called the *exit window*, since this defines the area of the image seen. The entrance window of a camera is thus the image of the plate-holder in the lens. The view-finder of a camera attempts to show the user a close approximation to the entrance window; it cannot be precise because the lens of the view-finder is not in the same place as that of the camera (except, of course, in a reflex camera).

We may now ask how the field stop is chosen. As we have already pointed out, it gives a sharp edge to the field of view. The choice of the position of this boundary is based upon two considerations: the simpler is that we do not want to have obvious variation in the brightness of the image, and it is usual to limit the field so that the brightness does not fall, at the edges, to less than half the value at the centre; the other is that the aberrations (Appendix II) necessarily increase with distance from the centre, and the image should be cut off before these aberrations become apparent.

There are other aspects also of the theory of stops – such as depth of field and of focus – but we shall not deal with them since they do not enter into the subsequent topics dealt with in this book.

9.2.4 *Illuminating systems.* For microscopes, the object must be illuminated in some way, either naturally or artificially. If the latter, the intensity of illumination must usually be as great as possible and therefore a condensing lens is used to focus light from a source onto the object. From the geometrical-optics point of view, the design of this component of a microscope may seem quite trivial; from the physical-optics point of view, however, we shall see later (§ 9.5.1) that a condenser can be comparable in complexity with a microscope objective if the highest possible resolution is to be obtained.

9.3 Diffraction theory of image formation: the Abbe theory

We know that the theory in the first two parts of this chapter cannot be complete; the description of image formation in terms of rays can be only an approximation, since light is a wave motion and cannot adequately be described in terms of rays. We must now ask 'What difference does the wave nature of light make to our conclusions?'

We shall try to answer this question in two ways. The first (§ 9.3.1) is a rather intuitive method which can be applied only to a periodic object, but shows clearly that the wave nature of light limits the resolution attainable with optical instruments. This method was first discussed by

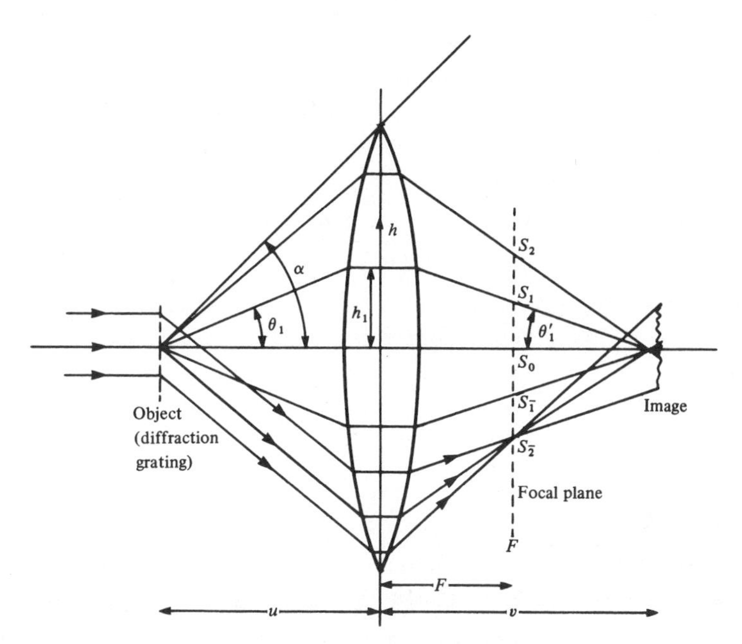

Fig. 9.7. Formation of image of diffraction grating. Five orders of diffraction are shown, producing five foci S in the plane F. The complete ray bundle is only shown for the $\bar{2}$ order. The angular semi-aperture of the lens is α.

Abbe in 1867. The second method (§ 9.3.3) is an analytical method based on the scalar theory of diffraction which shows how the exact relationship between object and image can be formulated in general as a double Fourier transform.

9.3.1 *The image of a periodic object.* We have seen in § 7.4.2 that if parallel light falls normally upon a diffraction grating several orders of diffraction are produced (Fig. 9.7). Each order can be considered as a plane-wave and the set of plane-waves can be refracted by a lens so that they converge individually to a set of points S in the focal plane F of the lens and then continue so that they all overlap in the plane I. Here they form a complicated interference pattern; this pattern is the image.

The advantage of taking a diffraction grating as an object is that the process of image formation can easily be seen to be divisible into two distinct stages. Firstly we have the stage between O and F. Here we have produced, in plane F, the Fraunhofer diffraction pattern of the object. Secondly we have the stage between F and I. The orders $S_2, S_1, \ldots, S_{\bar{2}}$ behave like a set of equally spaced point sources and the image is their diffraction pattern. Thus the process of image formation appears to be divisible into two diffraction processes, applied sequentially.

The second diffraction process in this example can be analysed without difficulty. Each pair of orders S_j and $S_{\bar{j}}$ produces Young's fringes in the plane I. If the object grating has spacing d, the order S_j appears at angle θ_j given, for small angles, by

$$\theta_j \approx \sin \theta_j = j\lambda/d \tag{9.1}$$

(the small-angle approximation will be seen later to be unnecessary: § 9.3.2). By simple geometry one can see from Fig. 9.7 that

$$\theta_j \approx \tan \theta_j = h/u, \tag{9.2}$$

$$\theta'_j \approx \tan \theta'_j = h/v, \tag{9.3}$$

and so

$$\theta'_j \approx u\theta_j/v. \tag{9.4}$$

The waves from the first orders, S_1 and $S_{\bar{1}}$ converge on the image at angles $\pm\theta'_1$ and thus form periodic fringes with spacing

$$d' = \lambda/\sin \theta' \approx \lambda v/\theta_1 u = vd/u. \tag{9.5}$$

Thus a magnified image has been produced; the magnification is v/u. Fringes from the higher orders produce harmonics of this periodic pattern, and contribute to determining the detailed structure of the image. The finest detail observable in the image is determined by the highest order of diffraction which is actually transmitted by the lens.

The zero order contributes a constant amplitude, the zero-order term of the Fourier series. This zero-order term is of crucial importance. Without it, the interference pattern of the first orders would appear to have half the period of the image, because we observe intensity, and not amplitude; the function $\sin^2 x$ has half the periodicity of $\sin x$. However, the addition of the constant restores the correct periodicity to the intensity, as is illustrated in Fig. 9.8.

9.3.2 *The Abbe sine condition.* The theory in the last section might suggest that a faithful image would be built up only if the angles of diffraction were kept small. This would be an intolerable condition for an efficient optical instrument and Abbe realized that the image would be faithful if the ratio $\sin \theta/\sin \theta'$ rather than θ/θ' were constant for all zones of a lens. If we had had, exactly:

$$\frac{\sin \theta_j}{\sin \theta'_j} = M \tag{9.6}$$

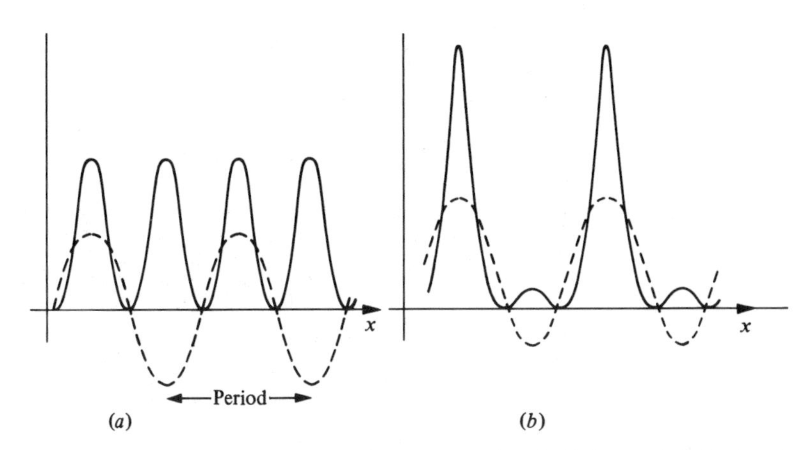

Fig. 9.8. (*a*) Squaring a sine curve. The square has half the period of the sine function. (*b*) Squaring the sine curve after addition of a constant: the function $\frac{1}{4}(1+2\sin x)^2$, corresponding to the image of a grating produced using the 0 and ±1 orders. The period of the image is the same as that of the sine function.

we should then have the period of the fringes in the image

$$d'_j = \lambda/\sin\theta'_j = M\lambda/\sin\theta_j = Md_j. \tag{9.7}$$

The harmonics would then have exactly the right periods to fit the fundamental d'_1 and the image would be as perfect as possible. One important system having this property is described in Appendix III, and forms the basis of high-power microscope objectives. The Abbe sine condition does *not* state that $\sin\theta/\sin\theta'$ is a constant in *any* imaging system, but requires that this condition be met if the system is not to produce aberrations when large angles θ and θ' are used. It is not satisfied, for example, by a simple lens.

9.3.3 *Image formation as a double process of diffraction.* In § 9.3.1 we introduced, in a qualitative manner, the idea that image formation can be considered as a double process of diffraction, and in § 9.3.2 we saw the Abbe sine condition to be necessary for its realization. In this section we shall formalize the approach mathematically in one dimension. There is no particular difficulty in the extension to two dimensions.

The calculation is based on the scalar-wave theory (§ 6.1.3) and will assume the object and image to be small compared to their distances from the lens.

The scalar-wave amplitude leaving the object is $f(x)$, which is taken to represent the object (Fig. 9.9). Then the amplitude of the wave reaching

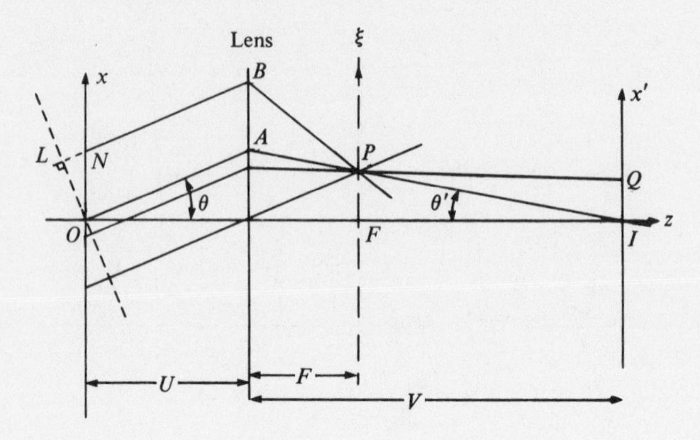

Fig. 9.9. Ray diagram for the demonstration of the image–object relationship.

the point P on the focal surface is, following the treatment and notation of § 7.3.1 in one dimension, the Fourier transform of $f(x)$:

$$a(u) = \exp(\mathrm{i}k_0\overline{OAP}) \int_{-\infty}^{\infty} f(x) \exp(-\mathrm{i}ux)\mathrm{d}x, \tag{9.8}$$

where u defines the point P:

$$u = k_0 l = k_0 \sin\theta. \tag{9.9}$$

Now the amplitude $b(x')$ at Q in the image plane can be calculated using Huygens' principle. The optical distance from P to Q is

$$\begin{aligned}\overline{PQ} = PQ &= (PI^2 + x'^2 + 2x'PI \sin\theta')^{\frac{1}{2}} \\ &\approx PI - x' \sin\theta' \end{aligned} \tag{9.10}$$

when $x' \ll PI$; this is our assumption of a small image.

If the Abbe sine condition is obeyed,

$$\sin\theta = M \sin\theta', \tag{9.11}$$

where M is a constant. We therefore write, from (9.9)

$$PQ = PI - x'u/Mk_0, \tag{9.12}$$

whence the amplitude at Q is

$$\begin{aligned} b(x') &= \int a(u) \exp(\mathrm{i}k_0 PQ)\, \mathrm{d}u \\ &= \int \exp(\mathrm{i}k_0 PI) a(u) \exp(-\mathrm{i}x'u/M)\, \mathrm{d}u. \end{aligned} \tag{9.13}$$

This is the second Fourier transform in the problem. Inserting (9.8) into (9.13) we write the relationship between the image $b(x')$ and the object $f(x)$:

$$b(x') = \int_{-\infty}^{\infty} \left[\exp \{ik_0(\overline{OAP} + PI)\} \int_{-\infty}^{\infty} f(x) \exp(-iux)\, dx \right] \times \exp(-iux'/M)\, du. \tag{9.14a}$$

The combined phase factor $\exp\{ik_0(OAP + PI)\}$ appears at first sight to be a function of the point P, and hence of the parameter u. This is indeed true if the planes of O and I are chosen arbitrarily. But if they are *conjugate* planes then by Fermat's principle (§ 8.7.2) the optical path from O to I is *independent* of the point chosen for P, and the factor can be written as a constant $\exp(ik_0\overline{OI})$ and can be taken outside the integral. We are left with the integral

$$b(x') = \exp(ik_0\overline{OI}) \int_{-\infty}^{\infty} \left\{ \int_{-\infty}^{\infty} f(x) \exp(-iux)\, dx \right\} \times \exp(-iux'/M)\, du, \tag{9.14b}$$

which can be evaluated by the Fourier inversion theorem. Substituting (3.35) we then have:

$$b(x') = \exp(ik_0 OI) f(-x'/M). \tag{9.15}$$

This equation represents the well-known fact that the image is an inverted copy of the object, magnified by the factor M. (If we employ Gaussian optics (Appendix I) we can put $M = V/U$, but this is not necessarily correct for a system employing large angles.)

The above result, first proved by Zernike, can be stated simply: an optical image can be represented as the Fourier transform of the Fourier transform of the object. It applies exactly only if the lens is well corrected; i.e. it obeys the Abbe sine rule and the optical path $\overline{OPI}$ is independent of the point P.

9.3.4 *Illustrations of the diffraction theory of image formation.* In the previous section we have shown theoretically that the imaging process can be considered as a double Fourier transform, when the image is illuminated coherently. We shall now describe some experiments to confirm this result. The experiments are carried out in an imaging system, Fig. 9.10, which allows the comparison of the intermediate transform and the final image.

A transparent object is illuminated with a parallel coherent beam. It is imaged by a converging lens. We observe the illumination in the focal

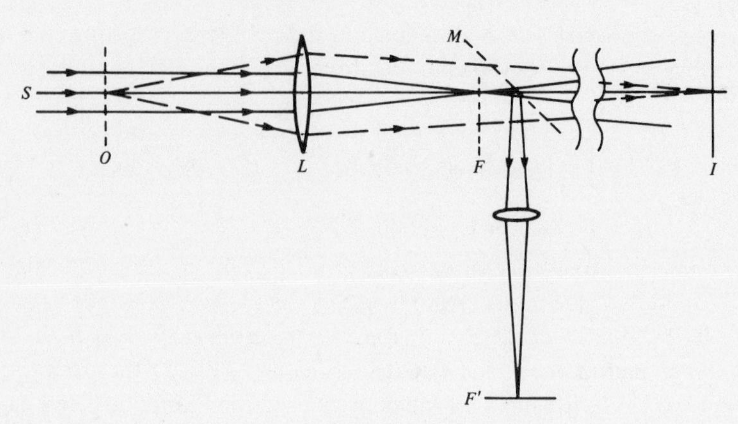

Fig. 9.10. Optical imaging system to illustrate the Abbe theory of image formation. The object O is illuminated by parallel coherent light from the laser source S. It is imaged by the lens L onto a distant screen I. The half-silvered mirror M is used to form a separate image of the focal plane F at F'. The various spatial filtering masks are inserted in the plane F, and F' and I are observed simultaneously (Fig. 9.11).

plane to be the Fraunhofer diffraction pattern of the object, and the image to be the Fourier transform of that diffraction pattern. The first stage, that of the formation of the Fraunhofer diffraction pattern, has been adequately illustrated in Chapter 7. To confirm that the second stage is also a Fourier transform, we can modify the transform in the focal plane by additional masks or obstacles and observe the resultant changes in the final image. Such processes are called 'spatial filtering' by analogy with the corresponding process in the time-domain in electrical circuits. Spatial filtering has some very important applications which will be discussed in detail in later sections. Let us consider first an object consisting of a piece of gauze. It is two-dimensional, and is basically periodic, although there are deviations from exact periodicity as well as defects such as blocked holes. We image it in the system of Fig. 9.10. The diffraction pattern (plane F) is shown in Fig. 9.11(a). It contains well-defined spots, corresponding to the periodic component of the gauze, and an additional light distribution corresponding to the non-periodic components (cf. § 7.7.3). The complete image of the gauze is shown in (b).

We now insert various masks into the plane F, and thereby cut out parts of the diffraction pattern. For example, if the mask transmits only orders on the horizontal axis (c) the image becomes a set of vertical lines (d); this is the object which would have given (c) as its diffraction pattern. Similarly, a mask which transmits only the orders $(0, \pm 1)$, $(\pm 1, 0)$, (e),

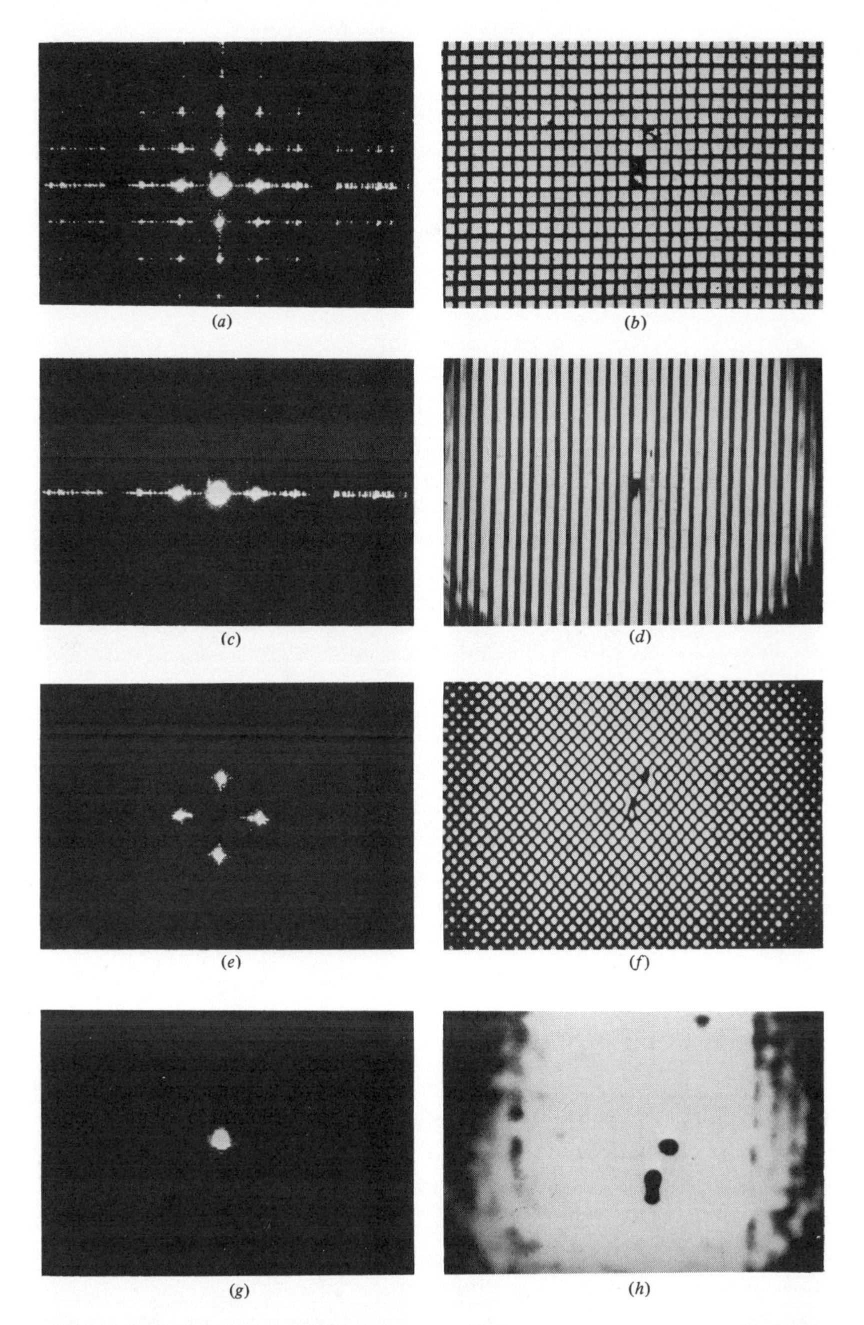

Fig. 9.11. Illustrating the Abbe theory of image formation. On the left are the selected portions of the diffraction pattern, and on the right the corresponding images. (Figure continued overleaf.)

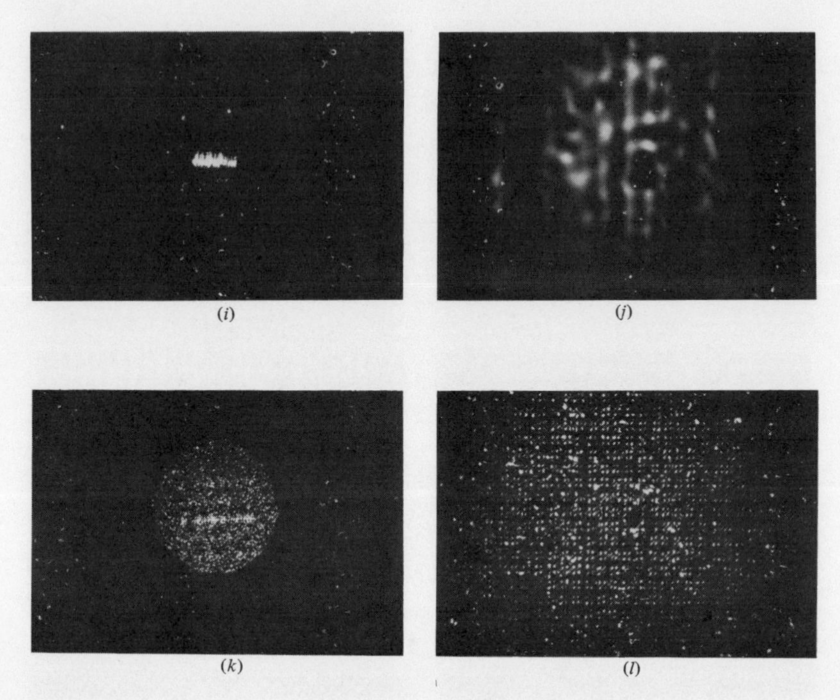

Fig. 9.11 (*continued*).

gives us a different gauze (*f*). But the irregularities are the same, because they contribute to the diffraction pattern at all points. The zero order alone, (*g*), gives us an image in which no gauze is visible, but only the irregularities – particularly the blocked holes. Finally, small regions of the diffraction pattern (*i*) and (*k*) remote from the centre emphasize different aspects of the deviations from exact periodicity, (*j*) and (*l*).

9.3.5 *The phase problem.* A question which is always asked at this point refers to the possibility of separating the two stages of the image-forming process. Suppose we were to photograph the diffraction pattern in the focal plane and in a separate experiment illuminate the photograph with coherent light and observe its diffraction pattern. Should we not have produced the diffraction pattern of the diffraction pattern and have reconstructed the image? The flaw in the argument concerns the phases of the diffraction pattern. The illumination $a(u)$ is a complex quantity containing both amplitudes and phases. Photography records only the intensity $|a(u)|^2$ and the phase is lost. A second diffraction process as suggested above would be carried out in ignorance of the phases, and

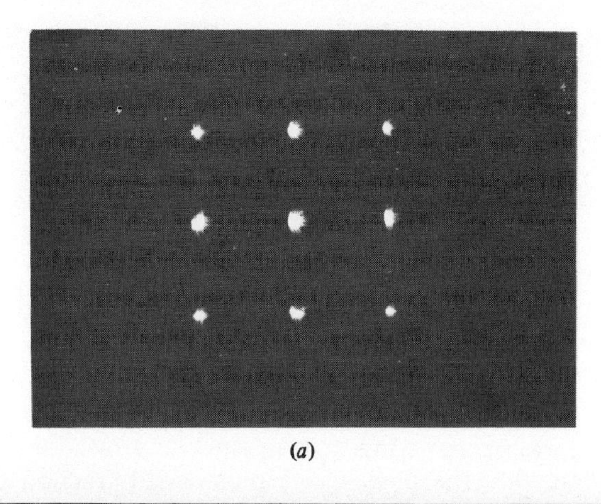

(*a*)

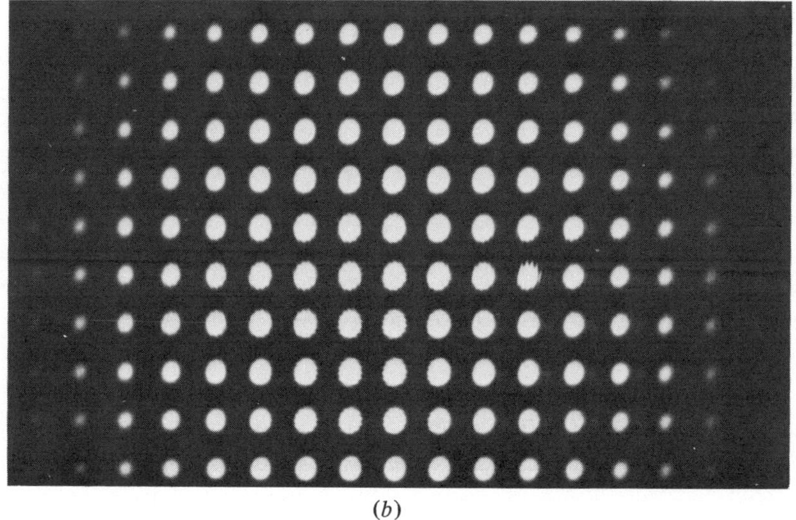

(*b*)

Fig. 9.12. The mask (*a*) has nine holes with areas proportional to the calculated amplitudes of the diffraction spots from the gauze. Its diffraction pattern is shown in (*b*).

therefore would be unlikely to give the right answer. In fact, the second process would assume the phases all to be zero, and would indeed give the correct image if this were so. For example, in the example of Fig. 9.11 it is quite easy to show that the centre nine spots in the diffraction pattern all have the same (zero) phase. Ignoring the rest of the spots, we can observe the diffraction pattern of a photograph of these nine spots. The nine spots and their pattern are shown in Fig. 9.12. The diffraction pattern is indeed a tolerable reconstruction of the gauze.

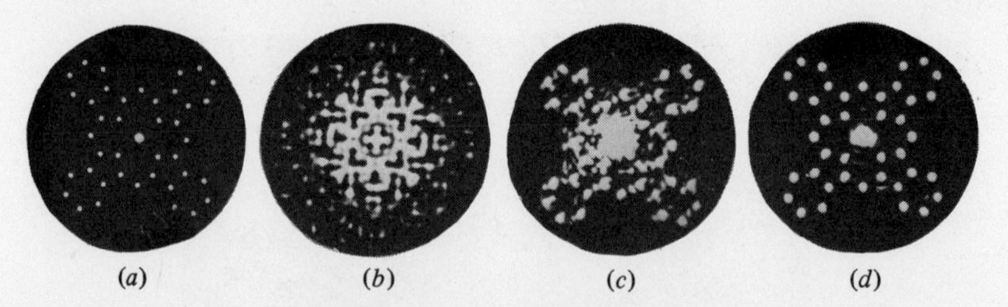

Fig. 9.13. (*a*) Mask representing chemical molecule with heavy atom at centre; (*b*) transform of (*a*); (*c*) transform of (*b*) without precautions to eliminate phase changes. The pattern of (*a*) is quite recognizable; (*d*) transform of (*b*) placed between optical flats with cedar-wood oil to eliminate phase changes. The pattern of (*a*) is quite clear.

The problem that the phases in a recorded diffraction pattern are generally unknown is called the 'phase problem', and its solution is of central importance to the interpretation of X-ray diffraction patterns, where one wants to create an image of a crystal whose diffraction pattern intensity has been recorded. Under certain circumstances the problem has a solution, and allows the image to be reconstructed by calculation. The diffraction pattern itself is the first Fourier transform in the imaging process; the calculation provides the second.

For example, consider a planar object (a chemical molecule) having a centre of symmetry and a strong scattering atom at that centre. The centre of symmetry gives the molecule the properties of a two-dimensional even function (§ 3.3.1) whose Fourier transform is all real, and therefore has phase zero or π, corresponding to positive or negative cosine terms. Now since the object has a strong positive scattering point at the centre, all the cosine terms must add at this point, and so they must all have zero phase. Thus the phases are all known, and in this case retransforming a photographic transparency of the diffraction pattern (which process *assumes* all the phases to be zero) gives a good reconstruction of the object. This method is used practically in X-ray crystallography for crystals in which heavy metal atoms can be introduced into known positions in the unit cell; it will be discussed further in § 11.3.2.

These conditions may be produced artificially by making a symmetrical arrangement of holes with a larger one at the centre (Fig. 9.13(*a*)) and covering the smaller ones with gauze to reduce their contributions to the optical transform. The resulting transform is shown in Fig. 9.13(*b*). This is placed between the lenses of the optical diffractometer, sandwiched between optical flats and immersed in cedarwood oil which has the same refractive index as the film emulsion; the resulting diffraction pattern (Fig. 9.13(*c*)) can be seen to be a good image of the object.

In visible optics the phase problem can be overcome by recording the diffraction pattern holographically. This technique will be discussed in § 9.7.

9.4. The effect of finite apertures

The light which forms the image in an optical system is limited angularly by the aperture stop (§ 9.2.1). In this section we shall use the Abbe theory of image formation in an attempt to understand how the size of the aperture stop and the coherence of the illumination affect the characteristics of the image, and in particular how they limit the resolution attainable. It will appear that the limits of perfect coherence and perfect incoherence of the illumination can be treated fairly clearly; the intermediate case of partially coherent illumination is complicated and the results can only be indicated in rather general terms.

9.4.1 *A completely incoherent object.* The simplest and most well-known resolution criterion is that of Rayleigh and applies to the case of a self-luminous and incoherent object; it is usually applied to an astronomical telescope, because stars certainly fulfil the requirements of self-luminosity and incoherence; but it equally well applies to a microscope observing, for example, a fluorescent object.

If we consider a single point on the object, we have seen in § 6.2.3 that we observe in the image plane the Fraunhofer diffraction pattern of the limiting aperture, on a scale determined by the distance between exit pupil and image. For a single lens the aperture stop and exit pupil coincide with the lens itself.

An extended object can be considered as a collection of such points, and each one produces a similar Fraunhofer diffraction pattern in the image plane; because the sources are incoherent we add intensities of the various patterns to get the final image.

The Rayleigh resolution criterion arises when we consider two neighbouring points on the object, separated by a small angle. If the exit pupil has diameter D, its diffraction pattern has intensity (from (7.45)) at deviation θ proportional to:

$$I(\theta)=\{J_1(\tfrac{1}{2}k_0D\sin\theta)/(\tfrac{1}{2}k_0D\sin\theta)\}^2. \tag{9.16}$$

Rayleigh considered that the two points on the object are distinguishable if the central maximum of one lies outside the first minimum of the other. Now the function (9.16) has its first zero when

$$\tfrac{1}{2}k_0D\sin\theta_1=\pi D\sin\theta_1/\lambda=3.832. \tag{9.17}$$

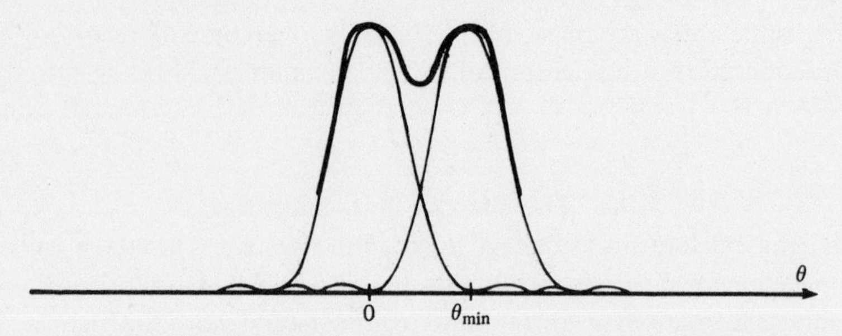

Fig. 9.14. Addition of the images of two pinholes incoherently illuminated. The thin lines show the intensity curves and the thick line their sum.

The angle θ_1 is the minimum resolvable angular separation of such incoherent sources: the resolution limit is thus

$$\theta_{\min} = \theta_1 = 3.832\lambda/\pi D = 1.22\lambda/D. \tag{9.18}$$

This criterion is a little coarser than what can actually be resolved in practice, but illustrates the principle adequately. Notice that only the angular separation of the sources enters the result.

9.4.2 *Extension of Rayleigh's method to coherent sources.* As an illustration of the difficulties arising when the sources are coherent, we can try to use the same method again. If the sources are coherent and have the same phase, we must add the amplitudes of the diffraction patterns

$$A(\theta) = J_1(\tfrac{1}{2}k_0D \sin \theta)/(\tfrac{1}{2}k_0D \sin \theta). \tag{9.19}$$

Again, one could use the Rayleigh criterion blindly and get the same result as (9.18) but we should look more carefully at the significance of the criterion. In Fig. (9.14) we show the intensity as a function of position on a line through the images of two incoherent sources separated by $\theta_{\min}$. It is the sum

$$I_{\text{incoh}} = \left[\frac{J_1(\frac{1}{2}k_0D \sin \theta)}{\frac{1}{2}k_0D \sin \theta}\right]^2 + \left[\frac{J_1\{\frac{1}{2}k_0D \sin (\theta - \theta_{\min})\}}{\frac{1}{2}k_0D \sin (\theta - \theta_{\min})}\right]^2. \tag{9.20}$$

This curve definitely has a minimum at $\theta = \frac{1}{2}\theta_{\min}$, indicating that the images are resolved. Now look at the equivalent result when the sources are coherent, and we add amplitudes before squaring to find the intensity. We then have

$$I_{\text{coh}} = \left[\frac{J_1(\frac{1}{2}k_0D \sin \theta)}{\frac{1}{2}k_0D \sin \theta} + \frac{J_1\{\frac{1}{2}k_0D \sin (\theta - \theta_{\min})\}}{\frac{1}{2}k_0D \sin (\theta - \theta_{\min})}\right]^2, \tag{9.21}$$

which is illustrated in Fig. 9.15. This curve has no minimum at the centre, indicating that the images are not resolved.

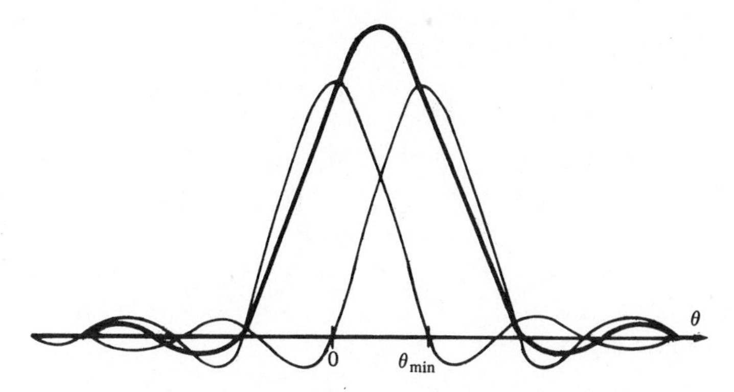

Fig. 9.15. Addition of the images of two pinholes coherently illuminated. The thin lines show the amplitude curves and the thick line their sum.

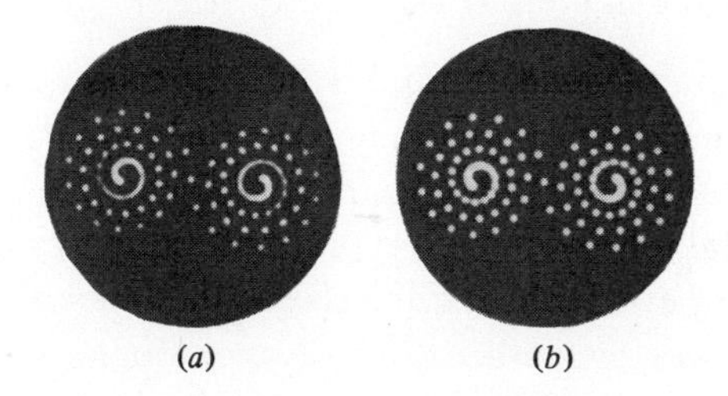

Fig. 9.16. Comparison between coherent and incoherent imaging, made with the apparatus of Fig. 9.17: (*a*) coherent; (*b*) incoherent. Note the better resolution in (*b*).

This argument is not general, in that we had to assume a given phase relation between the sources in order to demonstrate the result. However, in general it can indeed be said that incoherent illumination results in the better resolution. Fig. 9.16, in which images of a test object (which will become very familiar in what follows!) have been formed under identical conditions apart from the coherence of their illumination, demonstrates the improvement in resolution obtained with incoherent illumination.

9.4.3 *Application of the Abbe theory to resolution.* Most microscopes work with coherent or partially coherent illumination, because of the small dimensions of the object and the practical difficulties (§ 9.5.1) of producing truly spatially incoherent light. The Abbe theory discussed in § 9.3.1 applies to coherent illumination and is therefore fairly appropriate to a discussion of resolution by a microscope.

We therefore return to the model of a periodic object. The resolution that can be obtained with a given lens or imaging system is, as mentioned in § 9.3.1, limited by the highest order of diffraction which the finite aperture of the lens will admit. If the object has period d, the first order appears at angle θ_1, given by

$$\sin \theta_1 = \lambda/d, \tag{9.22}$$

and in order to image an object with such a period, the angular semi-aperture of the lens, α, must be at least as large as θ_1. Thus the smallest period which can be imaged is given by

$$d_{\min} = \lambda/\sin \alpha. \tag{9.23}$$

It is usual to anticipate the possible immersion of the object in a medium of refractive index μ, where the wavelength is λ/μ, and to write $d_{\min}$ in terms of the numerical aperture (§ 9.2.2):

$$d_{\min} = \lambda/\mu \sin \alpha = \lambda/\mathrm{NA} \tag{9.24}$$

is the coherent imaging resolution in this case. Now we have assumed in the above discussion that the illumination is parallel to the axis, and acceptance of the zero and two first orders is necessary to form an image with the correct period. In fact, the correct period will be imaged if the zero order and *one* first order alone pass through the lens. So we can improve the resolution by illuminating the object with light travelling at angle α to the axis, so that the zero order just passes through; then the condition for the first order on one side to pass through also is that

$$d_{\min} = \lambda/2\mu \sin \alpha = \lambda/2 \times \mathrm{NA}, \tag{9.25}$$

where we have used the result for Fraunhofer diffraction in oblique illumination from § 7.3.3. This result represents the best that can be achieved with a given lens, and is usually quoted as the ultimate resolution limit. It is interesting that it also applies, as will be shown below, to partially coherent illumination of a form which is much used in microscopy.

9.4.4 *Illustration of coherent resolution.* This theory may be illustrated by using a simple extension of the optical diffractometer; as shown in Fig. 9.17, an extra lens is used to form a direct image of the diffracting object. It might be asked why these illustrations are not also carried out with the apparatus described in Fig. 9.10. The reason is that the use of the diffractometer, whose light source is a conventional mercury arc, allows us to show the effects of partially, as well as completely, coherent

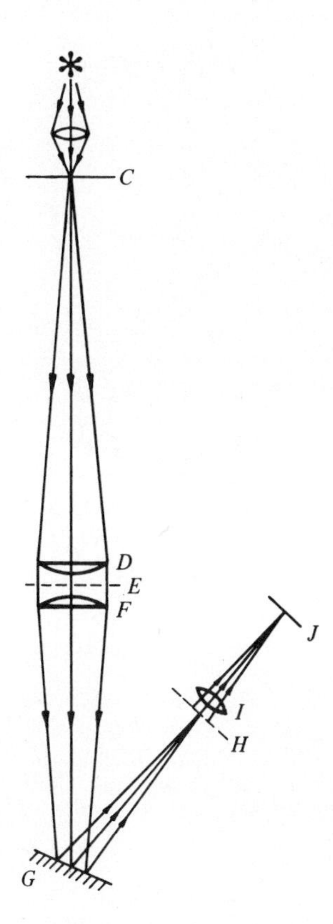

Fig. 9.17. An image of the pinhole C is formed in the plane H by the optical system DFG. The diffraction pattern of an object in the plane E is formed in the plane H and its image J is formed by the auxiliary lens I. (From Taylor & Lipson, 1964.)

illumination (Fig. 9.16). These effects are very important in practical microscopy. We can now investigate the changes that occur in an image if the optical transform is limited in some way.

For example, suppose that we have a general object as shown in Fig. 9.18(a); its transform is shown in Fig. 9.18(b). We may place a series of successively smaller holes over the transform, and we can then see how the image is affected. The succession of diagrams is self-explanatory.

The resolution limit imposed by a finite aperture can also be considered as an application of the convolution theorem (§ 3.6.4). As far as the reconstruction of the image is concerned, the optical transform of the object has been limited in extent by multiplying it by a finite aperture, a function which is zero outside certain limits. The reconstruction, which is

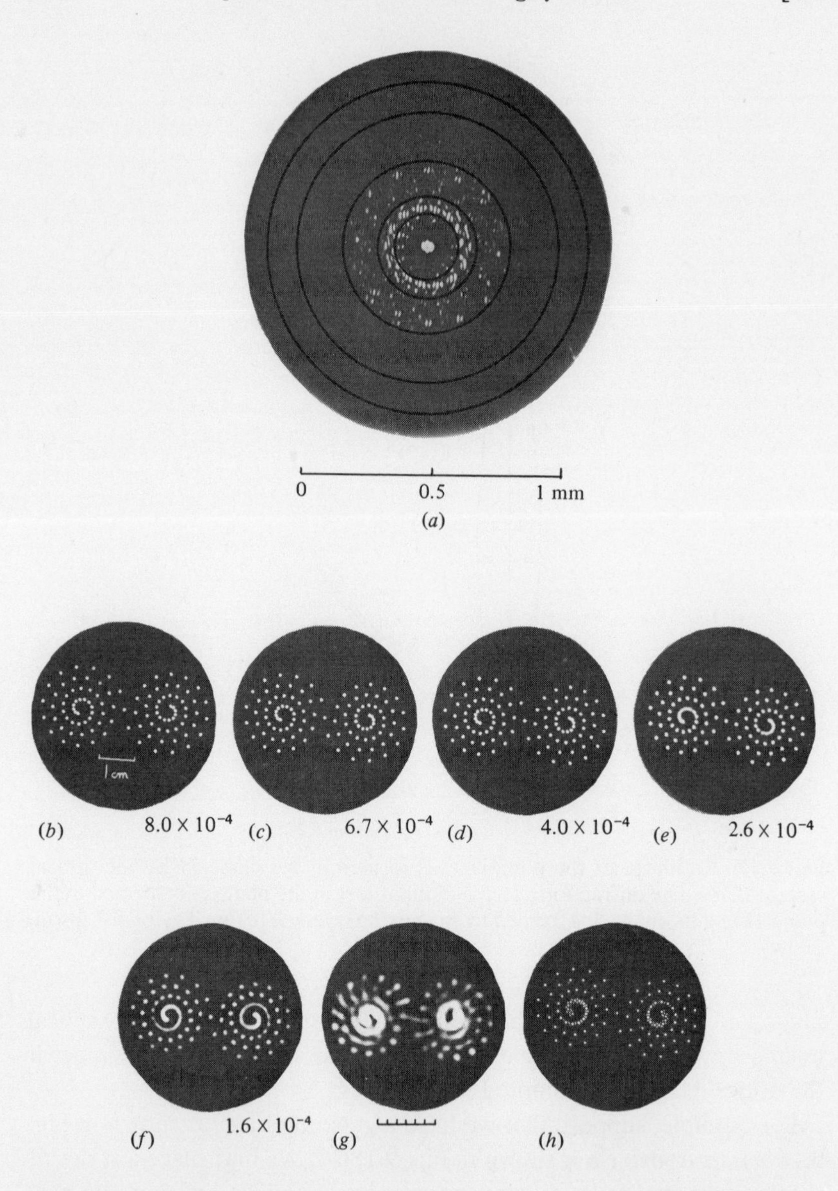

Fig. 9.18. (*a*) Diffraction pattern of set of holes shown in (*b*). The circles indicate the apertures used to limit the transform. (*b*)–(*g*) Images of object shown in (*b*), with different numerical apertures. The apertures used are shown as circles superimposed on the diffraction pattern shown in (*a*). The divisions of the line below (*g*) show the Rayleigh limit of resolution. (*h*) is the image formed by the part of the diffraction pattern between the second and third circle from the centre. This illustrates apodization (§ 9.6.7). The image is sharper than the object, but contains false detail.

the transform of the limited transform of the object, is therefore the convolution of the original object with the transform of the aperture. The latter transform exists for a certain region around zero, and therefore the convolution can be described roughly by saying that each point in the original sharp object is blurred into a region of this size.

9.4.5 *False detail.* It is often thought that the sole effect of reducing the aperture of an optical system is to reduce the resolution by blurring fine detail. There can be, however, a more serious defect – the production of false detail, which may be finer than the limit of resolution. The use of an optical instrument near to its limit of resolution is always liable to produce effects of this sort; when the Abbe theory (§ 1.7.1) was first announced, many microscopists adduced such effects as evidence that the theory was unacceptable. Even now, when the theory is fully accepted, it is sometimes forgotten in dealing with images produced by, for example, the electron microscope.

The formation of false detail can be conveniently illustrated in the framework of Fig. 9.11. Suppose that the aperture stop of the instrument limits the transform 9.11(*b*) to the centre five orders only (Fig. 9.19(*a*)). The image is then illustrated by Fig. 9.19(*b*). Notice the formation of bright spots on the crosses of the gauze wires. One can easily see the origin of these spots by considering one dimension of the pattern. If each wire has a thickness equal to half the period l, the centre row of diffraction spots (Fig. 9.11(*d*)) have amplitudes: zero order 1, first orders $2/\pi$, second orders 0, third orders $-2/3\pi$, etc. Reconstruction of the zero and first orders *only* gives an image amplitude

$$\begin{aligned} b(x) &= 1+\frac{2}{\pi}\exp(2\pi \mathrm{i}x/l)+\frac{2}{\pi}\exp(-2\pi \mathrm{i}x/l) \\ &= 1+\frac{4}{\pi}\cos(2\pi x/l), \end{aligned} \tag{9.26}$$

where the three terms are respectively the 0, +1 and −1 orders. Now $4/\pi$ has the value 1.28, so that $b(x)$ contains some negative regions. On squaring, to obtain the image intensity $|b(x)|^2$ we find positive (bright) regions just at the centres of the black lines. A two-dimensional analogue of this argument confirms the false detail.

9.5 Imaging with partially coherent illumination

9.5.1 *The importance of the condenser.* As far as geometrical optics is concerned, the condenser in a microscope merely serves to illuminate the

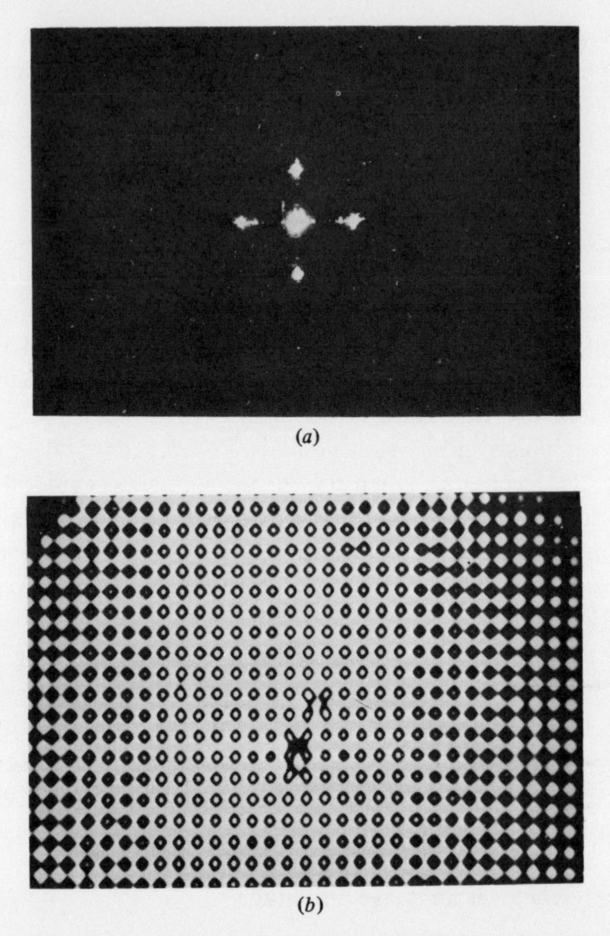

(a)

(b)

Fig. 9.19. False detail produced by imaging with a restricted region of the diffraction pattern (cf. Fig. 9.11).

specimen strongly. According to the wave theory, however, the coherence of the incident light is important, and the condenser has as much importance as any other part of the optical system. In practice, for producing the best results from a microscope used to its limits, the condenser must be as good as the objective, and usually has exactly the same design.

The reason for this requirement can best be expressed in terms of coherence. Ideally, as we shall show in § 9.5.2, the object should be illuminated in completely incoherent light, which we could obtain by a general external illumination from a large source such as the sky. But this would be very weak, and we increase it by using a lens to focus a source of light onto the object. An image, however, cannot be perfect, and each point on the source is represented by an area of finite size on the object. In other words, neighbouring points on the object are illustrated by partially coherent light.

In practice, two forms of illumination are widely used. The first is called critical illumination and is obtained by forming an image of a source (Fig. 9.20(*a*)) directly on the object by means of a condenser. This arrangement, however, has the defect that irregularities of the source can affect the image formed. An arrangement that does not have this defect is called Kohler illumination and is shown in Fig. 9.20(*b*). An extended source is

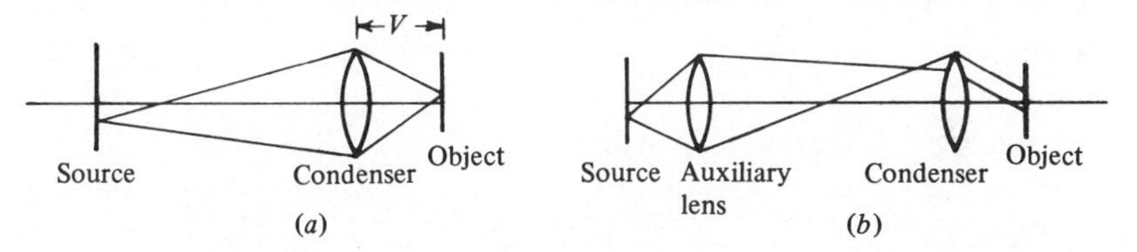

Fig. 9.20. Types of incoherent illumination: (*a*) critical; (*b*) Kohler.

used, and although any one point on the source gives parallel coherent illumination at a certain angle, the total illumination from all points on the source is indeed almost incoherent. This is because the individual coherent plane-waves have random phases and various directions of propagation and therefore add up with different relative phases at each point in the field.

For either of the above condensing systems the spatial coherence distance z_c is simply related to the numerical aperture as follows. In the case of critical illumination each uncorrelated point on the object produces an image which is at least the size of the Airy disc (§ 7.3.5) which has diameter $2.4\, V\lambda/D$. Spatial coherence exists over a region of about this size, so that the spatial coherence distance can be written in terms of the NA:

$$z_c \sim \lambda/\mathrm{NA}. \qquad (9.27)$$

If the condenser is not an ideal lens, aberrations increase the size of the coherence region above this value. Similarly in Kohler illumination, when coherent plane-waves of random phase are incident at angles in the range α, the coherence diameter (§ 8.6.1), within which the plane-waves add together with a given phase relation, is

$$z_c \sim \lambda/\sin\alpha = \lambda/\mathrm{NA}. \qquad (9.28)$$

It would seem that aberrations in the condenser are less critical for this type of illumination.

9.5.2 *Resolution with partially coherent illumination.* In § 8.4.4 we describe partially coherent illumination in a manner similar to Kohler as a set of incident plane-waves which have no phase relation between them and lie in a certain range of angles to the axis of the imaging lens. The resultant image is then a superposition of the coherent images produced by each plane-wave separately. These images do not interfere, because of the random phase differences between them. The plane-waves which are incident at angle α to the axis will allow the maximum resolution of $\lambda/2\times\text{NA}$ to be achieved; and so, if the illumination cone reaches out to this angle, we have created the right conditions for maximizing the resolution. Since there are waves in all directions at angle α around the cone, the high resolution is isotropic.

We can summarize this result by stating that the best resolution that can be achieved is

$$d_{\text{min}}=\lambda/2\times\text{NA} \quad \text{(Abbe)}, \tag{9.29}$$

which arises when the object is illuminated with a cone of light at least as steep as the acceptance cone of the imaging lens. Under these conditions the coherence distance z_c is approximately equal to the resolution limit, so that neighbouring resolution elements are substantially uncorrelated. As a result of this incoherence of neighbouring resolvable points, the resolution limit (9.29) should be very similar to that given by the Rayleigh criterion. The rewriting of the result (9.18) in terms of the numerical aperture indeed gives a result only 20 per cent larger:

$$d_{\text{min}}=0.6\lambda/\text{NA} \quad \text{(Rayleigh)}. \tag{9.30}$$

9.6 Applications of the Abbe theory: spatial filtering

Optical instruments can be used without more than a cursory knowledge of how they work, but a physicist should know more than this. He can then fully appreciate their limitations, can find the conditions under which they can be best used, and – most important – may find ways of extending their use to problems that cannot be solved by conventional means. Examples of such procedures will be described in the following sections.

The methods to be described are all operations carried out at the transform stage of an imaging system, i.e. in the focal plane of the imaging lens. The microscopic methods date from the 1930s; pattern recognition by coherent optical methods became popular in the late 1960s when continuous-wave lasers became available as common laboratory tools, and has subsequently been shown to be rather limited in its application to real problems.

In §§ 9.6.1–9.6.3 we shall give a description of the three most well-known spatial filtering systems – dark ground, schlieren and phase-contrast microscopy. In § 9.6.4 we shall discuss a new method – modulation contrast. In § 9.6.5 we shall carry out a calculation to compare the images produced by the three methods in the case of a simple one-dimensional phase object. In §§ 9.6.6 and 9.6.7 we shall discuss interference microscopy and apodization.

9.6.1 *Dark-ground illumination.* Suppose that we wish to observe a very small non-luminous object. If we use the ordinary method of illumination, in which the incident light bathes the specimen and enters the objective, it is likely that the amount of light scattered will be so small that it will be negligible compared with that contained in the incident beam and the object will not be seen. We can avoid this difficulty by arranging that the incident light is directed obliquely at the specimen so that it does not enter the objective; for observation of Brownian motion, for example, it is usual to direct the incident light perpendicularly to the axis of the microscope.

The method is adequate if we merely want to know whether a scattering object is present or not; it is equivalent to forming an image by using only a small, off-centre, part of the transform, and this will not give much information about the nature of the object. For a reasonable image of the object we must use as much as possible of the transform, and this is achieved in practice by cutting out the incident beam precisely and using as much of the scattered light as possible (Fig. 9.21). This procedure is useful for an object that produces little absorption, but because of variations in thickness or refractive index, introduces appreciable phase differences into the waves transmitted by various parts – a phase object (§ 7.2.5).

The method can be illustrated quite simply by producing the transform of an object and placing a small black spot over the central peak. We have chosen as object a pattern of holes punched in a thin transparent film (Fig. 9.22(a)); since the film is not optically uniform the transform (Fig. 9.22(b)) is rather diffuse. A *very* small ink-spot on a piece of glass – it need not be optically flat since only a small area is used – is then placed

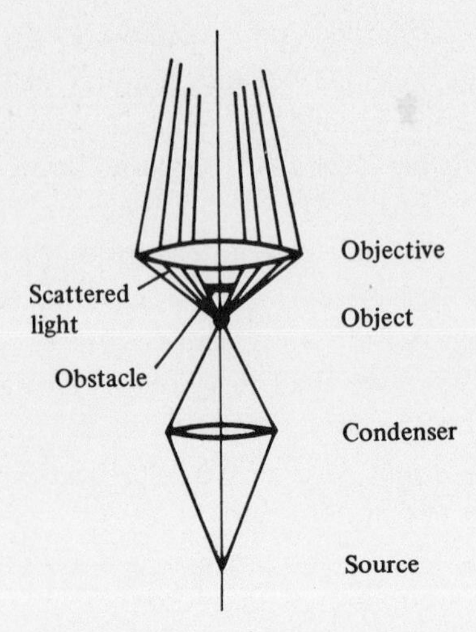

Fig. 9.21. Principle of dark-ground illumination.

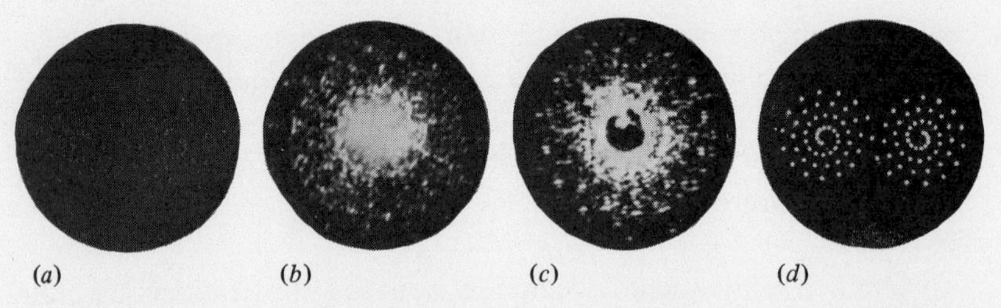

Fig. 9.22. (*a*) Image of a pattern of holes in cellophane sheet; (*b*) diffraction pattern of (*a*); (*c*) as (*b*) with small spot over centre; (*d*) image formed from (*c*).

over the centre of the transform (Fig. 9.22(*c*)) and the final image (Fig. 9.22(*d*)) compared with that obtained when the ink-spot is absent.

Since the original pattern consists of holes in a transparent sheet, it cannot easily be seen by the naked eye; in the instrument, however, the holes appear clearly because the edges scatter light, but the *intensity* inside and outside the holes is the same (Fig. 9.22(*a*)). But when the image is formed with the centre of the transform blacked out, there is considerable contrast (Fig. 9.22(*c*)) between intensities in the holes and the background.

9.6.2 *Schlieren patterns.* An alternative method is to cut off the central peak by a knife-edge, thereby cutting off half the transform as well. In practice the object to be studied is placed in an accurately parallel beam

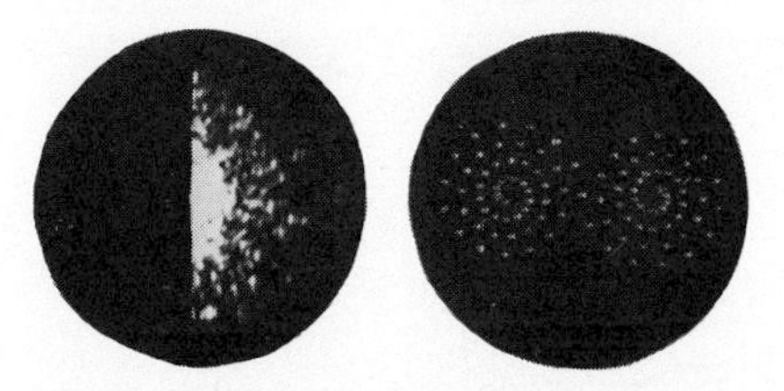

Fig. 9.23. Image formed by cutting out half the diffraction pattern in Fig. 9.22(*b*).

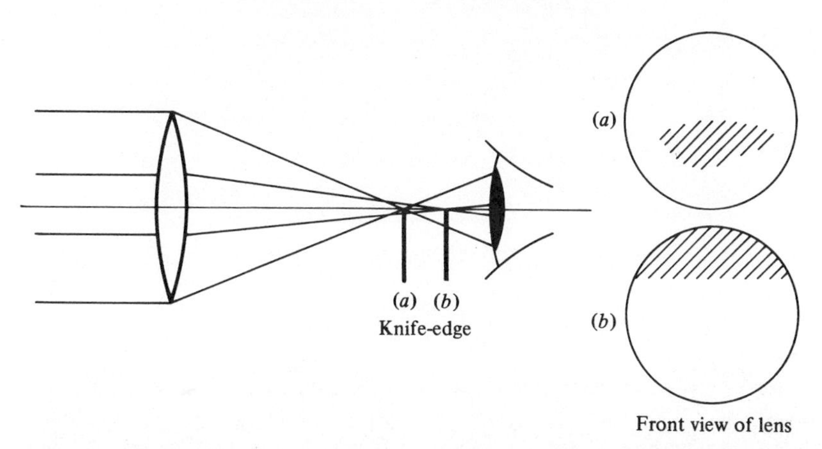

Fig. 9.24. Appearance of lens suffering from spherical aberration when subject to Foucault knife-edge test: (*a*) when knife-edge is in marginal focal plane; (*b*) when knife-edge is in paraxial focal plane.

which is brought to a focus by a lens accurately corrected for spherical aberration. A knife-edge is then translated in the focal plane of the lens until it just overlaps the focus. A clear image of the object can then be seen (Fig. 9.23). We have illustrated this method by the same object that we used for dark-ground illumination (Fig. 9.22(*a*)); the image now has some defects that will be discussed later.

The schlieren method has two by-products. First, it can be used to test a lens – the Foucault knife-edge test – for if a lens suffers from spherical aberration it will not be possible to put the knife-edge in a position to cut off only half the transform. Thus, if we observe the lens with the eye close to the knife-edge as it traverses the focal plane, the intensity of illumination across the surface will appear to change (Fig. 9.24); we can then deduce what modifications to make to the lens. The Foucault method can also be used to locate the focus of a lens extremely accurately.

A second use of the schlieren method is important in fluid dynamics. A wind tunnel in which the density of air is constant (and hence the

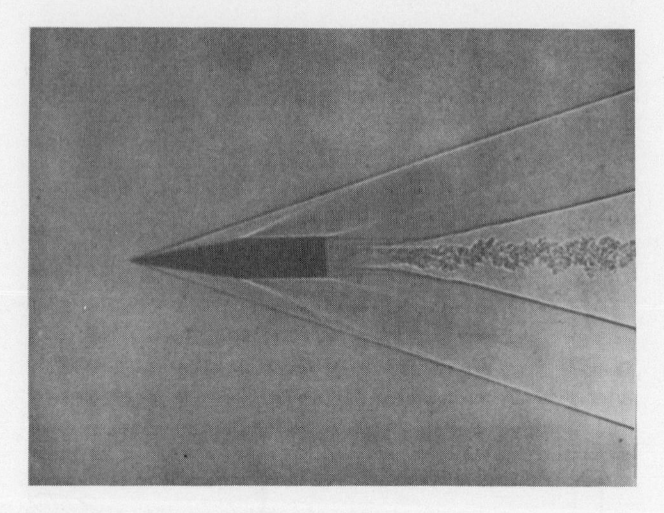

Fig. 9.25. Schlieren pattern of bullet-shaped object at Mach number 3.62. (From Binder, 1962, p. 245.)

refractive index is constant too) is an object with neither phase nor amplitude variations. The existence of waves or other disturbances in the tunnel will modify the density and refractive index in a non-uniform way, and thus produces an object with phase variation but still no amplitude variations. By using the schlieren technique, such variations can be studied visually as changes in intensity in the final image (Fig. 9.25).

An important difference between dark-ground systems and Schlieren systems is that the latter operate in one meridian only. The other difference in practice is that usually schlieren systems are adjusted to attenuate but not remove the zero-order light. This greatly increases the sensitivity by leaving a reference beam in the final image plane which the scattered light can interfere with.

9.6.3 *Phase-contrast microscopy*. Phase-contrast microscopy is another method of rendering visible the phase variations in a transparent object. It can be explained as follows.

Suppose that we represent the light amplitude transmitted by an object by a vector in the complex plane. In a phase object (§§ 7.2.5 and 9.6.1) the vectors representing the complex amplitude at various points on the object are all equal in length, but have different phase angles. In Fig. 9.26, OA_1, OA_2, OA_3 are typical vectors. The zero order of the diffraction pattern of the object is the sum of the vectors OA_1, OA_2, etc. This order

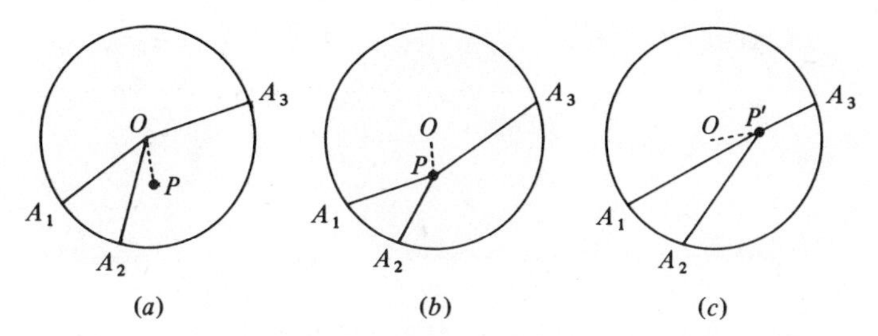

Fig. 9.26. Vector diagrams illustrating (*a*) normal imaging of a phase object, (*b*) dark ground, and (*c*) phase-contrast.

makes a uniform contribution to the image represented by the vector *OP*, which is the mean value of OA_1, OA_2, etc.

In the dark-ground method, we obstructed the zero order, and therefore subtracted the vector *OP* from each of the vectors $OA_1, OA_2, \ldots$. The resulting vectors $PA_1, PA_2, \ldots$ had different lengths and therefore intensity contrast was achieved (Fig. 9.26(*b*)).

The phase-contrast method, invented by Zernike in 1935, involves changing the phase of the vector *OP* by $\pi/2$, and therefore replacing it by the vector OP'. The new image-point vectors $P'A_1, P'A_2, \ldots$ once again have different lengths (Fig. 9.26(*c*)). This method has the advantage that all the light transmitted by the object is used in forming the image.

The phase-contrast method can be described simply, when the phase variations are small, by expanding the complex transmission function:

$$A\,\mathrm{e}^{\mathrm{i}\phi} \approx A + \mathrm{i}A\phi. \tag{9.31}$$

Changing the phase of the zero-order term by $\pi/2$ gives the function

$$\mathrm{i}A + \mathrm{i}A\phi \approx \mathrm{i}A\,e^{\phi}, \tag{9.32}$$

which has a real variation of intensity.

It is difficult to illustrate this device in any primitive way; putting a very thin plate of about 0.1 mm diameter over the centre of a transform is too difficult. We can, however, produce an arbitrary phase difference by putting a transparent sheet with a small hole in it over the transform, and the results of this are shown in Fig. 9.27.

In practice, the device is not as simple as we have described. As we have seen in § 9.5.1, we do not want to use completely coherent light, and if we use incoherent light there is no precise transform over the centre of which a thin plate – the phase plate – can be introduced. We must use a beam of finite angular dimensions in order to have sufficient intensity and also to produce a small depth of focus; otherwise the out-of-focus parts of the

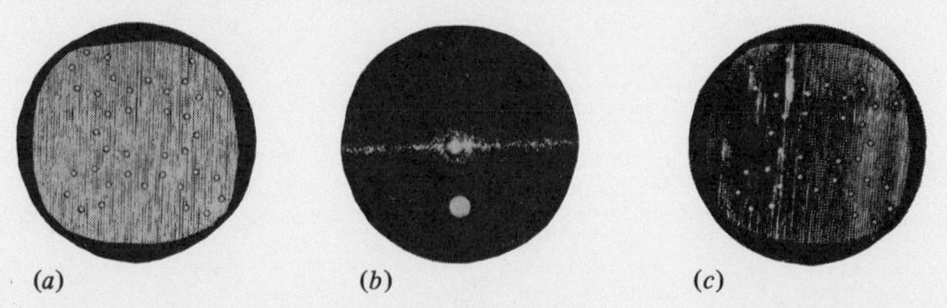

Fig. 9.27. (*a*) Pattern of holes punched in cellophane. (Note the irregularities produced by the method of production of the sheet.) (*b*) Fraunhofer diffraction pattern of (*a*). The white spot shows the size of hole in the thin transparent sheet placed over the centre of the pattern. (*c*) Image of (*a*) formed with hole in position. (Note the contrast between the holes and the background.)

image will be disturbing. A compromise is therefore necessary, and is effected as follows.

The beam from the source of light is limited by an annular ring (Fig. 9.28) placed between the source and the condenser; the condenser and the objective form a lens system which produces a real image of the ring above the objective. A phase plate is then inserted in the plane of this image to match its dimensions exactly. All undeviated light from the specimen must therefore pass through this plate. The final image is formed by interference between the undeviated light passing through the phase plate and the deviated light which passes by the side of it. The ideal conditions are only approximately obeyed, for some of the deviated light will also pass through the phase plate. The phase plate is constructed by the vacuum deposition of a dielectric material such as cryolite (Na_3AlF_6) onto an optically parallel glass support. The techniques of vacuum deposition have reached a level of perfection which makes control of the dimensions and thickness of such a plate a fairly simple process.

If the object produces only small differences of phase, the centre spot of the diffraction pattern is outstandingly strong and changing its phase produces too great a difference in the image. Therefore the phase plate is made to transmit only about 10–20 per cent of the light. It looks like a small dark ring on a clear background.

9.6.4 *Modulation-contrast microscopy.* It will be seen that phase-contrast microscopy, as practised, is greatly dependent upon the dark-ground principle as well, and it might therefore be asked whether the phase-contrast element is really justified. In fact, a new system called *modulation-contrast* has recently been devised; it is much simpler to produce and to use, and is claimed by its originators, Hoffman & Gross (1975), to be as

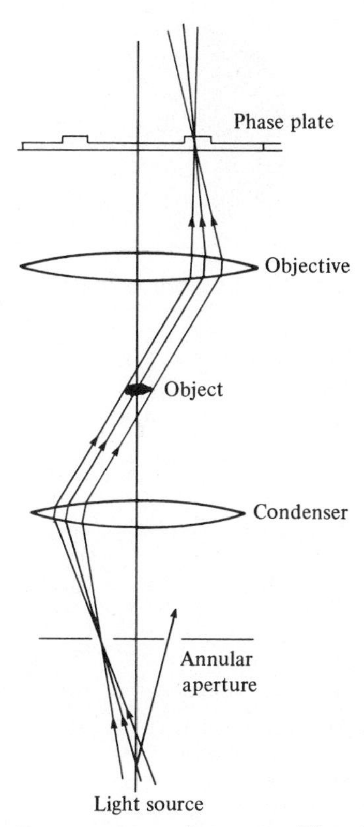

Fig. 9.28. Principle of phase-contrast microscope. The phase plate is coincident with the image of the annular ring formed in the condenser–objective lens system. Kohler illumination is used.

effective as phase-contrast. It is essentially based upon the idea of illuminating the specimen with an intensity gradient, and is therefore more general than the schlieren method.

In practice, a discontinuous function of intensity is used. The incident light is defined by a slit beneath the specimen, and in the plane conjugate to this slit is placed a screen with three different transmissions (Fig. 9.29) – dark (less than 1 per cent), grey (about 15 per cent) and bright (100 per cent) – with boundaries parallel to the slit. Published photographs of images show results comparable with those produced by phase-contrast. The method, unlike phase-contrast, is directional, but this property can sometimes be turned (literally) to advantage, when detail in some particular direction is being examined.

The components are easier to produce than those for phase-contrast, and it is possible to provide modulation plates with different properties to

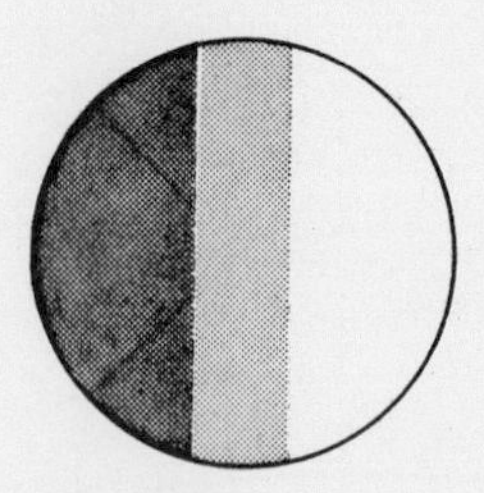

Fig. 9.29. The modulation-contrast plate.

suit different specimens; the bands of transmission (Fig. 9.29) can be of different extent or of different colours.

It can be seen, therefore, that the subject of microscopy is not stagnant; it is still capable of producing interesting new ideas.

9.6.5 *An analytical example illustrating dark-ground, schlieren and phase-contrast systems.* In this section we shall calculate the intensity distributions in the image of a simple one-dimensional phase object, when the filters discussed in §§ 9.6.1–9.6.3 are used. The object can be called a 'phase slit'; it is a transparent field containing a narrow strip of different optical phase from its surroundings. In one dimension, x, normal to the length of the strip we describe such an object by:

$$f(x) = \exp\{i\phi(x)\}, \tag{9.33}$$

where

$$\phi(x) = 0 \quad \text{when } |x| > a,$$
$$\phi(x) = \alpha \quad \text{when } |x| \leq a.$$

This function can be written as the sum of a uniform field and the difference in the region of the strip:

$$f(x) = 1 + (e^{i\alpha} - 1)g(x), \tag{9.34}$$

where $g(x) = 1$ in the region $|x| \leq a$, and is zero otherwise. (A normal transmitting slit of width $2a$ is simply represented by $g(x)$.) The transform of the function written this way is:

$$F(k) = \delta(k) + 2a(e^{i\alpha} - 1)\,\text{sinc}\,(ak). \tag{9.35}$$

First, let us consider the effect of dark-ground illumination (§ 9.6.1). Here we eliminate the $k = 0$ component; this is the $\delta(k)$ and a narrow region of negligible width at the centre of the sinc function. After such filtering of the transform is, to a good approximation,

$$F_1(k) = 0 + 2a(e^{i\alpha} - 1)\,\text{sinc}\,(ak), \tag{9.36}$$

and the resultant image, the transform of $F_1(k)$, is

$$f_1(x) = (e^{i\alpha} - 1)g(x). \tag{9.37}$$

Recalling the definition of $g(x)$ (equation (9.34)), we see that the slit appears bright on a dark background. In fact, the intensity of the image is dependent on the phase α:

$$I_1(x) = |f_1(x)|^2 = 2(1 - \cos\alpha)g(x). \tag{9.38}$$

It is reassuring to note that the image disappears when $\cos\alpha = 1$, i.e. $\alpha = 0, 2\pi, \ldots,$ because under those conditions there is no physical difference between the strip and its surroundings.

Using the same example we can illustrate the schlieren system (§ 9.6.2). In this case the filter cuts out the $\delta(k)$ and all the transform for $k \leqslant 0$, leaving us with

$$F_2(k) = 2a(e^{i\alpha} - 1)\,\mathrm{sinc}\,(ak)S(k), \tag{9.39}$$

where $S(k)$ is the step function: $S(k) = 1$ $(k > 0)$ and zero otherwise. Using the convolution theorem, the transform of this is:

$$f_2(k) = (e^{i\alpha} - 1)g(x) * s(x), \tag{9.40}$$

where $s(x)$ is the transform of the step function:†

$$s(x) = \int_{-\infty}^{\infty} S(k)\,e^{ikx}\,dk = \int_0^{\infty} e^{ikx}\,dk = 1/ix. \tag{9.41}$$

For the particular form of $g(x)$ (unity when $|x| \leqslant a$) the convolution then takes the simple form:

$$\begin{aligned} g(x) * s(x) &= \frac{1}{i}\int_{-\infty}^{\infty} \frac{g(x - x')}{x'}\,dx' \\ &= \frac{1}{i}\int_{x-a}^{x+a} \frac{dx'}{x'} = \frac{1}{i}\ln\left|\frac{x + a}{x - a}\right|. \end{aligned} \tag{9.42}$$

Thus the image is

$$f_2(x) = -i(e^{i\alpha} - 1)\ln|(x + a)/(x - a)|, \tag{9.43}$$

with intensity

$$I_2(x) = |f_2(x)|^2 = 2(1 - \cos\alpha)\{\ln|(x + a)/(x - a)|\}^2, \tag{9.44}$$

which is illustrated in Fig. 9.30. The schlieren method in this case clearly emphasizes the edges of the slit, which are discontinuities in object phase. In general it can be shown to highlight phase gradients in the direction normal to the knife edge; this effect can be observed in Fig. 9.23.

† $S(k)$ does not strictly have a Fourier transform since it does not decay to zero at infinity. By multiplying by a very slowly decaying function $e^{-\epsilon x}$, the function can be made transformable and ϵ can subsequently be taken to be zero. This 'trick' gives us the result stated.

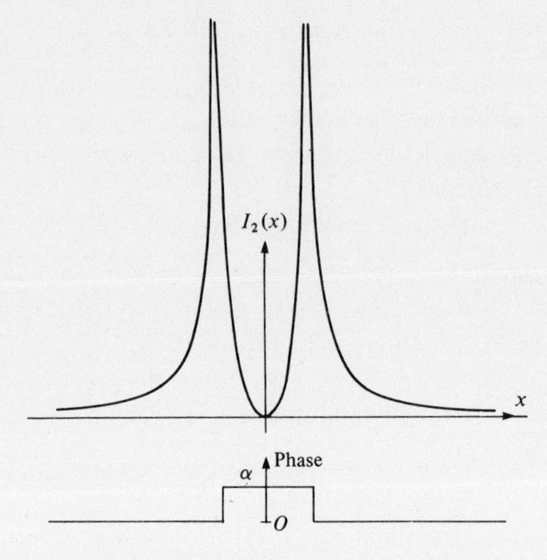

Fig. 9.30. Schlieren image of the phase slit.

Finally, this model can be used to illustrate the phase-contrast method (§ 9.6.2). The transform (9.35)

$$F(k)=\delta(k)+2a(e^{i\alpha}-1)\operatorname{sinc}(ak) \tag{9.45}$$

is filtered by the phase plate which in principle changes the phase of the $k=0$ component by $\pi/2$ (i.e. multiplication by i):

$$F_3(k)=i\delta(k)+2a(e^{i\alpha}-1)\operatorname{sinc}(ak). \tag{9.46}$$

The image amplitude is the transform of this:

$$f_3(x)=i+(e^{i\alpha}-1)g(x), \tag{9.47}$$

which has value i in the region $|x|>a$ and value

$$i+(e^{i\alpha}-1)=(\cos\alpha-1)+i(\sin\alpha+1) \tag{9.48}$$

within the slit. The respective intensities are unity ($|x|>a$), and

$$(\cos\alpha-1)^2+(\sin\alpha+1)^2=3-\sqrt{2}\cos(\alpha+\pi/4). \tag{9.49}$$

The maximum intensity ratio between inner and outer values is obtained when $\alpha=3\pi/4$ and is equal to $3+\sqrt{2}$.

Although the above examples treat the microscopic methods quantitatively, it should be stressed that each use is mainly qualitative. They are used to visualize phase-objects, but rarely to analyse them quantitatively.

9.6.6 *The interference microscope.* Interference microscopy allows quantitative analysis of phase-objects by including a two-beam interferometer as part of a microscope. It is not a spatial-filtering technique, but we describe it in this section because of its complementary relationship to the techniques described in §§ 9.6.1–9.6.4.

To observe the interferogram of the object, it is placed in one of the two beams of the interferometer. The variation of optical thickness from point to point then becomes visible in a quantitative manner (§ 7.2.5).

There are several different systems possible, each with advantages and disadvantages, and it is not possible to discuss them all here. We shall only describe the general principles and discuss one well-tried design in detail.

For producing the basic interference pattern two coherent beams are needed. Four different ways of producing these beams have been used, as follows:

(a) by using a semi-silvered mirror as in the Michelson interferometer (§ 7.8.2);

(b) by means of a doubly-refracting crystal (§ 5.5.1(a));

(c) by using two orders of diffraction from a diffraction grating (§ 7.7.1);

(d) by extracting two parts of a coherent wavefront.

The instrument that we shall describe was made by Lebedeff and was based upon the second of these methods (Fig. 9.31). The incident beam is polarized at 45° to the plane of the diagram and the first doubly-refracting crystal gives two beams of equal intensity polarized perpendicularly. These two beams then pass through a half-wave plate that turns the planes of polarization through 90°, so that, after the beams have passed through a second crystal exactly similar to the first, the two combine to form a beam polarized at 45° to the plane of the diagram. The half-wave plate is clearly necessary in order that the ordinary and extraordinary beams should interchange their roles in the second crystal and coincide on leaving it. Two beams polarized at right angles cannot produce interference effects, and therefore the analyser is necessary in order to extract two components that *can* interfere; it is crossed (§ 5.7.4) with respect to the polarizer.

If no object is present, the illumination should be perfectly uniform, since the two interfering beams are parallel. If white light is used, the field of view will be coloured since some wavelengths will be eliminated by interference. If now an object is placed in one of the beams, its image will be seen crossed by fringes which give quantitative information about the object. If the object covers both beams the interpretation of the image is clearly much more complicated.

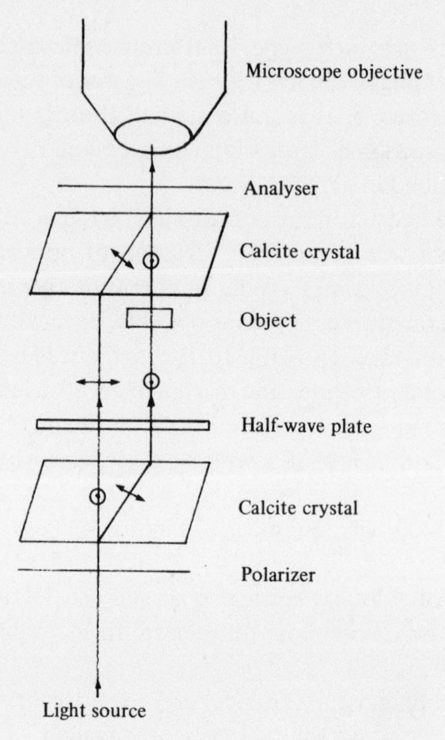

Fig. 9.31. Interference microscope.

9.6.7 *Apodization.* Apodization is a spatial filtering technique applied to the outer regions of the Fourier transform, the parts of highest spatial frequency k, to achieve two possible ends. One is to achieve higher resolution than the Rayleigh limit; the other is to suppress the diffraction ring images prominent in images with resolution close to the Rayleigh limit.

The first application, that of achieving high resolution, is carried out by masking out the central part of the Fourier transform, usually by simply covering the centre of the imaging lens with an opaque disc. The result of this is to form an image, by using only the highest spatial frequencies, which emphasizes highly resolved detail. As an example, consider the effect of leaving only a thin annular transparent ring of thickness b around the rim of an imaging lens of radius a (§ 7.3.2). If we are forming an image of a point source at infinity, the image in the focal plane will be the diffraction pattern of the annular ring; this is

$$\psi(\zeta, \eta) = \int_0^{2\pi} \exp\{-\mathrm{i}(a\zeta \cos\eta \cos\theta + a\zeta \sin\eta \sin\theta)\} ba \, \mathrm{d}\theta \quad (9.50)$$

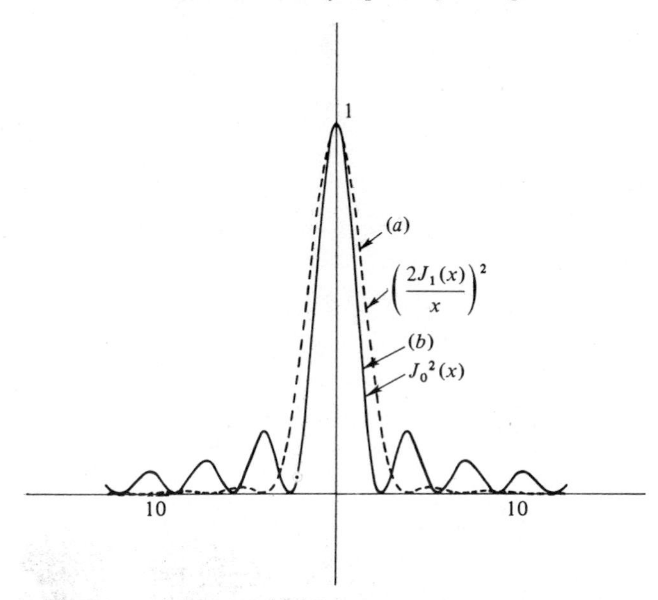

Fig. 9.32. Comparison between the diffraction rings around a point source imaged through (a) a circular aperture and (b) an annular aperture.

following the notation of (7.28) and realizing that the integral in that equation from 0 to a is now replaced, for the annulus, by the value of the integrand at the rim. The thickness of the ring is b. Equation (9.50) can be rewritten

$$\psi(\zeta, \eta) = \int_0^{2\pi} \exp\{-\mathrm{i}a\zeta\cos(\theta-\eta)\}ab\,\mathrm{d}\theta \tag{9.51}$$

$$= 2\pi ab J_0(a\zeta), \tag{9.52}$$

where $J_0(a\zeta)$ is the zero-order Bessel function for which the definition as an integral takes exactly the form of (9.51). This function has its first zero at $a\zeta = 2.405$. Since ζ is the radial component of $k = 2\pi\sin\theta/\lambda$, this gives us the angular value of the first dark ring:

$$\sin\theta = \lambda\zeta/2\pi = 2.405\lambda/2\pi a = 0.38\lambda/a, \tag{9.53}$$

compared to $\sin\theta = 0.61\lambda/a$ for the full lens.

This apparent improvement is very much negated by the greater intensity of the diffraction rings produced by the annular aperture. The two intensity functions $J_1^2(a\zeta)/(a\zeta)^2$ (for the full lens) and $J_0^2(a\zeta)$ (for the obstructed lens) are compared in Fig. 9.32. It is doubtful whether this method really has any practical application, although it can be considered as a two-dimensional analogue of the Michelson Stellar interferometer. An example is shown in Fig. 9.18(*h*).

The second application of apodization is exactly the opposite, and concerns reduction of the diffraction rings surrounding images with resolution near the diffraction limit. We consider the image of a point source, which is the Fraunhofer diffraction pattern of the imaging aperture, usually circular. For a lens of radius a, this is

$$\psi(\zeta, \eta) = 2\pi a^2 J_1(a\zeta)/a\zeta. \tag{9.54}$$

This image has the characteristic rings shown in Fig. 7.17. The rings can be annoying in high-quality images, and can even confuse important information. At the expense of efficiency and some resolution, the rings can be suppressed to any desired extent by covering the lens with a mask having a Gaussian transmission profile; if r is the radius in the plane of the lens, this would imply an amplitude transmission:

$$t(r) = \exp(-r^2/2\sigma^2). \tag{9.55}$$

The image of the point source now becomes the transform of the lens aperture multiplied by $t(r)$; this transform is therefore the convolution

$$\psi(\zeta, \eta) = 2\pi a^2 J_1(a\zeta)/a\zeta * 2\pi\sigma^2 \exp(-\zeta^2\sigma^2/2). \tag{9.56}$$

Without entering into details of the exact behaviour of this function, it is clear that it will influence the intensity of the diffraction rings significantly when σ is the order of a. For this value, however, the mask absorbs some 23 per cent of the incident intensity and increases the resolution limit to about $\sin\theta = 0.9\lambda/a$. The intensity of the first bright ring of the diffraction pattern is reduced from 1.75 per cent of the central intensity (without apodization) to 0.65 per cent.

9.7 Reconstruction of images from recorded diffraction patterns, and holography

Since all the information concerning the image of an object is contained in its diffraction pattern, it is tempting to ask whether this information can be recorded on a photographic plate and then used to reconstruct the image. The germ of this idea was suggested in 1948 and had some limited success; but the invention of the laser has now enabled the operation to be carried through completely successfully.

The difficulty involved is the recording of the relative phases of the different parts of the optical transform (§ 9.3.5); photographic film can record only intensities and a direct diffraction pattern of the intensity distribution, although it has some relation to the object, is not an image of it.

9.7.1 *Gabor's method.* A reason for trying to overcome this difficulty was an attempt by Gabor to solve the problem of image formation in the electron microscope. As we shall see later (§ 11.3.1), the resolution obtainable in the electron microscope is not limited by the wavelength (~0.1 Å) but by the aberrations (Appendix II) of the electron lenses: these cannot be corrected and therefore a very small aperture has to be used. This aperture is so small that the resultant image can be considered effectively as a Fresnel diffraction pattern (§ 6.3) of the object. Gabor thought that a better image might be reconstructed from this pattern with visible light, using a lens system that could correct for the aberrations of the electron lenses.

To demonstrate this idea, he made some optical diffraction patterns and reconstructed images from them. The phase problem was solved by using an object that consisted of a small amount of black detail on a large transparent background; the background would produce a uniform phase and the variation in phase of the diffraction pattern would be recorded as differences in intensity. The intensity would be greatest where the phase of the diffraction pattern was the same as that of the background and least where there was a phase difference of π.

The method did not work very well, even for optical patterns, as we can see from Fig. 9.33, and was never applied to electron-microscope pictures. But it had an idea in it that was later developed with complete success in another approach, which will be described in § 9.7.3.

9.7.2 *X-ray microscope.* A more successful approach in a limited field was made by Bragg in 1939; he used the fact that if a crystal had an atom of preponderant weight in the unit cell, that atom's diffraction pattern would decide the phases of the diffracted radiation. In particular, if the atom were at the origin of the unit cell the phases would all be the same and could be taken as zero. Therefore if one made a representation of a section of the X-ray diffraction pattern of a crystal in the form of holes in an opaque plate (Fig. 9.34(*a*)), the sizes of the holes being related to the corresponding intensity of the order of X-ray diffraction, then the Fraunhofer diffraction pattern of the plate should be an image of the structure (Fig. 9.34(*b*), (*c*)).

Other workers have taken the subject further, by extending the scope of the method to crystals that do not contain a heavy atom and for which the relative phases of the orders of diffraction can adopt general values between 0 and π. The resulting diffraction patterns can be regarded as photographs of atomic arrangements with magnification of the order of

Hologram

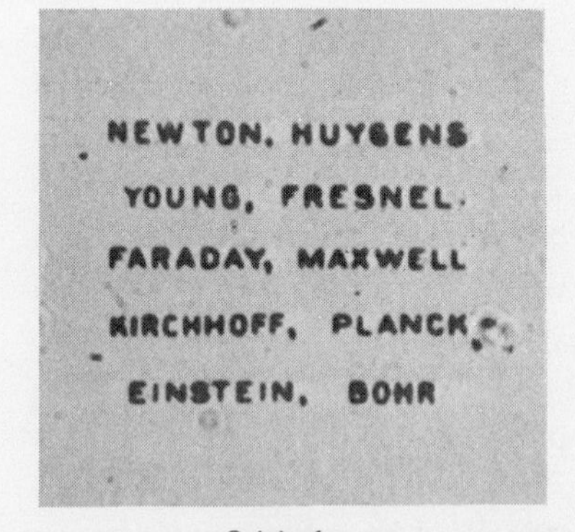

Original

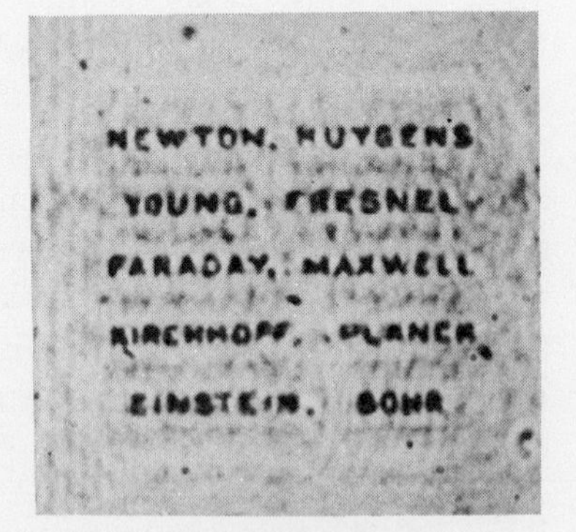

Reconstruction

Fig. 9.33. Gabor hologram and the image reconstructed from it. (From Gabor, 1948.)

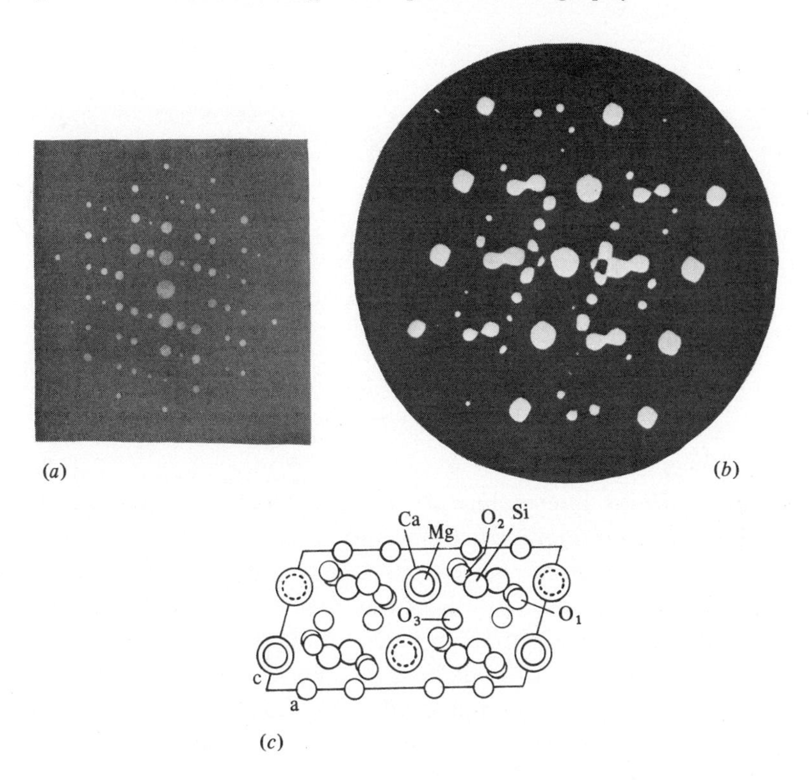

Fig. 9.34. (*a*) Set of holes representing orders of X-ray diffraction from diopside, $CaMg(SiO_3)_2$; (*b*) diffraction pattern of (*a*) representing atoms in crystal structure; (*c*) diagram of structure to compare with (*b*). (From Bragg, 1939.)

10^8. It must be pointed out, however, that these are not images in the proper sense of the word since extra information – the relative phases of the waves – has to be supplied artificially.

9.7.3 *Application of the laser.* The hologram approach has now been applied successfully, by Leith and Upatnieks, because of the intense beams of spatially coherent light that can be produced by a laser (§ 8.7). In principle, the same method can be used with ordinary sources of light, but the production of spatially coherent sources in this way usually results in such a small intensity that the experiments are hardly practicable.

The experimental set-up is quite simple (Fig. 9.35). A spatially-coherent laser beam is divided, either in wavefront or amplitude, so that one part falls directly on a photographic plate, and the other falls on the subject to be recorded, which scatters light on to the same plate. The two waves, called the 'reference wave' and the 'object wave' respectively,

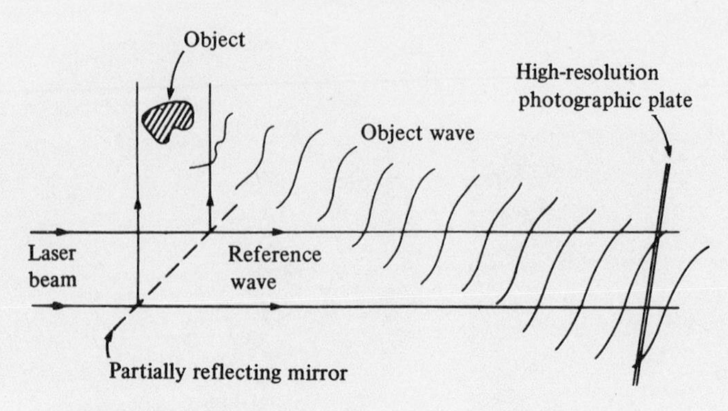

Fig. 9.35. Example of a holographic recording set-up.

interfere; and the interference pattern is recorded by the plate. It is necessary to reduce vibrations of the various components to amplitudes considerably less than one wavelength to avoid blurring the interference fringes. Reconstruction of the image is carried out by illuminating the developed plate with a light wave which is similar, if not identical, to the original reference wave. Two images are usually observed.

We shall first give a qualitative interpretation of the recording and reconstruction processes, and afterwards discuss them in a more quantitative manner.

The process can be described in general terms by considering the hologram as analogous to a diffraction grating (§ 7.7). Suppose that we photograph the hologram of a point scatterer (Fig. 9.36(*a*)). The point scatters a spherical wave, the object wave, and this interferes with the plane reference wave. The result is a set of curved fringes (Fig. 9.36(*b*)) like an off-centre part of a zone-plate (§ 6.3.4). The hologram is photographed and the plate developed.

To reconstruct the image we illuminate the hologram with a light wave identical to the original reference wave (Fig. 9.36(*c*)). We can consider each part of the hologram individually as a diffraction grating with a certain line-spacing. Illumination by the parallel reference wave gives rise to a zero and two first orders of diffraction, at angles θ_1, θ_{-1} depending on the local spacing of the fringes. It is not difficult to see that the -1 orders intersect and form a real image of the point scatterer, and the $+1$ orders form a virtual image in the position identical to the original point. The images are localized in three dimensions.

Two other important points are brought out by this model. First, the reconstructed point is more accurately defined in position if a large area

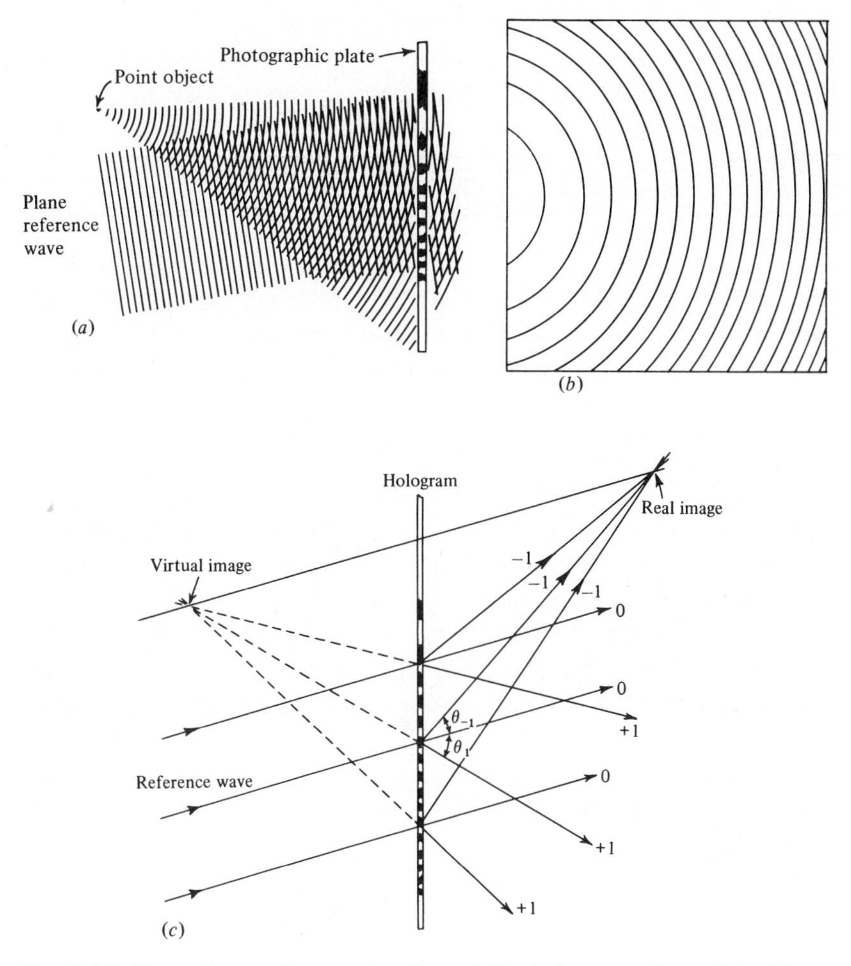

Fig. 9.36. Formation and reconstruction of the hologram of a point object: (*a*) spherical wave from the object interferes with plane reference wave; (*b*) fringes recorded on the photographic plate; (*c*) the plate behaves as a diffraction grating with non-uniform line spacing.

of the plate is used, and the reconstruction orders meet at a considerable angle; the resolution is therefore a function of the size of the hologram. Secondly, the fringes are sinusoidal, since only two waves interfere. If the plate records this function faithfully, only zero and first orders will be produced on reconstruction, and only the above two images are produced.

In a more quantitative fashion, we can now see how both the amplitude and the phase of the scattered light are recorded in the hologram.

Suppose that at a general point (x, y) in the plate the scattered light has amplitude $a(x, y)$ and phase $\phi(x, y)$. Furthermore, we shall assume that the reference wave is not necessarily a plane-wave, and has amplitude A and phase $\phi_0(x, y)$ at the general point. Then the total wave amplitude at (x, y) is

$$\psi(x, y) = A \exp\{i\phi_0(x, y)\} + a(x, y) \exp\{i\phi(x, y)\} \tag{9.57}$$

and the intensity:

$$I(x, y) = |\psi(x, y)|^2 = A^2 + a^2 + 2Aa \cos\{\phi(x, y) - \phi_0(x, y)\}. \tag{9.58}$$

One usually arranges in holography for a to be much smaller than A, in which case the term a^2 can be neglected and

$$I(x, y) \approx A^2 + 2Aa(x, y) \cos\{\phi(x, y) - \phi_0(x, y)\}. \tag{9.59}$$

This is the hologram. It consists of a set of interference fringes with sinusoidal profile and phase $\phi - \phi_0$. The visibility of the fringes is $2a/A$. Thus both $a(x, y)$ and $\phi(x, y)$ are recorded in the hologram. The need for coherent light to record the hologram should now be clear, since the phase *difference* $\phi - \phi_0$ is recorded in the interference pattern.

To deduce the form of the reconstruction, we assume that the hologram is now photographed with a plate whose amplitude transmission $t(x, y)$ after development is linearly related to the exposure intensity $I(x, y)$:

$$t(x, y) = 1 - \alpha I(x, y). \tag{9.60}$$

The hologram is illuminated by a wave identical to the original reference wave $A \exp\{i\phi_0(x, y)\}$ and so the transmitted amplitude is:

$$\begin{aligned}
&At(x, y) \exp\{i\phi_0(x, y)\} \\
&\quad = \{1 - \alpha I(x, y)\} A \exp\{i\phi_0(x, y)\} \\
&\quad = A(1 - \alpha A^2) \exp\{i\phi_0(x, y)\} && \text{(a)} \\
&\qquad - \alpha A^2 a(x, y) \exp\{i\phi(x, y)\} && \text{(b)} \\
&\qquad - \alpha A^2 a(x, y) \exp[i\{2\phi_0(x, y) - \phi(x, y)\}]. && \text{(c)}
\end{aligned}$$

The three terms in the above equation are interpreted as follows:

(a) an attenuated continuation of the reference wave (the zero order);

(b) the virtual image; apart from the constant multiplier αA^2, the reconstructed wave is exactly the same as the object wave and so the light appears to come from a virtual object perfectly reconstructed. This is the first order.

(c) The second image. This wave is the complex conjugate of the object wave if ϕ_0 is a constant, and is then a good reconstruction. Otherwise it is distorted. This is the -1 order.

The recording of holograms is almost unbelievably simple, although good results require careful work. There is a considerable literature on the subject including several textbooks (e.g. Collier, Burckhardt & Lin, 1971); we shall only mention a few points which arise directly from the above discussion.

The intensity ratio between the object beam and the reference beam, a^2/A^2, has been required to be small; in general a ratio of 1:5 is sufficient, although for some purposes even 1:2 can be tolerated. The result of too great a ratio is to put structure into the zero-order transmitted beam, which may interfere with the reconstructed image.

The reconstruction requires that the photographic plate faithfully record the light intensity. Quantitatively, we require a linear relationship between the amplitude transmission of the developed plate (i.e. the square root of the intensity transmission) and the exposure intensity. This can be obtained with normal photographic plates only under restricted conditions. However, the condition can be relaxed quite considerably for many purposes, since the main effect of non-linearity in the plate is to create second- and higher-order reconstructions which are usually separated in space from the main reconstructions.

Another obvious requirement is for high spatial resolution in the photographic plate. If the reference beam and the object beam are separated by angle θ, the period of the fringes in the hologram is just $\lambda/\sin\theta$. For, say, $\theta = 6°$ – a rather small separation – this period is 10λ, only about 6 μm with the common helium–neon laser. To record fringes on this scale, the plate must be capable of resolving less than 3 μm. This is a very stringent requirement, and needs special photographic plates to fulfil it.

These plates were produced originally for spectroscopic instruments; it is unlikely that holography alone would have been sufficient incentive for their development. Such fine-grain plates are almost incredibly insensitive compared with normal photographic materials.

9.7.4 *Achievements of holography.* Holographic reconstructions have two main advantages over ordinary photographs. They are three-dimensional, and they contain phase information (§ 9.6.3). On the other hand they are monochromatic (colour holography has been realized, but is more of a novelty than a useful method) and require both rigid support of the object and laser illumination for their production.

The three-dimensional aspect has few really valuable applications, although the great depth of field attainable in a single photograph is

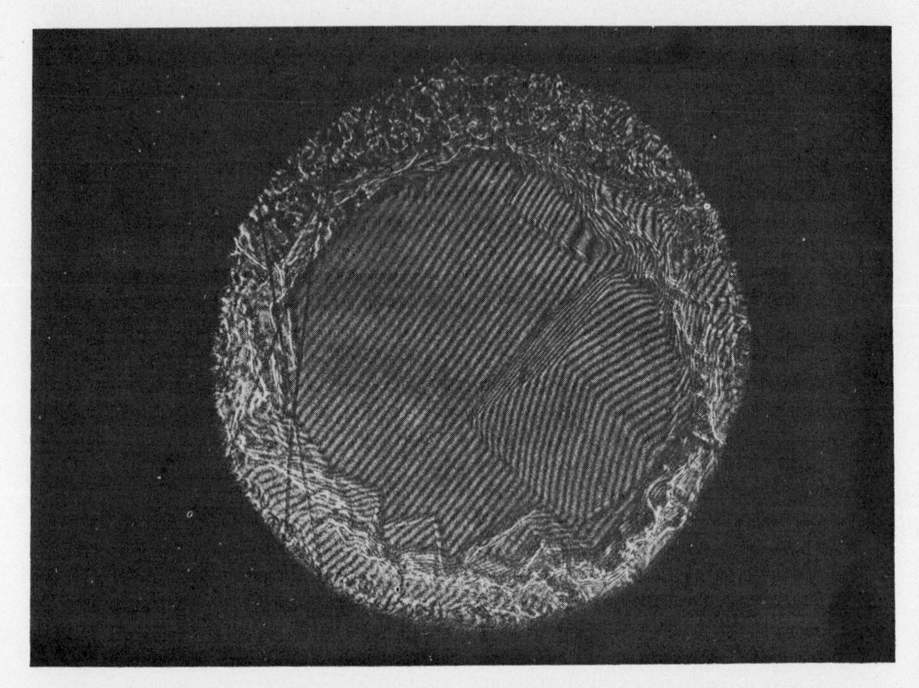

Fig. 9.37. Holographic interferogram showing a growing crystal of solid helium at a temperature of 0.5 K.

sometimes of importance. The recording of phase information is much more useful. It has allowed the development of *holographic interferometry* in which an object can be compared interferometrically with a holographic recording of itself. If any changes, of optical density or dimensions, for example, have occurred since the recording was made, the differences will be apparent as interference fringes. It is not our intention to go into practical details here; an example is shown in Fig. 9.37 in which the growth of a crystal of a transparent material (actually solid helium) is observed by recording the pattern of changes in optical density. The details of the exact shape of the experimental cell are irrelevant to an interpretation of the photograph, since only *changes* in the optical density are observed. Other applications of interferometric holography include vibration analysis and aerodynamic experiments. The interested reader is referred to the bibliography.

9.7.5 *Non-optical holography.* The idea of holography can of course be applied to any waves, not only to light waves, provided that a suitable recording medium exists. Acoustic holography, using ultrasonic waves, has been widely studied in recent years, and one is beginning to see its application to medical diagnosis. On the other hand, there have been suggestions that the method might be used with X-rays; by taking a hologram with X-rays and viewing it with visible light from a laser, one might be able to 'see' images of molecular structures directly. This project seems unlikely; making an X-ray laser is several orders of magnitude more difficult than making an optical laser, and controlling an object to a fraction of an X-ray wavelength would appear to be impossible. Moreover a crystal does not give all its diffraction pattern at one setting (§ 7.6), and a three-dimensional pattern cannot be recorded on a plate. The method would therefore be applicable only to non-crystalline matter. Success in this field would indeed be of inestimable value, but the problems of producing spatial coherence on the X-scale seem to be remote.

9.8 Applications of the Abbe theory involving holographic filters

The spatial filtering techniques discussed in § 9.7 involve very simple geometrical filters such as the phase plate, the schlieren knife-edge and so on. Holography allows us to record diffraction patterns including their phases, and the use of holographic filters adds another dimension to spatial filtering.

Suppose that the object $f(x)$ is imaged in the usual spatial filtering system (Fig. 9.10). In the focal plane of the lens we see the complex Fourier transform $F(u)$ and the final image is its transform $f(-x/m)$, where m is the magnification of the system, and will henceforth be assumed to be unity. In general we can insert any filter $G(u)$ in the focal plane; it multiplies the transform $F(u)$ and the resultant image is then the transform of $[F(u)G(u)]$ which is $f(-x) * g(-x)$. There are several interesting applications of this result, when $G(u)$ is complex. In practice, a complex $G(u)$ is recorded on a photographic plate holographically, in the system of Fig. 9.38, so that the filter is actually proportional to $|G(u)+A\,\mathrm{e}^{\mathrm{i}\alpha u}|^2$, where the term $A\,\mathrm{e}^{\mathrm{i}\alpha u}$ represents the amplitude of the reference beam incident at an angle β (α and β are proportional; both are constants and the exact relationship is not important here). The filter transmission is proportional to the hologram intensity

$$|G(u)+A\,\mathrm{e}^{\mathrm{i}\alpha u}|^2 = A^2 + AG(u)\,\mathrm{e}^{-\mathrm{i}\alpha u} + AG^*(u)\,\mathrm{e}^{\mathrm{i}\alpha u} + G^2. \quad (9.61)$$

Under the usual holographic recording conditions, G^2 is negligible (§ 9.7.3). The action of the filter on an image produced with the same lens in the same position is then to multiply $F(u)$ by the function (9.61); the

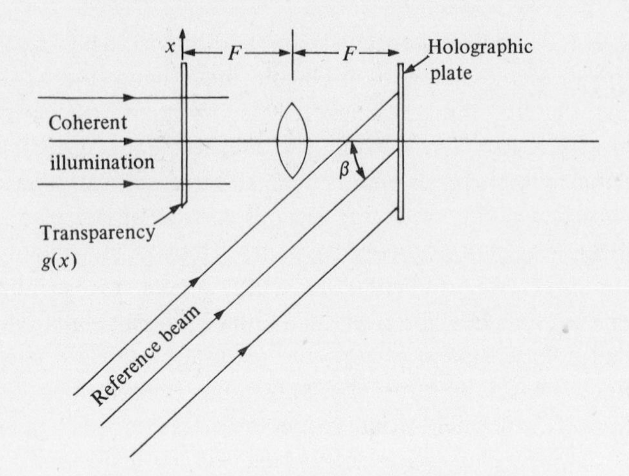

Fig. 9.38. Production of a holographic filter.

amplitude distribution emerging from the filter is thus

$$F'(u) = F(u)\{A^2 + AG(u)\,\mathrm{e}^{-\mathrm{i}\alpha u} + AG^*(u)\,\mathrm{e}^{\mathrm{i}\alpha u}\}. \tag{9.62}$$

The image is the transform of this function which is just

$$\begin{aligned} f'(-x) = {} & A^2 f(-x) + Af(-x) * g(-x) * \delta(-x+\alpha) \\ & + Af(-x) * g^*(x) * \delta(-x-\alpha). \end{aligned} \tag{9.63}$$

This expression contains three identifiable terms:

(a) $A^2 f(-x)$ is just the image of $f(x)$;

(b) the second term is the *convolution* of f and g, centred on the point $x = \alpha$;

(c) the third term is the *correlation function* of f and g on the point $x = -\alpha$.

The three image functions are physically separated in the image plane because they are centred on different points.

One application of such a system is to pattern recognition. Suppose that $g(x)$ corresponds to a known object, and $f(x)$ to an unknown object. If the two are identical the correlation function has a strong central spot, which is weaker or absent if the two are not identical. The method appears applicable to problems such as the recognition of fingerprints, by comparison of an unknown one with banks of known ones; this application has yet to be carried out in a satisfying manner. In fact the optics is just too pedantic, and is sensitive to all sorts of unreasonable differences such as variations in emphasis from point to point, slight rotations or scale changes.

Another application is to deblurring of photographs or other pictures. As was pointed out in § 3.6.1, blurring is equivalent to convolution by a 'blur function'. We call this function $b(x)$. Then a blurred image of $f(x)$ is $f(x) * b(x)$. The transform of this blurred image is $F(u)B(u)$. What could be simpler than multiplying $F(u)B(u)$ by a filter $1/B(u)$ to restore $F(u)$ and the sharp image $f(-x)$ in the image plane? This idea, which would allow deblurring to be carried out whenever the function $b(x)$ is known (for example, out-of-focus photographs, pictures of moving objects and clinical X-ray photographs, which are essentially shadows from a finite source), is deceptively simple.

What happens when $B(u)$ goes through zero? Almost all transforms have zeros, and their presence results in irretrievably lost information in the blur transform $F(u)B(u)$. Some little success has been achieved for simple blur functions by ignoring the amplitude of $B(u)$ and concentrating on its phase only. As we have pointed out many times, the phase of a transform is as important as its amplitude, so that at least some improvement should be noticeable. It is, but hardly seems worth the trouble taken.

Filters of this type are often produced by calculation and drawing, usually with the aid of a computer. The calculation is simply reproducing the optical interference, but can be programmed to ignore amplitude variations and make other changes. Such filters are known as 'computer generated holographic filters'.

10

The classical theory of dispersion

10.1 Why classical theory is usually adequate

Long before the advent of quantum theory, which is necessary for a complete description of the interaction between radiation and matter, many of the more striking dispersion effects in materials had been explained on very simple and none-too-unrealistic classical models of the structure of matter. The reason why classical and quantum results do not diverge much is that most of the interesting results are the effect of resonances, and it turns out that the classical description of a resonance is very similar in almost all respects to the quantum description.

We shall discuss dispersion effects arising from two distinct causes. The first effects, which result from sub-atomic processes, and whose explanation involves postulating a model of the structure of an atom, are resonance phenomena and occur at frequencies in and around the optical region; in quantum theory they are transitions between atomic energy levels. The second group of effects arises from interactions between neighbouring atoms or molecules and are generally characterized by being relaxation rather than resonance phenomena. Those we shall discuss are typical in having their greatest effects at frequencies lower than optical, extending over a band very broad compared with the resonance effects.

10.1.1 *The classical atom.* To describe the internal structure of an atom or molecule we shall use a model consisting of two parts of opposite charge connected by springs (Fig. 10.1).

The spring constant, together with the masses of the two parts, determines a resonant frequency for the system; it is a weakness of the model that only a single resonant frequency results, since we know that atoms and molecules have many modes of oscillation, which we call spectral lines. However, this turns out not to be very important since radiation incident on the system interacts with it only very close to a resonant

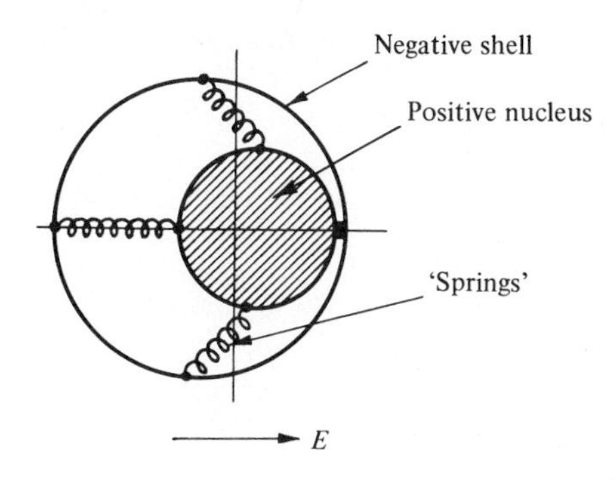

Fig. 10.1. The model 'classical atom'.

frequency, and so generally only a single resonance is involved. In addition to the resonance effects, a molecule may have a permanent dipole moment; the positive and negative charges may be separated even when the system is at rest. Its response to an external electric field is then a tendency to alignment.

10.1.2 *Properties of a forced harmonic oscillator.* The electric field in an electromagnetic wave will exert forces and couples on a molecule such as we have described, and since it has a resonant frequency we shall first give a brief résumé of the properties of a damped simple-harmonic oscillator subject to a periodic force.

Suppose that the oscillator has a natural frequency η and damping constant κ, so that its equation of motion when the driving force has the form $F_0 \exp(\mathrm{i}\omega t)$ is:

$$\ddot{x} + \kappa\dot{x} + \eta^2 x = (F_0/m)\exp(\mathrm{i}\omega t). \tag{10.1}$$

The equation can be shown to have two solutions. The first is a transient

$$x = x_0 \exp(-\kappa t/2)\exp(\mathrm{i}\eta t) \tag{10.2}$$

which dies away during a period $2\kappa^{-1}$ after the initiation of the driving force, and is of no particular interest in optics. The second is a steady response

$$x = \frac{F_0 \exp(\mathrm{i}\omega t)}{m(\eta^2 - \omega^2 + \mathrm{i}\kappa\omega)} = A(\omega)\exp(\mathrm{i}\omega t) \tag{10.3}$$

having frequency ω and a complex amplitude $A(\omega)$. The latter indicates that the oscillations follow the applied force, but with a phase difference. The modulus and phase of $A(\omega)$ are illustrated in Figs. 10.2 and 10.3; the form of the curves depends on the factor κ/η.

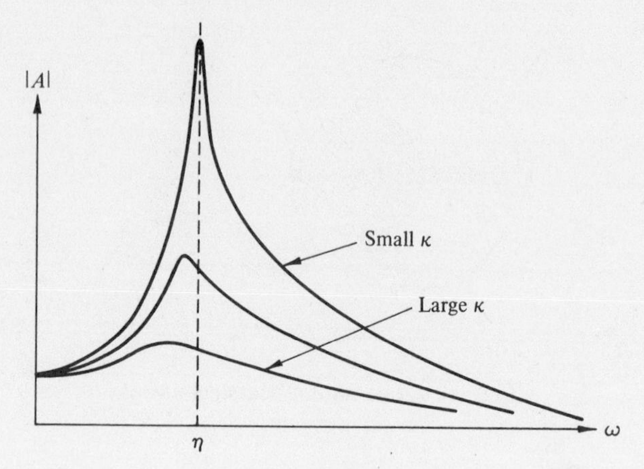

Fig. 10.2. Amplitude of response of a simple harmonic oscillator for various degrees of damping.

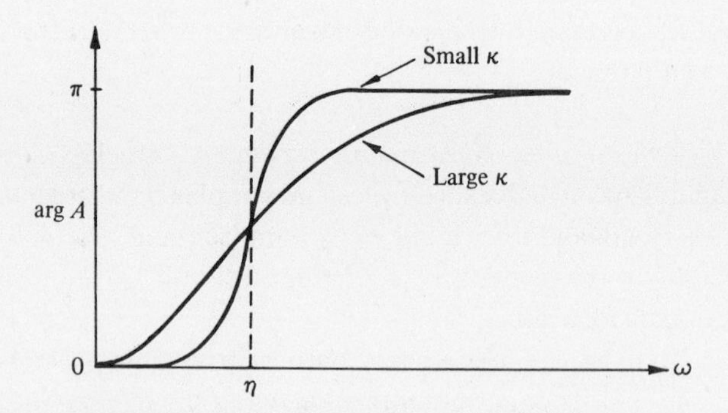

Fig. 10.3. Phase of the response in Fig. 10.2.

The important facts to notice are as follows:

(a) The lightly damped oscillator has a large amplitude only in a band with a width of the order of κ about the resonant frequency η (Fig. 10.2). At frequencies below η, $\arg A = 0$, implying that the oscillator moves in phase with the force, whereas above the resonant frequency the oscillator moves out of phase with the force (Fig. 10.3). At the resonant frequency η, the oscillator and force are in phase quadrature ($\arg A = \pi/2$).

(b) The heavily damped oscillator responds to the force over a very much larger frequency range, again $\pm\kappa$ about η, but never reaches as large an amplitude as the lightly damped one. The phase relations are the same as in (a), the transition being spread out over the larger frequency band.

(c) The maximum amplitude is not achieved exactly at the resonant $\eta = \omega$, but always at a slightly lower value of ω, depending on κ:

$$\omega \text{ (max amplitude)} = \left(\eta^2 - \frac{\kappa^2}{2}\right)^{\frac{1}{2}}.$$

10.1.3 *Polarizability.* When our model atom is subjected to an oscillating electric field, $E = E_0 \exp(\mathrm{i}\omega t)$, the positive and negative parts (charge q) experience opposed forces

$$F(t) = E_0 q \exp(\mathrm{i}\omega t). \tag{10.4}$$

Their relative movement $x(t)$ can then be calculated from (10.3). The result of this charge separation is to create an oscillating dipole moment which has the value

$$p(t) = qx(t) = qA(\omega) \exp(\mathrm{i}\omega t). \tag{10.5}$$

Substituting (10.3) and (10.4) we calculate the atomic or molecular polarizability

$$\alpha(\omega) = \frac{p(t)}{\varepsilon_0 E(t)} = \frac{q^2}{\varepsilon_0 m(\eta^2 - \omega^2 + \mathrm{i}\kappa\omega)}. \tag{10.6}$$

This quantity, which indicates the linear relationship between the applied wave-field and the resultant oscillating dipole moment, is real when $|\eta^2 - \omega^2| \gg \kappa\omega$ but becomes complex when ω is close to the resonant frequency η.

10.2 Rayleigh scattering

When an electromagnetic wave falls on an isolated particle, it is either absorbed or scattered. If the wave frequency ω is well removed from any resonant frequency η, the absorption of the wave is negligible, and only scattering need be considered.

Rayleigh scattering occurs when the particle is very small, very much smaller than the wavelength, so the wave field it experiences is essentially uniform. The result will be seen to be particularly useful for scattering by isolated atoms or molecules, although it is also applicable to very fine particulate matter and density fluctuations.

We write the instantaneous dipole moment (see (10.6))

$$p(t) = \alpha\epsilon_0 E(t). \tag{10.7}$$

Now if $E(t)$ has the form $E_0 \exp(\mathrm{i}\omega t)$, $p(t)$ behaves as an oscillating dipole. This, we know, radiates energy at a rate (equation 4.85))

$$W = \frac{\omega^4 p_0^2}{12\pi\epsilon_0 c^3} = \frac{\omega^4 E_0^2 \alpha^2 \epsilon_0}{12\pi c^3}. \tag{10.8}$$

If there are N *independent* scattering particles in a cube of unit volume, the total power scattered is just N times the result (10.8). But the radiant power incident on a face of the cube is the Poynting vector $\mathbf{\Pi}$ (§ 4.2.4) which has average value $\frac{1}{2}E_0^{\,2}\epsilon_0 c$. Therefore the loss of power per unit distance of propagation is

$$\frac{\mathrm{d}\Pi}{\mathrm{d}z} = -NW = \frac{-\Pi N\omega^4\alpha^2}{6\pi c^4}. \qquad (10.9)$$

This equation has the solution:

$$\Pi = \Pi_0 \exp\left(\frac{-N\omega^4\alpha^2 z}{6\pi c^4}\right) = \Pi_0 \exp(-z/z_0), \qquad (10.10)$$

where

$$z_0 = 6\pi c^4/N\omega^4\alpha^2. \qquad (10.11)$$

The quantity z_0 is a decay distance, telling us that the intensity of light travelling through the collection of scattering particles falls to exp (-1) of its initial value in a distance z_0.

The quantity z_0 is amenable to exact calculation for many systems, such as gases, where N and α are known. However, it is important to recall that the calculation has assumed the scattering from the individual particles to be independent, so that the scattered waves are incoherent and the intensities of the scattered waves are to be added. This assumption is very often untrue, and will be examined in more detail in § 10.2.3.

10.2.1 *Wavelength dependence of scattered radiation.* The most striking part of equation (10.11) is the fourth-power dependence on frequency: blue light is scattered about ten times more intensely than red light; this is the reason for the common observation that the sky is blue (weather permitting) during most of the day, but can appear red when one looks directly towards the sun at dawn or sunset. The mid-day appearance involves radiation scattered by air molecules at all heights and is therefore predominantly blue; the redness occurs at daybreak and sundown because at those times the sun's light passes horizontally through the atmosphere, and the very long air-passage results in the scattering away of a much greater fraction of the blue light than of the red.

Rayleigh scattering is also responsible for other everyday effects, such as the colours of diluted milk and cigarette smoke, when either the transmitted or scattered light is observed.

10.2.2 *Polarization of scattered radiation.* The dipole moment produced in the atom is parallel to the electric vector in the incident light and will

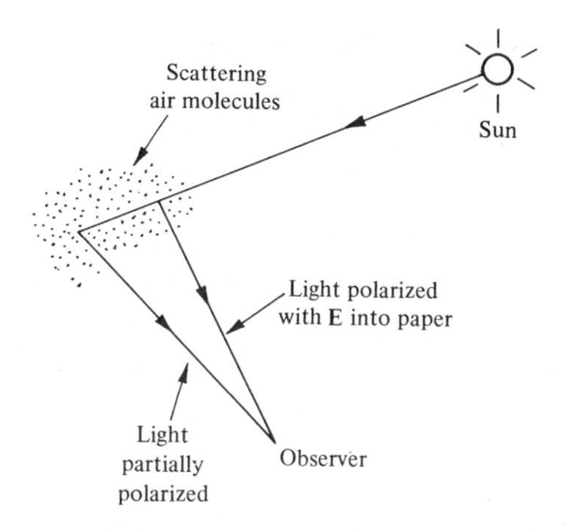

Fig. 10.4. Polarization of atmospherically scattered sunlight.

re-radiate with a radiation polar diagram as described in § 4.5.1 (Fig. 4.19). The intensity radiated along the axis of the dipole is zero. It therefore follows that scattered radiation along a line perpendicular to the incident light is plane-polarized normal to the plane containing the incident and scattered light. In other directions the light will appear partially polarized. With the aid of a single polaroid sheet (§ 5.7.3), these conclusions can easily be tested using ordinary sunlight (Fig. 10.4), although polarization is far from complete because of multiple scattering.

10.2.3 *Incoherent and coherent scattering.* We can use equation (10.10) to calculate the decay distance for clean air at atmospheric pressure. The atomic polarizability α is related to the refractive index μ of the gas via the dielectric constant ε:

$$\mu = \epsilon^{\frac{1}{2}} = (1 + N\alpha)^{\frac{1}{2}} \approx 1 + \tfrac{1}{2}N\alpha. \tag{10.12}$$

Thus z_0 can be written, substituting the wavelength $\lambda = 2\pi c/\omega$:

$$z_0 = \frac{6\pi c^4}{N\omega^4\alpha^2} = \frac{3}{32}\frac{N\lambda^4}{\pi^3(\mu-1)^2}. \tag{10.13}$$

Using the values at atmospheric pressure $\mu - 1 = 3\times 10^{-4}$ and $N = 3\times 10^{25}\ \text{m}^{-3}$, we find, for green light of $\lambda = 5000$ Å, $z_0 = 65$ km.

At first sight this figure seems surprisingly low, particularly as molecular scattering is unfortunately not the only factor which limits visibility through the atmosphere. One can easily find places where the meteorological visibility often exceeds 100 km and even reaches 200 km. However, one should remember that z_0 corresponds to an attenuation factor of $1/e = 0.37$, and factors of $1/e^2$ (at $2z_0$) or even $1/e^3 = 0.05$ (at $3z_0$) can be tolerated before a distant view of snow-capped mountains against an azure sky merges into the haze.

Nevertheless, we have already remarked that the above theory is correct only for particles which scatter incoherently. Now, the mean distance between air molecules under atmospheric conditions is two orders of magnitude less than a wavelength of light, so that almost completely coherent scattering would be expected. For such a medium of uniform density, we shall see in § 10.3.1 that there is no net scattering at all. It is the *deviations from uniform density* which then give rise to scattering.

The subject of scattering by density fluctuations can be treated fully by thermodynamics (see e.g. Landau & Lifshitz, 1960) but we can get an idea of the results by a very simple argument. One would expect incoherent Rayleigh scattering to result mainly from independent 'blocks' of material separated by distances of order λ, each one therefore having volume V of order λ^3. (Larger blocks are not 'small' compared with the wavelength, and smaller ones will not scatter incoherently.) Now, in such a volume there are on average NV molecules. If the molecules do not interact with one another we have a perfect gas, and the exact number of molecules in the volume V will be governed by Poisson statistics (§ 8.2.2). For such statistics, the r.m.s. fluctuation in the number is $(NV)^{\frac{1}{2}}$, and it is these fluctuations which should be considered as the scattering 'particles'. Thus, with $\alpha(NV)^{\frac{1}{2}}$ as the 'particle' polarizability, and $1/V$ as the 'particle' density we substitute in (10.11) to find

$$z_0 = 6\pi c^4 \Big/ \left(\frac{1}{V}\right) \omega^2 \{\alpha(NV)^{\frac{1}{2}}\}^2.$$

On cancelling the Vs, this reverts to (10.11); (10.12) and (10.13) then follow as before. Thus the scattering by density fluctuations in a perfect gas is just the same as if all the molecules were to scatter incoherently, and our estimate of z_0 is correct.

Under what conditions will Rayleigh scattering differ from the incoherent case? If the material is relatively incompressible the motions of the particles are correlated in such a way as to avoid one another. This suppresses density fluctuations; the scattering is less than in the incoherent case and approaches zero at the uniform density limit (§10.3.1). On the other hand, near a critical point there is a tendency towards local condensation and enhanced density fluctuations. We then see excess scattering and the phenomenon of *critical opalescence*.

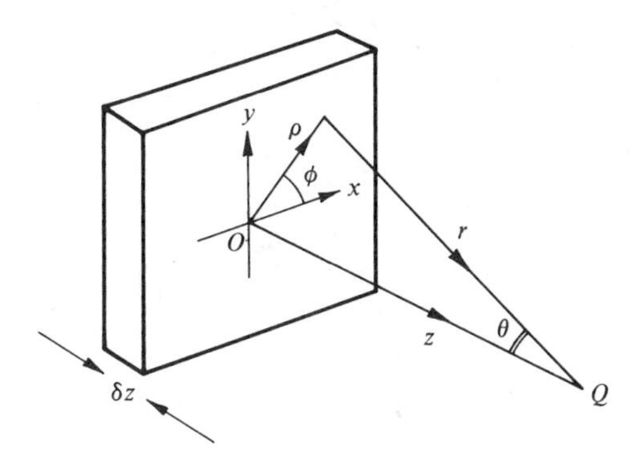

Fig. 10.5. Coherent scattering by a dense medium.

10.3 Coherent scattering and dispersion

We shall now consider the problem of scattering by a uniform dense medium, where the molecules are much closer than a wavelength and therefore the scattered waves are correlated in phase. In this problem we have to sum the *amplitudes* of the scattered waves. It turns out that a real polarizability α results in no net scattering whatsoever; the material simply refracts the incident wave. But when α is complex, absorption of the incident light occurs.

10.3.1 *Refraction as a problem in coherent scattering.* Consider scattering by a thin slab of thickness δz in the plane $z=0$, where z is the axis of propagation of the radiation (Fig. 10.5). In this slab are N molecules per unit volume, each having polarizability α. Now the oscillating dipoles will radiate waves and we can calculate their combined effect at the point Q $(0, 0, z)$. An atom in the slab at point $(x, y, 0)$ behaves as an oscillating dipole of magnitude:

$$p(t) = \alpha\epsilon_0 E_0 \exp(\mathrm{i}\omega t) \tag{10.14}$$

and its radiation field at Q is (from (4.81))

$$e(t) = \frac{\omega^2 \alpha\epsilon_0 E_0 \exp\{\mathrm{i}(\omega t - kr)\}\cos\theta}{4\pi\epsilon_0 c^2 r}, \tag{10.15}$$

where r is the distance from $(x, y, 0)$ to $(0, 0, z)$, i.e. $(x^2+y^2+z^2)^{1/2}$, and θ is the angle between the vector $\mathbf{r}$ and the z-axis. The total field from all the molecules in an elementary volume $\mathrm{d}x\,\mathrm{d}y\,\delta z$ at this point is (10.15) multiplied by $N\,\mathrm{d}x\,\mathrm{d}y\,\delta z$. We can therefore write down the total

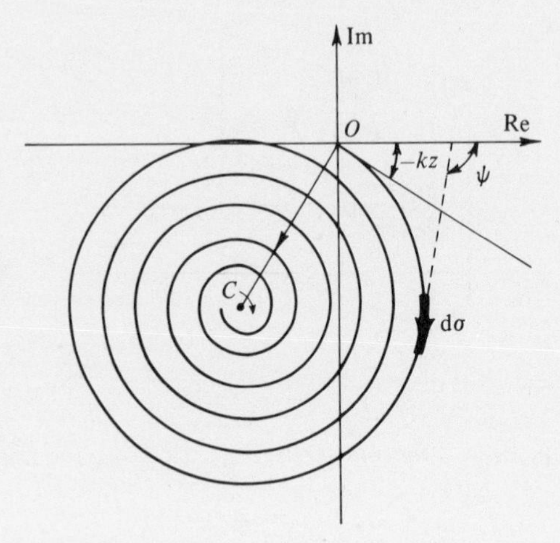

Fig. 10.6. Amplitude–phase diagram for the integral (10.17).

scattered field δE_Q at Q as the integral over the whole slab:

$$\delta E_Q = \frac{N\omega^2 \alpha E_0 \delta z \exp(\mathrm{i}\omega t)}{4\pi c^2} \iint_{-\infty}^{\infty} \frac{z \exp(-\mathrm{i}kr)}{r^2} \,\mathrm{d}x\,\mathrm{d}y, \quad (10.16)$$

where $\cos\theta$ has been replaced by z/r. Now we can write this integral in cylindrical polar coordinates (ρ, ϕ, z) as

$$\delta E_Q = \frac{N\omega^2 \alpha E_0 \delta z \exp(\mathrm{i}\omega t)}{4\pi c^2} \times 2\pi z \int_0^{\infty} \frac{\exp\{-\mathrm{i}k(\rho^2 + z^2)^{\frac{1}{2}}\}}{\rho^2 + z^2} \rho \,\mathrm{d}\rho. \quad (10.17)$$

Using the relationship $r^2 = \rho^2 + z^2$, the integral in (10.17) can conveniently be written as

$$\int_z^{\infty} \frac{\exp(-\mathrm{i}kr)}{r} \,\mathrm{d}r.$$

It can easily be evaluated by an amplitude–phase diagram, as explained in § 6.4.1. The curve (Fig. 10.6) is constructed from elementary vectors of length $\mathrm{d}\sigma = \mathrm{d}r/r$ at angle $\psi = -kr$ to the real axis. The curve has local radius of curvature $\mathrm{d}\sigma/\mathrm{d}\psi = 1/kr$, and when $kz \gg 1$ the turns approximate to circles with this radius. As the integral continues, and r increases, the turns slowly spiral in to the centre. The integral from $r = z$ to ∞ is thus the vector OC

$$OC = (kz)^{-1} \exp\{-\mathrm{i}(kz + \pi/2)\}. \quad (10.18)$$

Thus

$$\delta E_Q = \frac{N\omega^2 \alpha E_0 \delta z}{2c^2 k} \exp\{\mathrm{i}(\omega t - kz - \pi/2)\}$$
$$= \frac{-\mathrm{i}N\omega^2 \alpha E_0 \delta z}{2c^2 k} \exp\{\mathrm{i}(\omega t - kz)\}. \tag{10.19}$$

Replacing ω^2/c^2 by k^2 we have

$$\delta E_Q = -\tfrac{1}{2}\mathrm{i}Nk\alpha E_0 \delta z \exp\{\mathrm{i}\,(\omega t - kz)\}. \tag{10.20}$$

This scattered wave must be added to the unscattered wave which has reached Q; if δz is small, the unscattered wave is negligibly different from the incident wave $E_{Q0} = E_0 \exp\{\mathrm{i}(\omega t - kz)\}$, whence

$$\delta E_{Q0} = -\tfrac{1}{2}\mathrm{i}kN\alpha\delta z E_Q. \tag{10.21}$$

If α is real, the scattered amplitude is in phase quadrature with the direct wave and therefore does not alter its magnitude, but only its phase; in

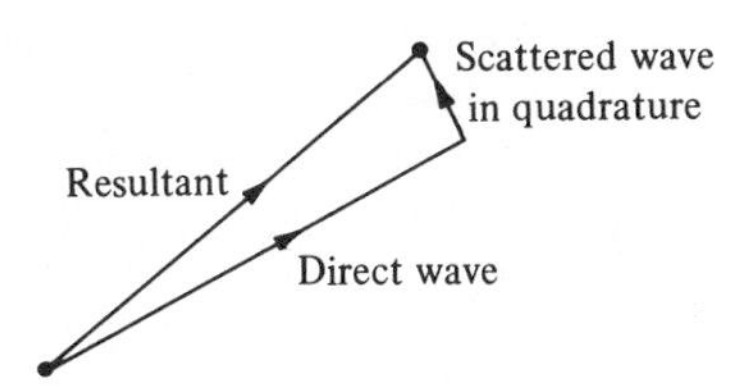

Fig. 10.7. Amplitude–phase diagram for coherent scattering when there is no absorption.

other words, the velocity of the wave is modified, but there is no attenuation (Fig. 10.7). Then

$$E_Q = E_{Q0} + \delta E_Q = (1 - \tfrac{1}{2}\mathrm{i}kN\alpha\delta z)E_{Q0}$$
$$\approx \exp(-\tfrac{1}{2}\mathrm{i}kN\alpha\delta z)E_{Q0}. \tag{10.22}$$

If we had inserted a transparent plate, with refractive index μ and thickness δz, into the beam, we should have increased the optical path by $(\mu - 1)\delta z$ and modified the wave E_{Q0} to

$$E_Q = E_{Q0} \exp\{-\mathrm{i}k\delta z(\mu - 1)\}. \tag{10.23}$$

So coherent scattering by the slab has resulted in an effective refractive index

$$\mu = 1 + \tfrac{1}{2}N\alpha. \tag{10.24}$$

This is just the refractive index we have used, for example in (10.12). Thus coherent scattering results in refraction, but not absorption.

If the medium is dense, so that μ is not close to unity, we must consider the field which polarizes the molecules as the local field, and not simply the applied field. This makes the treatment more complicated but does not introduce absorption.

10.3.2 *Resonance and anomalous dispersion.* Near the resonance frequency α is not real, and as a result the statement that the scattered wave is in quadrature with the direct wave is no longer correct. The

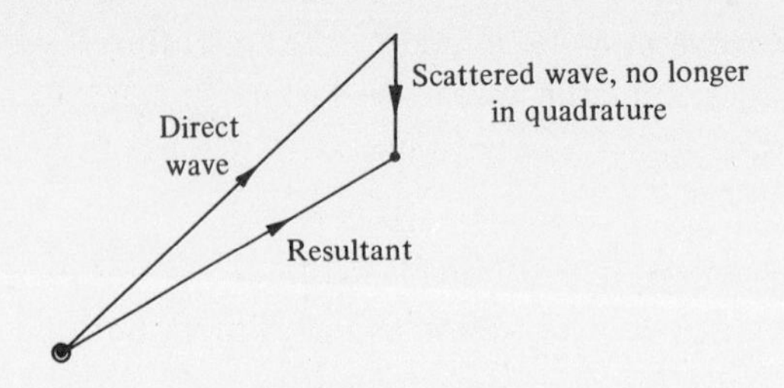

Fig. 10.8. Absorption occurs when the scattered wave is no longer in quadrature with the direct wave.

refractive index is still modified, but absorption may also occur (Fig. 10.8). Under the influence of the oscillating wave-field (10.4) we saw that the complex polarizability is given by (10.6):

$$\alpha = q^2/\{\epsilon_0 m(\eta^2 - \omega^2 + i\kappa\omega)\}$$

and thus the refractive index (10.24) is approximately

$$\mu = 1 + \tfrac{1}{2}N\alpha = 1 + \tfrac{1}{2}\Omega^2(\eta^2 - \omega^2 + i\kappa\omega)^{-1} \tag{10.25}$$

when $N\alpha \ll 1$. In (10.25), Ω is the plasma frequency (§ 10.3.5)

$$\Omega = (Nq^2/\epsilon_0 m)^{\frac{1}{2}}. \tag{10.26}$$

According to the approximation $N\alpha \ll 1$ (which is often untrue, but still indicates the physics of the problem correctly) the real and imaginary parts of μ are

$$\mu_1 = 1 + \frac{\tfrac{1}{2}\Omega^2(\eta^2 - \omega^2)}{(\eta^2 - \omega^2)^2 + \kappa^2\omega^2}, \tag{10.27}$$

$$\mu_2 = \frac{-\tfrac{1}{2}\Omega^2\kappa\omega}{(\eta^2 - \omega^2)^2 + \kappa^2\omega^2}. \tag{10.28}$$

These two quantities, μ_1 and μ_2, are illustrated in Fig. 10.9 which shows two important features:

(a) The refractive index becomes large at frequencies just below resonance, and sharply drops to a value less than unity just above the

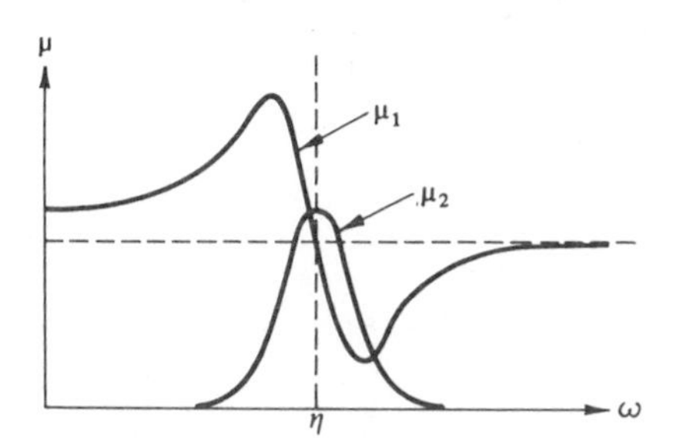

Fig. 10.9. Anomalous dispersion.

resonance. The region of sharp change, where $d\mu_1/d\omega$ is negative, is called the *anomalous-dispersion region*, because dispersion usually gives rise to positive $d\mu_1/d\omega$.

(b) The anomalous-dispersion region is accompanied by absorption. We shall show that this is necessary from very general considerations (§ 10.5).

This absorption is, of course, the absorption line corresponding to an emission line in the atomic spectrum. If the atom is excited, its oscillations at its natural frequency will be radiated and die away in the decay time κ^{-1} (the transient of § 10.1.2). Such a decaying oscillation gives rise to a spectral line of width κ corresponding to the absorption line at the same frequency and of the same width as that which we have calculated above. In fact, the correspondence between the emission and absorption line characteristics is sufficiently complete for it to lead to a convenient method of measuring linewidths which will be discussed shortly (§ 10.3.3).

A real atom has a series of spectral lines at various frequencies, and anomalous dispersion takes place in the region of each one of them. The classical theory cannot explain this multitude of resonances, but takes them into account by assigning a number N_j of model atoms to the jth resonance so that the refractive index is written

$$\mu_1 - 1 = \frac{e^2}{2\epsilon_0 m} \sum_{j=1}^{\infty} \frac{N_j(\eta_j^2 - \omega^2)}{(\eta_j^2 - \omega^2)^2 + \kappa_j^2 \omega^2}. \tag{10.29}$$

The N_js are called *oscillator strengths* and are related to the matrix elements derived from the quantum-mechanical description. Fig. 10.10 shows a typical refractive-index curve.

10.3.3 *Experimental investigation of anomalous dispersion.* The refractive index and absorption coefficient in an anomalous region can be

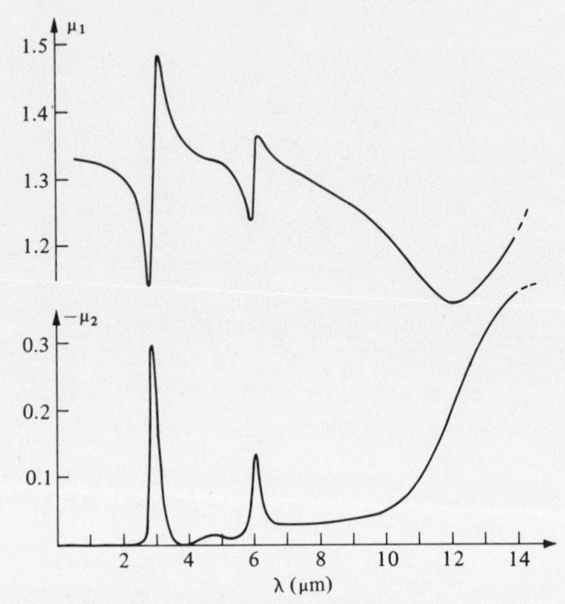

Fig. 10.10. Real and imaginary components of the refractive index of sea water at wavelengths between 0.6 and 14 μm.

determined by a number of methods, the common property being that they must all be sensitive to small changes in wavelength. For example, the reflectivity of a surface for oblique incidence and various polarizations can be used to calculate the real and imaginary parts of the refractive index. The method has been used to study dispersion in the region of the infra-red rotation-spectral lines in water (Fig. 10.10).

Probably the most ingenious of the methods that can be used in the visible region is the interference method due to Roschdestwenski. Sodium vapour with uniform density is introduced into one tube of a Jamin interferometer (§7.8.1) illuminated by white light. The resulting interference pattern is analysed on a spectrometer so that the components at various wavelengths are separated, and since a fringe shift proportional to the refractive index occurs in the interferometer the analysed interference pattern is bent into *hooks* (Fig. 10.11). This method has been used to investigate the shape of the absorption line, and hence the width of the corresponding emission line, the pattern being calibrated by the fringe separation. The line-shape can therefore be reliably measured.

10.3.4 *Dispersion remote from an absorption band: X-ray refractive index.* Equation (10.27) can be applied when η and ω are far from equality.

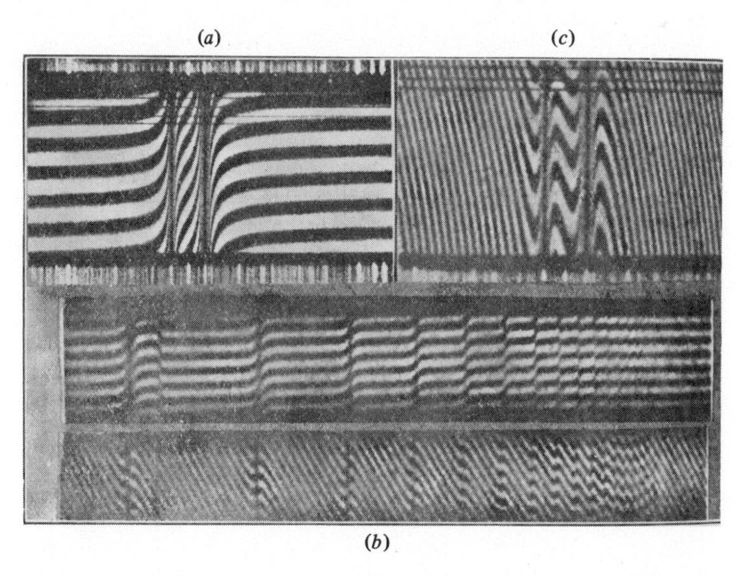

Fig. 10.11. Roschdestwenski's hooks, observed in sodium vapour. (From Wood, 1934.)

When $(\eta^2-\omega^2)\gg\kappa\omega$ absorption is negligible and we can calculate the normal dispersion

$$\mu_1=1+\frac{1}{2}\frac{\Omega^2}{(\eta^2-\omega^2)}. \tag{10.30}$$

At very high frequencies, very much greater than the η for the highest-frequency spectral line, we can write

$$\mu_1=1-\frac{1}{2}\frac{\Omega^2}{\omega^2}, \tag{10.31}$$

which shows that the refractive index in the X-ray region is less than unity, but only just so. Substitution of typical values gives

$$1-\mu_1\approx 10^{-7}. \tag{10.32}$$

10.3.5 *Refractive index of a free-electron assembly.* If the electrons in a medium are unbound, either as a plasma in the ionosphere or as conduction electrons in a simple metal, for example, we can work out the refraction properties as a function of frequency. Substituting zero for the resonant frequency η, we obtain from (10.6) and (10.26)

$$\epsilon=\mu^2=1+N\alpha=1+\frac{\Omega^2}{i\kappa\omega-\omega^2}. \tag{10.33}$$

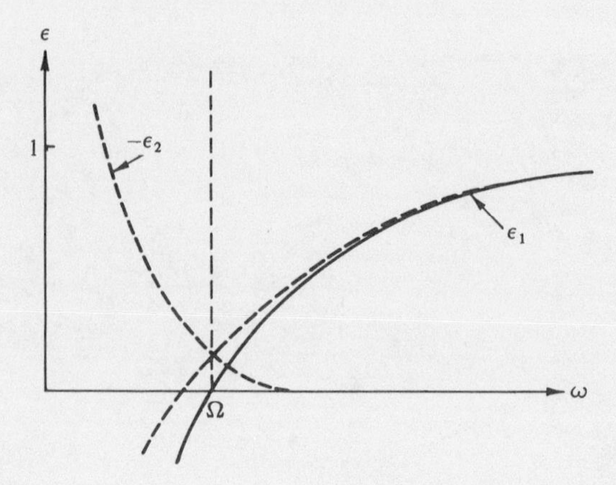

Fig. 10.12. Real and imaginary parts of the complex dielectric constant for a free-electron assembly. The solid line is for zero damping, the broken line for non-zero damping.

In the zero-damping case, $\kappa = 0$,

$$\epsilon = 1 - \frac{\Omega^2}{\omega^2}, \tag{10.34}$$

which shows that when ω is below the frequency Ω all modes of propagation are evanescent, since ϵ is negative. This can be seen immediately from the diagram (Fig. 10.12); Ω is called the *plasma frequency* because at that frequency collective oscillations of the whole plasma can occur. When $\epsilon = 0$, we have

$$\frac{\omega}{k} = \frac{c}{\mu} = \infty; \tag{10.35}$$

therefore, $k = 0$, $\lambda = \infty$. This corresponds to the whole plasma oscillating in phase. At a slightly higher frequency, phase differences will occur between neighbouring parts.

When $\kappa \neq 0$ the real and imaginary parts of the dielectric constant become

$$\epsilon_1 = 1 - \frac{\Omega^2}{\omega^2 + \kappa^2}, \tag{10.36}$$

$$\epsilon_2 = \frac{-\Omega^2 \kappa}{\kappa^2 \omega + \omega^3}. \tag{10.37}$$

The plasma frequency is thus shifted, zero ϵ_1 occurring at the lower frequency

$$\omega^2 = \Omega^2 - \kappa^2. \tag{10.38}$$

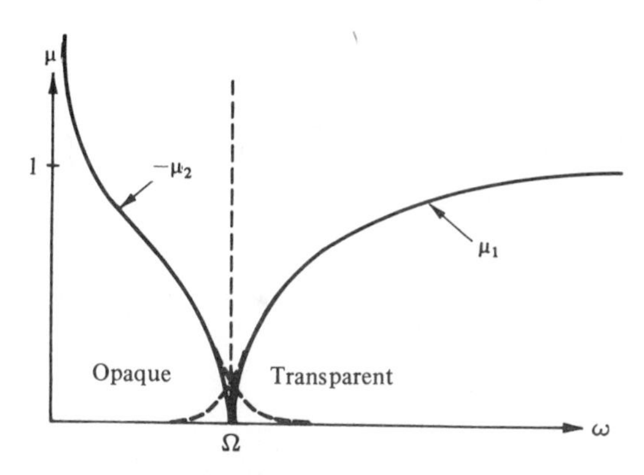

Fig. 10.13. The refractive index arising from Fig. 10.12. This figure should be compared with Fig. 10.14, in which experimental results for sodium are shown. Note that the plots in Figs. 10.13 and 10.14 are functions of ω and λ respectively.

The plasma oscillations are now damped, however, since at this frequency

$$\epsilon_2 = -\frac{\kappa}{\omega}. \tag{10.39}$$

The refractive index arising from (10.36) and (10.37) is sketched in Fig. 10.13. For comparison are shown some measurements made in sodium, which is a fairly good free-electron metal. Since it has a finite electrical conductivity, κ is not zero (Fig. 10.14).

10.4 Relaxation processes

It is well known that water and ice have a dielectric constant of about 80 for a static electric field, whereas their refractive index at optical frequencies is about 1.3, corresponding to a dielectric constant of about 1.7. The large change in refractive index occurs in a single frequency range, 10^9–10^{11} Hz for water and 10^3–10^5 Hz for ice; this strong dispersion is accompanied by an absorption band. It is also well known that the H_2O molecule has a permanent dipole moment. When an electric field is applied to water or ice, the permanent dipoles try to align themselves to the field to attain minimum potential energy, and the wholesale rotation of the molecules is naturally resisted by frictional forces. If the electric field is applied for long enough, the dipoles will eventually reach a state of equilibrium in which the alignment has resulted in a considerable net dipole moment, and this is responsible for the high static dielectric constant. Complete alignment will not be reached for two reasons. First, thermal agitation will tend to randomize

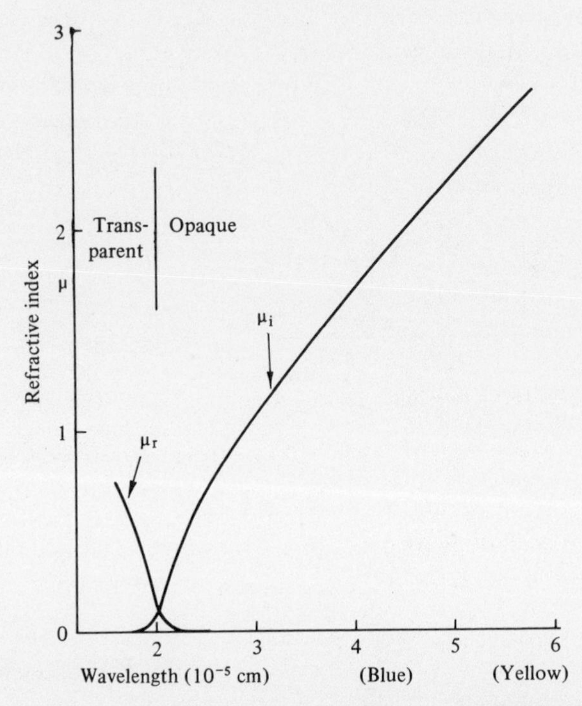

Fig. 10.14. Complex refractive index of sodium measured as a function of wavelength near the plasma wavelength. Calculated from the free-electron density, the plasma wavelength $\lambda_p = \Omega/c$ should be 2.1×10^{-5} cm.

the directions. Secondly, if the electric field alternates at a high enough frequency there will not be sufficient time, during a single half-cycle of the field, for the dipoles to align themselves. Therefore at sufficiently high frequencies the effects of permanent dipoles disappear.

The frequency dependence of the dielectric constant (and hence refractive index) resulting from the latter process gives rise to a type of dispersive effect which is qualitatively different from the resonant effects responsible for anomalous dispersion. The effects are known as 'relaxation processes' and, although they generally occur at frequencies lower than optical, will be discussed briefly below.

10.4.1 *The impulse response.* A convenient way of understanding the dynamic behaviour of a system which responds to an external field is to start by considering the effect of a single impulse of the field. This approach is by no means modern; Newton used it in his analysis of the Earth's motion in the Sun's gravitational field. The response to the field impulse can usually be deduced from a physical understanding of the working of the system. In our case the application of the impulse – an electric field applied for a very short time – starts an alignment of the molecules which cannot be completed during the pulse. Then the partially induced alignment decays back to zero. We shall suppose that this decay process is exponential, although there are many systems for which it is much more complicated. Writing $Q_I(t)$ for the polarization at time t we have, for a δ-function field impulse $\delta(t)$ at $t=0$,

$$\begin{aligned} Q_I(t) &= 0 \ (t<0), \\ Q_I(t) &= Q_0 \exp(-t/\tau) \ (t \geq 0), \end{aligned} \tag{10.40}$$

where τ is the time constant of the decay process.

The effect of an alternating electric field at frequency ω can now be calculated. The field $E_0 \exp(\mathrm{i}\omega t)$ can be represented by a series of impulses at intervals $\mathrm{d}t$ and of magnitude $E_0 \exp(\mathrm{i}\omega t)\,\mathrm{d}t$. If the system responds linearly, the response $P(t)$ to a series of impulses is just the superposition of the individual responses; this is the convolution

$$\begin{aligned} P(t) &= \int E_0 \exp(\mathrm{i}\omega t') Q_I(t-t')\,\mathrm{d}t' \\ &= E_0 \exp(\mathrm{i}\omega t) \int \exp\{-\mathrm{i}\omega(t-t')\} Q_I(t-t')\,\mathrm{d}(t-t') \\ &= E_0 \tilde{Q}(\omega) \exp(\mathrm{i}\omega t), \end{aligned} \tag{10.41}$$

where $\tilde{Q}(\omega)$ is the Fourier transform of $Q(t)$. Now the dielectric constant $\epsilon(\omega)$ at frequency ω is related to $P(t)$ and E_0 by the relation

$$\epsilon(\omega)\epsilon_0 E_0 \exp(\mathrm{i}\omega t) = \epsilon_0 E_0 \exp(\mathrm{i}\omega t) + P(t), \tag{10.42}$$

whence

$$\epsilon_0\{\epsilon(\omega)-1\} = \tilde{Q}(\omega). \tag{10.43}$$

From (10.40) we calculate $\tilde{Q}(\omega)$, following § 7.5.1, to be

$$\tilde{Q}(\omega) = Q_0(\mathrm{i}\omega + 1/\tau)^{-1} \tag{10.44}$$

and

$$\epsilon(\omega) - 1 = Q_0/\{\epsilon_0(\mathrm{i}\omega + 1/\tau)\} = \{\epsilon(0)-1\}/(1+\mathrm{i}\omega\tau), \tag{10.45}$$

where the value in a static field $\epsilon(0)$ has been used to replace Q_0. The real and imaginary parts of (10.45) are then:

$$\epsilon_1(\omega) - 1 = \{\epsilon(0)-1\}/(1+\omega^2\tau^2), \tag{10.46}$$

$$\epsilon_2(\omega) = -\{\epsilon(0)-1\}\omega\tau/(1+\omega^2\tau^2). \tag{10.47}$$

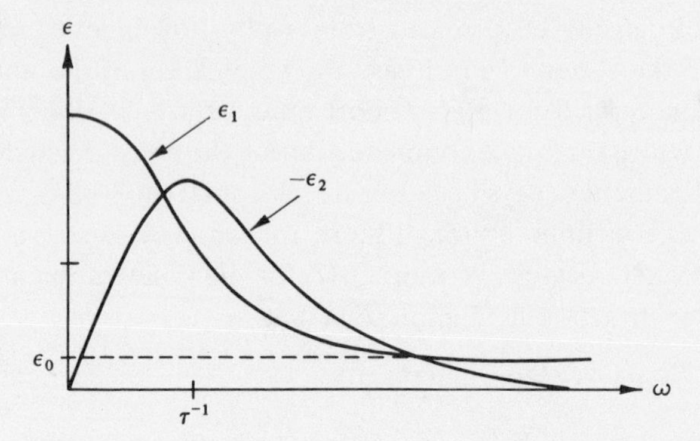

Fig. 10.15. Frequency dependence of the electric constant in a medium dominated by relaxation processes.

These functions are illustrated by Fig. 10.15. The behaviour is characterized by a broad absorption band and a Lorenzian dielectric-constant curve centred on zero frequency. This is typical of relaxation processes. It should be realized that the above calculation can easily be modified if the approach to alignment is not exponential. Any functional form $Q_I(t)$ can be written for (10.40) and its Fourier transform (10.44).

The high-frequency limit of (10.46) and (10.47) should be noted. When $\omega \gg 1/\tau$

$$\epsilon_1(\omega) \to 1, \qquad \epsilon_2(\omega) \to 0. \tag{10.48}$$

However, ϵ_1 approaches 1 from above, in contrast to the behaviour of resonant systems (§ 10.3.4). In Fig. 10.16 we show the observed behaviour of the dielectric constant of ice, which fits the above model very well.

10.5 Dispersion theory

In the two examples of classical dispersive processes discussed in this chapter, resonant and relaxation, the real and imaginary parts of the dielectric constant appeared together as parts of the same analytic function and are obviously related. However complicated the dispersion process may be on a microscopic scale, such relationships must exist, and in this section we shall explore, in a somewhat non-rigorous manner, some of the consequences of these relationships.

10.5.1 *The Kramers–Krönig relations.* The response $Q_I(t)$ introduced in § 10.4 can, as we mentioned, have many forms of which (10.40) is just a

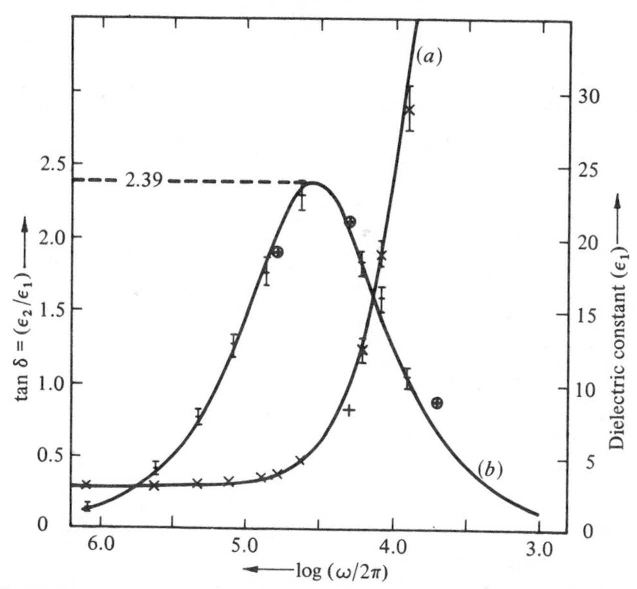

Fig. 10.16. Dielectric constant as a function of frequency in ice at -5 °C. The two curves show: (*a*) the real part and (*b*) the ratio of real to imaginary parts $\tan\delta = \epsilon_2(\omega)/\epsilon_1(\omega)$ which is called the loss factor. The points show experimental measurements (Lamb, 1946) and the curves are the best fit to equations (10.46) and (10.47).

simple example. For instance, in a resonant system $Q_I(t)$ would be oscillatory. But it must always obey one simple rule: $Q_I(t)$ must be zero when $t<0$; otherwise the response would precede its cause, the impulse of the electric field at $t=0$. This behaviour is called *causal*, and leads to some general relationships between dispersion and absorption.

We can use causality to relate the real and imaginary parts of the refractive index and hence of the dielectric constant. Let us define a unit step function $D(t)$;

$$D(t) = \begin{cases} \lim_{s\to 0} [\exp(st)] \quad (\sim 1) & \text{when } t<0 \\ 0 \quad \text{when } t \geqslant 0. \end{cases} \tag{10.49}$$

We have defined it via the limiting process so that its Fourier transform can be calculated for the reasons explained in § 3.4.4; its transform is

$$\tilde{D}(\omega) = \lim_{s\to 0} (s - \mathrm{i}\omega)^{-1}. \tag{10.50}$$

Now since $Q_I(t)$ only starts at $t=0$, and $D(t)$ finishes by then, we can write

$$Q_I(t)D(t) = 0. \tag{10.51}$$

Taking the Fourier transform of this equation

$$0=\tilde{Q}(\omega)*\tilde{D}(\omega)=\lim_{s\to 0}\int_{-\infty}^{\infty}\frac{\tilde{Q}(\omega')}{s-\mathrm{i}(\omega-\omega')}\,\mathrm{d}\omega'. \tag{10.52}$$

Thus from (10.43) we have

$$0=\lim_{s\to 0}\int_{-\infty}^{\infty}\frac{\epsilon(\omega')-1}{s-\mathrm{i}(\omega-\omega')}\,\mathrm{d}\omega'. \tag{10.53}$$

We can divide this integral into two parts: the integral from $\omega-s$ to $\omega+s$ and the rest. The first integral can be evaluated straightforwardly:

$$\lim_{s\to 0}\int_{\omega-s}^{\omega+s}\frac{\{\epsilon(\omega')-1\}\,\mathrm{d}\omega'}{s-\mathrm{i}(\omega-\omega')}=\{\epsilon(\omega)-1\}\int_{\omega-s}^{\omega+s}\frac{\mathrm{d}\omega'}{s-\mathrm{i}(\omega-\omega')}$$
$$=\pi\{\epsilon(\omega)-1\}. \tag{10.54}$$

The rest is called the *principal part* of the integral denoted by $\mathscr{P}$ and the small s can be neglected:

$$\lim_{s\to 0}\left(\int_{-\infty}^{\omega-s}+\int_{\omega+s}^{\infty}\right)\frac{\{\epsilon(\omega')-1\}\,\mathrm{d}\omega}{s-\mathrm{i}(\omega-\omega')}\equiv\mathscr{P}\int_{-\infty}^{\infty}\frac{\epsilon(\omega')-1}{-\mathrm{i}(\omega-\omega')}\,\mathrm{d}\omega'. \tag{10.55}$$

Using the above three equations we see that

$$\pi\{\epsilon(\omega)-1\}=\mathscr{P}\int_{-\infty}^{\infty}\frac{\epsilon(\omega')-1}{\mathrm{i}(\omega-\omega')}\,\mathrm{d}\omega'. \tag{10.56}$$

We can equate real and imaginary parts of (10.56) separately and obtain two integral relationships between $\epsilon_1(\omega)$ and $\epsilon_2(\omega)$, where $\epsilon(\omega)=\epsilon_1(\omega)+\mathrm{i}\epsilon_2(\omega)$. Notice from (10.28) and (10.47), for example, that absorption corresponds to negative $\epsilon_2(\omega)$.

$$\pi\{\epsilon_1(\omega)-1\}=\mathscr{P}\int_{-\infty}^{\infty}\frac{\epsilon_2(\omega')}{(\omega-\omega')}\,\mathrm{d}\omega\quad\left[=2\mathscr{P}\int_{0}^{\infty}\frac{\omega'\epsilon_2(\omega')\,\mathrm{d}\omega'}{\omega^2-\omega'^2}\right], \tag{10.57}$$

$$\pi\epsilon_2(\omega)=\mathscr{P}\int_{-\infty}^{\infty}\frac{\epsilon_1(\omega')-1}{(\omega-\omega')}\,\mathrm{d}\omega'\quad\left[=2\mathscr{P}\int_{0}^{\infty}\frac{\omega\{\epsilon_1(\omega')-1\}\,\mathrm{d}\omega'}{\omega^2-\omega'^2}\right]. \tag{10.58}$$

In the bracketed forms of (10.57) and (10.58) we have in each used the properties $\epsilon(\omega)=\epsilon^*(-\omega)$, resulting because $\epsilon(\omega)$ is the Fourier transform of a real response to an applied electric field. Equations (10.57) and (10.58) are known as the Kramers–Krönig relations.

10.5.2 *Qualitative implications of the Kramers–Krönig relations.* We shall consider two examples to show qualitatively how the relations (10.57) and (10.58) can be used to deduce one ϵ from the other.

Suppose that a material is known to have the same real dielectric constant in two separated frequency regions around ω_1 and ω_2. We can immediately deduce that ϵ_2 must be zero between these two frequencies, so that there is no absorption band between them. For, if $\epsilon_2$were not zero in this range, (10.57) shows that the contribution it would make would oppositely affect $\epsilon_1(\omega_1)$ and $\epsilon_1(\omega_2)$ because the signs of $\omega_1^2-\omega^2$ and $\omega_2^2-\omega^2$ are opposite when ω is in this region. Thus no absorption can occur between ω_1 and ω_2. It is important to notice that there must be *regions* of equal ϵ_1 around ω_1 and ω_2 because it is always possible that isolated frequencies in regions of rapidly changing ϵ_1 will have the same value. This occurs in the anomalous regions.

Similarly, at very high frequencies, we can see that ϵ_1 approaches unity from below, when all absorption has taken place at lower frequencies. The imaginary susceptibility ϵ_2 is always negative, representing absorption, and (10.57) shows that $\epsilon_1(\omega)-1$ will also be negative when ω is well above the absorption frequencies.

10.5.3 *A wave-group propagated in a dispersive medium.* The way in which a wave-group propagates in the presence of dispersion can be described formally in a fairly straightforward manner. If we know, either by theory or by experiment, the complex refractive index $\mu(\omega)$, we can write a propagating sine wave as:

$$\exp\{\mathrm{i}(\omega t-kx)\}=\exp[\mathrm{i}\{\omega t-\mu(\omega)\omega x/c\}]. \tag{10.59}$$

From the initial form of the wave-group $f(x, t)$, i.e. $f(0, t)$, we can calculate its Fourier transform; then we consider the propagation of each component separately. Thus if

$$f(0, t)=\int_{-\infty}^{\infty} a(\omega)\exp\{\mathrm{i}\omega t\}\,\mathrm{d}\omega \tag{10.60}$$

then

$$f(x, t)=\int_{-\infty}^{\infty} a(\omega)\exp[\mathrm{i}\omega\{t-\mu(\omega)x/c\}]\,\mathrm{d}\omega. \tag{10.61}$$

The question then arises whether the integral (10.61) can be calculated, and what is its physical content.

Now in a reasonably behaved material for which the dispersion is normal and $\mu(\omega)$ is a monotonically increasing function, there are no

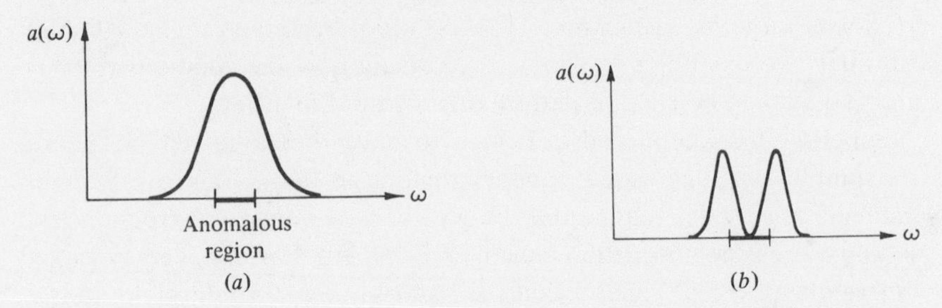

Fig. 10.17. Spectrum of a wave-group (a) before and (b) after passage through a medium exhibiting anomalous dispersion.

particular surprises. For example (problem 10.1), if we consider a Gaussian wave-group $f(0, t)$ we find that the peak of the envelope propagates at the group velocity $\partial\omega/\partial k$ (§ 2.4) and that the width of the envelope increases steadily at a rate depending on the second differential $\partial^2\omega/\partial k^2$. This result is very important in communication theory because the spreading of a wave-group sets a limit to the rate at which pulses can be transmitted by the medium. We shall refer to this result again in § 11.9.5 in connexion with the transmission of pulses along optical fibres.

Propagation of a wave-group in the region of anomalous dispersion poses some more interesting problems. The group velocity can be written in terms of the real part μ_1 of the refractive index $\mu(\omega)$:

$$v_g = \frac{\partial\omega}{\partial k} = c\left(\mu_1 + \omega\frac{\partial\mu_1}{\partial\omega}\right)^{-1}. \tag{10.62}$$

In the anomalous region $\mu_1 \sim 1$ and $\partial\mu_1/\partial\omega$ is negative, so that v_g becomes greater than c. Does this mean that signals can be propagated faster than the velocity of light in free space, in contradiction to the theory of relativity?

Before answering the above question we must stress the inevitability of absorption in the anomalous region. Obviously a wave-group composed mainly of components in this region could never travel far, and it is questionable if any relativity-defying result could ever be observed; it is also clear that the wave-group becomes seriously distorted because of the selective absorption of components in a narrow frequency range (Fig. 10.17). It is therefore necessary to specify the question rather more exactly.

10.5.4 *Signal velocity.* A necessary concept is the *signal velocity*. We start with a wave-group $f(0, t)$ which is zero for all times $t < 0$, and then at point

x we wait for the first vestiges of signal to arrive. The arrival at time T of the first signal, however weak, defines the signal velocity x/T. The question is whether *this* velocity can exceed c.

An elegant and complete answer can be given in terms of integrals in the complex-frequency plane, and is discussed very fully by L. Brillouin in his book *Wave Propagation and Group Velocity* (1960). We shall only give a qualitative treatment here, and attempt to show where the difference between group and signal velocity lies.

In § 2.4 we derived the group velocity by requiring that the phase relation between the various components of a wave-group remain unchanged as the wave progresses. If $\phi = \omega t - kx$ is the phase of a component, then the group velocity relates x and t when $\partial\phi/\partial\omega = 0$. The idea was then applied to a wave-group for which the Fourier components all had the same phase; any *symmetrical* group such as a Gaussian wave-group satisfies this condition.

Another way of describing the same calculation invokes the method of stationary phase which we have already introduced in § 6.4.5. The integral (10.61) is dominated by the region for which $\partial\phi/\partial\omega = 0$, and therefore has its peak value at the point $x(t)$ for which this relationship applies.

Now let us consider the definition of signal velocity. We have defined the initial wave-group to be *unsymmetrical* by requiring it to have zero value at all times less than zero. For such a wave-group the group-velocity is then no longer the velocity at which the envelope progresses. If we write the transform $a(\omega)$ as $|a(\omega)| \exp\{i\psi(\omega)\}$ we can substitute in (10.61) to get:

$$f(x, t) = \int_{-\infty}^{\infty} |a(\omega)| \exp[i\{\omega t - \mu(\omega)\omega x/c + \psi(\omega)\}] \, d\omega. \quad (10.63)$$

Then the method of stationary phase tells us that the dominant contribution to the integral comes from the region where

$$\frac{\partial\phi}{\partial\omega} = \frac{\partial}{\partial\omega}\{\omega t - \mu(\omega)\omega x/c + \psi(\omega)\} = 0 \quad (10.64)$$

and the $x(t)$ relationship implied by this equation gives the signal velocity. Now the equation (10.64) involves particulars of the wave-group itself, so that the signal velocity is not determined solely by the medium. But the requirement that $f(0, t)$ be zero for $t < 0$ can easily be shown to result in $\partial\psi/\partial\omega$ being negative, which in turn constrains the signal velocity to be less than the group velocity, and in general less than c. We shall illustrate this result by a single example.

Suppose that the propagating wave-group has the initial form

$$f(0, t) = \exp(-st)\exp(\mathrm{i}\eta t) \quad (t \geqslant 0) \tag{10.65}$$

and is zero for $t < 0$. Then its Fourier transform is

$$\begin{aligned} a(\omega) &= -\{s + \mathrm{i}(\omega - \eta)\}^{-1} \\ &= -\{s^2 + (\omega - \eta)^2\}^{-1/2} \exp[-\mathrm{i}\tan^{-1}\{(\omega - \eta)/s\}]. \end{aligned} \tag{10.66}$$

Thus in (10.63) we substitute

$$\psi(\omega) = -\tan^{-1}\{(\omega - \eta)/s\}. \tag{10.67}$$

The relationship (10.64) then gives us

$$\begin{aligned} 0 &= t - \left\{\omega \frac{\partial \mu}{\partial \omega} + \mu(\omega)\right\} \frac{x}{c} + \frac{\partial \psi}{\partial \omega} \\ &= t - \frac{x}{v_\mathrm{g}} + \frac{\partial \psi}{\partial \omega} \end{aligned} \tag{10.68}$$

$$t = \frac{x}{v_\mathrm{g}} + \frac{s}{s^2 + (\omega - \eta)^2}. \tag{10.69}$$

This equation does not define a constant velocity, but it does tell us that the time t at which the wave-group has significant amplitude is later than we should have calculated from the group velocity alone. In other words, the group travels more slowly than the group velocity. In particular, suppose that we construct the ideal wave-group for testing the relativity-defying result, by letting η be the exact resonance frequency of the material, and letting $s \to 0$ so as to define the frequency of the group as closely as possible. But then the term $s/\{s^2 + (\omega - \eta)^2\}$ has a high peak, of value $1/s$ just at $\omega = \eta$, and the signal velocity is correspondingly small.

11

Some applications of optical ideas

11.1 Introduction

Optics has played such a large part in the development of physics that it might almost be said that the history of optics is the history of physics. The point is not so much that optical instruments have always been important in experimental research, but that optical ideas and optical results have often influenced physics and physicists in unexpected ways which have turned out to be of fundamental significance.

The reason for this strong influence is that on the whole problems in optics can be clearly formulated, and are usually soluble – in principle, if not in detail. The existence of a solved problem in one field has often been a stimulus to the solution of a similar but more elusive problem in another field, perhaps completely unrelated at first sight.

But for the same reason – that optical problems can generally be formulated and solved, with more or less difficulty – optics has been through a long period of stagnation, when it was considered as a subject unworthy of the attention of serious research workers striving for the boundaries of knowledge. The names, however, remained: witness, for example, the optical model of the nucleus, and dispersion relations in elementary particle physics. The re-emergence of optics, stimulated by the invention of the laser in 1960, has brought with it a host of new developments, and the stigma of a 'worked out' subject has largely disappeared.

This chapter is intended to illustrate, with a few examples, how the ideas which formed the theme of this book have been applied both to subjects outside optics and also to present-day advances within optics itself.

11.2 Some general uses of optics

11.2.1 *Basic units.* A considerable number of the fundamental physical constants accepted today have been determined by optical methods.

Moreover, the constants determined optically are, almost without exception, the most accurately known ones. Of the basic standards, only that of length (1 metre = 1 650 753.73 wavelengths of a certain ^{86}Kr emission line) is now defined optically; the old astronomical definition of the second was replaced in 1967 by a new definition based on an atomic resonance frequency in cesium. Amongst the physical constants, optical determinations play a leading role. The velocity of light, c, has been determined by a number of methods, including optical time-of-flight methods and microwave interferometry. Avogadro's number, N_A, is determined by an accurate measurement of the unit-cell dimensions of a crystal whose structure is known; this involves measuring the wavelength of X-rays absolutely by diffraction at glancing angle off a ruled diffraction grating. Accurate spectroscopic measurements are necessary for determining the fine-structure constant $\alpha = e^2/\hbar c$ and the Rydberg constant $R_\infty = me^4/4\pi\hbar^3 c$.

The type of indefatigability and patience needed for such measurements is illustrated by Michelson's first determination of the metre in terms of the wavelength of a spectral line. This standard can be reproduced identically in laboratories throughout the world, in contrast to the standard metre rod which is preserved in holy isolation in an underground vault.

Michelson's experiment was based on his interferometer (§ 7.8.2) illuminated simultaneously by the monochromatic source and by white light. The white light serves to label the zero-order fringe. Several etalons (§ 7.7.6), each having two steps, were prepared, and each one was measured as precisely as possible by conventional means with respect to the standard metre. Each etalon was approximately twice the length of the previous one. The interferometer was used first to determine the number of fringes corresponding to the shortest etalon (about 0.4 mm, of the order of 600 fringes) to an accuracy of 0.1 fringe. From this, the number of fringes corresponding to the second etalon could be estimated to an accuracy of about 0.2 fringe. But the odd fraction of a fringe could be measured with the interferometer to an accuracy of 0.1 fringe with no further need for counting fringes; Michelson simply needed to observe the match between the fringes crossing the two steps. The process was repeated until finally the longest etalon – one metre – was finally measured, also to an accuracy of 0.1 fringe (Michelson, 1927).

11.2.2 *Surface topography.* Interferometry is also of great value in applied physics – to the testing of surfaces. Here the use of optics in

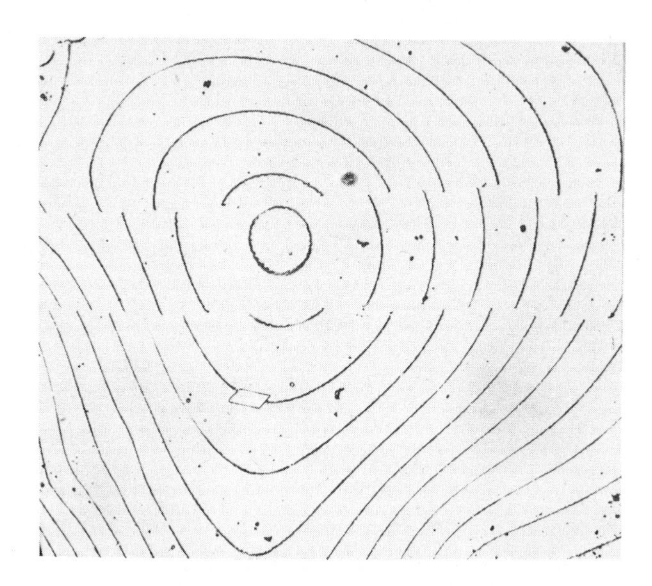

Fig. 11.1. Cleavage surface of mica as revealed by multiple-beam interference. (By courtesy of Professor S. Tolansky.)

maintaining accuracy is of outstanding importance, and the mass production of machinery is dependent upon it.

To test the flatness of a surface we can use fringes of equal thickness (§ 7.1.6). We must have a standard flat and this can be tested with two others; if all of the three possible pairs give straight fringes when they form small-angle wedges they must all be flat. These flats can then be used to test the flatness of other surfaces to about $\lambda/10$ – about 10^{-5} cm. This is better than can be obtained by mechanical means.

The accuracy can be still further improved by multiple-reflexion methods; surfaces can then be examined down to molecular dimensions. For example, steps in mica of the order of 2×10^{-7} cm have been measured (Fig. 11.1) (Tolansky, 1960).

There is also a feed-back to optics itself. Lens surfaces can be tested to a high degree of accuracy by forming Newton's rings with a known surface; small departures from circular symmetry can easily be detected. Such tests will not, however, give any information about aberrations, for which more appropriate methods, such as the knife-edge test (§ 9.6.2), should be used.

Fringe methods also enable other properties to be measured. For example, if Newton's rings are formed between a lens and a flat plate, the curvature of the plate when a uniform bending moment is applied can be measured by the change in the diameters of the rings. The rings will be no longer circular and the diameters measured across the bending give the transverse curvature; thus Poisson's ratio can also be derived.

Coefficients of thermal expansion of quite small crystals can be measured by counting the fringes that pass through the centre of a Newton's ring system as the crystal is heated; a crystal face provides the flat surface. Moreover, the expansion can be measured in different directions, and the method therefore enables the three principal coefficients of an anisotropic crystal to be measured – one of the few methods available for this type of measurement.

11.3 Examination of matter

11.3.1 *The electron microscope.* We have seen in §§ 1.7.2 and 1.7.3 that the limit of resolution of a microscope can be extended by the use of radiation of shorter wavelengths than those of visible light. By far the most successful of such radiations is the electron beam; the principle of the electron microscope is exactly that of the light microscope, but it uses electrostatic or magnetic fields to produce focused images.

The basic fact that allows electrons to be focused is that their trajectories can be varied so as to simulate refraction. In fact, the theory of refraction is exactly that which was put forward by the supporters of the corpuscular theory of light (§ 1.2.3); if an electron is accelerated through a potential difference V_1 it will acquire a velocity $(2eV_1/m)^{\frac{1}{2}}$ and this will be constant in the region of constant potential; if it now enters another region at potential V_2 the velocity will be $(2eV_2/m)^{\frac{1}{2}}$. Thus the refractive index of the second region with respect to the first is $(V_2/V_1)^{\frac{1}{2}}$.

The exploitation of this result is in some ways more difficult and in other ways more easy than designing glass lenses. Curved equipotential surfaces arise naturally and so some focusing is bound to occur. But the surfaces do not have the right configuration for *accurate* focusing and so aberrations (Appendix II) are always present.

As a result of the aberrations, only very small angular apertures can be used; the resolution limit (§ 9.4.3) $\lambda/2 \sin \alpha$ therefore falls two or three orders of magnitude short of what might theoretically be possible.

An electron lens must have cylindrical symmetry. The simplest type of electrostatic lens is shown in Fig. 11.2. The theory of even this simplest type of lens cannot be worked out exactly, and it may be said that, while theory is necessary to see what possibilities exist, ultimately the precise design of electron lenses is based upon empiricism.

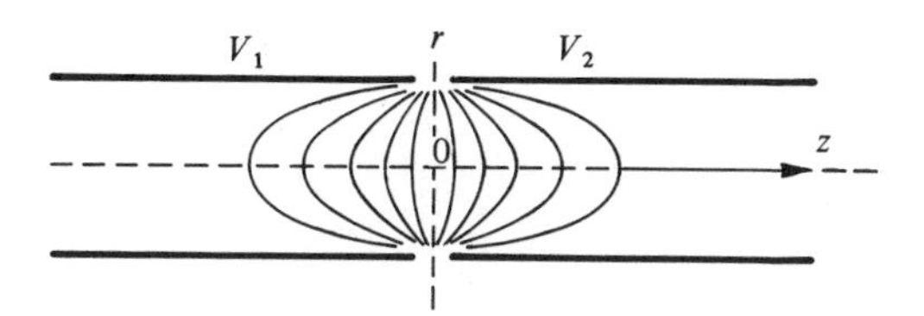

Fig. 11.2. Electron lens formed between two coaxial cylinders. (From Cosslett, 1950.)

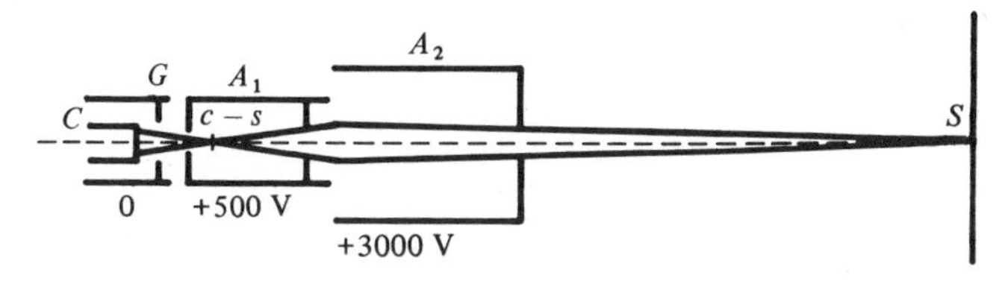

Fig. 11.3. Electron gun. (From Cosslett, 1950.)

Electrostatic fields are now not much used in electron microscopes because the higher fields necessary for short focal lengths may cause ionization in the residual gas; magnetic fields produced by electromagnets are almost always used. The theory is much more complicated because the motion of an electron in a magnetic field can be described only in three dimensions; in a uniform magnetic field electrons move in helical paths.

The type of electromagnetic field necessary for high magnification is that produced by a short solenoid enclosed in a soft-iron sheath, the best dimensions being chosen empirically.

How do we set about making an electron microscope? The first requirement is a source which must produce a beam of *monochromatic* electrons; it is called an *electron gun.* A typical form is shown in Fig. 11.3; it can use an accelerating voltage V of 100 kV, giving $\lambda = 0.04$ Å, since

$$\lambda = \frac{h}{(2meV)^{\frac{1}{2}}}. \tag{11.1}$$

This beam is focused on the specimen by a lens that corresponds to the microscope condenser. After traversing the specimen – which must be thin enough not to cause a spread of velocities of the electrons – the beam passes through a stronger lens, which corresponds to the objective (Fig. 11.4); an image with a magnification of perhaps 100 is produced. This is further magnified by a factor of up to about 200 by a third lens – the

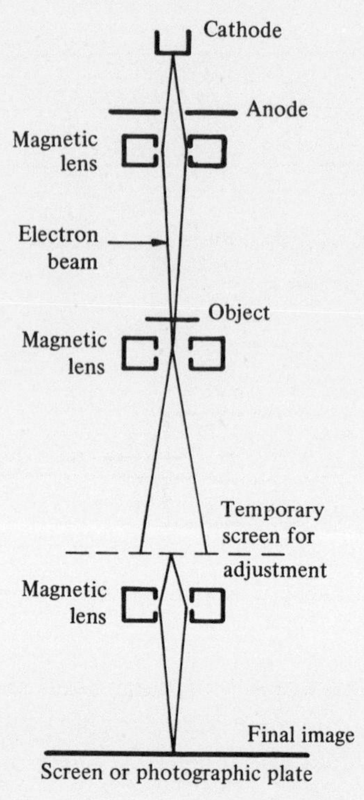

Fig. 11.4. Transmission electron microscope. (From Thomas, 1948.)

projection lens. The final image, as shown in Fig. 11.4, then falls on a fluorescent screen or on to a photographic plate.

The whole apparatus must be highly evacuated, and thus only specimens that are not distorted by being placed in a vacuum can be examined. Moreover, they must be very thin, in order to have negligible absorption of electrons. Metals, for example, can normally be examined only by producing a plastic replica of the surface, and methods of producing such replica have received a great deal of attention. Very fine microtomes can be used to produce very thin sections.

If a specimen is so thin that it does not absorb electrons, there will be no contrast in the image. Some sort of staining must be used. The most successful procedure is called *shadowing*; a beam of heavy atoms, such as palladium, is evaporated at a glancing angle onto the specimen and this can show detail in most convincing relief. An example of an electron micrograph is shown in Fig. 11.5.

Spatial filtering techniques (§ 9.6) have also found their place in electron microscopy. For images of crystalline or polycrystalline objects,

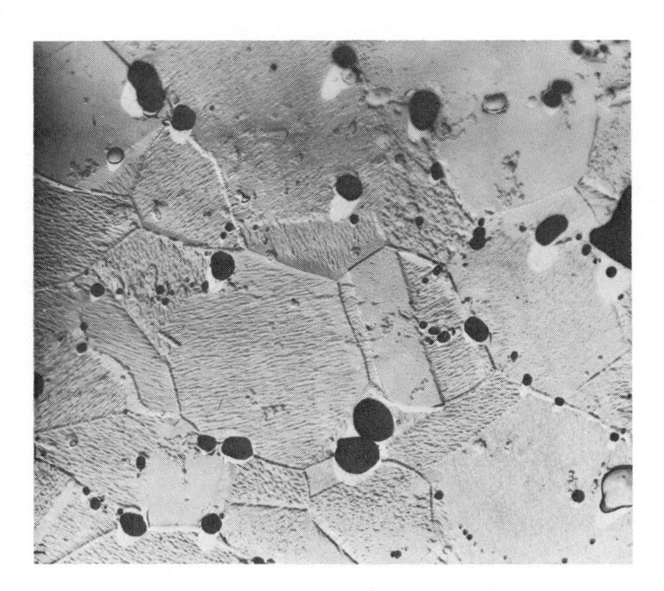

Fig. 11.5. Electron electromicrograph of nickel surface with particles of thoria on it, showing shadows cast by shadowing. (By courtesy of Dr D. W. Ashall.)

the numerical aperture of the microscope is often insufficient to allow more than a single diffraction order to contribute to the image. If only the zero order and its surroundings contribute, we have what is called *diffraction contrast*; if a non-zero order is used the technique is called *dark field*. The latter particularly emphasizes crystallites diffracting strongly into that order. Phase contrast can also be used; the phase plate is a graphite layer about 200 Å thick, or it can be simulated electrostatically.

The experiment described in § 9.3.4 to illustrate the Abbe theory of image formation is ideally suited to modelling the electron microscope, and some simulations are shown in Fig. 11.6. The object consists of a number of small pieces of the gauze of Fig. 9.11 used to represent crystallites oriented randomly. Its diffraction pattern is shown in Fig. 11.6(*a*). A diffraction-contrast image (zero order only) is shown in (*b*). When the aperture is large enough to transmit a few orders of diffraction, images like (*c*) (including the zero order) and (*d*) (excluding the zero order) can be obtained. In the last two examples, as in many high resolution electron microscopes, the relative phases of the various diffraction orders have not been preserved exactly, because of aberrations or simply because of the thickness of the sample. So, although the periodicity of the image is correct, the 'atomic' detail within the individual

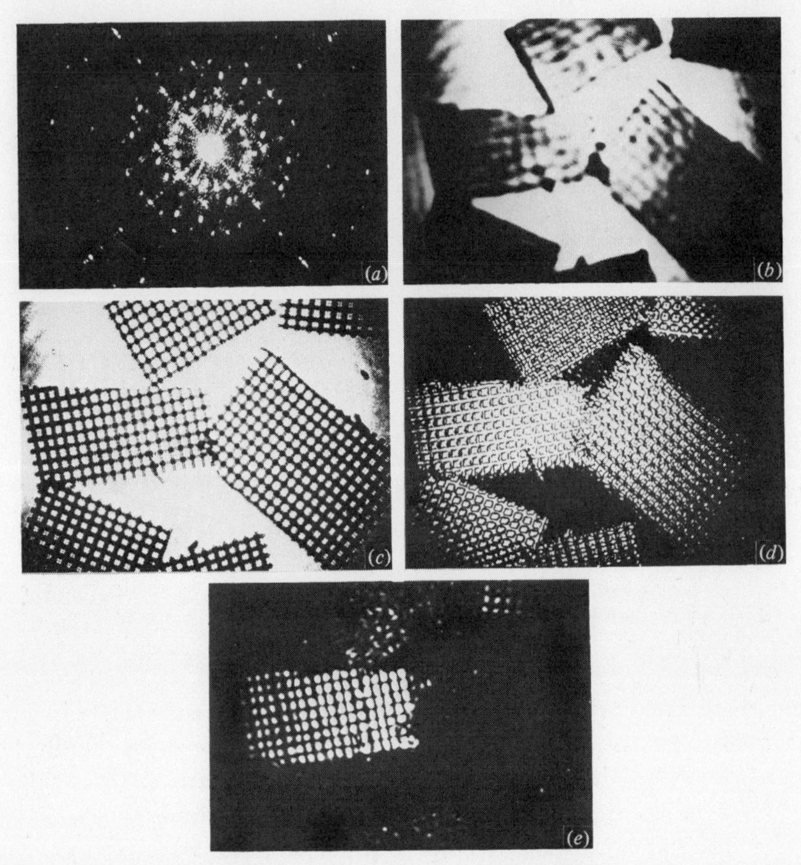

Fig. 11.6. Simulation of electron images of a polycrystal, using coherent optics.

crystallites is quite false. In Fig. 11.6(*e*) we show a dark-field image. Only one of the crystallites appears prominently.

There seems little possibility of any extreme breakthrough in resolving power of electron microscopes. The limit of 7 Å so far claimed seems to be set by a large number of practical factors such as accuracy of machining and control of currents and voltages. Much credit will accrue to anyone who can design a corrected electron lens and so lead the way to production of micrographs with detail of atomic dimensions. A fuller account of electron microscopy, which emphasizes analogies with optical microscopy, is given by Grivet (1972).

Another type of electron microscope which is widely used is the *scanning electron microscope*. Although it is a much more versatile instrument, requiring specimens which are neither thin nor situated in a vacuum, its imaging principle has no analogue in optical microscopy and we shall not discuss it here. The resolution that can be obtained is also considerably inferior to that of the transmission electron microscope.

11.3.2 *X-ray diffraction.* X-rays have wavelengths comparable with interatomic distances and are much more manageable than electron beams; they are little absorbed in air and can be diffracted by specimens of reasonable size. The theoretical basis of such diffraction (§ 7.6) is well understood. X-rays lack only one property: they cannot be refracted in any appreciable sense and thus image formation by means of lenses is impossible. If we wish to find out what the object is, we have to recognize it solely from its Fraunhofer diffraction pattern.

We know, however, that an image is the Fourier transform of the vector-amplitude distribution in the diffraction pattern of the object (§ 9.3.3), and if such a transformation could be effected we could produce a representation of the object. We have seen in § 7.6 that the procedure required for a crystal is the summation of a three-dimensional Fourier series, each coefficient being an order of diffraction. We cannot, however, carry out this procedure because we know only the scalar amplitudes of the orders of diffraction and not the phases. This difficulty is known as the phase problem (§ 9.3.5).

There is no general solution to the phase problem. For crystals that are centrosymmetric – i.e. for any atom at position (x, y, z) there is another similar atom at $(-x, -y, -z)$ – the relative phases are either 0 or π; this statement corresponds to the result that an even function is expressible in terms of cosine curves only, positive and negative at the origin corresponding to phase angles of 0 and π respectively. But even with this limitation the phase problem is not generally solvable.

Over the years, however, since the original discovery of X-ray diffraction (§ 1.7.3), experience has accumulated and methods of approach have been devised that enable quite complicated crystal structures to be solved, culminating in the structure of the proteins, haemoglobin and myoglobin, determined by Perutz and Kendrew. An outline of the physical basis of this type of work is given below.

Suppose that the crystal structure that we are concerned with has a number of atoms in the unit cell, of which a few (one to four) are of predominant weight. These few atoms may be found easily because their effect on the diffraction pattern will also be predominant. We can calculate the phases given by atoms in these positions; a Fourier summation with these phases should show the other atoms. The most

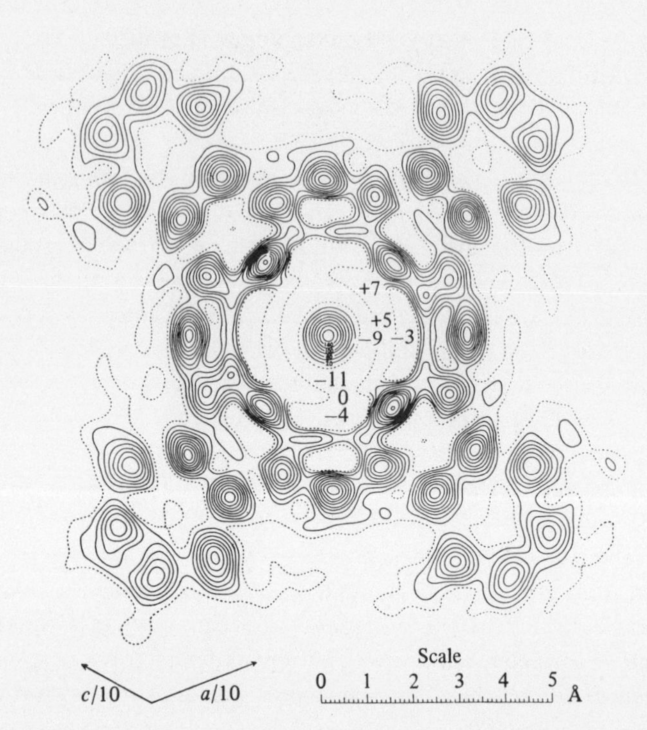

Fig. 11.7. Electron density in projection of a platinum phthalocyanine molecule. Contours are at intervals of 1 electron per $Å^2$ except near the centre where some values of electron densities are indicated. (From Lipson & Cochran, 1966.)

well-known example of successful work of this sort is the determination of the structure of phthalocyanine (Fig. 11.7), the phases being all fixed as 0 by a heavy platinum atom at the centre of the molecule. In effect, this method is equivalent to having a clear point on which to focus, and when this focus is found, the rest of the detail also becomes visible. The similarity to the hologram method of image reconstruction (§ 9.7) is very great.

If it is not possible to find an atom heavy enough to enable this procedure to be carried out successfully, a more powerful method may be used; the heavy atoms may be replaced by others of different weight, and the amplitude of the various orders compared. An early investigation involved the compounds $CuSO_45H_2O$ and $CuSeO_45H_2O$. The *changes* of intensity give clearer evidence than the effect of one heavy atom alone. This method is called the *isomorphous replacement* method. If the crystal is not centro-symmetric, so that general phase angles are involved, two different atoms have to be replaced.

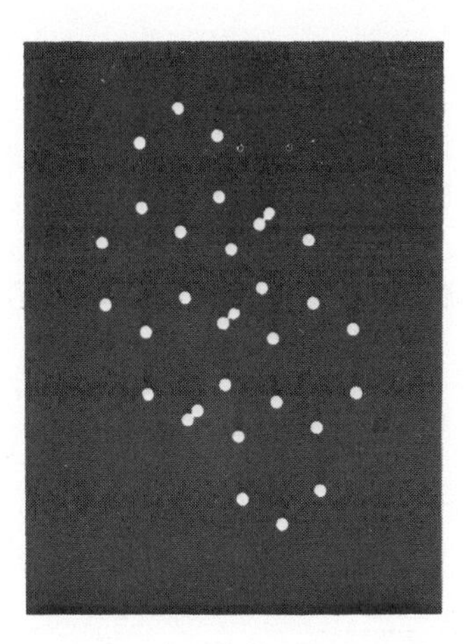

Fig. 11.8. Mask of holes representing two molecules of pyrene, the contents of the unit cell of the crystal.

One can sometimes arrange for a single atom to exhibit a change in scattering by using radiations of different wavelengths. If the wavelength is just short enough to excite the K electrons in an atom, the phase of the scattered wave is affected, and the scattering factor has to be represented by a complex quantity. Comparison of the intensities of the same order of diffraction with two different X-radiations can give some information about the phase angle. This is called the *anomalous-scattering* method. Further discussion of these methods is given by North (1963).

A completely different approach that may be of particular interest to readers of this book involves the use of optical transforms (two-dimensional diffraction patterns, § 7.3). Suppose that we have some idea of the positions of the atoms in the unit cell and wish to see if the calculated intensities of the orders of diffraction agree with those observed. We can make a mask (Fig. 11.8) with holes representing atoms, and prepare its optical transform by means of the optical diffractometer (Appendix IV). We may superimpose upon this a representation of the diffraction pattern of the crystal as a reciprocal-lattice section (§ 7.6.2) in which each point has a black spot whose size is roughly related to the observed intensity (Fig. 11.9). One can see immediately whether the sampling (§ 7.4.4) of the transform by the reciprocal lattice produces intensities which have some relation to those observed.

11.3.3 *Electron diffraction.* Electron-diffraction patterns (Fig. 11.10) are observed over much larger angles than can be used for image formation in

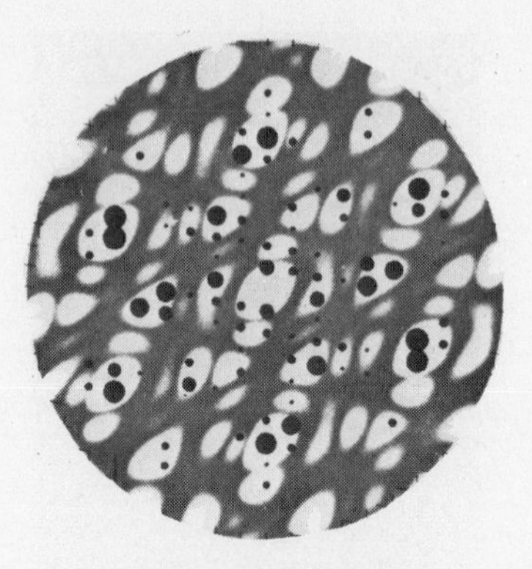

Fig. 11.9. Transform of Fig. 11.8 superimposed on the representation of the X-ray diffraction pattern. Note how each strong order of X-ray diffraction lies upon a strong part of the transform.

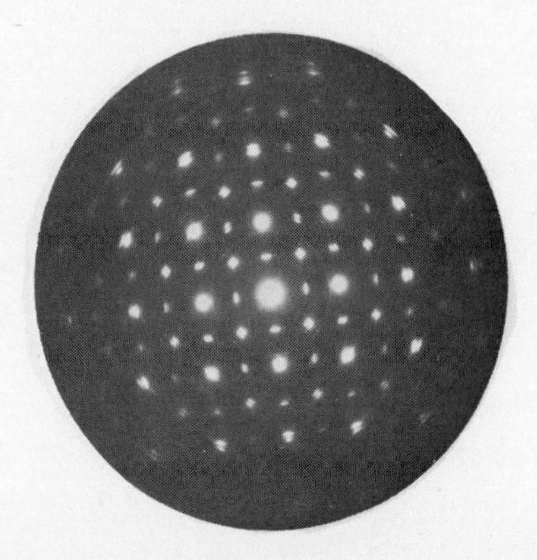

Fig. 11.10. Electron-diffraction pattern of mica crystal.

the electron microscope, and therefore the resolving-power limitation is reduced. But all the problems associated with X-ray diffraction occur, and in addition there are the difficulties introduced by having to put the specimen and recording film in a vacuum, and by requiring the specimen to be very thin. Why then is electron diffraction worth while?

First, it gives information that is complementary to X-ray information; X-rays are scattered by electrons whereas electrons are scattered by electric fields, and a light atom such as hydrogen, which hardly scatters X-rays, can produce appreciable effects on electrons. Secondly, the structure of matter in the form of a thin film may be different from that in bulk. Thirdly, some materials are obtainable only in the form of fine powder, but an electron beam may be able to pick out single crystals.

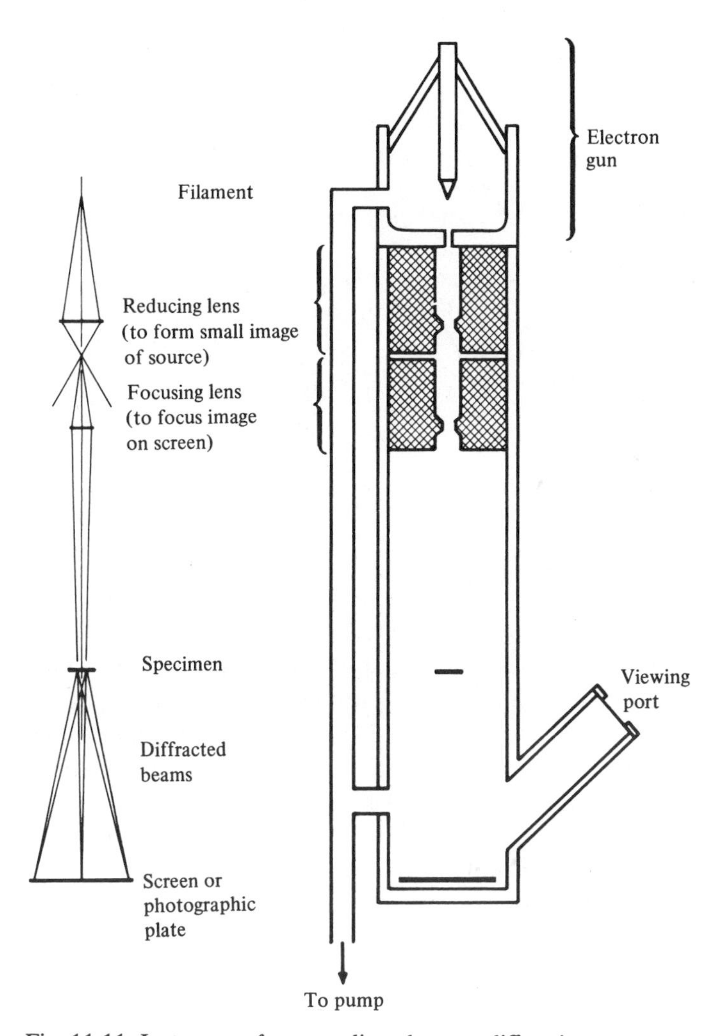

Fig. 11.11. Instrument for recording electron-diffraction patterns.

Finally, for imperfect structures, electrons may be able to provide information by selecting a much smaller volume of a specimen than X-rays can.

Although the first electron-diffraction experiments were rather crude, considerable improvement occurred when electron-beam focusing (§ 11.3.1) became possible. A typical instrument for electron-diffraction experiments is shown in Fig. 11.11; it can be seen that it has much in common with the electron microscope (Fig. 11.4) and usually the same instrument is used for both purposes.

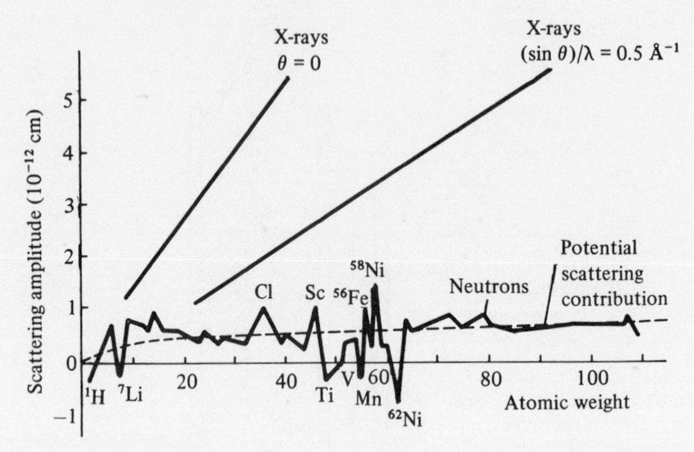

Fig. 11.12. Scattering cross-sections of atoms. (From Bacon, 1962.)

11.3.4 *Neutron diffraction.* Neutrons (§ 1.6.3) are scattered mainly by nuclei and so provide still further information that is complementary to that provided by X-rays and electrons. For neutrons there is no simple relation between scattering and atomic number (Fig. 11.12) and light atoms such as hydrogen scatter as strongly as most other elements.

It will be noted from Fig. 11.12 that some scattering amplitudes are negative; in other words, such nuclei scatter exactly out of phase with most others. Scattering amplitudes may also be quite different for isotopes of the same element. Moreover, the magnetic moment of neutrons may interact with the moments associated with the incomplete electronic shells of the transition elements. All these facts enable information of a completely new type to be obtained; the last, in particular, has cast new light upon the subject of magnetism and this will be discussed in greater detail later in this section.

Until recently, the only source of neutrons for diffraction experiments was a nuclear reactor. Recently a new process, called spallation, has been realized, in which a neutron beam is produced as the result of aiming a pulse of energetic electrons from a linear accelerator at a uranium target. These produce gamma rays, which in turn generate a pulse of neutrons. The neutron flux is considerably greater than that obtainable from a reactor, and has the great experimental advantage of being pulsed.

The neutrons produced by either method are usually in thermal equilibrium with either the reactor core or a moderator, and are therefore moving with considerable thermal velocities. The kinetic energy contribution from a selected direction x, that of the collimated output beam, is

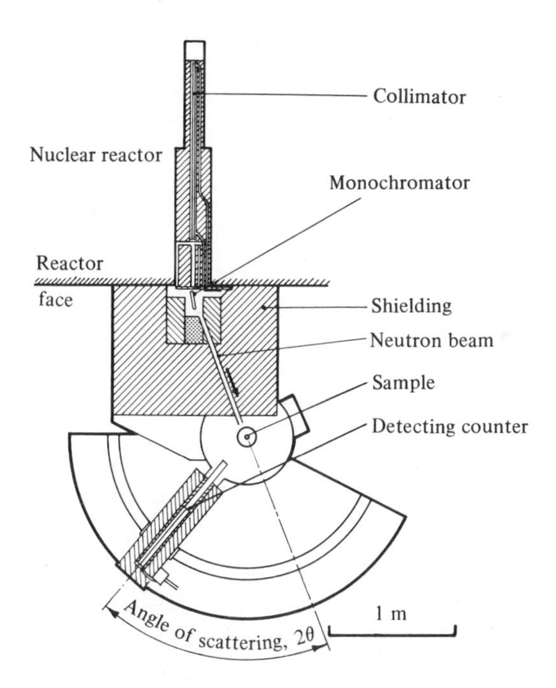

Fig. 11.13. Neutron diffractometer (DIDO at Harwell). (From Bacon, 1954.)

$\frac{1}{2}mv_x^2$. The neutrons at the centre of the thermal distribution at temperature T therefore satisfy:

$$\tfrac{1}{2}mv_x^2 = \tfrac{1}{2}k_{\mathrm{B}}T \tag{11.2}$$

and have de Broglie wavelength:

$$\lambda = \frac{h}{mv} = \frac{h}{(mk_{\mathrm{B}}T)^{\frac{1}{2}}}. \tag{11.3}$$

If T is of the order of 300 K, λ is of the order of 10^{-8} cm – about the same as the wavelength of X-rays.

The neutrons are allowed to leave the reactor through a channel in absorbing material, as shown in Fig. 11.13, thus producing a well-defined beam. It still has a non-uniform distribution of velocities, and to produce a 'monochromatic' beam it is reflected from a crystal (Fig. 11.13) so that Bragg reflexion (§ 7.6.1) occurs for wavelengths near to the maximum of the distribution.

Alternatively, the beam can pass through two or more discs rotating on the same axle, and having slits at certain angles. The only neutrons which can pass through all the slits then have a well-defined velocity. The monochromatic beam so selected can then be used in the same way as an

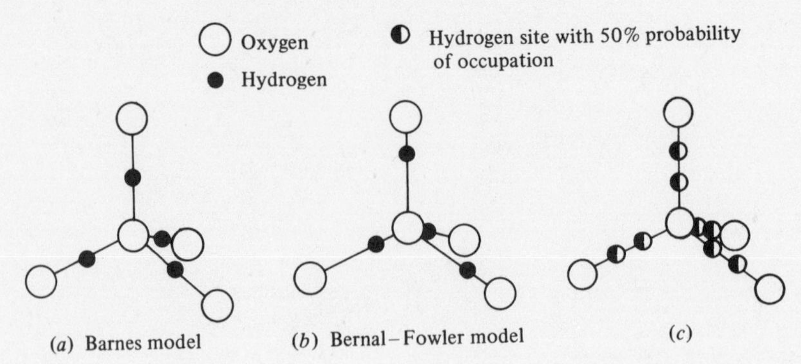

Fig. 11.14. Two incorrect structures ((*a*) and (*b*)) and the correct structure (*c*) of ice.

X-ray beam, with one exception: neutrons do not affect photographic film and thus diffraction patterns can be obtained directly only by counter methods.

Nearly all the information obtained by neutrons has been concerned with crystal structures already derived by X-rays. For example, X-ray work on the structure of ice had clearly established the positions of the oxygen atoms, but the hydrogen atoms were left open to speculation; were they midway between the oxygen atoms or were two hydrogen atoms closely attached to each oxygen atom forming clear molecules of H_2O (Fig. 11.14)? A neutron-diffraction investigation of heavy ice, D_2O, shows that neither possibility is correct. The deuterium atoms are statistically distributed round each oxygen atom in a way that can be most effectively described by the formula $(\frac{1}{2}D)_4O$.

But the most remarkable result of neutron diffraction is the extension of our knowledge of magnetism. To the classical subdivision of materials into diamagnetics, paramagnetics and ferromagnetics, we must now add antiferromagnetics, ferrimagnetics and other types that are not readily classified. The existence of the subdivisions had been suspected by Néel around 1932 from anomalies in specific-heat measurement, and by others from odd changes in symmetries of certain crystal structures; but the complete elucidation of the subject had to wait for the advent of neutron diffraction.

The reason for this is that the spin of the neutron interacts with the resultant spin of the electrons in an atom. If the resultant spin is zero, there is no interaction, but if it is finite there is an interaction which results in the atom's having a scattering factor that varies with angle; this scattering factor is the Fourier transform of the density of unpaired

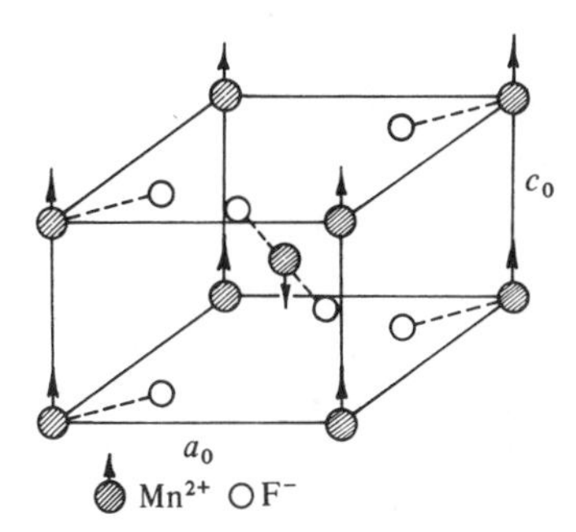

Fig. 11.15. The unit cell of MnF_2, showing the antiferromagnetic arrangement of the magnetic moments of the Mn^{2+} ions. (From Erickson, 1953.)

electrons (compare § 7.6). Neutron diffraction is therefore the ideal method for studying these electrons.

In an ideal paramagnetic the moments of the individual atoms are randomly oriented and the scattering is incoherent; the intensity of scattering is therefore a continuous function and is not confined to reciprocal-lattice points (§ 7.6.2). In a ferromagnetic, however, the spins are aligned and thus coherent crystal-type scattering occurs.

The existence of the new types of magnetic materials appeared when the X-ray diffraction patterns and the neutron-diffraction patterns of certain compounds were compared; it was found that the latter contained orders of diffraction that should not occur. A simple example is given by MnF_2 (Fig. 11.15). It will be seen that the manganese atoms lie at the corners of the unit cell and at the centre; accordingly they give no contribution to the order of diffraction for which $h+k+l$ is odd. The neutron-diffraction results, however, indicated that the atoms at the corners were different from those at the centres; the obvious difference is that their moments are oppositely aligned (Fig. 11.15). This is the anti-ferromagnetism predicted by Néel in 1932.

Since the spins cancel, the compounds would appear by ordinary magnetic measurements to be paramagnetic. Nevertheless, the ordering of the spins is as perfect as in a ferromagnetic and the variation of the degree of order follows the same law as ferromagnetism. In the same way that ferromagnetism disappears at a particular temperature (the Curie temperature), antiferromagnetism also disappears at a particular temperature (the Néel temperature) and it was the specific-heat and latent-heat anomalies associated with this temperature that first led Néel to the conclusion that the materials that he was examining were antiferromagnetic.

A ferromagnetic material is one in which the spins of two different atoms are opposite but unequal; such a material would therefore appear to be weakly ferromagnetic. Other types of magnetism exist because spins are not confined merely to two different directions; differently related, but still ordered, directions may occur, and it would take too long to discuss here the structures that have been found.

Another device remains to be discussed. Neutrons are scattered from magnetic materials selectively according to their spins. It is usual therefore for an order of diffraction from a magnetic crystal to have a distribution of spins that is not random, and such a beam is said to be partially polarized. (It is rather remarkable that this is exactly the type of polarization envisaged by the corpuscularists, such as Newton, in the seventeenth century.) The polarization may be complete, and the beam is then said to be plane-polarized. The 220 order of diffraction from magnetite (Fe_3O_4) is an example.

This discovery opens up a new field. Plane-polarized neutrons can be obtained by replacing the crystal in Fig. 11.13 by a magnetic crystal, and with such beams experiments can be carried out more precisely. A great deal has been discovered about the magnetic properties of crystals in this way (Bacon, 1962; Squires, 1978).

11.4 Some more complicated applications of Fraunhofer diffraction theory

We shall next discuss some extensions of the simple Fraunhofer diffraction theory in Chapter 7 which will enable us to deal with two particular types of problem – an object whose properties are statistical, and a moving obstacle.

11.4.1 *The auto-correlation function and its relation to the diffraction pattern.* If we could observe the amplitude and phase at each point on a diffraction pattern it would merely be a matter of computation for us to work out to an accuracy of about one wavelength, which corresponds to the finest detail that diffracts at a real angle, a complete description of the object. Unfortunately, all observing and measuring instruments are sensitive only to intensity, which contains no information about the phase, and, unless we compare the pattern with a coherent wave that has not been diffracted (§ 9.7), it is impossible to deduce this information. The question of what is the maximum amount of information that can be derived unambiguously from the observed intensity pattern is therefore important; we cannot, of course, make any estimate of what can be achieved as a result of intelligent guesswork and preconceived ideas of the answer!

We shall consider diffraction in two dimensions again, to avoid the necessity of repeated statements about which part of the Fourier transform can be observed. The intensity of I of the diffracted beam ψ in the direction (u, v) is the product of the amplitude with its complex conjugate, and the transform of this is therefore the convolution (§ 3.6) of the object with its complex conjugate;

$$I(u, v) = \psi\psi^*(u, v), \tag{11.4}$$

is the transform of

$$F(x, y) = \int\int f(x', y')f^*(x'-x, y'-y)\,\mathrm{d}x'\,\mathrm{d}y', \tag{11.5}$$

where $f(x, y)$ is the transmission function of the object. This function is the *auto-correlation* function, (§ 3.6.7) and must necessarily have a real positive Fourier transform (11.4). The concept is particularly useful when applied to objects for which the self-correlation is an important parameter; for example, we shall discuss the effect of thermal disorder of a lattice in § 11.4.3.

11.4.2 *Simple auto-correlations.* Let us first consider a very simple example. The idealized Young's interference experiment, with pin-holes separated by a distance a, has

$$f(x, y) = \delta(y)\tfrac{1}{2}\{\delta(x - a/2) + \delta(x + a/2)\} \tag{11.6}$$

and gives

$$\psi(u, v) = \cos\frac{au}{2}. \tag{11.7}$$

As $f(x, y)$ is a product we can consider correlation in the x and y directions independently. In the y-direction, the function is concentrated entirely at $y = 0$, and therefore the integrand of (11.5) is zero unless $y = 0$. The quantity $F(x, y)$ thus has y-dependence $\delta(y)$. In the x-direction, f consists of two δ-functions A and B (Fig. 11.16(a)). The integrand of (11.5) is thus zero unless (a) $x = a$, when it is the product of A and B, which is a δ-function of strength $\frac{1}{4}$; (b) $x = -a$, when it is the same product; or (c) $x = 0$, when it has two contributions, one from $A \times A$ and the other from $B \times B$. The self-correlation function is therefore as illustrated in Fig.

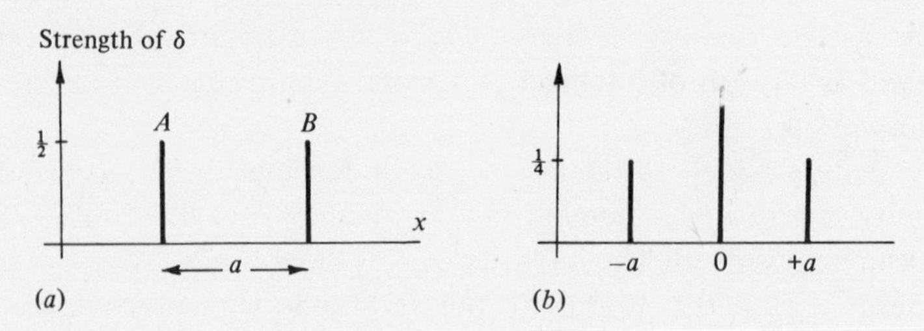

Fig. 11.16. (*a*) Two δ-functions separated by distance *a*; (*b*) the self-correlation function of (*a*).

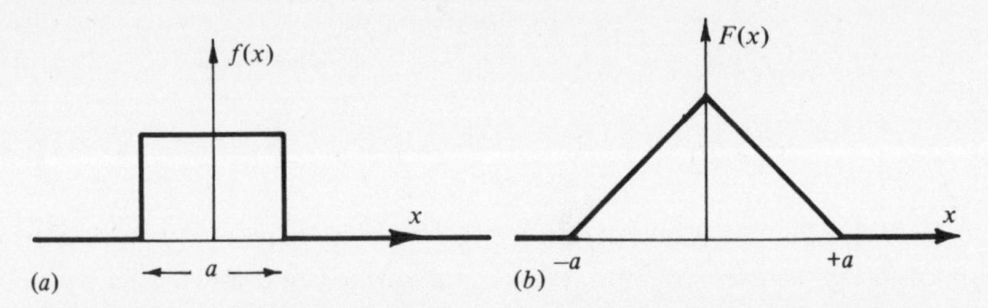

Fig. 11.17. (*a*) A square pulse; (*b*) the self-correlation function of (*a*).

11.16(*b*), the centre δ-function having twice the strength of the outer ones. The transform of this function is

$$\begin{aligned} I(u, v) &= \tfrac{1}{4}\{\exp(-iua)+2+\exp(iua)\} \\ &= \tfrac{1}{2}(1+\cos au) \\ &= \cos^2 \frac{au}{2} \\ &= \psi^2(u, v). \end{aligned} \tag{11.8}$$

A second example we can consider is that of a slit of width *a*. It is a simple matter to show that the self-correlation function is as shown in Fig. 11.17 and, as was shown in the examples in §§ 7.2.1 and 7.2.2, the transforms of $f(x, y)$ and its self-correlation function are respectively,

$$\frac{\sin ua/2}{u} \quad \text{and} \quad \frac{\sin^2 ua/2}{u^2}. \tag{11.9}$$

Such examples are of academic interest only, but we shall now consider an example which is of real importance.

11.4.3 *Thermal disorder in a lattice.* A crystal is never exactly perfect. At any temperature the atoms are vibrating about their true lattice positions, and as the frequency of vibration is very much less than that of the X-rays

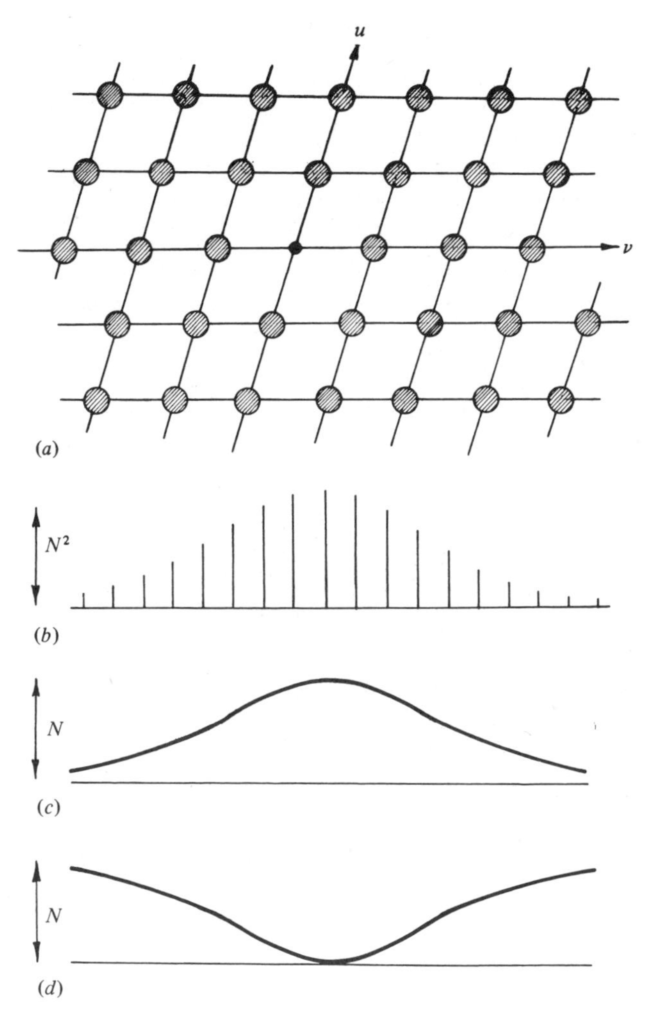

Fig. 11.18. (*a*) The self-correlation function of a structure with randomly displaced atoms. All peaks have a strength *N*. (*b*) One-dimensional representation of the transform of (*a*) assuming that the origin peak were as broad as the rest. (*c*) Transform of the origin peak alone. (*d*) The difference between the transform of a δ-function at the origin and the transform of a broadened peak.

which are used to investigate them the diffraction pattern observed is that of a stationary crystal whose atoms are displaced by random small distances from their true lattice positions. We shall consider a two-dimensional model of this.

The auto-correlation of such a model is as shown in Fig. 11.18(*a*); it forms a two-dimensional lattice of which all the 'points' are broadened except that at the origin, which is sharp because all the atoms are at exactly zero distance from themselves. The intensity distribution in the required diffraction pattern is the Fourier transform of this self-convolution. We can derive this transform by taking the transform of a lattice with *all* the 'points' broadened, subtract the transform of the broadened central 'point' and add the transform of a true point at the origin.

The transform of a lattice of broadened 'points' is a perfect lattice in which the heights of the peaks fall off in some sort of quasi-Gaussian way (Fig. 11.18(*b*)), the central peak having a height N^2, where N is the number of atoms in the crystal. The transform of the broadened central 'point' is a continuous curve (Fig. 11.18(*c*)) with the same form, the height at the centre being N. This curve must be subtracted from the transform of the central point, which, being a true point, is now a uniform distribution of height N. The difference (Fig. 11.18(*d*)) gives a continuous distribution that is zero in the centre and increases with distance outward.

We have thus reached the conclusion that the diffraction pattern of a lattice with thermal disorder is a lattice of sharp points with a background that is zero in the centre and increases in intensity with angle of diffraction.

This is an idealized treatment. In practice, thermal displacements are not independent of each other, and so the peaks in the self-correlation function are not all equal. Moreover, the dependence of the displacements on each other is a function of direction and so the peaks may have odd cross-sections. Although these facts do not alter the conclusion that all the diffraction peaks remain sharp, they do lead to structural features in the background (Fig. 11.19). Consideration of this subject would take us far out of the realm of this book.

11.4.4 *A moving system: neutron diffraction by lattice vibrations.* In the previous section we assumed that the X-rays used in a crystal-diffraction experiment saw an instantaneous snapshot of the vibrating crystal, because the X-ray frequency was much greater than that of the atomic vibrations. A different situation arises when a wave of frequency comparable with that of the atomic vibrations is incident on a crystal.

We then have to take into account the way in which the atoms are moving; their motion essentially Doppler-shifts the frequency of the incident wave. This effect is most important for the diffraction of a coherent beam of slow neutrons, which have velocities comparable with the velocity of sound (§ 11.3.4).

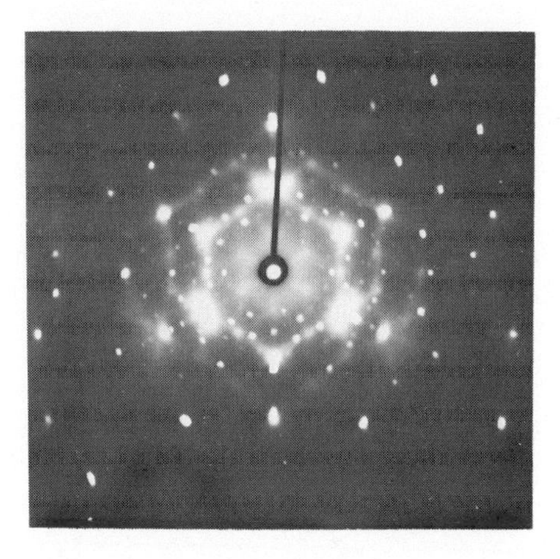

Fig. 11.19. X-ray diffraction photograph of benzil $(C_6H_5CO)_2$. (By courtesy of Professor K. Lonsdale.)

It is useful to represent the thermal vibrations of the crystals by the superposition of sinusoidal waves called phonons, having various wave-vectors **K**, frequencies Ω and polarizations. The relationship between Ω and **K** is generally rather complicated; the function $\Omega(\mathbf{K})$ is intimately related to the nature of the interatomic forces in the crystal, and its determination is an important tool in the study of solids. An example of $\Omega(\mathbf{K})$ will be illustrated later (Fig. 11.22). It is important to realize that phonon wavelengths $2\pi/|\mathbf{K}|$ smaller than the interatomic spacing have no physical meaning. At very low frequencies, i.e. for small Ω and **K**, the two are related by the sound velocity $V_s = \Omega/|\mathbf{K}|$ but this velocity may itself be a function of the direction of propagation in an anisotropic crystal.

Just to illustrate the elements of the problem, we shall discuss the diffraction of a neutron wave incident in the z-direction:

$$\psi = \psi_0 \exp\{i(\omega_0 t - k_0 z)\} \tag{11.10}$$

by a longitudinal phonon travelling in the x-direction. This phonon (Fig. 11.20) can be written by representing the displacement $\zeta(x)$ of the atom originally at position $x = na$:

$$\zeta(x) = \zeta_0 \cos(\Omega t - Kx). \tag{11.11}$$

This formulation reminds us of the treatment in § 7.7.3 of periodic errors in a diffraction grating. But in this case we take into account the dynamic

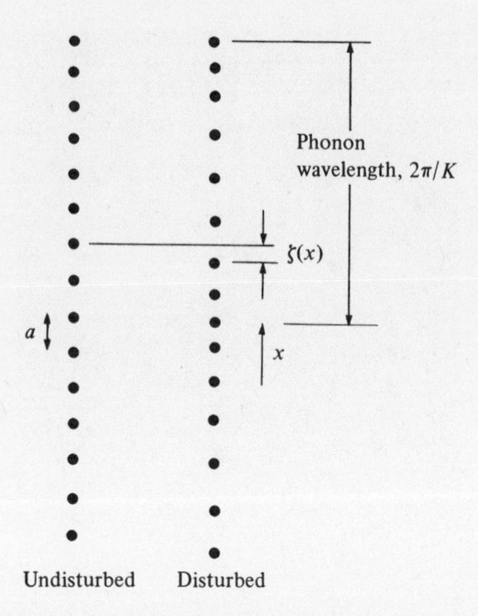

Fig. 11.20. A longitudinal phonon in a crystal.

nature of the problem by explicitly including the time-variations; the diffracted wave must then be of the form:

$$\begin{aligned}\psi &= \psi_0 \exp\{\mathrm{i}(\omega t - kz\cos\theta - kx\sin\theta)\}\\ &= \psi_0 \exp\{\mathrm{i}(\omega t - kz\cos\theta - ux)\}.\end{aligned} \tag{11.12}$$

We can then write the diffraction pattern as a Fourier transform $\psi(\omega, u)$ integrated over both position and time. Regarding the atoms as coherent δ-function sources of amplitude $\exp(\mathrm{i}\omega_0 t)$ in the plane $z=0$ at positions $x = na + \zeta(na)$ we have:

$$\begin{aligned}\psi(\omega, u) &= \int_{-\infty}^{\infty} \mathrm{d}t \sum_{n=-\infty}^{\infty} \delta\{x - na - \zeta(na)\} \exp(\mathrm{i}\omega_0 t) \exp(-\mathrm{i}\omega t + \mathrm{i}ux)\\ &= \int_{-\infty}^{\infty} \mathrm{d}t \sum \exp\{\mathrm{i}(\omega_0 - \omega)t\} \exp[\mathrm{i}u\{na + \zeta(na)\}].\end{aligned} \tag{11.13}$$

Assuming the amplitude of the phonon ζ_0 to be small ($\ll 1/u$) we can expand one of the exponentials and have

$$\begin{aligned}\psi(\omega, u) = \int_{-\infty}^{\infty} \mathrm{d}t \sum_n \exp\{\mathrm{i}(\omega_0 - \omega)t\} \exp(\mathrm{i}una)\\ \times[1 + \tfrac{1}{2}\mathrm{i}u\zeta_0 \exp\{\mathrm{i}(\Omega t - Kna)\}\\ + \tfrac{1}{2}\mathrm{i}u\zeta_0 \exp\{-\mathrm{i}(\Omega t - Kna)\}].\end{aligned} \tag{11.14}$$

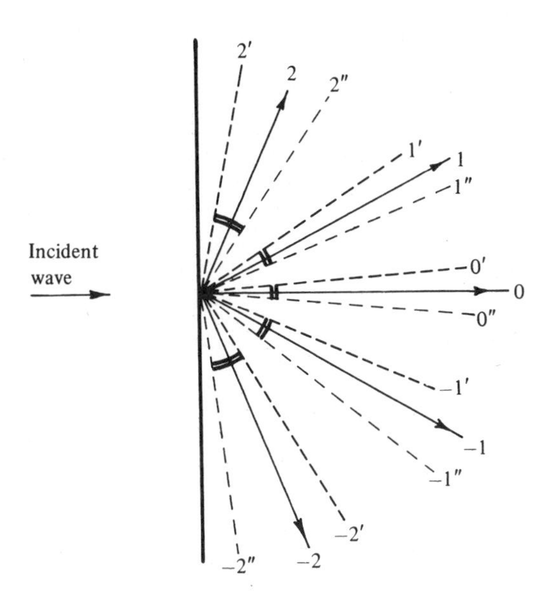

Fig. 11.21. Extra orders of diffraction produced by a sinusoidally disturbed crystal.

The integral and summation then gives us three series of δ-functions (cf. (7.108)):

$$\psi(\omega, u) = \psi_1 + \psi_2 + \psi_3, \tag{11.15}$$

where

$$\psi_1(\omega, u) = \delta(\omega_0 - \omega) \sum_m \delta(u - 2\pi m/a), \tag{11.16}$$

$$\psi_2(\omega, u) = \tfrac{1}{2}\mathrm{i}u\zeta_0\delta(\omega_0 - \omega + \Omega) \sum_m \delta(u - K - 2\pi m/a), \tag{11.17}$$

$$\psi_3(\omega, u) = \tfrac{1}{2}\mathrm{i}u\zeta_0\delta(\omega_0 - \omega - \Omega) \sum_m \delta(u + K - 2\pi m/a). \tag{11.18}$$

The δ-functions in ψ_1 represent the diffraction orders from the undisturbed crystal; those in ψ_2 and ψ_3 represent ghost orders, displaced from the main orders by $\Delta u = \pm K$ and by frequency $\Delta\omega = \pm\Omega$ (Fig. 11.21). The factor $u\zeta_0$ in ψ_2 and ψ_3 shows the ghost orders to be weak, particularly at small angles ($u \to 0$). This analysis shows what happens when a one-dimensional crystal, disturbed by a single sinusoidal phonon, diffracts a slow neutron wave. A real crystal has a continuous spectrum of phonons, and each main diffraction order is surrounded by ghosts which differ from it by frequency Ω and wave-vector $\mathbf{K}$. The effects are most marked when ω_0 is a few times greater than the maximum value of Ω and k_0 is a few

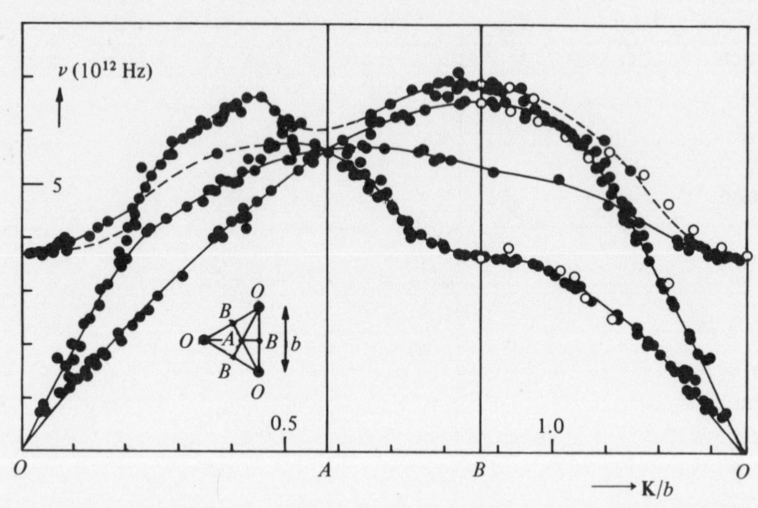

Fig. 11.22. Experimental results for the phonon dispersion relation in magnesium, obtained by slow-neutron diffraction. The frequency ν is $\Omega/2\pi$, and the vector **K** lies along symmetry directions in the basal plane of the hexagonal unit cell. The inset diagram shows the basal plane of the reciprocal-lattice unit cell, and the **K**-values investigated follow the lines *OABO*. Notice that both acoustic modes ($\Omega \to 0$ as $K \to 0$) and optic modes ($\Omega \not\to 0$ as $K \to 0$) of various independent polarizations are present. (By courtesy of Dr G. L. Squires.)

times greater than the maximum value of $|\mathbf{K}|$; together these conditions indicate a neutron wave having velocity similar to the velocity of sound and wavelength somewhat smaller than the interatomic spacing. These correspond to the peak of a neutron assembly in thermal equilibrium at about 25 K. An example of the results of an analysis of the phonon spectrum of magnesium deduced from analysis of the energy of the ghost spectrum as a function of angle is shown in Fig. 11.22.

11.4.5 *Diffraction considered as a collision*: *energy and momentum conservation.* In the same way as the neutron, generally considered as a particle, has been treated here as a wave, it is possible to consider the phonon as a particle. The diffraction process then becomes a collision in which energy E and momentum $\mathbf{p}$ are conserved. Since the energy and momentum of a non-relativistic particle of mass m are related to ω and $\mathbf{k}$ by the relations:

$$\mathbf{p} = m\mathbf{v} = \hbar\mathbf{k}, \tag{11.19}$$

$$E = \tfrac{1}{2}mv^2 = \frac{\hbar^2 k^2}{2m} = \hbar\omega, \tag{11.20}$$

we can write the interaction with the crystal in two forms:

(a) *elastic scattering*, in which the neutron loses no energy but is diffracted by the crystal lattice:

$$\mathbf{k}-\mathbf{k}_0=\mathbf{g}, \qquad \omega-\omega_0=0; \tag{11.21}$$

(b) *inelastic scattering*, in which the neutron gains or loses energy and momentum to a phonon:

$$\mathbf{k}-\mathbf{k}_0=\mathbf{g}\pm\mathbf{K}, \tag{11.22}$$

$$\omega-\omega_0=\frac{\hbar}{2m}(k^2-k_0{}^2)=\pm\Omega. \tag{11.23}$$

Now equations (11.21) for the elastic process can be interpreted as the geometrical construction (§ 7.6.1) for the sphere of observation and the reciprocal lattice in crystal diffraction, and correspond to the term ψ_1 (11.16). Similarly, the inelastic process can be seen to correspond to ψ_2 (11.17) when the neutron gains energy or ψ_3 (11.18) when the neutron loses energy, since $u=k\sin\theta$ is the x-component of the vector $\mathbf{k}$. This can be represented by a more complicated form of observation-sphere diagram; but it is more useful practically to draw a diagram which allows us to see what frequency Ω will be observed for diffraction with particular directions of observation and incidence. Such a diagram is illustrated schematically in Fig. 11.23. A fuller discussion is given by Squires (1978).

11.4.6 *Light waves scattered by phonons: Brillouin scattering and the acousto-optic effect.* A longitudinal phonon travelling through a medium represents a pattern of periodic compression and rarefaction. The refractive index of the material, being related to the density, also varies periodically; the material therefore behaves like a three-dimensional phase grating and diffracts light. Since the wavelength of visible light is far larger than the interatomic spacing, the medium can be considered as a continuum.

Now it is clear, when we describe the periodic compression and rarefaction as a phase grating, that only phonons having wavelengths longer than one-half the wavelength of the light can interact with it. The process of light scattering by phonons is therefore restricted to phonons of very long wavelength, compared to the interatomic spacing. So the frequency Ω is extremely small compared to ω_0.

We write equations (11.22) and (11.23) as:

$$\omega=\omega_0\pm\Omega, \tag{11.24}$$

$$\mathbf{k}=\mathbf{k}_0\pm\mathbf{K}. \tag{11.25}$$

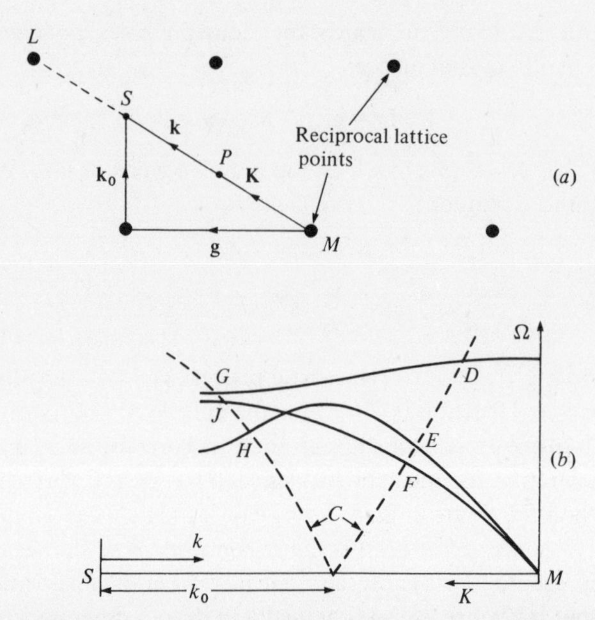

Fig. 11.23. Investigation of the phonon dispersion curve using inelastic neutron scattering. (*a*) Vector diagram in reciprocal space for equation (11.22); $\mathbf{k}_0$ is chosen so that its end-point *S* lies on the line *LM* in (*a*), scattered neutrons are observed whenever the curve *C* (equation (11.23)) intersects a branch of the dispersion curve, i.e. at *D, E, F* (absorption of a phonon) or *G, H, J* (creation of a phonon).

The frequency change on scattering is small, but measurable by spectroscopic techniques of very high resolution. The direction of the scattering gives the wave-vector $\mathbf{K}$; since ω and ω_0 are so closely equal it is easy to show that, on scattering through angle θ, $\mathbf{K}$ must bisect the angle between $\mathbf{k}$ and $-\mathbf{k}_0$ and then

$$|\mathbf{K}| = |\mathbf{k} - \mathbf{k}_0| = \frac{2\omega_0}{c} \sin \tfrac{1}{2}\theta. \tag{11.26}$$

When a laser beam is incident on a material (solid, liquid or gas) and is scattered by phonons the process is called *Brillouin scattering* and can be used to investigate the properties of very-long-wavelength density waves. The scattering is very weak and high-intensity lasers, or very elaborate detection techniques, are necessary to observe it. The frequency shift is often measured by creating beats between the scattered light and the incident light (§ 8.4). When we observe the light scattered at a certain angle θ, and analyse its spectrum, we find intensity peaks at the values of

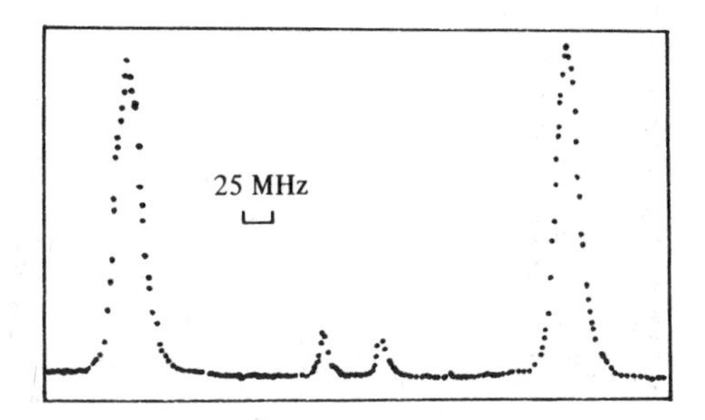

Fig. 11.24. Brillouin scattering by superfluid helium. The incident laser frequency is at the centre of the graph. The two pairs of peaks indicate two types of sound wave, first and second sound. (From Vinen, 1978.)

$\omega = \omega_0 - \Omega$ corresponding to **K** in (11.26). There will be a number of peaks, because Ω can be positive or negative for the same $|\mathbf{K}|$ and moreover a number of different types of density wave may exist. An example is shown in Fig. 11.24.

In other experiments a phonon of precisely known Ω and **K** is created by an electronic transducer on the surface of a block of transparent material. The observed scattering of a light wave is then known as the *acousto-optic effect* and under suitable conditions in many materials a diffraction efficiency (§ 7.7.4) close to unity can be achieved when the vectors satisfy (11.25) and (11.26). The acousto-optic effect is becoming widely used as a method of modulating and steering light beams; particularly suitable materials are lead molybdate and, surprisingly enough, water.

11.4.7 *Non-linear optics.* We have become used to regarding the polarization of an atom by an external field as a linear process, i.e. that the dipole moment produced is proportional to the applied field. This must, of course, be true for small fields, as we can always expand the dipole moment P per unit volume as a Taylor expansion

$$P(E) = P(0) + E\left(\frac{\partial P}{\partial E}\right)_0 + \tfrac{1}{2}E^2\left(\frac{\partial^2 P}{\partial E^2}\right)_0 + \dots \qquad (11.27)$$

For a molecule with no static dipole moment $P(0) = 0$, and the second term, the linear induced moment, gives

$$P(E) \approx E\left(\frac{\partial P}{\partial E}\right)_0 = \chi E. \qquad (11.28)$$

If we take a material which needs an expansion like (11.27) to do it justice and apply a wave $E = E_0 \exp(i\omega t)$ to it we get

$$P(E) = \chi E + \tfrac{1}{2}\chi' E^2 + \ldots$$
$$= \chi E_0 \exp(i\omega t) + \tfrac{1}{2}\chi' E_0^{\ 2} \exp(2i\omega t) + \ldots. \qquad (11.29)$$

$P(E)$ has a harmonic content; a frequency 2ω has been generated and will be re-radiated by the oscillating dipole. Higher terms in the expansion will also occur, giving frequencies $n\omega$ in general, but we shall assume for the moment that these effects are negligible. What governs the intensity of the observed harmonic waves? First of all the intensity of the 2ω component in P is proportional to $E_0^{\ 2}$ so that the high intensity from a laser will be a great help in producing observable effects. Secondly, the value of χ' must not be zero. The crystal symmetry of the dielectric material may often put conditions on χ'. If the crystal has a centre of symmetry, for example, χ' and all odd derivatives must be zero, for there can be no difference in the effect produced by electric fields $+E$ and $-E$ if r and $-r$ are identical. But there is one further point of importance, and that is that matching of the propagation velocities of the ω and 2ω waves is necessary. The 2ω wave is produced coherently with the ω wave at its point of origin. It then propagates at the phase velocity $v(2\omega)$. If the 2ω waves are being produced at all points within the material they can add constructively only if the phase velocities of the 2ω and ω waves are equal. Once this condition is satisfied, it becomes quite easy to observe harmonic generation. Consistent with the requirement that χ' should not be zero are some biaxial crystals, and the anisotropy of the propagation velocities can then be used to find a direction in which the velocities of ω and 2ω waves are equal (Fig. 11.25). One of the most well-known examples is potassium di-hydrogen phosphate.

A second, associated, problem concerns the possibility of mixing light waves of different frequencies ω_1 and ω_2. If the two light beams are incident simultaneously on a non-linear material we have:

$$E = E_1 \exp(i\omega_1 t) + E_2 \exp(i\omega_2 t) \qquad (11.30)$$

$$P = \chi E_1 \exp(i\omega_1 t) + \chi E_2 \exp(i\omega_2 t)$$
$$+ \tfrac{1}{2}\chi' E_1^{\ 2} \exp(2i\omega_1 t) + \tfrac{1}{2}\chi' E_2^{\ 2} \exp(2i\omega_2 t)$$
$$+ \chi' E_1 E_2 \exp\{i(\omega_1 + \omega_2)t\}. \qquad (11.31)$$

The three lines in (11.31) correspond, respectively, to the fundamental frequencies, the harmonics and the sum frequency. Had we written *real* expressions, $E = E_1 \cos \omega_1 t$, etc., for the fields, both sum and difference frequencies would have been evident.

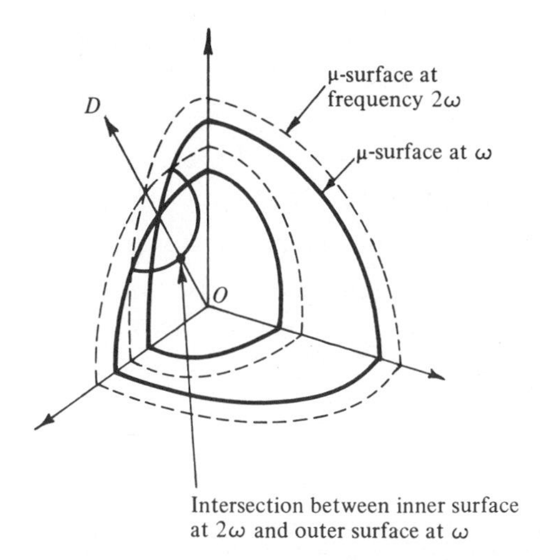

Fig. 11.25. To illustrate the matching of refractive indices or phase velocities, at frequencies ω and 2ω in a biaxial crystal. In all directions such as OD lying on a cone around the optic axis, $\mu_i(2\omega) = \mu_o(\omega)$, the suffices i and o referring to the inner and outer branches of the surface.

11.4.8 *Observations on Raman spectra.* Raman scattering of photons is a type of inelastic scattering which is most simply explained by saying that a photon exchanges a quantum of rotational or vibrational energy with the scattering atom. Looked at from a wave-mechanical point of view the process is as follows.

A light wave of frequency ω is incident on an atom which has, say, angular momentum states with energy differences $\hbar\omega_1$, $\hbar\omega_2$, etc. The atom becomes polarized by the wave and re-emits most of its energy by elastic scattering (§ 10.2) at the original frequency ω. However, during the interaction, the perturbation of the light wave has put the atom into a mixed quantum state, which contains real charge oscillations at frequencies ω_1, ω_2, etc. Depending on the phase relations between ω and these oscillations, there is a slight probability of re-emission at combination frequencies

$$\omega \pm \omega_1, \qquad \omega \pm \omega_2, \text{ etc.}$$

However, because they do not conserve energy, such processes are rather unlikely. If the atom can lose the spare energy, emission at the lower frequencies $\omega - \omega_1$, $\omega - \omega_2$ can occur; emission at the higher frequencies requires an extra source of energy, such as thermal energy, and is less likely. The emission lines at $\omega - \omega_1$, etc., are called *Stokes lines*; those at $\omega + \omega_1$, etc., are *anti-Stokes* lines, and are generally very weak.

If the incident light is very intense, however, there becomes possible an alternative method of conserving energy and momentum. This is by the simultaneous (or nearly simultaneous) incidence of two photons ω. Energy can thus be conserved if re-emission takes place simultaneously at

$$\omega + \omega_1 \quad \text{and} \quad \omega - \omega_1.$$

Moreover, for certain angles of scattering the momentum $\hbar\mathbf{k}$ can also be conserved, and thus the intensity of emission can be large. We can draw a vector diagram (Fig. 11.26) for conservation of $\mathbf{k}$ in which we represent

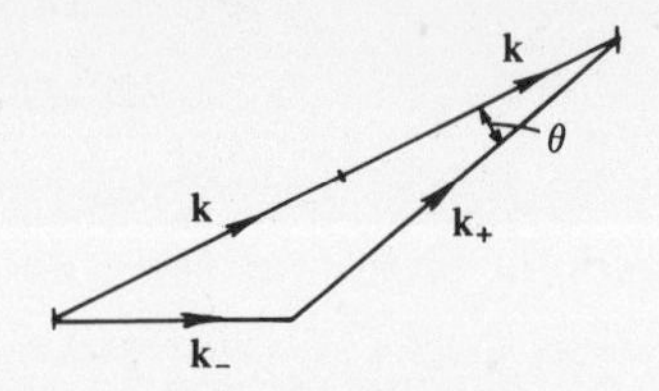

Fig. 11.26. Conservation of **k**-vector for Raman-assisted double-photon transition.

the wave-vector for the ω waves by the vector $\mathbf{k}$, and those of $\omega \pm \omega_1$ by $\mathbf{k}_\pm$ respectively. We write $|\mathbf{k}_+|$ and $|\mathbf{k}_-|$ by Taylor expansions:†

$$k_+ = K \pm \omega_1 \frac{\partial K}{\partial \omega} + \tfrac{1}{2}\omega_1{}^2 \frac{\partial^2 K}{\partial \omega^2} + \dots \tag{11.32}$$

and using the cosine rule for the triangle we find that

$$\cos\theta = \frac{\omega_1{}^2(\partial^2 K/\partial\omega^2)K}{2\{K^2 + \omega_1(\partial K/\partial\omega)K\}}. \tag{11.33}$$

For small angles θ this gives:

$$\begin{aligned} \theta &= \omega_1\left(\frac{1}{K}\frac{\partial^2 K}{\partial\omega^2}\right)^{\frac{1}{2}} \\ &= \omega_1\left\{\frac{1}{\mu}\left(\frac{\partial^2\mu}{\partial\omega^2} + \frac{2}{\omega}\frac{\partial\mu}{\partial\omega}\right)\right\}^{\frac{1}{2}}, \end{aligned} \tag{11.34}$$

since $k = \mu\omega/c$. The strong Raman scattering in the directions satisfying these equations is evident as rings around the direct beam $\theta = 0$ (Fig. 11.27); it has been observed particularly well when a collimated laser beam from a ruby laser induces Raman scattering in nitrobenzene.

† The abbreviation $\partial\mathbf{K}/\partial\omega$ means $\partial\mathbf{k}/\partial\omega$ evaluated at Ω where $\mathbf{k} = \mathbf{K}$.

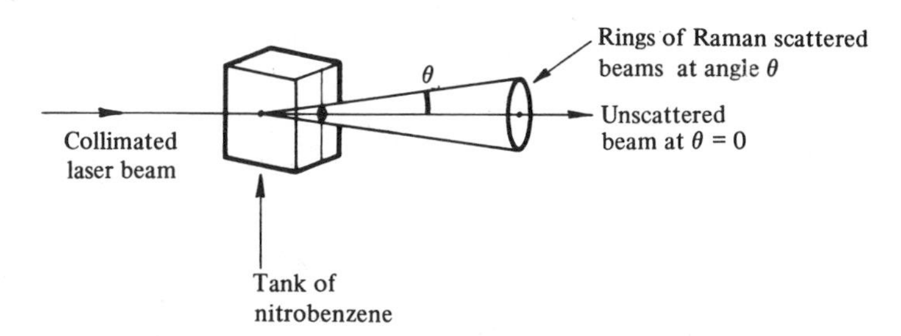

Fig. 11.27. Observation of Raman rings in nitrobenzene.

11.5 Astrophysics

Until recently, our knowledge of the universe had been obtained solely from optical measurements; visible light had been the only means of communication with the rest of the universe, and all that we knew about it had depended on our ability to interpret this light. But a whole spectrum of electromagnetic waves is presumably emitted by any stellar object; the reason that we have been so limited to visible light until the last few decades is that we are shielded from most of them by the Earth's atmosphere. Two major 'windows' exist in the spectral absorption of the atmosphere: visible light, and radio waves. The former is the basis of this section, and the latter the basis of the next.

Two optical instruments are of paramount importance. The first is the telescope, which has to be able to resolve the stellar object of interest to us from other objects in its immediate neighbourhood. It must also collect enough light for the object to be photographed or its image analysed spectroscopically. We shall discuss the problem of resolution in § 11.5.2.

The second instrument is the spectrograph which is attached to the telescope. Most of our conjectures about stellar constitution and structure are based on spectroscopic information.

11.5.1 *Spectroscopy in astrophysics.* As far as instrumentation is concerned, we have little to add to earlier chapters. The diffraction grating (§ 7.7) and the Fourier transform spectrometer (§ 8.5) are the main instruments used for analysis, and although much work has been done to adapt them specifically to astronomical purposes, no new principles have evolved.

Once the lines in a stellar spectrum have been identified, high-resolution spectroscopy allows other observations to be made on them. First, the lines may be broadened. Following the discussion in § 8.3, it is possible to determine the temperature and maybe the pressure of the emitting gas. Secondly, the lines may be shifted from their usual positions in the spectrum. If all the lines from a given object have the same relative shift in wavelength, $\delta\lambda/\lambda$, it is assumed that it arises from a Doppler shift due to motion along the lines of sight; observations of this sort have led to one of the most extraordinary and puzzling features in the whole of astrophysics – the galactic red shift. It appears that the vast majority of stellar objects are receding from us, some of them at fantastic velocities approaching the speed of light. Moreover, when there exists any reliable information as to the distance of the objects, there appears to be an approximate proportionality between the distance and the velocity of recession. This relationship is called *Hubble's law*. It is not difficult to convince oneself that this law does not put the earth or the solar system in any privileged position! The constant of proportionality seems to be about 0.024 m s^{-1} (light year)$^{-1}$. The simplest explanation of the observation is that the universe is expanding uniformly, and at present there seems to be no other reasonable interpretation of the Doppler shift.

When a stellar object is large enough for spectroscopic measurements to be made on isolated parts of it, a differential Doppler shift between the edges indicates rotation of the body. In the same way, the doubling of each spectral line indicates a rotating double star, generally called a 'spectroscopic binary'.

11.5.2 *Astronomical speckle interferometry.* The theoretical resolution limit of a telescope, $\theta_{\min} = 1.22\lambda/D$, cannot be achieved by any Earth-based instrument because of the presence of non-uniformities in the atmosphere. Local pressure and temperature variations result in the atmosphere's having a rather poor optical quality and its properties vary widely as a function of the weather, the time and the azimuth angle.

Just to get some idea of the parameters involved we can quote some typical deviations from the mean optical thickness of the whole atmosphere. The r.m.s. fluctuation amplitude is between two and three wavelengths of visible light, and changes randomly in a time of the order of 10 ms. In the spatial dimension, fluctuations are correlated within transverse distances of about 0.1 m. It should be pointed out that the optical thickness is not a local property of the atmosphere, but a value integrated as a light wave travels from outer space to the Earth's surface. Most of the fluctuations arise within a few hundred metres of the surface, and are substantially affected by local conditions such as water and trees. The reduction of atmospheric turbulence is one of the conditions affecting the choice of a site for an astronomical observatory. The atmospheric fluctuations are responsible for the twinkling of small stars.

The general effect of atmospheric fluctuations on a stellar image is called by the astronomer 'the seeing'. Numerically, this might be 3 arcsec on a poor night and 1 arcsec on an exceptionally good, still night. It is rarely better than 0.5 arcsec. This figure can be related to the correlation distance in the wavefront since we have seen in § 8.6 that the distribution of intensity in the image formed by an instrument is the Fourier transform of the spatial correlation function of the light in the entrance pupil. If a typical spatial correlation distance x_c is 0.1 m in the incident wavefront, the resultant image will have an angular radius of order

$$\theta_{atm} \approx \lambda / x_c = 5 \times 10^{-6} \text{ rad} \approx 1 \text{ arcsec.}$$

This should be compared with the Rayleigh resolution limit for, say, a 1 m telescope which is 0.05 arcsec. The resolution that can be achieved with a very large telescope therefore seems to be no better than that from a telescope of diameter 10 cm; only the brightness of the image is greater with the larger telescope.

Two major inventions attempted to overcome the resolution limit set by the atmosphere: the Michelson stellar interferometer (§ 8.6.6) and the Brown–Twiss interferometer (§ 8.6.7). Both of these exhausted all of the stellar objects which were bright enough for their operation. Recently two new techniques have been introduced in an attempt to overcome the problem of atmospheric degradation – speckle interferometry and adaptive optics – and in addition technical advances have made the Michelson stellar interferometer applicable to far weaker and smaller sources than Michelson and Pease were able to consider. Because of its relationship to the main theme of this book we shall describe the principle of speckle interferometry in its most elementary form (Labeyrie, 1976).

The technique was first used by Labeyrie, Gezari and Stachnik in 1970 and arose from careful observation of 'instantaneous' photographs of stellar images. With the introduction of image-intensifier tubes it had become possible to photograph images through a narrow-band filter at a magnification sufficient to see detail at the Rayleigh resolution limit using an exposure less than the 10 ms correlation time of the atmospheric fluctuations. These images have an overall size of the order of seconds of arc (as estimated above) but contain a wealth of detail. The detail changes with time, so that a long exposure (greater than say 0.1 s) shows *only* the overall size.

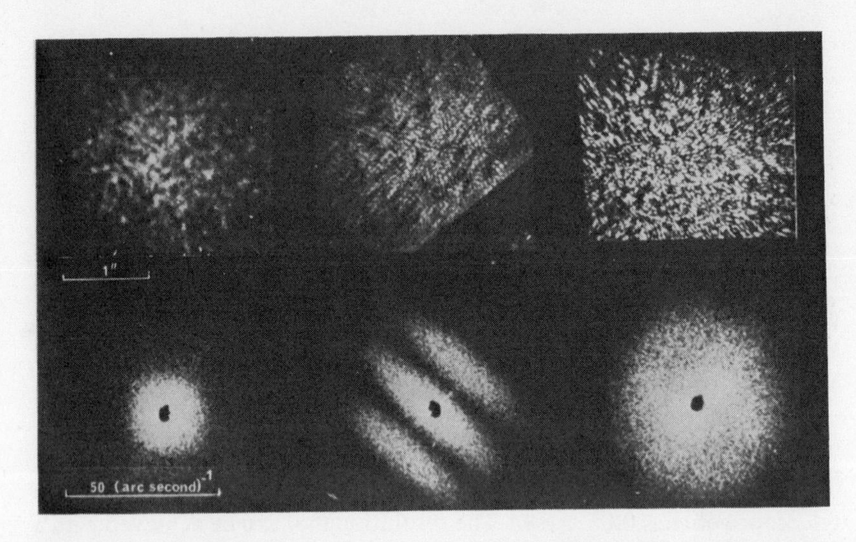

Fig. 11.28. Speckle images (above) and corresponding spatial power spectra $\tilde{O}(u, v)$ (below). From left to right: (*a*, *d*) Betelgeuse (resolved disc), (*b*, *e*) Capella, (resolved binary) and (*c*, *f*) an unresolved reference star. (From Labeyrie, 1976.)

Three examples of 'instantaneous' photographs are shown in Fig. 11.28. There are obvious differences in their detailed structure and these differences represent real differences in the objects. The method of speckle interferometry separates the atmospheric and object contributions to these images by using a series of exposures during which the atmosphere changes from exposure to exposure, but the star remains invariant. The analysis essentially follows § 7.4.6, where we discussed the diffraction pattern of a random array of identical apertures.

Suppose, first, that the telescope was used to observe an ideal point star. The image, photographed instantaneously through the atmosphere, has an intensity distribution $R(x, y, t)$ at time t. Now the point star illuminated the atmosphere from above with a plane-wave, so that the light distribution $R(x, y, t)$ is just the intensity of the Fourier transform of the atmosphere considered as a phase object. It is called the atmospheric speckle pattern.

Now consider the extended star as an *incoherent* collection of such point stars. If the atmosphere had been homogeneous we should have seen an ideal image of intensity $O(x, y)$ with resolution limited only by the finite aperture of the telescope. In the presence of the real atmosphere each ideal point in $O(x, y)$ gives rise to an identical atmospheric speckle pattern $R(x, y, t)$ and so the composite image is the convolution:

$$I(x, y, t) = O(x, y) * R(x, y, t). \tag{11.35}$$

This image is photographed at time t_i, under conditions such that the photographic film has amplitude transmission proportional to the exposure intensity (§ 9.7.3).

Subsequently the developed photograph is used as a diffraction obstacle in a diffractometer. One records on a second film the intensity of the Fourier transform of $I(x, y, t_i)$ (by virtue of the linear relationship between amplitude transmission and exposure intensity in the original film). The process is repeated for a series of exposures at times t_i, the transforms being superimposed on the second film. Now the transform of $I(x, y, t_i)$, (11.35), is

$$\tilde{I}(u, v, t_i) = \tilde{O}(u, v)\tilde{R}(u, v, t_i), \tag{11.36}$$

and its intensity

$$|\tilde{I}(u, v, t_i)|^2 = |\tilde{O}(u, v)|^2 |\tilde{R}(u, v, t_i)|^2. \tag{11.37}$$

The summation for a long series of t_is gives:

$$\sum_i |\tilde{I}(u, v, t_i)|^2 = |\tilde{O}(u, v)|^2 \sum_i |\tilde{R}(u, v, t_i)|^2. \tag{11.38}$$

And since $|\tilde{R}(u, v, t_i)|^2$ is a random function in which the detail is continuously changing, the summation becomes smoother and smoother as more terms are added. Finally, we have, when enough terms have been added to make $\sum |\tilde{R}(u, v, t_i)|^2$ smooth enough

$$\sum_i |\tilde{I}(u, v, t_i)|^2 = |\tilde{O}(u, v)|^2 \times \text{(a slowly varying function)}. \tag{11.39}$$

In this way the intensity of the Fourier transform $|\tilde{O}(u, v)|^2$ has been measured, and stellar structure can be deduced by a retransformation, as far as the Rayleigh diffraction limit will allow. A true stellar image cannot be deduced, because we do not record the phase of $\tilde{O}(u, v)$ (§ 9.3.5) but straightforward structural details can be derived.

In Fig. 11.28 we show three examples of speckle transforms obtained by this technique. One, of which (*c*) is a single exposure from the series of some hundred speckle patterns, corresponds to an unresolvable point star. The diameter of the transform disc (*f*) shows the star to have angular diameter less than $\frac{1}{50}$ arcsec which is the Rayleigh limit of the telescope. The second series, of which (*b*) is an example, is a binary with separation about $\frac{1}{20}$ arcsec. Close observation of the speckle pattern shows this to be true, but the transform (*e*) showing Young's fringes makes the observation quantitative. Finally the series typified by (*a*) shows in (*d*) a resolvable single star of diameter about $\frac{1}{15}$ arcsec.

The calibration scale 50 arcsec^{-1} is calculated absolutely from a knowledge of the geometry and wavelength used in the diffractometer, and the magnification of the telescope.

Finally, we should briefly mention some further points connected with the technique. If the telescope mirror contained aberrations, these would be apparent even in the transform disc of a point star such as Fig. 11.28(f); as a result one can show easily that dividing the intensity in (d) or (e) by that in (f) allows one to see diffraction-limited detail even with a rather poorly constructed telescope. (One can carry out the division as long as the intensity in (f) does not become zero.) It was therefore suggested that it would be useful to build an extremely large telescope using several mirrors on a rigid support. Of course the method as discussed here is not capable of reconstructing complete images and its use so far has been limited to stars with simple structures that can be recognized from their transform intensities. Some recent work has developed methods of processing the speckle images digitally in a way by which the phase of $\tilde{O}(u, v)$ can also be recovered and as a result the complete high-resolution image can be constructed. The method is at present being applied to solar and planetary images, but is too involved to be discussed here.

11.6 Radioastronomy

Although all electromagnetic waves can travel through empty space, astronomy from Earth-based observatories must use those wavelengths which are capable of penetrating the atmosphere completely. This leaves us essentially with two spectral regions: the visible, and the radio region. One can argue that it has been possible to make astronomical observations recently from balloons, rockets and satellites, and thus to use radiations which are either completely or partially absorbed by the atmosphere, but on the whole the resolution of such observations is rather poor and their thoroughness cannot be compared with that available from the ground. It is probable that the planned large space telescope will change this situation, but at present our knowledge of the universe is based on visible and radio observations.

It was therefore an important event for astronomy when Jansky, in 1932, discovered cosmic radiation of a few centimetres wavelength. Based on this discovery, radioastronomy developed rapidly after the Second World War. After the initial discovery that radio waves, with wavelengths of the order of centimetres, were reaching the Earth, some form of telescope was clearly necessary. Metals reflect radio waves almost ideally, and therefore a reflecting telescope can easily be built. Accuracy is not difficult to attain; a tenth of a wave is a few millimetres. The real problem is simply that of size. As we shall see in § 11.6.3 even the largest aperture that has been constructed is optically very small, and therefore one tries to see whether optical principles suggest any devices that can improve resolution. These will be discussed in the following sections.

11.6.1 *The radio receiver as an optical instrument.* A radio receiver and a telescope have much in common. Each of them receives electromagnetic waves, albeit of very different frequencies, and converts them by some means into electric currents in a circuit; the aerial may be connected to a radio receiver in which the electric currents are converted into audible sounds, whereas the telescope is connected to an eye, which turns visible light into wave-impulses which are eventually interpreted by the observer's brain. As well as this functional similarity, the designers of both telescopes and radio aerials are often faced with similar problems of sensitivity and resolution. Since the solutions illustrate optical principles which should by now be familiar to the reader, it seems appropriate to discuss them briefly (Cole, 1977).

11.6.2 *The resolving power and directional properties of telescopes and aerials.* The resolving power of a telescope is a measure of its directional properties. To cast the problem in the same mould as would a radio-receiver designer, we should ask the following question: If we fix the position of the observer's eye (assumed to be a point), how does the signal received by it depend on the angular position of a point object which is being viewed? For a telescope of finite aperture the answer is quite simple. The point object produces an Airy-disc image in the focal plane of the eyepiece, and so, as the subject changes its angular position, the image will sweep past the observer's eye point, being visible over the angular diameter of the Airy disc. The Airy disc is, of course, the diffraction pattern of the objective lens of the telescope (§ 7.3.5) and depends on its dimensions. We can illustrate the directional properties of a telescope by a curve showing a signal S received as a function of the angular position (θ, ϕ) of the source, as in Fig. 11.29.

Now consider the same problem as applied to a dipole aerial. We have discussed the properties of an oscillating dipole as a radiator in § 4.5.1, and as a receiver it behaves in an identical manner because the relationship between aerial-current and electric field is independent of which one

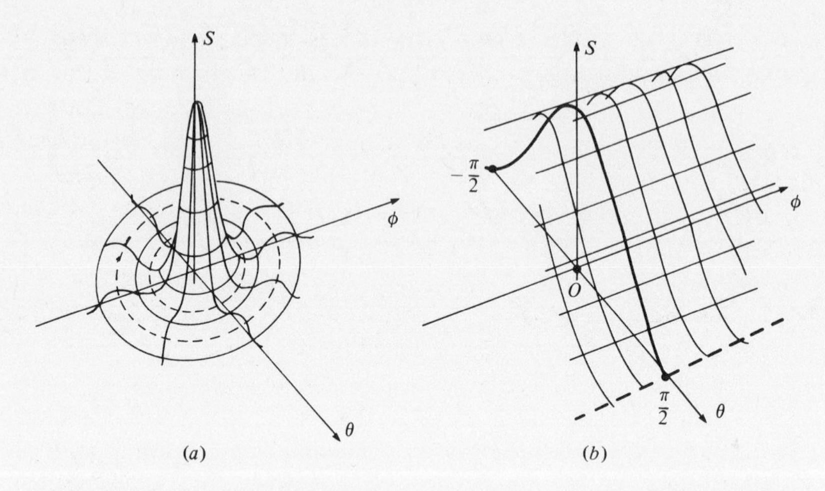

Fig. 11.29. (a) Directional properties of a telescope, the relationship between received-signal, S, and angular position of the source. (b) Directional properties of a dipole aerial.

causes the other. Following Fig. 4.19 we can plot the relationship $S(\theta, \phi)$ the same way as in Fig. 11.29(a), except that in this case there is a difference between the directional properties in the plane perpendicular to the dipole (ϕ varies) and in a plane containing the dipole (θ varies) (Fig. 11.29(b)). Clearly the dipole is behaving as if it had an aperture which is infinitely thin and of length $\lambda/2$, which is a fairly accurate description of the average radio aerial! In order to increase the selectivity we need to increase the dimensions of the aperture, or effective aperture.

11.6.3 *Large-aperture dish aerials.* The simplest way of increasing the effective aperture of a radio aerial is to back it by a parabolic reflector, which collects waves over a large area and presents them at a central dipole aerial with phase changes independent of position in the aperture. A point-source object now produces a signal spread over an Airy disc at the aerial position, and the directional properties are the same as those illustrated in Fig. 11.29(a), since we are now restricting our use of the dipole aerial to a small angular range around $\theta = 0$ (normal incidence).

The orders of magnitude are worrying, however. A telescope of aperture diameter D gives rise to an angular resolution (§ 9.4.1)

$$\delta\theta = \frac{1.22\lambda}{D}. \tag{11.40}$$

If the telescope is optical, with a 10 cm diameter lens and wavelength 5×10^{-5} cm, this gives

$$\delta\theta = \frac{1.2 \times 5 \times 10^{-5}}{10} = 6 \times 10^{-6} \text{ radians}$$
$$= 4 \times 10^{-4} \text{ degrees.} \tag{11.41}$$

A radio telescope, however, with dimensions 250 ft or 7.5×10^3 cm, working at a wavelength of 20 cm, has a limit of resolution

$$\delta\theta = \frac{1.2 \times 20}{7.5 \times 10^3} = 3 \times 10^{-3} \text{ radians}$$
$$= 2 \times 10^{-1} \text{ degrees.} \tag{11.42}$$

This is one order worse than the resolution achieved by the human eye, unaided by any optical instrument! And the Jodrell Bank telescope, whose diameter is used in the above calculation, represents a maximum order of magnitude for the practical construction of adjustable parabolic dish aerials.

Of course, the difference between the 2×10^{-1} degrees of the dish aerial and the 90 degrees of the unaided dipole is considerable, and the intensity of the signal is, of course, much larger. However, with the growing interest in more and more distant, and therefore smaller, objects, it has become of prime importance to increase the resolving power of radio telescopes as much as possible.

11.6.4 *The interferometer approach.* It will be remembered from § 8.6.6 that Michelson succeeded in increasing the resolving power of a telescope by using it as part of an interferometer. This he did at the expense of simplicity of the image, by producing an interference pattern out of the simple image of a star and observing the properties of this interference pattern. The same approach can be used with a radio telescope. Suppose we consider two dipole aerials separated by a distance L. They can be connected to an electronic device which measures the sum of their signals at a point midway between them, so that the phase lags between aerial and electronics are equal. This system has a response corresponding to a pair of infinitely thin Young's slits, and is illustrated in Fig. 11.30. It has a resolution angle

$$\delta\theta = \frac{\lambda}{2L}. \tag{11.43}$$

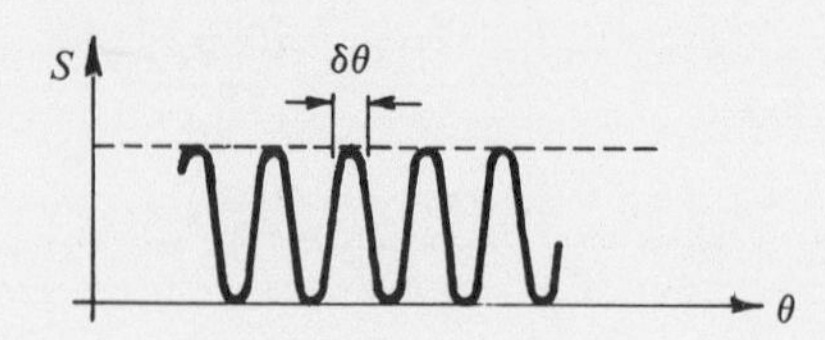

Fig. 11.30. Directional properties of two dipole aerials.

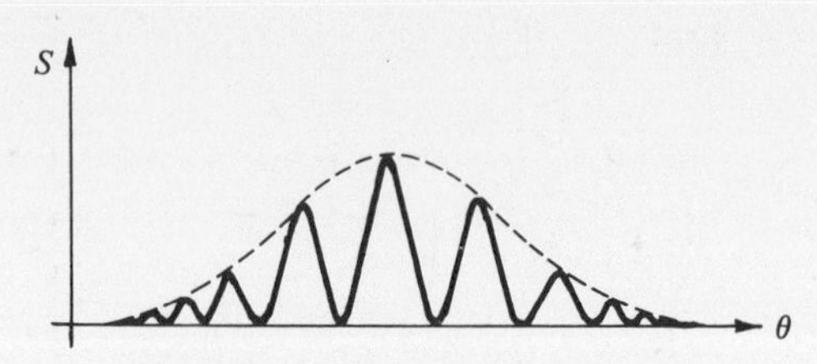

Fig. 11.31. $S(\theta)$ for a two-aerial interferometer observing a star emitting waves with a coherence-length of about 5λ.

By using a large value of L, the resolving angle can be reduced as much as required: for example, a base-line $L = 1$ km gives, for a 20 cm wavelength

$$\delta\theta = 10^{-4} \text{ radian} = 6 \times 10^{-3} \text{ degrees,} \tag{11.44}$$

which is as good as a 1 cm optical telescope. However, the observations of anything but a monochromatic point star are likely to be quite difficult to interpret by this method. The situation may not be quite as bad as the above discussion would indicate, because of the finite coherence-time (§ 8.2.4) of the radio-emission. If the path difference between the waves arriving at the two aerials is greater than the coherence-length of the radiation, no interference pattern will be seen. In fact, the coherence-time can be adjusted at will by controlling the bandwidth $\delta\lambda$ of the receiver. As a result the $S(\theta)$ curve for the interferometer would appear as in Fig. 11.31. Thus the signals received from well-separated stars can easily be resolved, and the position of each one determined to the accuracy of the interferometer. An example of the signals received from two sources, Cygnus A and Cassiopeia A, is shown in Fig. 11.32; the angle θ is swept by the rotation of the Earth. The process of adjusting the number of visible fringes by altering the receiver bandwidth is exactly equivalent to introducing a coloured glass filter into a white-light fringe system. With white light, only a few fringes can be seen; the average piece of coloured glass increases this to ten or so.

Just as in the Michelson stellar interferometer, the angular diameter of a source will also modify the interference pattern. This leads to a

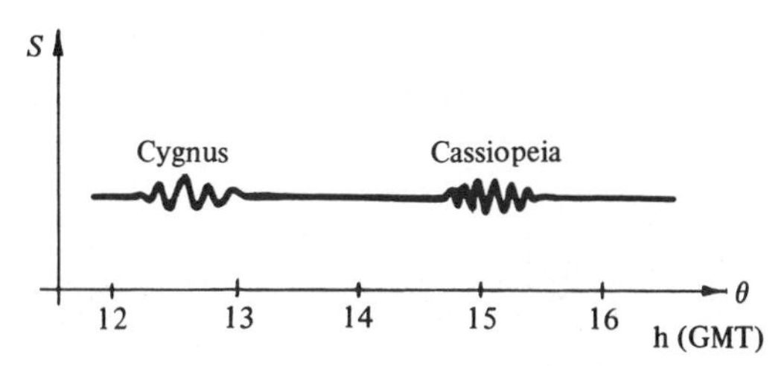

Fig. 11.32. Signals from two sources with finite coherence lengths – or observed within a finite bandwidth. (From Smith, 1962.)

reduction of the amplitude of the oscillations in $S(\theta)$ as the base-line L is lengthened, as a result of the finite coherence area of an extended source, and it is easy to see that the interference vanishes just when L becomes large enough for the interferometer to be able to resolve the star as other than a point source. In recent years this idea has been extended to its practical limit. The pair of receivers are situated on continents at opposite sides of the Earth; their signals, synchronized to atomic clocks, are stored on magnetic tape and subsequently analysed.

11.6.5 *Diffraction gratings and aerial arrays.* To return, however, to the achievement of the highest possible resolving power, we can follow the progress of interferometry from Young's slits to the diffraction grating (§ 7.4.2). By arranging a line of n regularly spaced aerials, all connected to a summation device by transmission lines introducing equal delays, we can increase the angular resolution to

$$\delta\theta = \frac{\lambda}{2l(n-1)}, \tag{11.45}$$

where l is the spacing of the aerials. We thus achieve the same resolution as two aerials at the ends of the line, with separation $(n-1)l$, but the introduction of the extra aerials between them simplifies the interference pattern to a series of spikes at a spacing of $\sin^{-1} \lambda/l$ (Fig. 11.33). Of course, by placing the aerials at intervals of less than λ we would achieve an $S(\theta)$ curve consisting only of a zero-order peak, but this is usually impracticable.

A further improvement uses the principle of the blazed grating (§ 7.7.5) by employing two or more dish aerials beamed in a certain direction.

11.6.6 *The Mills cross.* The one-dimensional array of aerials has only a limited use. It can sweep the sky as the Earth rotates, and has a high

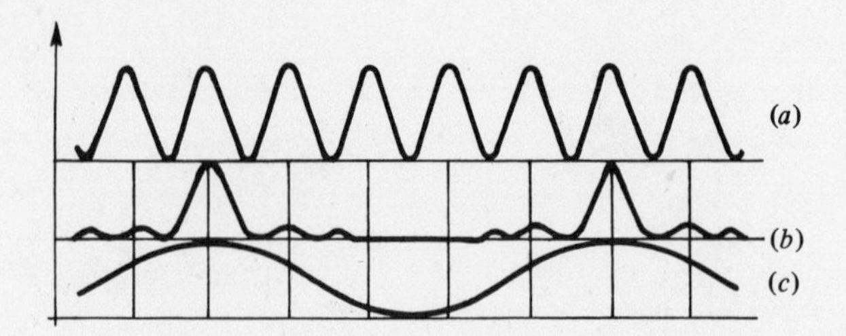

Fig. 11.33. Comparison between (a) two aerials with separation $(n-1)l$; (b) an array of aerials at separation l; and (c) two aerials with separation l.

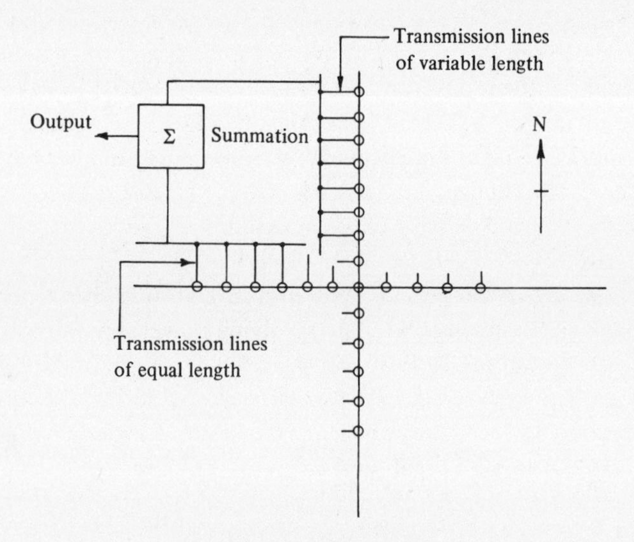

Fig. 11.34. Mills cross aerial.

angular resolution along its length, say east–west, but no resolution at right angles. The *Mills cross* is an extension of the same idea to two dimensions. By using two aerial arrays at right-angles, a high resolution in both north–south and east–west directions can be achieved (Fig. 11.34). Once again the east–west sweeping is achieved by using the Earth's rotation, but to sweep north–south requires the introduction of progressively increasing phase changes between the individual aerials.

A phase difference progressing by ψ radians from aerial to aerial is equivalent to tipping the whole array through an angle θ given by

$$\sin\theta = \frac{\psi}{2\pi} \times \frac{\lambda}{l}. \tag{11.46}$$

Scanning the sky therefore requires the introduction of variable delays into the signals from all the aerials in the north–south direction, or alternatively introducing the delays mathematically during the processing of the signals.

11.6.7 *Aperture synthesis.* Another method, called *aperture synthesis* uses the Fourier-transform relationship between the spatial correlation function of the received signal and the intensity distribution in the source – the van Cittert–Zernike theorem (§ 8.6.3). If the Earth were stationary, we could build up a picture of the radio universe by studying the correlation between signals V at pairs of points on the Earth's surface. By averaging over a long time the product signal from two aerials separated by vector **r**, we could calculate the correlation function:

$$\gamma(\mathbf{r}) = \frac{\langle V_1(0)\, V_2(\mathbf{r})\rangle}{(\langle V_1{}^2\rangle\langle V_2{}^2\rangle)^{\frac{1}{2}}} \tag{11.47}$$

and by Fourier-transforming this function we could then derive the brightness of the radio sky as a function of position. Of course, each evaluation of $\gamma(\mathbf{r})$ would involve observations of V_1 and V_2 over a long period in order to make an accurate assessment of the mean value. The fineness of structure observed in the sky depends only on the range of **r** employed. The larger the maximum values that can be achieved, the finer the detail observable.

In practice this process can be made simpler by employing the rotation of the Earth to sweep in an east–west direction, so that correlation measurements are only necessary with **r** in a north–south direction. The aerial array used is therefore as illustrated in Fig. 11.35; the east–west array is used as an ordinary interferometer, and the north–south pair as a correlation interferometer.

11.6.8 *Some results of radioastronomy.* It would be out of place here to devote much space to a discussion of the results of radioastronomical research. It is sufficient to say that the sky observed in centimetre waves appears quite different from that observed in visible light; although the former picture is crude by optical standards, it is clear that there are few stars that are both visible and radio emitters.

It is not possible to find the distance of radio emitters directly; even the red-shift cannot act as a guide since the only characteristic emission is a faint hydrogen line at 21 cm. The only evidence comes from the few radio emitters which can be identified with visible objects. Quite remarkably, it

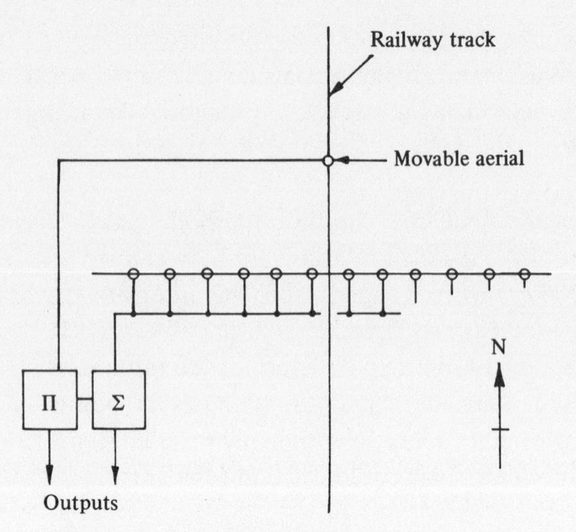

Fig. 11.35. Aperture synthesis aerial array.

is found that the radio-emission from these is very strong and it can be detected by means of fairly crude instruments. Some of the bodies are at extreme distances, and detectable only with several hours' exposure on the largest telescope. Since there are many much fainter radio emitters it must be concluded that radio telescopes can probe much deeper into space than optical telescopes can.

There are some strange objects, discovered in 1963, that must be near the extremities of the observable universe. Because of their remarkable properties they cannot be stars of ordinary types and therefore called *quasars* (quasi-stellar objects). Their red-shifts are remarkably high, indicating velocities up to 0.8 of the velocity of light. To be detected at the distance that this velocity corresponds to according to Hubble's hypothesis, they must be emitting an enormous amount of light, yet they are much more compact than would be expected.

The originally identified quasars were strong radio emitters; this was the reason why attention was focused upon them. But now a large number of visible quasars has been found, also with very large red-shifts, which move ultra-violet hydrogen lines into the visible region. These stars do not fit into the ordinary system of classification, and their presence is giving a great deal of scope for discussion amongst astrophysicists.

It must be remembered, however, that all these results depend upon the interpretation of the red-shift as a Doppler effect. If some other explanation of the observation were to be found the major problem would disappear; the Universe would stop expanding, and the quasars might be ordinary stars relatively near to our galaxy!

11.7 An introduction to magneto-ionic theory, and its relationship to optical activity

Rather nearer home than the far reaches of the Universe, radio waves are used to investigate the outer ionized layers of the Earth's magnetic field. Were the magnetic field absent, electromagnetic waves would propagate in the ionized plasma according to a refractive index of the type which was derived for the simplest example (a free-electron plasma) in § 10.3.5. The presence of a magnetic field makes the plasma anisotropic, and the anisotropy results in a great variety of different types of behaviour. The important parameters are the ratios between three frequencies: the plasma frequency Ω (10.26), the cyclotron frequency, eB/m, and the frequency of the propagating wave, ω. The observed effects are related to optical activity, which was described in § 5.5.4; this phenomenon, and the method of analysis developed in Chapter 5, are the basis of the following short discussion of the propagation of radio waves in the ionosphere.

11.7.1 *Optical activity.* In our discussion of optical activity we pointed out that the result of the helical structure of certain molecules is to give rise to a complex dielectric tensor of the form

$$\boldsymbol{\epsilon}=\begin{pmatrix}\epsilon_1 & ia & 0\\ -ia & \epsilon_2 & 0\\ 0 & 0 & \epsilon_3\end{pmatrix}, \tag{11.48}$$

where the axis of the helical structure is the z-axis. If we consider a crystal such as quartz, which is uniaxial and also has a helical structure (Fig. 5.24), we put $\epsilon_1=\epsilon_2$ in (11.48) and diagonalize the matrix. We then have:

$$\boldsymbol{\epsilon}=\begin{pmatrix}\epsilon_1-a & 0 & 0\\ 0 & \epsilon_1+a & 0\\ 0 & 0 & \epsilon_3\end{pmatrix} \tag{11.49}$$

and the principal axes are found to be (1, i, 0), (1, −i, 0) and (0, 0, 1). The meaning of the complex directions is not difficult to see. They have R-values (§ 5.2.3) of i and −i respectively, and correspond to two circularly polarized waves of opposite senses travelling along the z-axis. The normal modes are therefore these two circularly polarized waves and any wave travelling normal to z, polarized in the z-direction.

The μ-surface for the birefringent dielectric tensor (11.49) is just like that for any biaxial crystal, as in Fig. 11.36 (based on Fig. 5.20). Now

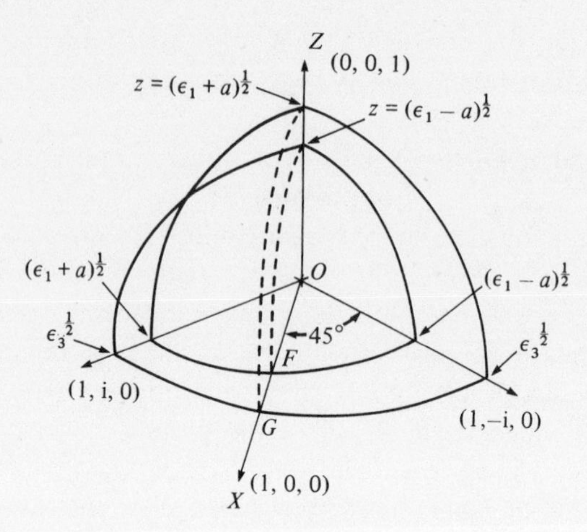

Fig. 11.36. μ-surface with complex axes for an optically active material.

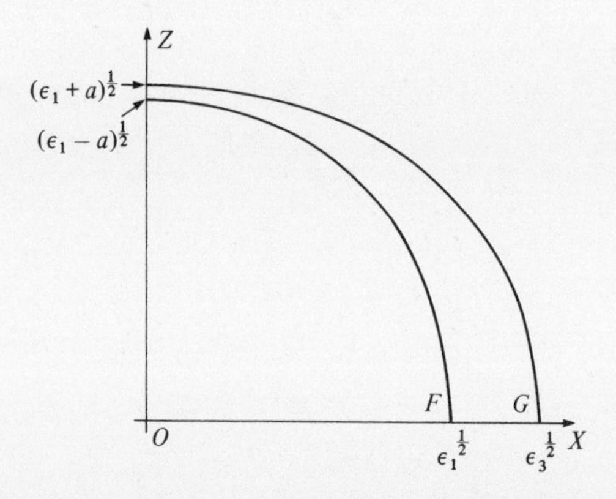

Fig. 11.37. Section of the μ-surface with real axes.

the z-direction is straightforward, since it is one of the principal axes. The x-direction can also be identified because it is half-way between (1, i, 0) and (1, −i, 0) since the linear superposition

$$\tfrac{1}{2}\{(1, \mathrm{i}, 0) + (1, -\mathrm{i}, 0)\} = (1, 0, 0). \tag{11.50}$$

This is marked on Fig. 11.36 by the axis OX. Then any direction in the plane OZX corresponds to a real direction. Now because the crystal is uniaxial all the information about its properties is contained in the plane OZX since y and x are degenerate. Thus the section of the μ-surface shown in Fig. 5.23(b) is just the section of Fig. 11.36 in this plane. We reproduce it in Fig. 11.37. The radius OF is $(\epsilon_1 - a^2/\epsilon_1)^{\frac{1}{2}}$ which is $\epsilon_1^{\frac{1}{2}}$ to a very good approximation since a is very small (§ 5.5.4).

11.7.2 *Magneto-ionic theory.* In a rather similar way the magneto-ionic problem can be illustrated. When an electron of mass m and charge e at position $\mathbf{r}$ moves in fields $\mathbf{E}$ and $\mathbf{B}_0$ we have an equation of motion

$$\mathbf{F} = m\mathbf{r} = \mathbf{E}e + e\dot{\mathbf{r}} \times \mathbf{B}_0. \tag{11.51}$$

Using the operator $\partial/\partial t \equiv \mathrm{i}\omega$ (§ 2.2.1) for sinusoidally oscillating quantities we have

$$-m\omega^2\mathbf{r} = \mathbf{E}e + \mathrm{i}\omega e\mathbf{r} \times \mathbf{B}_0. \tag{11.52}$$

Let us suppose that the charge in question is one of n per unit volume of a plasma, and that the effects of any other heavy ions can be neglected. Then the electric polarization $\mathbf{P}$ is related to $\mathbf{r}$ by

$$\mathbf{P} = ne\mathbf{r}. \tag{11.53}$$

By writing (11.53) in Cartesian coordinates with $\mathbf{B}_0$ along the z-axis, and using the relationship

$$\mathbf{D} = \epsilon\epsilon_0\mathbf{E} = \epsilon_0\mathbf{E} + \mathbf{P} \tag{11.54}$$

one calculates the dielectric tensor to be:

$$\epsilon = I - \frac{\Omega^2}{\omega^2 - \omega_0^2}\begin{pmatrix} 1 & \mathrm{i}\omega_0/\omega & 0 \\ -\mathrm{i}\omega_0/\omega & 1 & 0 \\ 0 & 0 & 1 - \omega_0^2/\omega^2 \end{pmatrix}, \tag{11.55}$$

where I is the unit tensor, Ω is the plasma frequency and ω_0 the electron cyclotron frequency,

$$\Omega^2 = ne^2/\epsilon_0 m, \tag{11.56}$$

$$\omega_0 = eB_0/m. \tag{11.57}$$

To give an idea of the orders of magnitude involved, the ionosphere at 60 km height contains of the order of 10^4 ions cm^{-3}. This gives a value of the plasma frequency:

$$\Omega \sim 10^6\ \mathrm{s}^{-1}.$$

In a magnetic field of 10^{-1} gauss, the order of the Earth's field, the electron cyclotron frequency

$$\omega_0 \sim 2 \times 10^6\ \mathrm{s}^{-1}.$$

Plasma propagation effects therefore take place at frequencies of the order of megacycles and less.

This complex tensor is formally just like the tensor we have studied in the case of the optically active medium. It diagonalizes on being referred to the axes (1, i, 0), (1, −i, 0) and (0, 0, 1) and has the diagonal form:

$$\epsilon = I - \frac{\Omega^2}{\omega^2}\begin{pmatrix} (1 + \omega_0/\omega)^{-1} & 0 & 0 \\ 0 & (1 - \omega_0/\omega)^{-1} & 0 \\ 0 & 0 & 1 \end{pmatrix}. \tag{11.58}$$

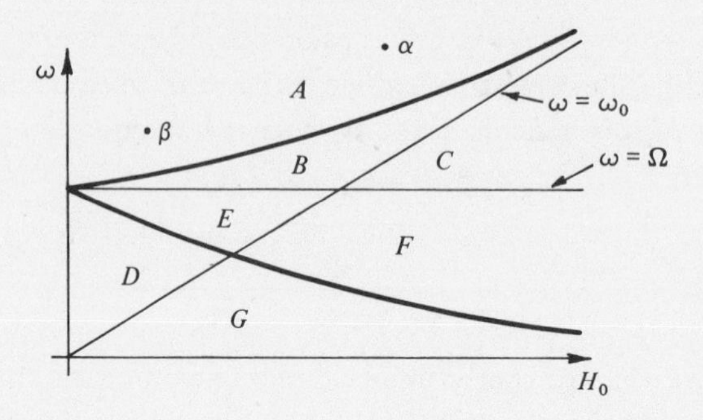

Fig. 11.38. Behaviour of waves with principal polarizations

Polarization:	(1, i, 0)	(1, −i, 0)	(0, 0, 1)
Mode	Clockwise	Anticlockwise	Plane polarized
	k‖z	**k**‖z	**E**‖z
Region *A*	Propagated	Propagated	Propagated
B	Propagated	Evanescent	Propagated
C	Propagated	Propagated	Propagated
D	Evanescent	Evanescent	Evanescent
E	Propagated	Evanescent	Evanescent
F	Propagated	Propagated	Evanescent
G	Evanescent	Propagated	Evanescent

What distinguishes the magneto-ionic medium from the optically active one is that the components of ϵ are highly dispersive. Since it is necessary that ϵ should be positive for real propagation (the wave is evanescent if $\mu=\epsilon^{\frac{1}{2}}$ is imaginary) we can place lower limits on ω for each of the three waves to be propagated. This is done in Fig. 11.38 in which ω is plotted vertically and B_0 horizontally. The boundary lines between regions are simply the solution of the equations:

$$\frac{\Omega^2}{\omega^2(1\pm\omega_0/\omega)}=1;\qquad \Omega=\omega;\qquad \omega=\omega_0(B_0). \tag{11.59}$$

Let us now ask ourselves what changes we must make in the formalism leading to Fig. 11.36 when one of the principal values of ϵ becomes negative, so that $\epsilon^{\frac{1}{2}}$ is imaginary. We know that this change converts an elliptical section into a hyperbolic section. In region *B* of Fig. 11.38, for example, we can redraw Fig. 11.36 replacing the ellipses involving the semi-axis $(\epsilon_1-a)^{\frac{1}{2}}$ (which is now imaginary) by hyperbolae. We must also notice that the value of ϵ_3 is always smaller than the other real principal value. The intersections are shown in Fig. 11.39 and it does not take much imagination to see that the inner surface is the region around *O* and the

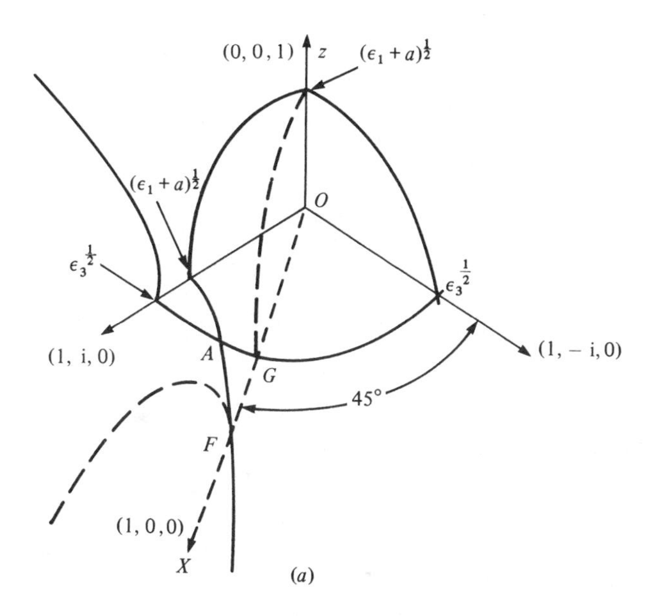

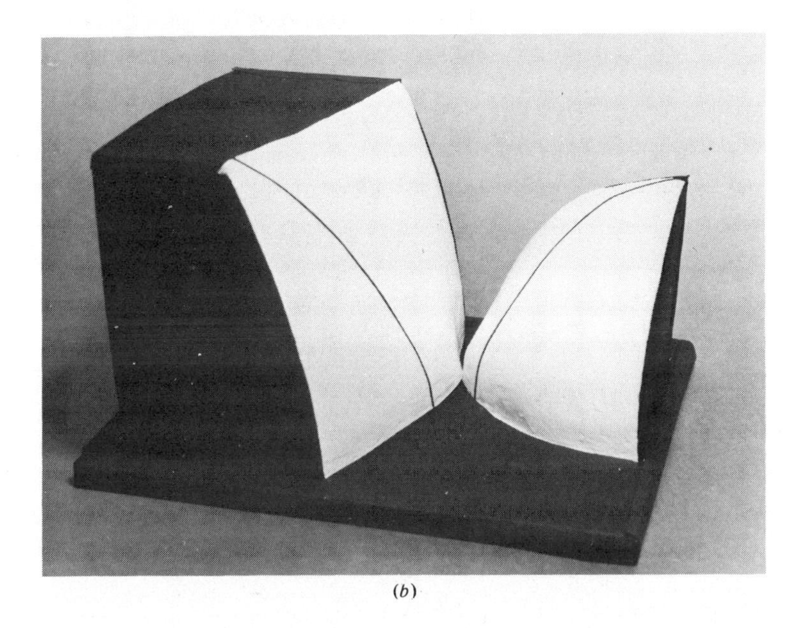

(b)

Fig. 11.39. (*a*) μ-surface with complex axes for a magneto-plasma in region *B* of Fig. 11.38, where $\epsilon_1 - a < 0$. Corresponding points on this figure and on Fig. 11.36 are labelled similarly. (*b*) Photograph of a model of the surface.

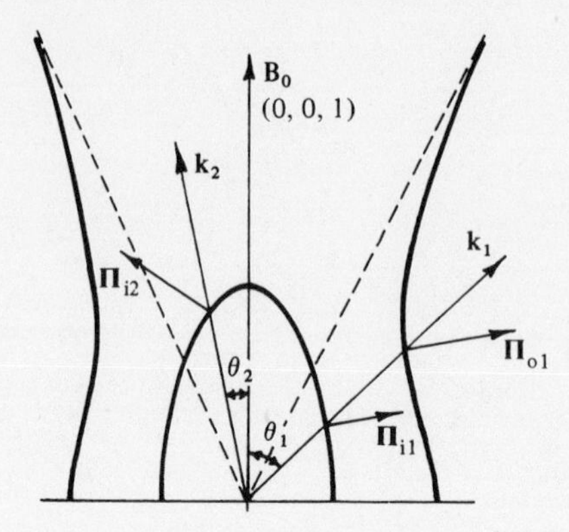

Fig. 11.40. μ-surface for propagation in region *B*.

outer surface touches it at *A* and diverges away from it. Now the real directions of propagation in the (x, z) plane lie, as explained in § 11.7.1, on the plane *ZOX* bisecting the angle between $(1, \mathrm{i}, 0)$ and $(1, -\mathrm{i}, 0)$ and so we find a μ-surface for propagation in real directions whose (x, z) plane is as shown in Fig. 11.40. Since the propagation depends only on the angle between the magnetic field $\mathbf{B}_0$ and the wave-vector $\mathbf{k}$, the medium is again uniaxial and information on propagation in all directions is contained in this diagram. The ray vector is once again normal to the surface, as is shown by the vectors on the diagram. Notice that $\mathbf{k}_1$ has two propagating characteristic waves, while $\mathbf{k}_2$ has only one. The shape shown in the figure was, admittedly, deduced by the liberal use of imagination; but the result can be demonstrated rigorously of course (see Budden, 1967).

11.7.3 *Whistlers.* Another phenomenon is that of whistlers, which are very-low-frequency waves which have been demonstrated experimentally to travel along the lines of the Earth's magnetic field. That is, the Poynting vector, $\mathbf{\Pi}$, is always very close to the z-direction in our formalism. Whistlers cause interference to radio communications when a thunderstorm occurs at the far end of the line of force passing through the receiving set (Fig. 11.41). They have the characteristic of a descending whistle because the phase velocity is proportional to ω, and so the highest frequencies arrive first.

Very low frequencies bring us into region *G* of Fig. 11.38. Here two of the principal values are imaginary and the birefringent μ-surface analogous to Figs. 11.36 and 11.39 is shown in Fig. 11.42. Only one sheet exists and the shape is like a hyperboloid of one sheet, but with different

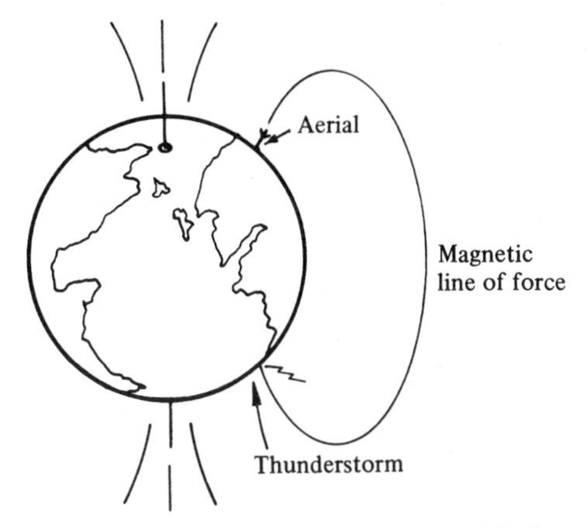

Fig. 11.41. Propagation of whistler along Earth's line of force.

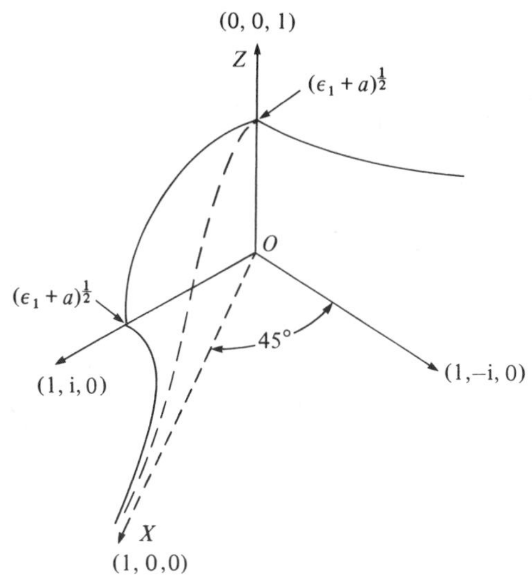

Fig. 11.42. μ-surface with complex axes for a magneto-plasma in region G of Fig. 11.38 (whistler mode), where $\epsilon_1 - a$ and ϵ_3 are less than zero. Labelling and orientation of the diagram are consistent with Figs. 11.36 and 11.39.

hyperboloidal sections in different planes containing the (1, –i, 0)-axis. The section in the ZOX plane is shown in Fig. 11.43, and we can see why the Poynting vector is almost parallel to the field $\mathbf{B}_0$ for all propagating directions of $\mathbf{k}$.

11.8 The optical transfer function

This book has not tried in any way to cover the applications of optics – particularly of geometrical optics – which are the basis of the optical

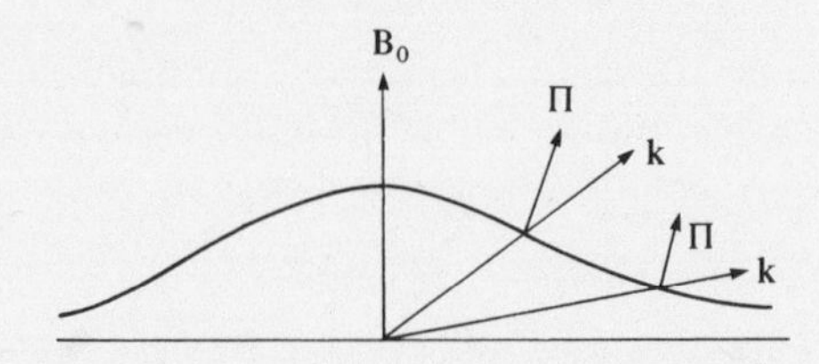

Fig. 11.43. μ-surface for whistler propagation.

industry. However, the final two sections will be used to discuss briefly two examples of industrial importance which are intimately linked to matter in earlier chapters.

The first subject is concerned with the characterization of imaging instruments, and shows the way in which Fourier analysis has provided a convenient way of condensing and presenting information on the quality of image formation in instruments.

11.8.1 *Fourier representation of image quality.* Suppose that one wants to specify the quality of the image formed by an imperfect imaging instrument, such as a telescope. One way would be to look at a point source on the axis and give a complete functional description of the image $I(x, y)$, which is called the *point spread function.* To provide complete information, the point source would then have to be moved around the field of view and the measurement repeated at selected off-axis points. This approach is not often used in practice, but has given way to a description in terms of its Fourier transform. Instead of using a point source one uses an extended incoherent sinusoidal object and measures the contrast of the image of the target at various points in the field of view. The resultant measurement of image contrast as a function of the spatial frequency of the image is called the *optical transfer function* (OTF), and has become a standard way of expressing the performance of real optical systems.

We should emphasize that the optical transfer function contains no more information than the function $I(x, y)$ – sometimes less – and still needs to be measured in a variety of off-axis regions to give a useful specification.

The optical transfer function, as we shall describe it below, strictly applies to linear systems alone, since the development assumes the linear superposition necessary to write a Fourier transform. However, the concept has also been applied to photographic films and electro-optic devices which are not linear, with consequent complications which will not be dealt with here.

Development of the concept is simply a mathematical exercise. We describe the sinusoidal target, which is incoherently illuminated, by the intensity function:

$$T_0(x, y) = \tfrac{1}{2}(1 + \cos ux), \tag{11.60}$$

where the spatial frequency, u, is 2π divided by the period of the target. This target has unit contrast or visibility (§ 8.2.6). For simplicity suppose it is magnified with unit magnification. The image is blurred to a certain extent and so its contrast is no longer unity; the image is also sinusoidal and has intensity

$$T(x, y) = \tfrac{1}{2}[a(0) + a(u) \cos \{ux + \phi(u)\}]. \tag{11.61}$$

The contrast of this image is

$$M(u) = a(u)/a(0) \tag{11.62}$$

and is called the *modulation transfer function* (MTF). The OTF is $M(u) \exp \{\mathrm{i}\phi(u)\}$, which contains information on a possible phase shift of the image; but $\phi(u)$ is rarely measured and the terms 'optical transfer function' and 'modulation transfer function' are usually synonymous. According to this definition, the MTF is unity when the spatial frequency u is zero.

The relationship between $I(x, y)$, the point spread function, and $M(u)$ is straightforward. Consider one dimension for simplicity. Each point on the target is an incoherent point source which produces an image $I(x)$. Then, according to the basic idea of convolution (§ 3.6),

$$T(x) = T_0(x) * I(x). \tag{11.63}$$

Taking the Fourier transform of each side as defined by (11.60) and (11.61) we see that

$$\begin{aligned} &\tfrac{1}{2}[a(0)\delta(k) + \tfrac{1}{2}a(u)\delta(k-u) \exp \{\mathrm{i}\phi(u)\} \\ &\qquad + \tfrac{1}{2}a(u)\delta(k+u) \exp \{-\mathrm{i}\phi(u)\}] \\ &\quad = \tfrac{1}{2}\{\tilde{I}(0)\delta(k) + \tfrac{1}{2}\tilde{I}(u)\delta(k-u) + \tfrac{1}{2}\tilde{I}(u)^*\delta(k+u)\}. \end{aligned} \tag{11.64}$$

In this equation $\tilde{I}(k)$ is the Fourier transform of $I(x)$, and $\tilde{I}(k) = \tilde{I}(-k)^*$ since $I(x)$ is a real function. Equating corresponding terms in (11.64) we then have:

$$\tilde{I}(u)/\tilde{I}(0) = M(u) \exp \{\mathrm{i}\phi(u)\}. \tag{11.65}$$

Thus, the OTF is the Fourier transform of the point spread function, normalized to unity at zero spatial frequency.

The transfer functions can be related simply to the properties of the imaging system itself. We saw in § 7.3 that, when observing a point source, what we actually see in the image plane is the Fraunhofer

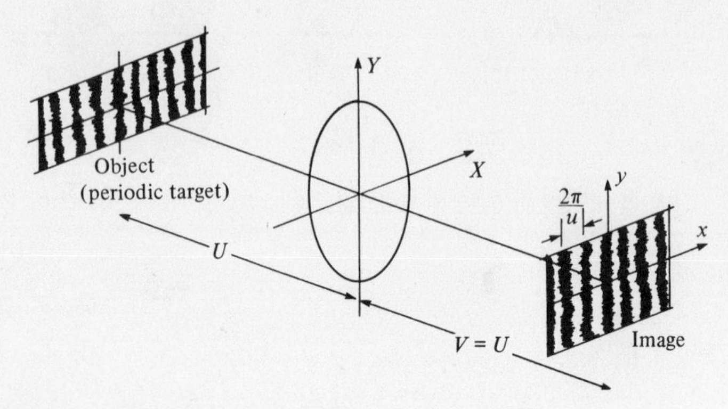

Fig. 11.44. Geometry of the OTF.

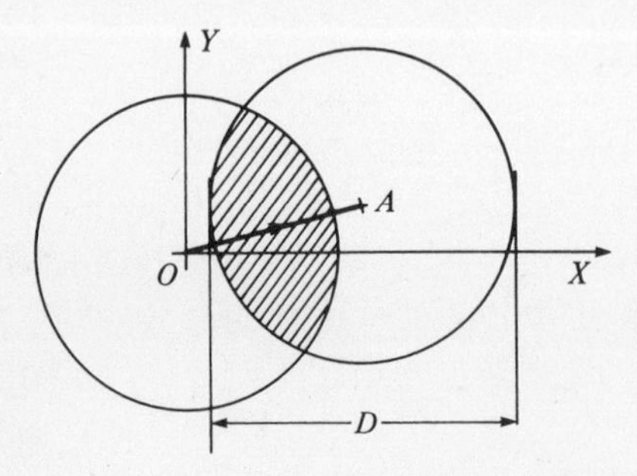

Fig. 11.45. Overlap of two circular apertures. The auto-correlation function is equal to the overlap areas shaded. The vector $OA=(u, v)V\lambda/2\pi$.

diffraction pattern of the entrance pupil. The intensity $I(x, y)$ is thus the intensity of the Fraunhofer diffraction pattern, and its Fourier transform $\tilde{I}(u, v)$ is therefore the self-correlation function of the complex amplitude distribution in the entrance pupil. One can work through the calculation in detail and show that the correlation function is written explicitly:

$$\tilde{I}(u, v)=\iint f(X, Y)f^*(X-uV\lambda/2\pi, Y-vV\lambda/2\pi)\,\mathrm{d}X\,\mathrm{d}Y \quad (11.66)$$

integrated across the entrance pupil in the (X, Y) plane, which is described by the transmission function $f(X, Y)$. The image distance is V (Fig. 11.44). Now, for a perfect lens, the function $f(X, Y)$ is unity inside the lens aperture, and zero outside it. The correlation function (11.66) is therefore the overlap area between two discs of diameter D (Fig. 11.45) with centres separated by the vector $(u, v)V\lambda/2\pi$ since outside this region at least one of f or f^* is zero. One can work out the value of the overlap area as a function of u (putting $v=0$) and find the transfer function $M(u)$ to be as in Fig. 11.46.

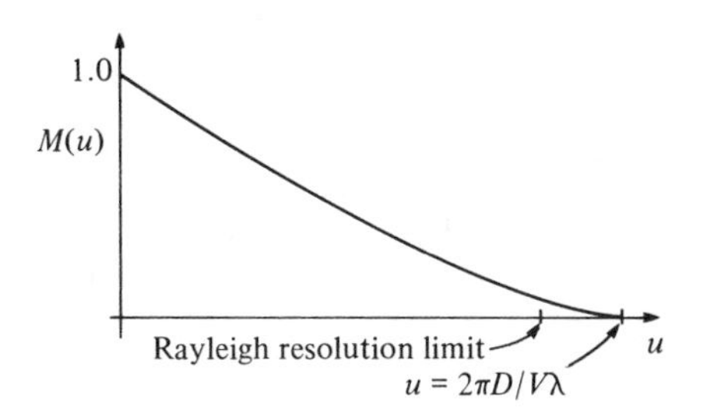

Fig. 11.46. OTF of an ideal circular lens.

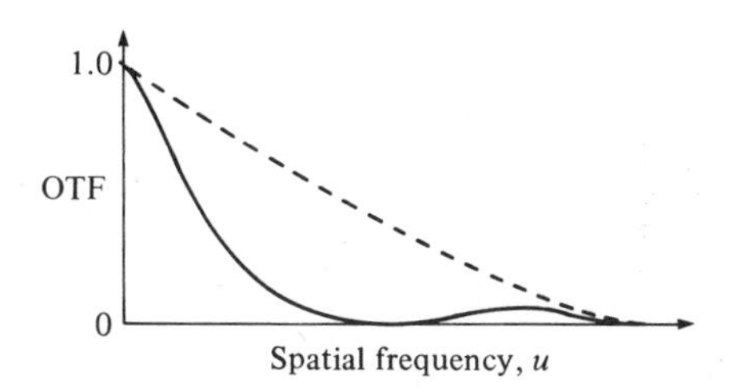

Fig. 11.47. Example of the OTF in the presence of astigmatism (full line) compared with the ideal lens (broken line).

The transfer function falls to zero at $u_m = 2\pi D/V\lambda$. This indicates that targets with a period less than $2\pi/u_m = V\lambda/D$ will be imaged with zero contrast; this agrees well with the Rayleigh criterion for resolution (§ 9.4.1) which gives a limiting spacing $V\theta_{min} = 1.22\, V\lambda/D$. When aberrations (Appendix II) are present, we make the transmission function $f(X, Y)$ complex to indicate phase errors in the emerging wavefront. Equation (11.66) can then be integrated – usually numerically – to give curves such as Fig. 11.47. For practical purposes, the resolution limit is considered to correspond to the *smallest* value of u for which $M(u)$ falls below 0.05.

11.8.2 *Measurement of the OTF.* Measurement of the OTF is carried out in several ways. First, it can be measured directly by photometrically scanning the image of a sinusoidal target. Secondly, the image of a point or slit, giving $I(x, y)$ or $\int I(x, y)\, dy$ can be scanned photometrically, and $M(u)$ calculated. But the most detailed method is to use a Twyman–Green interferometer (§ 7.9.4) to measure the wavefront amplitude and phase, $f(X, Y)$, when a point source illuminates the system, and to use (11.66) to calculate the transfer function numerically.

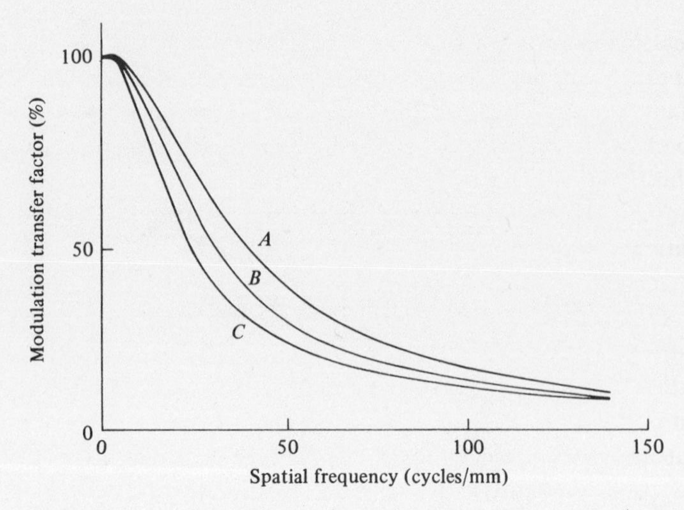

Fig. 11.48. Example of the MTF for a common photographic film in red, green and blue light (*A*, *B* and *C*, respectively). (From Thomas, 1973.)

11.8.3 *OTF of optical systems in series.* When several optical systems are placed in series, one might expect that the OTF of the combined system will be the product of the transfer functions of the components. This is usually not true. The reason is that the transfer function as defined above is valid only for an incoherently illuminated object. Now the first system will necessarily restrict the possible angles at which light can enter the second, and therefore (§ 8.6.4) the illumination at the second system is partially coherent. The discussion of transfer functions in partial coherence is too involved to be entered upon here. There are some restrictive conditions under which transfer functions of systems in tandem *can* be multiplied, but they are not very general; they do, however, include the pair consisting of a lens followed by photographic film.

As pointed out earlier, the concept of OTF can also be applied to a non-linear material such as photographic film. Provided that the contrast of the sinusoidal object is low, the image is also sinusoidal and the definition can be applied. The falling-off of MTF with spatial frequency depends on the grain statistics in the film, and there is no sharp cut-off as there is in a lens, but a Gaussian-like tail (Fig. 11.48). A discussion of the MTF of photographic films can be found in Thomas (1973).

11.9 Optical fibres

Transmission of light along a rod of transparent material by means of repeated total internal reflexion at its walls must have been observed countless times before it was put to practical use. Since the 1960s the subject of fibre optics, based on this principle, has grown to become an important industry. Optical fibres have two great uses, which will be described qualitatively in this section. The first is for transmitting images, either faithfully or in coded form, without the use of lenses; the second is for transmitting time-dependent optical signals, i.e. for optical communication.

11.9.1 *Geometrical theory of fibres.* The principle of the optical fibre can be illustrated by a two-dimensional model (corresponding really to a long strip rather than a fibre). The fibre has refractive index μ_2 and thickness d, and is immersed in a medium of lower refractive index μ_1. A plane-wave incident inside the fibre at angle $\hat{\imath}$ is reflected completely at the wall (§ 4.3.2) if $\hat{\imath}$ is greater than the critical angle $\hat{\imath}_c = \sin^{-1}(\mu_1/\mu_2)$. If the two sides of the fibre are parallel the wave is then reflected to and fro at the same angle repeatedly, ideally with no losses (Fig. 11.49(a)). According

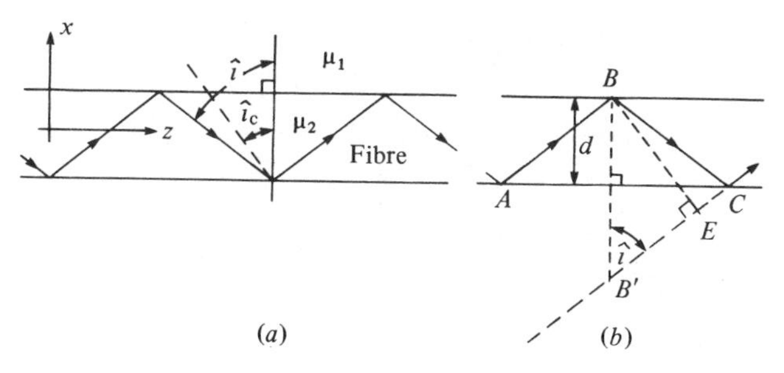

Fig. 11.49. Geometrical optics of propagation along a fibre.

to geometrical optics, any ray with $\hat{\imath} > \hat{\imath}_c$ can propagate in this way. However, physical optics requires us to look at the sum of all the waves travelling in the same direction, and to ensure that they do not interfere destructively. If we do this we naively calculate the phase difference between adjacent waves travelling parallel to one another to be (Fig. 11.49(b), cf. Fig. 7.4):

$$\Delta\phi = \mu_2 k_0(\overline{BC} - \overline{EC}) = \mu_2 k_0 \overline{B'E} = 2\mu_2 k_0 d \cos\hat{\imath}, \qquad (11.67)$$

where k_0 is 2π divided by the wavelength in free space. The requirement for constructive interference is then

$$\Delta\phi = 2\mu_2 k_0 d \cos\hat{\imath} = 2n\pi. \qquad (11.68)$$

Each integer value of n defines an allowed mode of propagation. There will always be at least one solution to (11.68) given by $n = 0$, $\hat{\imath} = \pi/2$. The

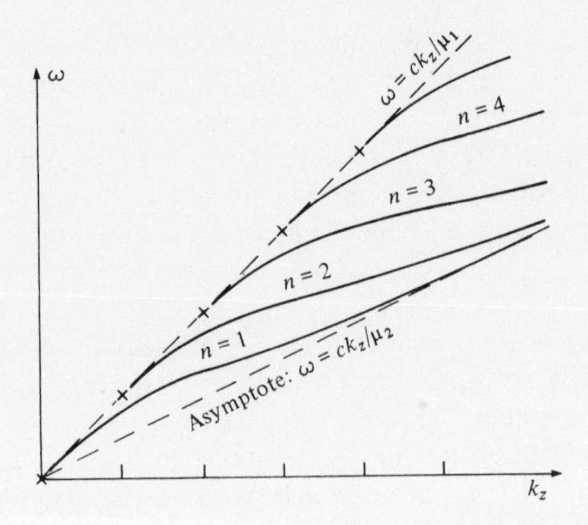

11.50. Dispersion curves $\omega(k_z)$ for propagating modes. The mode number is n, and the k_z axis is marked off in units of $\pi/\{2a(\mu_2^2/\mu_1^2-1)^{\frac{1}{2}}\}$.

number of additional solutions having $\hat{\imath} > \hat{\imath}_c$ is easily seen to be the integer part of:

$$\frac{\mu_2 k_0 d}{\pi} \cos \hat{\imath}_c = \frac{\mu_2 k_0 d}{\pi}(1-\mu_1^2/\mu_2^2)^{\frac{1}{2}}. \tag{11.69}$$

Now unfortunately the calculation is not quite as simple as this, because we have neglected to take into account the phase change $\alpha(\hat{\imath})$ which occurs on critical reflexion (§ 4.3.5). We should then write for (11.68)

$$\Delta\phi = 2\mu_2 k_0 d \cos \hat{\imath} + 2\alpha(\hat{\imath}) = 2n\pi. \tag{11.70}$$

The solution $n = 1$, $\hat{\imath} = \pi/2$ is now the first solution because $\alpha(\pi/2) = \pi$. However, since $\alpha(\hat{\imath}_c) = 0$, there may be one less mode than suggested by (11.69). The modes for the two principal polarizations will differ slightly because of the difference between $\alpha_{\parallel}$ and $\alpha_{\perp}$.

The characteristic property of a mode which describes the propagating wave is the value of $\mathbf{k}_0$ resolved along the fibre, i.e.

$$k_z = \mu_2 k_0 \sin \hat{\imath}. \tag{11.71}$$

By substituting the value of $\alpha(\hat{\imath})$ into (11.70) the dispersion curve $\omega(k_z) = ck_0(k_z)$ can be plotted for each mode (Fig. 11.50).

The modes described above, with $\hat{\imath} > \hat{\imath}_c$, are theoretically loss-less modes, and can propagate along an ideal fibre as far as the absorption coefficient of the medium will permit. In addition there are lossy modes with $\hat{\imath} < \hat{\imath}_c$, which die away after a certain number of reflexions, and are important only for very short fibres.

11.9.2 *The wave equation for a fibre.* We shall just mention that the fibre transmission problem can be formulated as a wave problem. We write down Maxwell's equations for the fibre, and satisfy the boundary conditions at the interface between the two materials. The solution is identical with what we have just described, the value of $\alpha(\hat{\imath})$ entering in a natural manner. The wave equation obtained in this way is, for the electric field vector,

$$\nabla^2 \mathbf{E} = \{k_z^2 - k_0^2 \mu^2(x)\}\mathbf{E}, \tag{11.72}$$

where $\mu(x)$ is the refractive index μ considered as a function of x, and the axis of the fibre lies along the z-direction. This equation will be recognized as analogous to Schrödinger's wave equation in quantum mechanics with $k_0^2\mu^2(x)$ taking the place of the potential energy function and the field $\mathbf{E}$ being analogous to the wave function ψ. The eigenvalue is then k_z^2. Many useful solutions of this equation can be found by analogy with quantum mechanical systems.

11.9.3 *Imaging applications.* For imaging, the mode structure of light transmission in a fibre is unimportant; we are only concerned that the light be transmitted from one end to the other. A bundle of fibres is arranged in an organized hexagonal array and the end is cut across cleanly. At the other end the fibres are arranged in the same way. What happens in-between is unimportant. An image projected on one end is then seen at the other end. This type of device is invaluable as a method of transmitting images from inaccessible regions; one important medical application is to the internal examination of patients. The resolution of the image is just determined by the diameter of each fibre, which is typically 50 μm. By changing the ordering of the fibres at the far end, an image can be coded, for example changed from a circular field of view to a slit-like field. Or the reordering may be simply an inversion; inverting an image by means of a fibre bundle is cheaper, is less bulky and introduces less aberration than would a lens system.

11.9.4 *Efficiency of light transfer by fibres.* Because all the light transmitted by a fibre must strike the surfaces at angles greater than the critical, the efficiency for energy input from a conventional light source is rather low. It is necessary to concentrate light onto a fibre or a bundle using a lens of rather small numerical aperture in order to put all the light into loss-less modes (Fig. 11.51) particularly when μ_2/μ_1 is close to unity. Input from a laser is considerably more efficient.

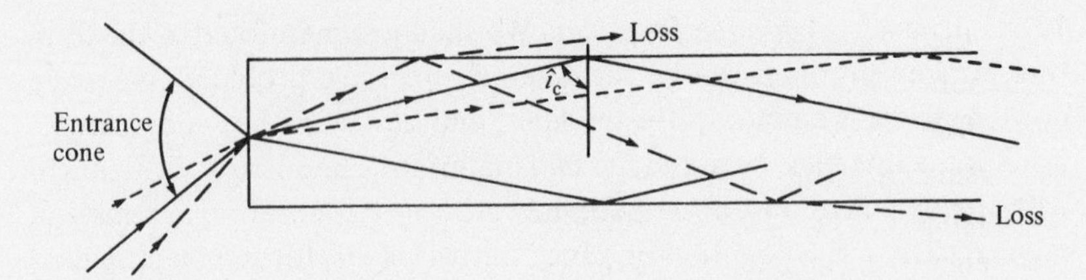

Fig. 11.51. Cone of angles allowing light to enter propagating modes.

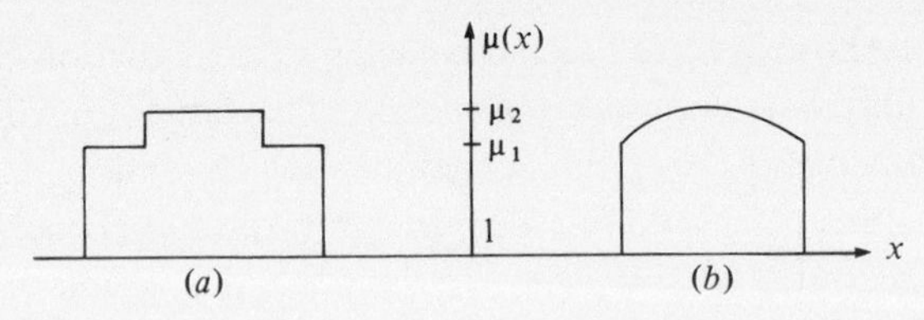

Fig. 11.52. Refractive-index profiles in (*a*) cladded fibre, (*b*) graded-index fibre.

11.9.5 *Communications through optical fibres.* The possibility of propagation of loss-less modes along fibres holds out great promise for the communications industry, and optical fibres are now serious candidates for a new generation of telephone lines. In practice, the losses in a fibre are eventually determined by the following factors:

(a) imperfections in the reflecting surfaces;

(b) absorption and scattering in the material of the fibre (glass or fused silica);

(c) losses at the input and output ends of the fibre;

(d) losses caused by the fibre's not being ideally straight, i.e. from bending.

It turns out that (d) is negligible under almost practical conditions, and (c) is tolerable provided that very long single fibres can be manufactured, and few joints are necessary. A tremendous amount of work has been put into (b), and types of glass have been developed for which absorption and scattering (Rayleigh scattering by inhomogeneities in the glass) are exceptionally small at certain wavelengths. It seems that the lowest total losses occur at about 1.3 μm wavelength although more efficient light sources and detectors are available in the region 0.8–1.0 μm, where losses are also tolerably low. And the solution to (a) is found in cladded fibres, where the reflecting surface is inside the fibre and is not exposed to possible damage (Fig. 11.52(*a*)), or in graded-index fibres (Fig. 11.52(*b*)), where reflexion takes place gradually as in a mirage (§ 4.4.2).

But the greatest attraction of optical fibres is in the potential number of 'telephone conversations' which can be transmitted simultaneously on a single fibre. If we suppose that the light transmitted has frequency ω, and that a single conversation covers a band of frequencies of width ω_1 then, by mixing various conversations each with a different intermediate frequency, in principle ω/ω_1 conversations can all be used to modulate the light wave simultaneously. If $\omega = 10^{15}\,\mathrm{s}^{-1}$ and $\omega_1 = 10^5\,\mathrm{s}^{-1}$, this number is 10^{10}. Technology is nowhere near capable of using this enormous potential, but the attraction remains. This estimate has in fact neglected several points of fundamental importance, one of which is dispersion (§ 11.9.6). This reduces the potential severely although it still leaves an impressive residue; at the present state of the technology some thousands of 'conversations' can be transmitted simultaneously on a glass fibre compared with twenty-four on conventional cable.

11.9.6 *Dispersion in fibres.* If we look at Fig. 11.49 again, and consider transmission of a signal along a fibre supporting a number of modes, we see that the time taken for the signal to traverse the fibre depends on the mode, since the total light path along a fibre of length L is $L/\sin\hat{\imath}$. As a result, an incident pulse will be received at the far end at times between $\mu_2 L/c$ and $\mu_2 L/c \sin\hat{\imath}_c = \mu_1 L/c$. This effect represents a spreading of the pulse or, in the terms of Chapter 10, dispersion. It is clearly reduced to a minimum if μ_2 and μ_1 are very close; under such conditions (11.69) suggests that the fibre will only have a single propagating mode. Such a fibre, satisfying

$$\mu_2 k_0 d\{1-(\mu_1/\mu_2)^2\}^{\frac{1}{2}}/\pi < 1 \tag{11.73}$$

is called a 'single-mode fibre', and this is what is necessary for the most demanding communications applications. Nevertheless, dispersion still remains a problem, because the very existence of a modulated signal implies a spectrum of frequencies, and we have seen (or at least mentioned) in § 10.6.1 that the derivative $\partial^2 k/\partial\omega^2$ of the group velocity in a material can also cause pulse distortion.

In the fibre, the value of k relevant to this derivative is the component k_z (see (11.71)) parallel to the axis, and the value of ω is just ck_0. So the dispersion is represented by the derivative:

$$\frac{\partial^2 k_z}{\partial\omega^2} = \frac{1}{c^2}\frac{\partial^2 k_z}{\partial k_0{}^2}. \tag{11.74}$$

Reference to Fig. 11.51 shows that $\partial^2 k_z/\partial k_0{}^2$ is not zero, and so dispersion can still occur in single-mode fibres, and this dispersion

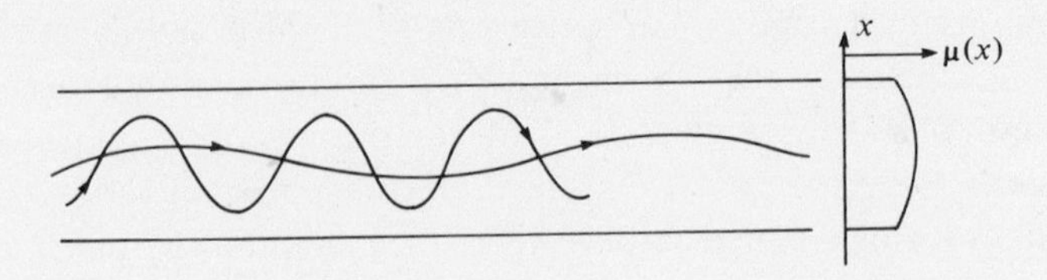

Fig. 11.53. Propagation of two modes in a graded-index fibre.

eventually sets an upper limit to the possible modulation frequency. The problem is much less serious in graded-index fibres, and for this reason they have been developed intensively.

One can see qualitatively the origin of the lesser dispersion in a graded-index fibre from Fig. 11.53. In the figure, the mode with the shorter path length is confined to the region where $\mu(x)$ is largest, while the mode with the longer path length enters regions of smaller refractive index. The two parameters – refractive index and path length – partially compensate, and give a dispersion which is less than in a step-index fibre. The dispersion of the glass itself must also be taken into account; in the normal dispersion region (§ 10.3.2), $\partial^2\omega/\partial k^2$ for glass conveniently has the opposite sign to that for the fibre dispersion, and further compensation is possible.

The subject of fibre optics is comprehensively covered by review articles and books at all levels. At present they are becoming out-of-date very rapidly, but the interested reader should have no difficulty in locating one of the many monographs available (e.g. Gloge, 1979).

APPENDIX I

Geometrical Optics

1 Fermat's principle

The elementary principles of optical instruments can best be considered in terms of rays rather than waves, a ray representing the direction in which a wave travels. That rays can be used is the consequence of Fermat's principle (§ 2.7) that the path of a ray from one point to another is either of maximum or minimum optical length (actual length multiplied by refractive index).

The law of rectilinear propagation in a uniform medium follows directly from Fermat's principle, and so do the laws of reflexion. The second law of refraction can be seen to result from the following reasoning. Let PQ (Fig. I.1) be a plane separating media of refractive indices μ_1

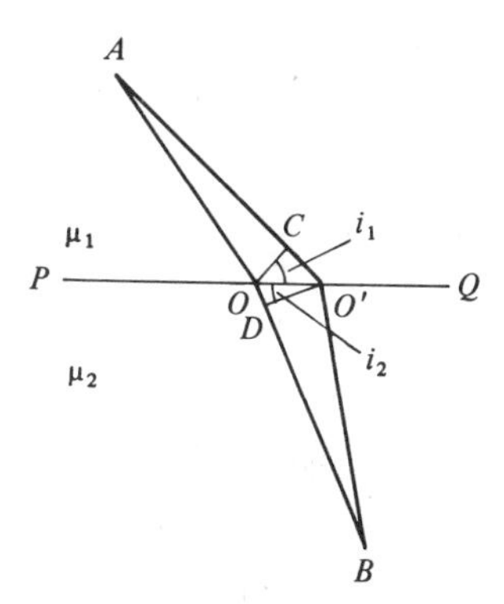

Fig. I.1. Explanation of refraction in terms of Fermat's principle.

and μ_2, and let AOB and $AO'B$ be two possible close paths between two points A and B. The optical length of the path through O is $\mu_1 AO+\mu_2 OB$, and that through O' is $\mu_1 AO'+\mu_2 O'B$. We require to find the condition that these two lengths are equal. They are obviously equal if $\mu_1 CO'=\mu_2 OD$, where OC and $O'D$ are perpendicular to AO' and OB

respectively. But $CO' = OO' \sin i_1$ and $OD = OO' \sin i_2$. Therefore the condition

$$\mu_1 CO' = \mu_2 OD$$

is re-stated as

$$\mu_1 \sin i_1 = \mu_2 \sin i_2,$$

which is the second law of refraction.

2 Sign conventions

It is helpful to have a consistent convention for the use of positive and negative quantities in optics. In this book, we use a simple Cartesian convention, as shown in Fig. I.2. Other conventions – based upon

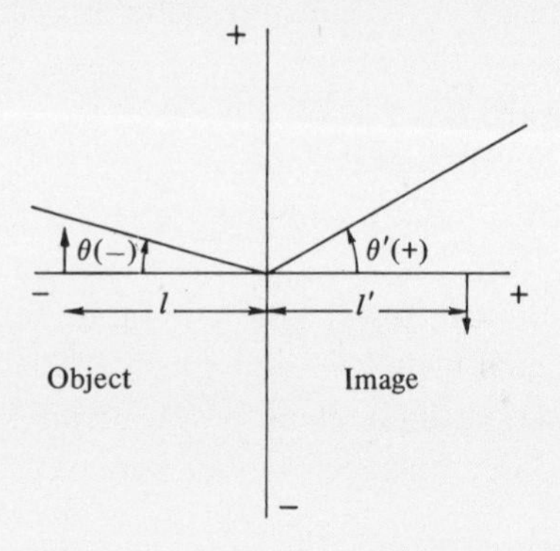

Fig. I.2. Sign convention for distances and angles.

whether real or virtual images or objects are involved – cannot be extended to two dimensions. Although there are no such direct applications in this book, anyone who wishes to extend his knowledge of lens aberrations, for example, will have to use a Cartesian system.

3 Thin lenses and mirrors

An image-forming system can be regarded as a device for equalizing a *limited* group of optical paths. Mirrors make use of the property of an ellipsoid of revolution – that the sum of the paths from a point on the surface to the two foci is constant; all mirrors can be regarded as parts of such an ellipsoid. The parabolic mirror in particular (Appendix III) can be regarded as part of an ellipsoid for which one focus is at infinity.

A lens equalizes optical paths by interposing a longer length of medium of high refractive index in those paths that are shorter. The lens formula

$$\frac{1}{l'} - \frac{1}{l} = (\mu - 1)\left(\frac{1}{r_1} - \frac{1}{r_2}\right)$$

can be proved from this approach. In this equation l and l' are object and image distances, and r_1 and r_2 are the radii of curvature of the two surfaces of the lens.

4 Thick-lens systems

A coaxial lens system is one for which the centres of curvature of all the refracting surfaces lie on a straight line – the axis. When such a system, which may be a single thick lens, has a thickness comparable with the focal length, we can no longer talk of measuring 'from the lens' (Fig. I.3(a)). It can be shown, however, that the thin-lens formula can still be

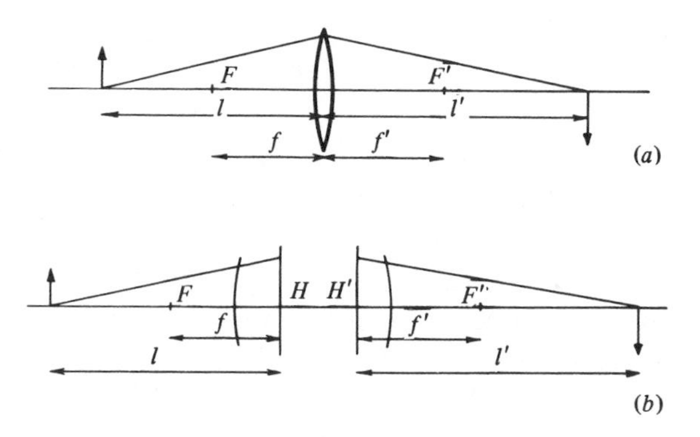

Fig. I.3. Comparison of thin-lens and thick-lens systems, showing significance of principal points H and H'.

used if object distances are measured from one point on the axis, and image distances from another. These two points (H and H' in Fig. I.3(b)) are called *principal points* and the planes through them perpendicular to the axis are called *principal planes.* Principal points are formally defined as the conjugate points for which the magnification is unity. The two focal lengths of the system f and f' are the distances from the principal points to the principal foci F, F'.

The two focal lengths are equal only if the image is formed in the same medium as that containing the object. In general,

$$f/\mu = f'/\mu',$$

where μ and μ' are the refractive indices of the initial and final media.

In thin-lens theory it is often convenient to make use of the concept of the optical centre of a lens; for a thick-lens system this point is replaced by two points, called the nodal points N and N'. These have the property that the ray along any line containing N will emerge along a parallel line

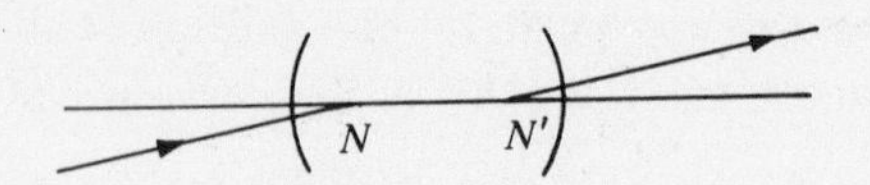

Fig. I.4. Nodal points of a thick-lens system. The incident and emergent rays are parallel.

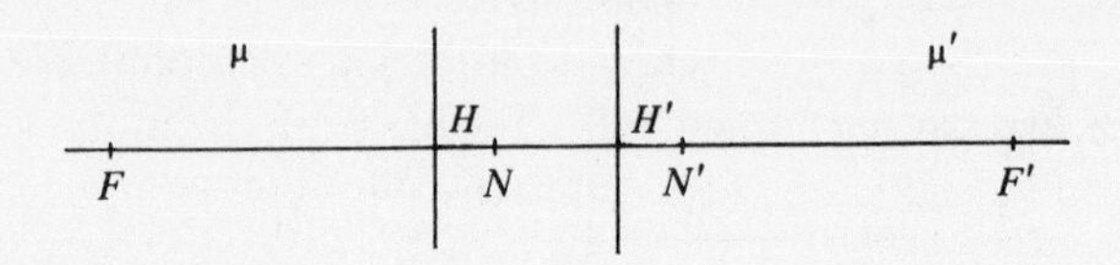

Fig. I.5. Cardinal points of a thick-lens system, F, F' = focal points; N, N' = nodal points; H, H' = principal points. $NH = N'H'$; $FN = H'F'$: $FH = N'F'$; $FH/H'F' = \mu/\mu'$.

containing N' (Fig. I.4), so that the nodal points can be defined as conjugate points for which the angular magnification is unity.

The points H, H', F, F', N, N' (Fig. I.5), are called the cardinal points of the lens system and suffice for solving any problem concerned with paraxial rays. There are certain relationships between them: e.g. $FN = f'$, and $F'N' = f$. Thus the distance between the nodal points equals the distance between the principal points. If the initial and final media are the same, the nodal points coincide with the principal points. Since the nodal points are normally more easily located, they often provide a means of finding the principal points.

5 Determination of cardinal points

To make use of the thick-lens formulae we need to locate the cardinal points. The principal foci can be determined theoretically by treating each refraction separately, using the equation

$$\frac{\mu'}{l'} - \frac{\mu}{l} = \frac{\mu' - \mu}{r},$$

the sign convention described in § 2 above, being used for the signs of l, l' and r. After each refraction the position of the image is used as object position for the next refraction. By taking the initial object and the final image at infinity, the two foci can be found. The principal points and nodal points cannot be located so directly.

It is usually best, however, to locate the cardinal points experimentally. The principal foci F and F' can be found by locating the image of a distant object, or by finding the position of an object that produces an image in the same place when a plane mirror is placed behind the lens system. We can then find a set of pairs of conjugate positions, each pair being an object position and its corresponding image position. If the distances of these points from F and F' are p and p', we can then use Newton's equation

$$pp' = -f^2$$

to find the focal length. If the initial and final media are not the same, the equation is

$$pp' = -ff'$$

and we need to know the two values of μ in order to find the two focal lengths.

Having found the principal foci and the focal lengths, we can find the principal points and the nodal points from the relationships given in § 4 above.

APPENDIX II

Aberrations

No image-forming system can form a perfect image of a finite object. Every image has defects, called aberrations, and the best that can be done in practice is to see that the aberrations that would be most noticed in a particular experimental arrangement are made as small as possible, sometimes at the expense of making other aberrations larger.

The classification of aberrations now usually accepted was first introduced by von Seidel about 1860. It is as follows.

1. Spherical aberration

Lenses and mirrors with spherical surfaces cannot form an accurate image of a point object on the axis; usually the more extreme rays are too greatly deviated (Fig. II.1) resulting in a defect known as spherical aberration. This defect can be removed in several ways.

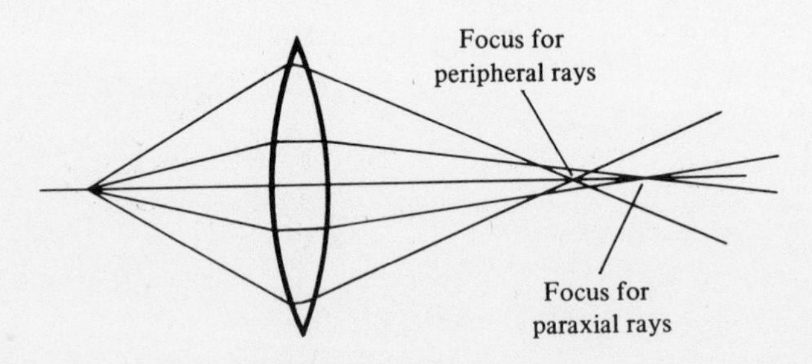

Fig. II.1. Spherical aberration.

(a) The refracting or reflecting surface can be made aspheric and the aberration removed empirically, the Foucault knife-edge test (§ 9.6.2) being used to assess the accuracy of the results. This is usually worth while only for large lenses.

(b) The aberration produced by a converging lens can be corrected by that produced by a diverging lens made from a different glass; the combination is usually in the form of a cemented doublet, and chromatic aberration (§ 6 below) can be eliminated at the same time.

(c) For simple systems use can be made of the rough rule that spherical aberration is reduced if the deviations are spread over several surfaces and is a minimum if the deviations of a particular ray are the same at each surface. Separated combinations – particularly eyepieces (Appendix III, § 3) – make use of this principle.

2 Coma

Even if spherical aberration is eliminated, we may still find that the image of a point slightly off the axis is not perfect. This defect is particularly important in astronomy, since the image of a star may look like a comet – hence the name *coma.* It can be reduced or eliminated in several ways.

(a) Coma can be eliminated by choosing radii of curvature appropriate to a given object position – the so-called 'bending' of the lens (Fig. II.2). This usually also leads to minimum spherical aberration.

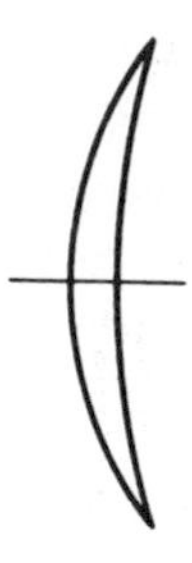

Fig. II.2. A 'bent' lens for eliminating coma.

(b) Coma and spherical aberration are absent in any system that obeys the Abbe sine condition (§ 9.3.2). Such a system is said to be *aplanatic,* and the two conjugate points are called *aplanatic points.* Aplanatic systems form the basis of the most powerful microscope objectives (Appendix III, § 2).

3 Astigmatism

Astigmatism can be regarded as an extreme form of coma; it occurs for very oblique rays when a bundle cannot be considered as even approximately passing through a point. Instead it forms a tapering wedge-shaped pencil, the edge of the wedge (Fig. II.3) forming what is known as a *focal line*: all the rays pass through this line and then diverge; but the taper

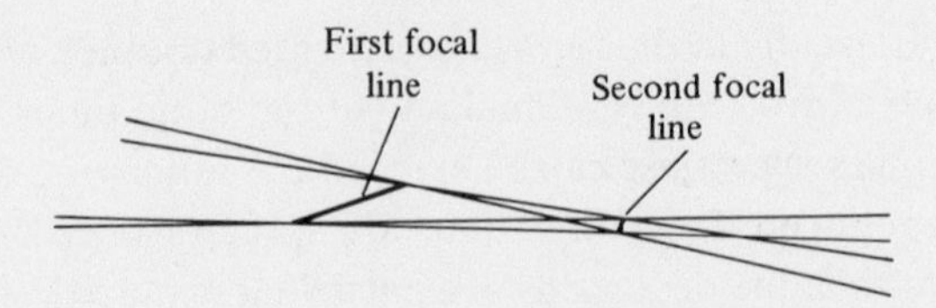

Fig. II.3. Astigmatic pencil, showing formation of two focal lines.

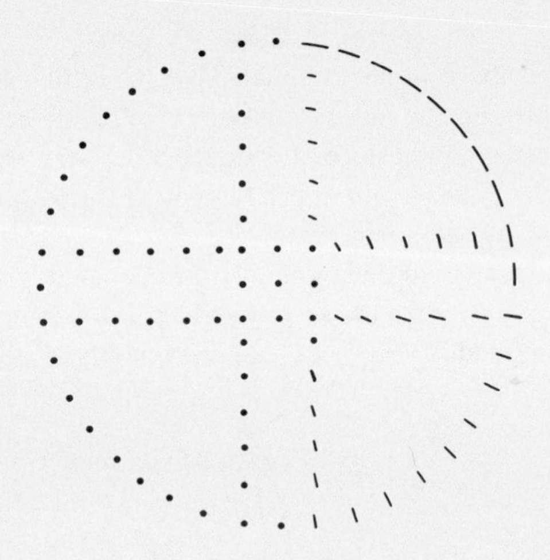

Fig. II.4. Astigmatism. The left-hand side shows the perfect image; the upper right-hand quadrant shows the image in the tangential focal line; the lower right-hand quadrant shows the image in the radial focal line.

continues, to cause the rays to pass through another focal line, perpendicular to the first.

The consequence of this defect is that it is impossible to produce a focused image of the outer parts of a plane object perpendicular to the axis. Radial detail – for example, the spokes of a wheel – may appear to be in focus in one focal line and tangential detail in the other (Fig. II.4). Moreover, the surfaces in which these apparent foci occur may not be even approximately plane, thus leading to the fourth aberration – curvature of field.

4 Curvature of field

The removal of astigmatism can be regarded as causing the two focal-line surfaces to coincide, and the removal of curvature of field as making these surfaces plane. The defects are particularly important in cameras in which

Fig. II.5. Pin-cushion and barrel distortion.

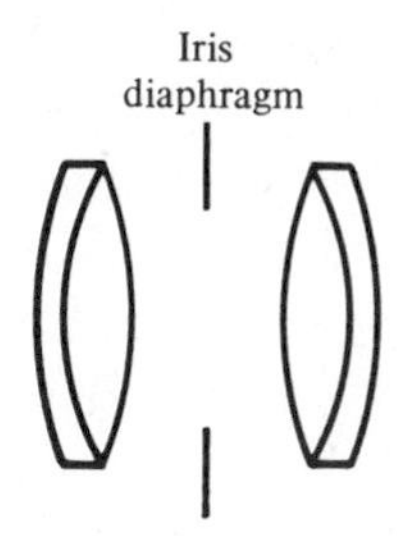

Fig. II.6. A symmetrical camera lens for the elimination of distortion.

the image has to be focused on a flat plate. Their removal, however, is rather difficult, and the principles cannot be simply explained.

A lens which is corrected for astigmatism and curvature of field is called an *anastigmat.*

5 Distortion

An image in which the previous aberrations have been corrected is still not necessarily a perfect representation of the object; if the magnification varies with the obliquity of the rays then the shape of the object may not be reproduced and the image is said to be distorted; if the magnification increases with obliquity the image of a square will have the shape of a pin-cushion and if it decreases with obliquity the image will be barrel-shaped (Fig. II.5).

It can be seen quite easily that the effect cannot exist for a symmetrical system forming a real image of the same size as the object ($l = -l'$). For since paths of rays are reversible, object and image can be interchanged and thus one cannot be larger than the other. Although this reasoning does not apply to other object distances, symmetrical lens systems have the property of producing little distortion for *any* object position. Most good camera lenses are therefore symmetrical, with the iris diaphragm in the plane of symmetry (Fig. II.6). Such a lens is said to be *orthoscopic* or *rectilinear.*

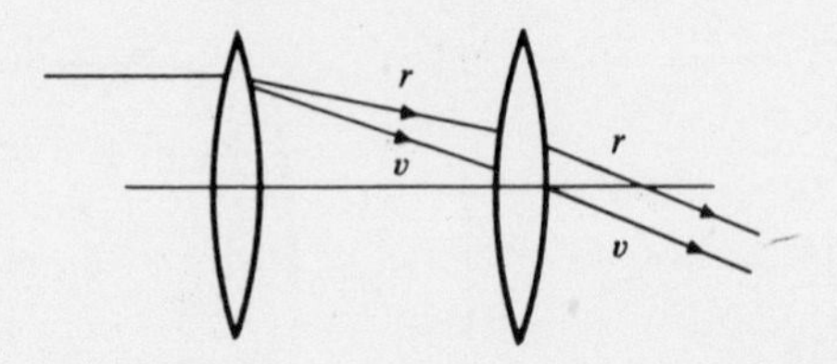

Fig. II.7. Corrections for lateral chromatic aberration by separated doublet. The red ray, r, is deviated less than the violet ray, v, at the first lens, but is deviated more at the second.

6 Chromatic aberration

Because the refractive index of glass is a function of wavelength, an image produced in white light usually has coloured edges. This defect is known as chromatic aberration.

A rough correction can be made by using two displaced lenses (Fig. II.7). The violet ray is more deviated than the red and so reaches the second lens nearer to the axis than the red ray; it is therefore less deviated by the second lens. The deviations can be arranged to combine in such a way that the red and violet rays emerge parallel to each other.

A more precise correction can be achieved by cemented doublets of two different glasses; such a lens is known as an achromatic doublet. The two glasses must have different *dispersive powers*, ω, which is defined as

$$\omega = \frac{\mu_1 - \mu_2}{\mu - 1};$$

μ_1 and μ_2 are refractive indices for two different wavelengths λ_1 and λ_2, and μ is their mean. If the two glasses are characterized by primed and unprimed symbols respectively, it is shown in elementary textbooks that the combination of two lenses is achromatic if

$$\frac{\omega}{f} + \frac{\omega'}{f'} = 0.$$

Such combinations work well in practice, but there is always some residual error resulting from the fact that ω is not strictly a constant for a given material; it depends upon the particular wavelength λ_1 and λ_2 chosen to define it. For a better correction a combination of three glasses can be made; this is called an *apochromat.* Such lenses are found to be perfectly satisfactory for even the most exacting work.

A thorough treatise on aberrations would have to deal with the chromatic variation of all the other errors. We shall merely state here that, for thick-lens systems, there are two kinds of chromatic aberration – axial (or longitudinal) and lateral. For the former the images formed by the different colours are the same size, but are not in the same plane; for the latter they are in the same plane, but are not of the same size. The system shown in Fig. II.7 obviously suffers from axial chromatic aberration, but not from lateral. It is not possible to correct for both these aberrations simultaneously except by making the lens system entirely of cemented doublets.

7 Summary

A lens system cannot be produced for which all aberrations are corrected. For any particular use, one must decide which errors are important and concentrate on eliminating these at the expense of possibly making the others worse. For example, the diffractometer described in Appendix IV has to form a point image of a pinhole on the axis; therefore spherical aberration is the only error that is important and is eliminated by using corrected doublets. Microscopes for measuring cosmic-ray tracks, however, must produce an image free from distortion and astigmatism, which might make the tracks difficult to measure accurately. It can be seen, therefore, that, although the design of lens systems is largely a technical matter, physicists must be aware of the problems involved in order that they can specify their requirements for particular researches.

APPENDIX III

Some important optical components

1 Telescope objectives

The simplest telescope objective consists of an achromatic combination (Appendix II, § 6), which can also be corrected for spherical aberration (Appendix II, § 1). For bigger telescopes, it becomes easier to use mirrors, and all large telescopes now have such objectives, usually of paraboloidal shape (Appendix II, § 1).

It is a property of a paraboloid that all rays parallel to the axis pass, after reflexion, precisely through the focus; i.e. spherical aberration is absent. But rays not parallel to the axis do not pass through a single point after reflexion; i.e. coma (Appendix II, § 2) is present. If, however, the mirror were spherical all directions would be equivalent, since they are all parallel to radii. A system that takes advantage of both these properties was devised by Schmidt in 1932 and is thus free from both spherical aberration and coma.

The Schmidt system is based upon a spherical mirror, combined with a transmitting plate that has the form of the difference between the paraboloid and the sphere (Fig. III.1). If the plate were in contact with the spherical mirror the two would be equivalent merely to a paraboloidal mirror. If the plate is placed at any other position it maintains this property of correcting the spherical mirror; if it is placed with its centre at the centre of curvature of the mirror it also has the property of treating alike (apart from an obliquity factor) beams of all directions (Fig. III.2). The system is therefore free from coma.

2 Microscope objectives

Low-power microscope objectives pose no great problem. For high powers, producing a limit of resolution near to the theoretical maximum (§ 9.4.1), a special system based upon the aplanatic points (Appendix II, § 2) of a sphere is universally used.

It can easily be shown (Fig. III.3) that rays diverging from a point distant R/μ from the centre of a glass sphere of radius R will, after

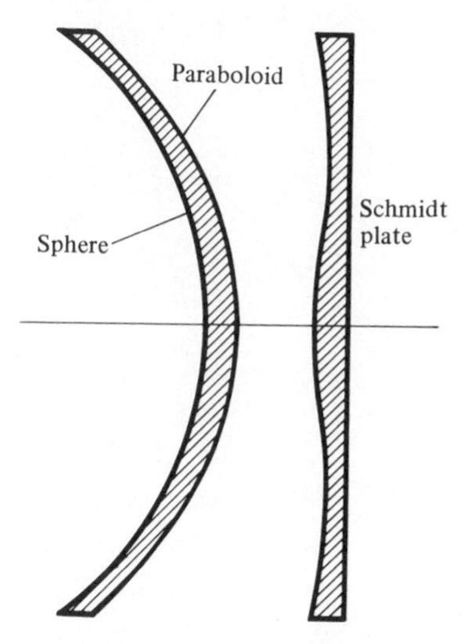

Fig. III.1. The Schmidt correcting plate as the difference between a paraboloid and a sphere.

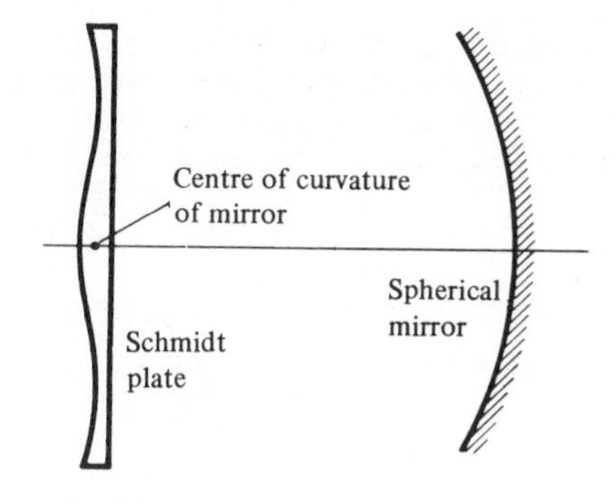

Fig. III.2. Schmidt objective system.

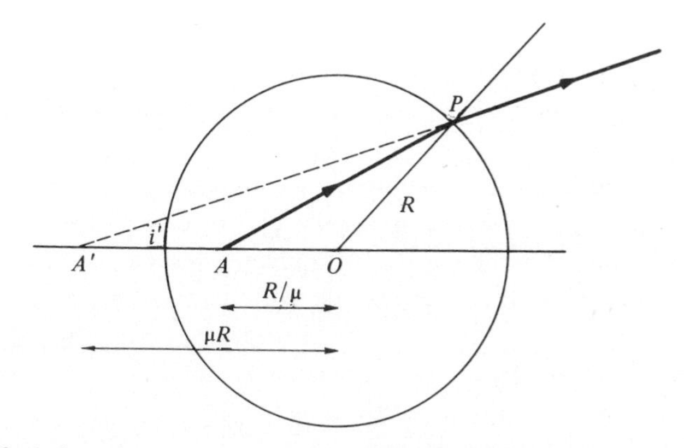

Fig. III.3. Aplanatic points of a sphere. Triangles AOP and POA' are similar.

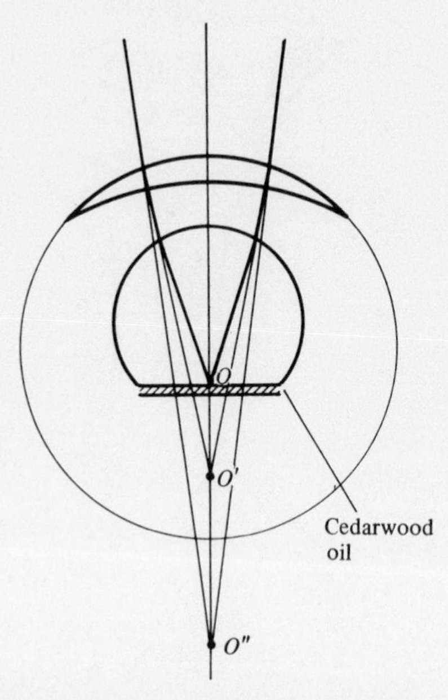

Fig. III.4. Microscopic objective with two stages, each making use of the aplanatic points of a sphere. O is inner aplanatic point of the small sphere; O' is outer aplanatic point. O' is also the centre of curvature of the lower surface of the upper lens and the inner aplanatic point of its upper surface. O'' is the outer aplanatic point of the upper surface.

refraction, appear to be diverging from a point distant μR from the centre; since no approximations are involved, the result is correct for all angles. A beam with a semi-angle of, say, 64° (sin 64° = 0.90) will thus emerge as a beam with a semi-angle of 37° (sin 37° = 0.60), if $\mu = 1.50$.

This property of a sphere can be used in two ways. First, the sphere can be cut by a section passing through its internal aplanatic point; the specimen is placed near this point and immersed in a liquid (cedarwood oil is used) of the same refractive index as the glass (Fig. III.4). The system is known as *oil-immersion* and is used universally for microscopes of the highest resolution.

The second way in which the principle can be used involves putting the object at the centre of curvature of the first face of a lens, making this point the inner aplanatic point of the second surface; all the deviation then occurs at the second surface and the image is formed at the outer aplanatic point. The system illustrated in Fig. III.4 uses both applications of the aplanatic principle in two successive stages. The semi-angle of the emergent beam is then reduced from 37° to 24°.

The freedom from coma of this system can be seen from the fact that *all* points at distance R/μ from the centre of the sphere are aplanatic points. Thus, if we ignore the curvature of the surface on which these points lie, we can say that all points on a plane object will form a plane image that is free from spherical aberration and coma.

3 Eyepieces

Although there is little noteworthy about eyepieces, they are used so frequently in scientific work that their principles should be well understood. It is quite remarkable that these principles should have been worked out so soon after the invention of the microscope; Huygens – who has given his name to a prototype eyepiece – lived at the time when the microscope was still quite new. Eyepieces consisting of pairs of lenses are used rather than single lenses because they can be better corrected for both spherical aberration and chromatic aberration. In addition, they can also carry other devices such as cross-wires and scales that can be put in focus with the final image. The lens nearer the eye is called the eye lens and the other is called the field lens.

There are two useful and easily proved rules to remember in dealing with simple lens systems: first, the deviation of a ray incident on a lens at a distance y from the axis is y/f; and, secondly, the chromatic aberration (Appendix II, § 6) of a single lens – the distance between the foci for two different wavelengths – is ωf, where ω is the dispersive power with respect to the same two wavelengths.

The equation for the focal length f of a combination of two lenses separated by a distance a is

$$\frac{1}{f}=\frac{1}{f_1}+\frac{1}{f_2}-\frac{a}{f_1 f_2}.$$

If f_1 and f_2 have a chromatic variation, then

$$\begin{aligned}\delta\left(\frac{1}{f}\right) &= \left(1-\frac{a}{f_2}\right)\delta\left(\frac{1}{f_1}\right)+\left(1-\frac{a}{f_1}\right)\delta\left(\frac{1}{f_2}\right)\\ &= \left(1-\frac{a}{f_2}\right)\frac{\omega f_1}{f_1^2}+\left(1-\frac{a}{f_1}\right)\frac{\omega f_2}{f_2^2}\\ &= \left(1-\frac{a}{f_2}\right)\frac{\omega}{f_1}+\left(1-\frac{a}{f_1}\right)\frac{\omega}{f_2}.\end{aligned}$$

This is zero if $f_1+f_2=2a$, whatever the value of ω. A combination with this property will be free from axial but not lateral (Appendix II, § 6) chromatic aberration.

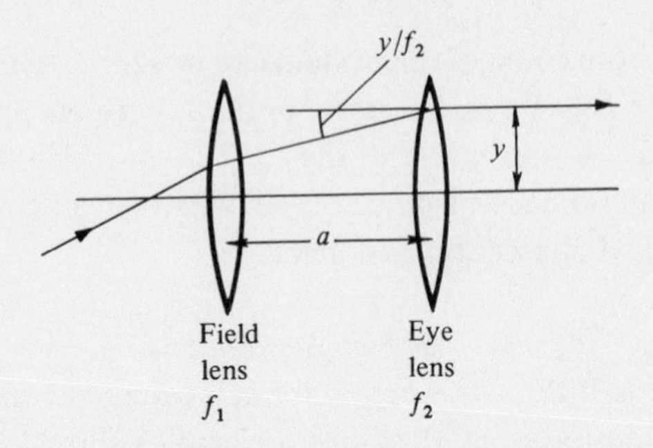

Fig. III.5. Principle of an eyepiece.

To correct for spherical aberration, we try to equalize the deviations produced by the two lenses. The equality can be only approximate, since rays are passing through at all angles, but we take as typical a ray emerging parallel to, and at a distance y from, the axis (Fig. III.5). This must be inclined at an angle y/f_2 between the lenses and must therefore be incident at the first lens at a distance from the axis $y-a(y/f_2)$; the deviation is therefore

$$\frac{y-a(y/f_2)}{f_1}.$$

Equating the two deviations, we have

$$\frac{y-(ay/f_2)}{f_1}=\frac{y}{f_2},$$

or

$$f_2-f_1=a.$$

If we combine this condition with the previous one, $f_1+f_2=2a$, we arrive at the result

$$f_1=a/2, \qquad f_2=3a/2.$$

This specifies the Huygens eyepiece.

Although it is theoretically the best combination, it has the disadvantage that the focus is between the lenses and so images of cross-wires or scales are not corrected. A more usual combination is the Ramsden eyepiece, for which

$$f_1=f_2=3a/2,$$

which does not have this disadvantage, and which still gives a reasonable performance.

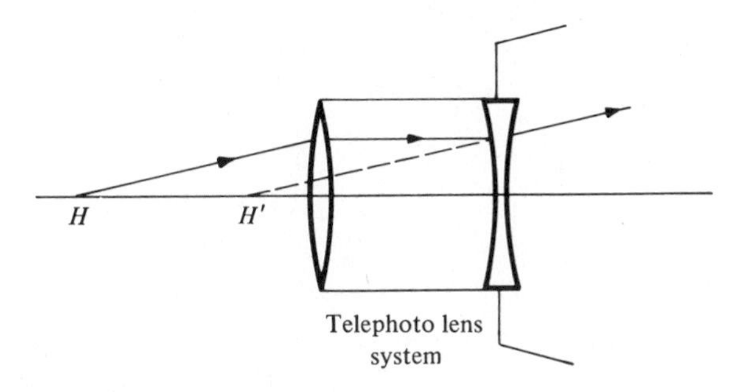

Fig. III.6. Positions of principal points for a telephoto lens.

4 Camera lenses

Camera lenses can vary from simple 'bent' lenses (Appendix II, § 2) with displaced stops in cheap cameras, to very complicated symmetric combinations (Appendix II, § 5) in expensive cameras. Only two special types will be discussed here, since they introduce some simple physical principles.

The first type is the telephoto lens, used for taking photographs of distant objects; a large focal length is needed if the image is to be of reasonable size, but it would be inconvenient for the camera to be correspondingly long. Therefore a separated combination of a converging lens and diverging lens is used, with principal planes (Appendix I, § 4) well outside the combination (Fig. III.6).

The second type is the 'zoom lens', which has resulted from the requirements, particularly for television, of a lens that would give a focused image while the magnification is changed. In principle this requirement is not difficult to meet; we need a separated combination for which the separation can be varied, and at the same time the lenses must be displaced in such a way that the second focal plane (Appendix I, § 4) remains fixed.

Let us consider a telephoto lens (Fig. III.6) for which the two lenses have equal and opposite focal lengths, f_1. The first principal point H must be at a distance of f_1 from the first lens, since the intermediate ray is then parallel to the axis, and the deviation at the second lens is therefore equal to that at the first; H and H' are thus nodal points and, since there is no change in medium, they are also principal points. The focal length f of the combination is, from the first equation in § 3 above, f_1^2/a. We can thus draw a diagram (Fig. III.7) showing the positions of the two lenses for various separations, in order to give a fixed second focal plane. All we

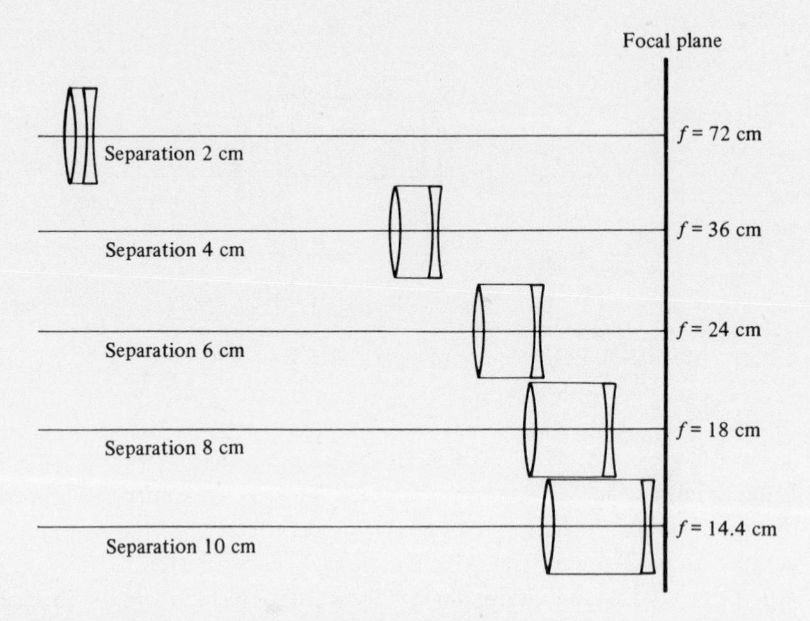

Fig. III.7. Simple zoom lens made from two components of equal and opposite focal length (12 cm).

now require is a mechanical system that, by means of one adjustment, can move the two lenses as required.

Zoom lenses are, of course, much more complicated than this, as the various aberrations require to be corrected, and allowance must be made for the fact that the object is not always at infinity.

APPENDIX IV

The optical diffractometer

The optical diffractometer was originally called the X-ray microscope by W. L. Bragg, who used it in 1939 to carry out the experiments described in § 9.7.2. Later workers found that it would serve other purposes as well in the study of the X-ray diffraction patterns of crystals, and a more precise instrument was built in 1949 in the Physics Department of the University of Manchester Institute of Science and Technology. It was then realized that it formed an ideal instrument for teaching physical optics; Fraunhofer diffraction patterns of quite coarse objects – even several centimetres – could be readily observed, and students could carry out experiments that they usually experience only in terms of illustrations in textbooks.

The instrument is in constant use in Manchester, and its existence is the main reason why we were prompted to write this textbook. We strongly recommend it as a teaching instrument and are therefore including details of it here; a more complete description is given by Taylor & Lipson (1964).

1 Construction

The instrument is essentially a spectrometer; as shown in Fig. 6.3 (p. 132), the top part is a collimator – using a pinhole, B, in place of a slit – and the bottom part is a telescope, with a microscope in place of an eyepiece. The plane mirror, E, at the bottom merely serves to direct the light into a convenient direction for viewing; in addition, it enables the viewer to be close enough to the lenses to allow him to manipulate the diffracting mask if he wishes, as described in § 4 below.

The instrument is basically simple, but since it is to be used to the limit of its potentialities several precautions must be taken in constructing it and in using it. First, the lenses, C and D, must be accurately made and corrected for spherical aberration (Appendix II, § 1); none of the other

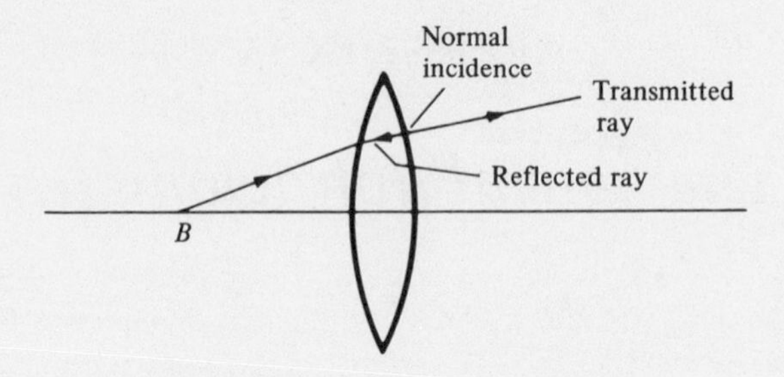

Fig. IV.1. The Boys point, *B*, of a lens.

aberrations is important. The lenses must include as few defects as possible; no glass can be made completely free from defects, but the odd bubble or foreign particle that could be tolerated even for very good ordinary lenses cannot be accepted for this purpose.

Secondly, the instrument must be very rigid. The basic component is a steel girder and the framework for supporting the optical apparatus is also made of steel; different metals may lead to variations with temperature because of different coefficients of expansion. To insulate the instrument from vibration of the building in which it is housed, it is supported on the walls by means of flexible rubber mounts.

Thirdly, the instrument must be accurately adjusted. One cannot depend upon the precise setting of a lens in its mount, and methods – to be described briefly in § 2 below – have been devised to be independent of any initial assumptions.

2 Adjustment

The requirements of the diffractometer are few: the lenses must be accurately coaxial, and the pinhole must lie accurately upon this axis; the viewing telescope must also be coaxial and the axis must also pass approximately through the centre of the plane mirror, *E*, at the bottom. The complete adjustment of a diffractometer is an excellent example of the application of geometrical optics; it is described in full by Taylor & Lipson (1964) and only a basic outline will be given here.

First, how does one find experimentally the axis of a lens? One of the simplest ways is to locate the *Boys points* – the points that are self-conjugate for reflexion from the back surface of the lens (Fig. IV.1), and which can be used, as shown in elementary textbooks, to determine the radius of curvature of the surface. The line joining the two Boys points – one on each side of the lens – is the axis of the lens. The method of adjusting the diffractometer does not make use of the Boys points themselves, but it uses the principle of these reflexions, which can be quite numerous if the lens is compound and if the source is bright enough for multiple reflexions to be seen.

The lower lens, *D*, is set first and its axis – which must be set roughly in the right position – then defines the axis of the whole system. When the second lens, *C*, is inserted more multiple reflexions are possible and help in setting the two axes coincident. The pinhole is then placed on this axis. The microscope can then be set coaxially by observing the reflexions from the various components of the objective on a translucent plate, *F*, placed in the focal plane of the lower lens. (This plate can be replaced by a maximum-resolution photographic film for recording the diffraction patterns.)

Some of these operations are also tests of the components themselves; for example, if the two components of a doublet are not mounted coaxially, it will not be possible to line up the various multiple reflexions.

Finally, it is necessary to focus the microscope accurately. For many simple purposes it is merely necessary to form as small an image as possible of the illuminated pinhole, but for precise purposes, particularly for photographing patterns with fine detail, a less subjective procedure is needed. One method is to form the diffraction pattern of the largest circular aperture that will give a clear diffraction pattern (§ 7.3.5); if the telescope is then raised or lowered, a succession of maxima and minima at the centre of the diffraction pattern will be observed, and the correct plane of focus is that which lies half-way between the two minima on either side of the approximate focus. In this way, the focal plane can be fixed to a fraction of a millimetre. (Anyone who has used an ordinary optical bench may find it difficult to believe that the focus of a lens of 150 cm focal length can be found so accurately; but such sceptics should verify for themselves that an error of 0.01 cm produces a difference of about $\frac{1}{5}\lambda$ between the central and marginal rays for a lens of 10 cm diameter, and this difference should begin to produce observable effects.)

3 Illumination

It is important that the light falling on the pinhole at *B* should be as intense as possible. When the diffractometer was first devised, the best illumination available was, as shown in Fig. 6.3 (p. 132), a compact-source mercury-vapour lamp with a large condenser lens. The invention of the laser provided an alternative intense source and this is now considered the most convenient form of illumination; the parallel beam from the laser is focused with a lens onto the pinhole.

Table IV.1

Diameter of mask (cm)	Size of pinhole (cm)
6	5×10^{-4}
3	10×10^{-4}
1	30×10^{-4}
0.3	100×10^{-4}

When the mercury-vapour lamp is used, it is necessary to select a single emission line by means of a gelatine or interference filter; the yellow line ($\lambda = 5970$ Å) is recommended, particularly if the patterns are to be photographed. The laser is, of course, monochromatic and no filter is needed.

It is found in practice that the extreme monochromaticity of the laser, in contrast with a spectral line from the mercury lamp, has accompanying disadvantages. Multiple reflexions from the lenses C and D produce unfocused waves which interfere in the plane of the diffraction pattern and which produce unwanted rings which are quite difficult to eliminate. Since long path differences are responsible for these effects, they do not occur with the broader band source. Such rings can be seen in Fig. 9.12, for example.

The size of pinhole is important. It is true that the smaller the pinhole the higher the degree of spatial coherence (§ 8.6), but it is unwise to have a pinhole so small that the patterns can be seen only with difficulty. It is best to use a pinhole that gives about 90 per cent coherence over the mask; thus the larger the mask the smaller the pinhole needs to be. Table IV.1 gives a rough guide of pinhole sizes for masks of given dimensions, for lenses of focal length 150 cm.

4 Some possible experiments

It is easily possible to make standard masks out of thin cardboard: old unused X-ray films are particularly suitable because they cut cleanly. For real permanency masks should be made out of sheet brass.

Static demonstrations, however, are never as inspiring as dynamic ones, and the following paragraphs give some suggestions for masks that can be varied as they are viewed by the observer (§ 1 above).

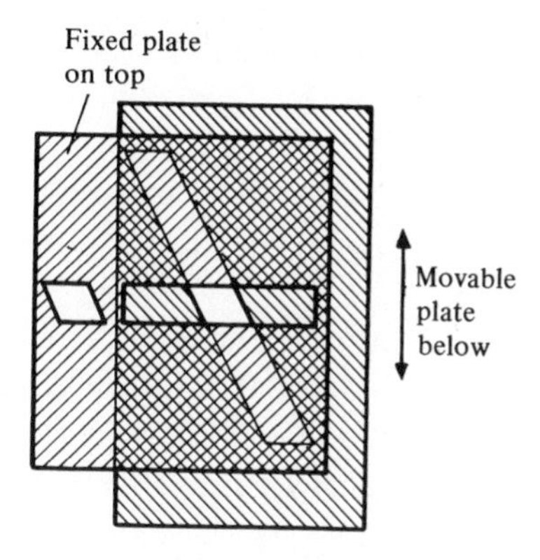

Fig. IV.2. Device for showing fringes from two similar apertures of variable separation.

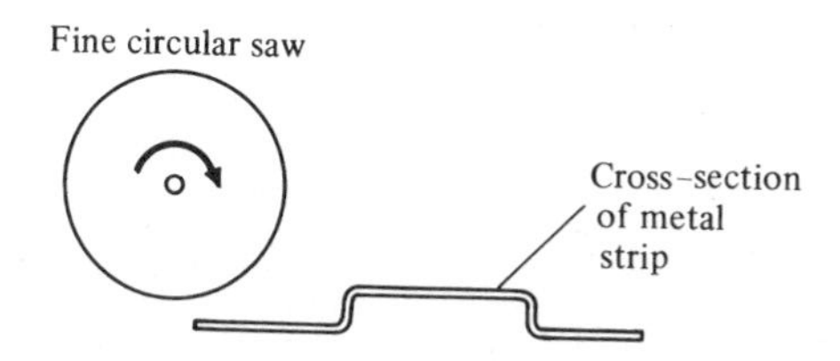

Fig. IV.3. Method of cutting slits for diffraction gratings.

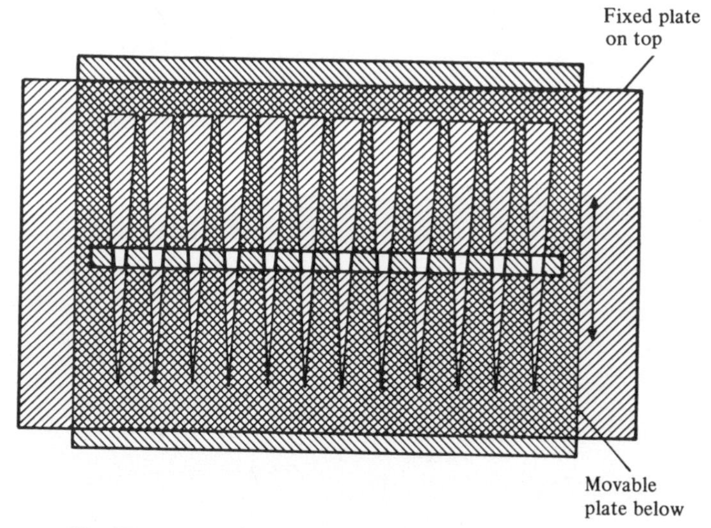

Fig. IV.4. Diffraction grating with slits of variable width.

(a) The diffraction patterns of two holes. Two holes about 5 mm in diameter and 2 cm apart give clear fringes, which disappear when one of the holes is covered.

(b) Variable fringes. The variation of the fringe separation with the separation of two similar apertures can be illustrated by the device shown in Fig. IV.2.

(c) Principal and subsidiary maxima for a diffracting grating. An accurate grating with, say, six lines about 0.2 mm wide and 1 mm apart can be produced by means of a fine slitting saw on a milling machine: a raised piece of metal (Fig. IV.3) is very suitable for this purpose. A sliding shutter can be caused to cover up the slits one by one and the changes in the secondary maxima (§ 7.4.2) noted.

(d) Effect of slit width of a diffraction grating. The effect of changing the width of the slit of a grating, keeping the spacing constant, can be shown by the type of device shown in Fig. IV.4.

No doubt other ideas will arise in the mind of the keen teacher of optics or, better still, in the minds of the students themselves. A liberal supply of cards and of unused razor blades can lead to a considerable increase of understanding of the basic principles of Fraunhofer diffraction.

APPENDIX V

Practical Fourier computations

We have emphasized the importance of Fourier theory in optics by devoting a complete chapter (3) to it. Nevertheless, although it is necessary to be able to cope with the basic mathematics, the number of problems that can be solved purely by mathematical methods is severely limited; if one wishes to deal with general problems one must be able to use numerical methods. This appendix is therefore devoted to a brief description of two practical methods of carrying out Fourier analysis numerically; one method is manual, and the second is designed for computer calculation.

1 Fourier strips

The first practical method of Fourier analysis of an arbitrary one-dimensional or two-dimensional function was introduced in 1934 by Beevers and Lipson. The method is based on a set of 'Fourier strips' which are card strips which record the values of $A \cos nx$ and $A \sin nx$ at intervals of 3° in the range $0° < x < 90°$. The set of cards covers values of A in units between -100 and $+100$, and in hundreds from -900 to $+900$; the values of n are from 0 to 30. Two typical strips are shown in Fig. V.1. Fourier synthesis or Fourier analysis is carried out for a periodic function by choosing the strips necessary to represent the integrals or sums of the form

$$a(k) = \int_0^1 f(u) \cos 2\pi uk \, du \approx c \sum_{m=0}^{30} f(3m°) \cos (3mk°), \qquad \text{(V.1)}$$

$$b(k) = \int_0^1 f(u) \sin 2\pi uk \, du \approx c \sum_{m=0}^{30} f(3m°) \sin (3mk°). \qquad \text{(V.2)}$$

The strips are laid out one under the other and the columns added mentally or with a calculating machine. Considerable work can often be saved by employing any symmetry properties of the function; for example, if the origin can be chosen so that the function is even, $b(k)$ is

76 $C7$	76	56	8	$\overline{45}$	$\overline{74}$	$\overline{66}$	$\overline{23}$	31	69	72	38	$\overline{16}$	$\overline{61}$	$\overline{76}$	$\overline{51}$	0
19 $S6$	0	11	18	18	11	0	$\overline{11}$	$\overline{18}$	$\overline{18}$	$\overline{11}$	0	11	18	18	11	0

Fig. V.1. Two Fourier strips. The first figure is the amplitude; C and S represent cos and sin. The next figure indicates the harmonic and the entries give the values of the ordinates for the first quarter of a period at intervals of $2\pi/60$. (From Lipson & Cochran, 1966, Fig. 90(i).)

automatically zero. For many years this method was the mainstay of Fourier analysis and synthesis in crystal-structure work. However, the ubiquity of digital computers has now made it obsolete, although it is still the fastest method for one-dimensional work.

2 The fast Fourier transform

The Fourier-strip method is called a 'direct Fourier transform' in the sense that it is based on a direct evaluation of the sums (V.1) and (V.2). Obviously a digital computation can replace the manual calculation and save a great deal of effort, but still the number of mathematical operations involved is large and consequently the price of a direct Fourier analysis is high. The scene was essentially revolutionized in 1965 with the invention of an algorithm called the 'Fast Fourier transform' by Cooley and Tukey, which reduced drastically the number of calculation steps involved in a Fourier analysis. This method has made Fourier analysis a practical possibility for very detailed functions defined at thousands of points in a single period, and in particular has brought the analysis of two-dimensional functions defined on a fine array (of say 1000×1000 points) within the bounds of reasonable computation time.

Although the method is available to anyone having access to a computer with a sub-routine library without his having to know how the algorithm works, we have tried to answer in this book the good physicist's need to understand the basic principles of his everyday tools. So we shall give a very brief account of the basic fast-Fourier-transform method so that the reader can see why it is so efficient in computation time.

The basic method is applied to a function defined at $N = 2^{\gamma}$ points in a cycle. It can be extended to other numbers, but the principle is the same. The transform (V.1) and (V.2) can be written in complex notation:

$$a(k) = \sum_{n=0}^{N-1} f(n) \exp(2\pi \mathrm{i} nk/N) = \sum_{n=0}^{N-1} f(n) W^{nk}, \qquad \text{(V.3)}$$

where $W = \exp(2\pi \mathrm{i}/N)$.

The straightforward evaluation of $a(k)$ from (V.3) would be carried out as follows. First, the various values of W^j are calculated and stored. This is not as much work as it might at first seem, because for $N=2^\gamma$ many of the values are related; for example:

$$W^j = W^{j \,\mathrm{modulo}(N)}; \qquad W^{j+N/2} = -W^j; \qquad W^{N/4-j} = \mathrm{i}(W^j)^*.$$

If $N=16$, there are only two distinctly different pairs of $(\cos 2\pi j/N + \mathrm{i}\sin 2\pi j/N)$ to be calculated.

After this step, the values of $a(k)$ from $k=0$ to $k=N-1$ would be calculated. Each one involves $N-1$ complex multiplications and $N-1$ complex additions. The computation time is mainly determined by multiplications, of which $(N-1)^2$ are necessary.

The fast Fourier transform is based on factorizing the calculation process. We shall simply illustrate it by writing down the equations for $N=8$, and reducing the number of calculation steps to a minimum. (It would be quicker to illustrate the method for $N=4$, but it then looks almost trivial!) The method can also be described rather compactly as a product of γ $(N\times N)$ matrices, but the meaning is the same. For $N=8$ we have:

$$W^0=1, \qquad W^4=-1, \qquad W^6=-W^2, \qquad W^{10}=W^2, \quad \text{etc.}$$

We first write the equations (V.3) in the following way:

$$\begin{aligned}
a(0) &= f_0+f_1+f_2+f_3+f_4+f_5+f_6+f_7\\
&= \{(f_0+f_4)+(f_2+f_6)\}+\{(f_1+f_5)+(f_3+f_7)\};\\
a(4) &= f_0+W^4f_1+W^8f_2+\ \dots\\
&= f_0-f_1+f_2-f_3+f_4\dots\\
&= \{(f_0+f_4)+(f_2+f_6)\}-\{(f_1+f_5)+(f_3+f_7)\};\\
a(2) &= f_0+W^2f_1+W^4f_2+\ \dots\\
&= f_0+W^2f_1-f_2-W^2f_3+f_4+W^2f_5-\ \dots\\
&= (f_0+f_4)-(f_2+f_6)+W^2\{(f_1+f_5)-(f_3+f_7)\};\\
a(6) &= f_0+W^6f_1+W^{12}f_2+\ \dots\\
&= f_0-W^2f_1-f_2+W^2f_3+f_4-\ \dots\\
&= (f_0+f_4)-(f_2+f_6)-W^2\{(f_1+f_5)-(f_3+f_7)\}.
\end{aligned}$$

To calculate the above coefficients efficiently, it is clear that we first evaluate the pairs (f_0+f_4), (f_1+f_5), etc. These are simply additions. From these pairs we next calculate the quartets $\{(f_0+f_4)\pm(f_2+f_6)\}$, $\{(f_1+f_5)\pm(f_3+f_7)\}$. If we multiply *one* of the quartets, $\{(f_1+f_5)-(f_3+f_7)\}$, by W^2

we have the ingredients of all the above equations; $a(k)$ can therefore be calculated for all the even values of k with *one complex multiplication* only.

Continuing, we write down in a similar manner for the odd values of k:

$$a(1) = \{(f_0 - f_4) + W^2(f_2 - f_6)\} + W\{(f_1 - f_5) + W^2(f_3 - f_7)\};$$
$$a(5) = \{(f_0 - f_4) + W^2(f_2 - f_6)\} - W\{(f_1 - f_5) + W^2(f_3 - f_7)\};$$
$$a(3) = \{(f_0 - f_4) - W^2(f_2 - f_6)\} + W^3\{(f_1 - f_5) - W^2(f_3 - f_7)\};$$
$$a(7) = \{(f_0 - f_4) - W^2(f_2 - f_6)\} - W^3\{(f_1 - f_5) - W^2(f_3 - f_7)\}.$$

To calculate the components $a(k)$ with k odd we need to evaluate the pairs $(f_0 - f_4)$, $(f_1 - f_5)$; $W^2(f_2 - f_6)$, $W^2(f_3 - f_7)$, and then the quartets $\{(f_0 - f_4) \pm W^2(f_2 - f_6)\}$, $\{(f_1 - f_5) \pm W^2(f_3 - f_7)\}$. Now the latter two pairs involve multiplication by W^2; thus two complex multiplications have to be performed. Finally the evaluation of the four values of $a(k)$ from the quartets involves two further multiplications, by W and W^3 respectively. Thus the odd-k components require another four multiplications. In all, *five* multiplications are needed to evaluate all eight components. This number should be compared to $(N-1)^2 = 49$ multiplications by the direct method.

In general one can show that rather less than $\gamma N/2$ complex multiplications are necessary to evaluate the transform. It is this saving, which amounts to a factor of $2N/\gamma$ compared with the direct method, which has made high-resolution Fourier computation a practical method in many areas of physics. The fast Fourier transform is described in some detail by Brigham (1974).

APPENDIX VI

The Michelson–Morley experiment

No textbook on optics would be complete without an account of the Michelson–Morley experiment which was carried out around 1887. Since the theme of Chapter 7 would have been interrupted if the account had followed the section (7.8.2) on the Michelson interferometer, we have decided to devote an appendix to it.

The idea of the experiment arose because Michelson realized that his interferometer was versatile enough to make measurements that had never before been possible. He was concerned by the fact that to explain the aberration of light – the apparent change of the direction of light from a star that occurs because the Earth is in motion around the Sun – Fresnel had had to assume that the aether must be at rest as an opaque body moves through it. He therefore set himself the task of measuring the velocity of the Earth with respect to the aether.

The problem had been considered earlier and it had been concluded that it was not possible to construct a piece of apparatus sensitive enough to make the necessary measurement. Michelson showed that, if the velocity of the Earth could be considered as the velocity in its orbit (he had to start with some assumption), his interferometer could make the measurement with reasonable certainty.

The difficulty is that the effect to be measured is a second-order one: the velocity of light can be found only by measuring the time taken for a light signal to return to its starting point; the difference between the time for a journey *up and down* the path of the Earth and that *across* the paths – to take the two extremes – is a second-order quantity derived as follows.

According to classical physics, the time t_1 for the up-and-down journey of a path L is

$$
\begin{aligned}
t_1 &= \frac{L}{c+v} + \frac{l}{c-v} \\
&= \frac{L}{c}\left[\left(1 - \frac{v}{c} + \frac{v^2}{c^2} - \ldots\right) + \left(1 + \frac{v}{c} + \frac{v^2}{c^2} + \ldots\right)\right],
\end{aligned}
$$

where v is the velocity of the Earth. Thus

$$t_1 = \frac{2L}{c}\left(1+\frac{v^2}{c^2}\right),$$

if higher-order quantities are neglected.

For the transverse passage, since the light would have effectively to cover a longer path $2L[1+(v^2/c^2)]^{\frac{1}{2}}$, the time taken would be

$$t_2 = \frac{2L}{c}\left(1+\frac{1}{2}\frac{v^2}{c^2}\right).$$

The time difference is thus $(L/c)(v^2/c^2)$ which corresponds to a path difference of Lv^2/c^2. If v is small compared with c, it would appear that the measurement of this quantity would not be possible.

But *is* it too small? The orbital velocity of the Earth is about 10^{-4} of the velocity of light. If L is about 100 cm – a value that Michelson regarded as normal for his experiments – the path difference is 10^{-6} cm, or about $\lambda/50$; this is certainly too small for measurement, but is large enough to suggest that with some modification a measurable distance might be expected.

The chief factor in producing a measurable path difference was an increase in the path L; the interferometer was mounted on a stone slab of diagonal about 2 m (Fig. VI.1) and the light reflected so that it traversed this diagonal several times, giving a total distance L of 1100 cm. Since there was no *a priori* knowledge of what might be the direction of the path of the Earth, the whole apparatus was caused to rotate and the maximum difference in path should be twenty-two times that previously calculated – just under half a fringe. Michelson and Morley were confident that they could measure this to an accuracy of about 5 per cent.

This experiment is described in some detail because it is one of the most important experiments in optics. It illustrates the importance of not accepting the impossibility of measuring second-order quantities; and the complete account of the care taken in avoiding spurious effects is well worth reading in the original.

The result was most surprising and disappointing; no certain shift greater than 0.01λ was found (Fig. VI.2). The broken line in Fig. VI.2 shows $\frac{1}{8}$ of the displacement expected from the orbital velocity of the Earth. It appeared the velocity of the Earth was zero!

There is just the possibility that the orbital velocity of the Earth happened at the time of the experiment, to cancel out the drift velocity of the solar system. This could not happen at all seasons of the year and

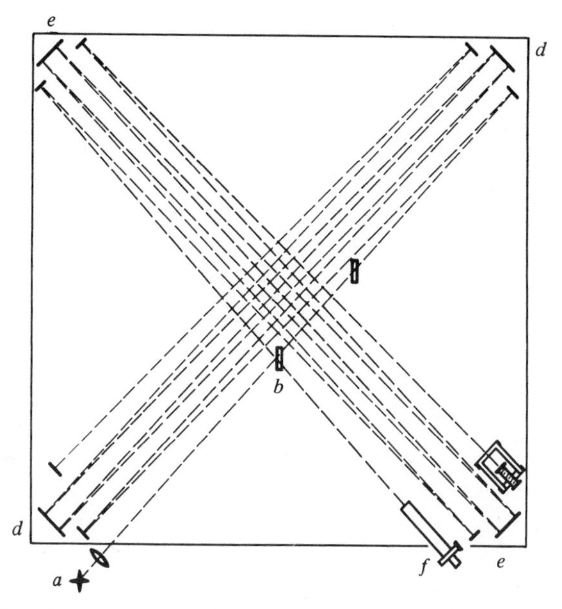

Fig. VI.1. Interferometer used in the Michelson–Morley experiment (From Michelson, 1927.)

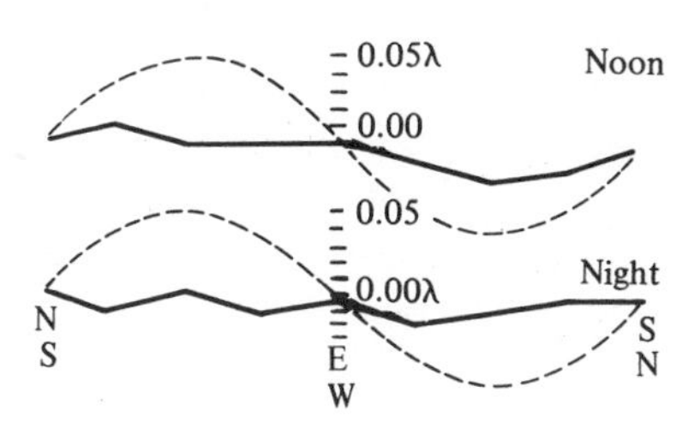

Fig. VI.2. Typical diurnal variation of the fringe shift. (From Michelson, 1927.)

therefore more measurements were made at intervals of several months. The result was always zero.

This result was one of the mysteries of nineteenth-century physics. It was perplexing and disappointing to Michelson and Morley, whose skill and patience seemed to have been completely wasted. But now we know that it was not so; it excited Einstein in 1905 to come forward with a new physical principle – relativity – the main assumption of which is that the velocity of light is invariant whatever the velocity of the observer. Thus out of an apparently abortive experiment, a new physical principle arose and a new branch of physics had its beginning.

A few suggestions for student projects

These suggestions intentionally lack specific practical details, but have all been carried out in recent years either as student laboratory projects or as lecture demonstrations.

1 Optical tunnelling can be investigated by means of the experiment suggested in § 4.3.3. One should confirm the decay of the fields in the gap, as calculated in problem 4.4.
2 Confirm conservation of energy in a Michelson interferometer (problem 7.10).
3 Demonstrate and compare dark ground, phase contrast and schlieren images of a phase object, using the set-up of Fig. 9.10. A convenient object is a smear of transparent glue on a microscope slide.
4 Make and reconstruct a computer-generated hologram of a simple object, such as letter of the alphabet. The computer must print out a pattern resembling the fringes of the hologram. The print-out is then photo-reduced onto a slide of suitable size for the reconstruction.
5 Use a Michelson interferometer to measure the spatial and temporal coherence functions of a multi-mode laser; one exciting several transverse modes is ideal.
6 Investigate the diffusive mixing of two liquids with the aid of holographic interferometry (see Bochner & Pipman, 1976).
7 Grow a crystal of $NaNO_3$ from aqueous solution and investigate its bi-refringence.
8 Use the photoelastic effect to investigate a plastic model of a simple loaded structure, and compare your results with calculations of the stress distribution (see problem 5.8).

Problems

Chapter 2

2.1 Find the wavelength and direction of the two-dimensional wave

$$\zeta = a \cos k_0(lx + my - ct).$$

2.2 Show that the sum of three similar waves with phase differences of $2\pi/3$ is zero.

2.3 Find the wavelength and velocity of the three-dimensional wave

$$\zeta = a \sin (Ax + By + Cz - Dt).$$

2.4 It is found that the refractive index of a medium can be expressed by the relation

$$\mu = \frac{\mu_0}{1 + B(\lambda - \lambda_0)}.$$

Find how the group velocity is related to the wave velocity.

2.5 A series of masses m are situated on the x-axis at equally spaced points $x = na$. They are connected by springs having spring-constants K, and are restrained to move only along the x-axis. Show that longitudinal waves can propagate along the chain, and that the dispersion relation is:

$$\omega = 2(K/m)^{\frac{1}{2}} |\sin \tfrac{1}{2}ka|.$$

Explain in physical terms why the dispersion relation is periodic.

2.6 For the example in problem 2.5, calculate the phase and group velocities when (a) $\omega \to 0$, (b) $\omega = 2(K/m)^{\frac{1}{2}}$.

2.7 Draw a sketch of $\mu(\omega)$ and hence deduce qualitatively the behaviour of the group velocity as a function of ω when

$$\mu(\omega) = 1 + \frac{A(\eta^2 - \omega^2)}{(\eta^2 - \omega^2)^2 + \kappa^2\omega^2}.$$

A, η and κ are all constants.

Chapter 3

3.1 To show that rounding the discontinuities of a sharp-edged function reduces the high-frequency (high k) components of its Fourier transform (cf.

§ 7.2.4), transform the following three functions and compare their behaviour at large k:

(a) $f(x)=1$ when $-\frac{1}{2}b<x<\frac{1}{2}b$; otherwise 0;

(b) $f(x)=\cos(\pi x/b)$ when $-\frac{1}{2}b<x<\frac{1}{2}b$; otherwise 0;

(c) $f(x)=\cos^2(\pi x/b)$ when $-\frac{1}{2}b<x<\frac{1}{2}b$; otherwise 0.

3.2 Find the Fourier transform of a decaying series of δ-functions:

$$f(t)=\sum_{u=0}^{\infty}\delta(t-nt_0)\,\mathrm{e}^{-\alpha n}.$$

How can this function be used to illustrate the properties of a Fabry–Pérot etalon?

3.3 Evaluate the convolution $f(t)$ of $f_1(t)$ and $f_2(t)$, where $f_1(t)$ is two δ-functions

$$\delta\left(t-\frac{b}{2}\right) \quad \text{and} \quad \delta\left(t+\frac{b}{2}\right)$$

and $f_2(t)$ is a square pulse as in problem 3.1(a). Show that the transform of $f(t)$ is the product of the transforms of $f_1(t)$ and $f_2(t)$.

3.4 Calculate the auto-correlation function of the slit function as defined in problem 3.1(a). Show by direct calculation that its transform is the square of the transform you calculated for 3.1(a).

3.5 The period of a square wave is b, the displacement being 1 for time c and zero for time $b-c$. How does the Fourier transform change with the ratio c/b? In particular, what happens when $c/b\to 1$?

3.6 In a pulsed laser, a δ-function-like excitation is reflected backwards and forwards within the resonator, growing with a gain G in every cycle. This process can be represented by a series of δ-functions:

$$f(t)=\sum_{n=0}^{\infty}\delta(t-nt_0)\exp(n\ln G).$$

This series diverges, but in practice the drain on the upper energy levels eventually becomes so great that G falls to zero. Approximately, we can write this fall-off as

$$G=G_0\exp(-\beta t).$$

Find the resultant transform for $f(t)$ and thus relate the linewidth of the laser output to the parameters β and t_0.

3.7 Derive the Fourier coefficients for a zigzag line as an odd function and as an even function. Why is the first coefficient predominant?

3.8 Calculate the two-dimensional Fourier transform of the function $f(\mathbf{r})=1/r$.

3.9 The convolution operation has some very odd properties. For example, a group of three δ-functions can be represented as an infinite periodic set of δ-functions $\Sigma\,\delta(x-na)$ multiplied by *any* symmetrical slit function having width b between $2a+\epsilon$ and $4a-\epsilon$, where ϵ is a very small quantity. Show by direct calculation that the transform of the product is indeed independent of the width of the slit function provided that it lies between these limits.

Chapter 4

4.1 A sandwich consists of three transparent materials with refractive indices μ_1, μ_2 and μ_3. The central layer has thickness l. If a wave of free-space wavelength λ enters normally through the first layer, show that the reflexion coefficient is

$$\text{(a)}\ \frac{\mu_3-\mu_1}{\mu_3+\mu_1} \qquad \text{when } \mu_2 l=\frac{\lambda}{2};$$

$$\text{(b)}\ \frac{\mu_2^2-\mu_3\mu_1}{\mu_2^2+\mu_3\mu_1} \qquad \text{when } \mu_2 l=\frac{\lambda}{4}.$$

In (a) the reflexion is unaffected by the second layer. Is this also true when μ_2 is complex? (b) illustrates the principle of lens blooming (§ 7.9.1).

4.2 Show that the following system satisfies Maxwell's equations and that it represents elliptically polarized light.

$$E_x = AZ_0 \cos(\omega t-\kappa z); \qquad H_x=-\epsilon^{\frac{1}{2}}B \sin(\omega t-\kappa z).$$

$$E_y = BZ_0 \sin(\omega t-\kappa z); \qquad H_z=\epsilon^{\frac{1}{2}}A \cos(\omega t-\kappa z).$$

$$E_z=0; \qquad H_z=0.$$

4.3 Calculate the reflexion coefficient at the surface of a dielectric medium for light incident at the Brewster angle.

4.4 Two 45°–90°–45° glass prisms are placed next to one another with their hypotenuses almost in contact, but separated by a small distance d (of the order of $\frac{1}{2}\lambda$). A plane-wave of unit intensity and wavelength λ is incident normally on one of the small glass faces of one prism. Calculate the amplitude and intensity of the plane-wave which leaves the parallel face of the second prism, after tunnelling through the air gap. Ignore reflexions at the small faces.

Chapter 5

5.1 Show that two equal coherent beams of circularly polarized light, when superimposed, produce linearly polarized light.

5.2 Mica has refractive indices 1.5998 and 1.5948 for propagation normal to the cleavage plane. A sheet of mica, observed between crossed polarizers, has a purple colour, i.e. red and blue light are transmitted, but not green. Estimate the thickness of the sheet. How does the colour of the mica change (*a*) as the mica is rotated in its own plane, (*b*) as the analyser is rotated in its plane, the mica and polarizer remaining in their original orientations?

5.3 A ray of unpolarized monochromatic light falls at normal incidence on a quartz plate having parallel faces and thickness t. The optic axis makes an angle of 45° with the normal to the plate. If $\mu_e = 1.533$ and $\mu_0 = 1.544$ find the separation of the emerging rays.

5.4 How is a quarter-wave plate used to convert linearly polarized light into circularly polarized light? How can the sense of polarization be reversed?

5.5 A beam of light is known to be a mixture of unpolarized and circularly polarized light. How would you find the proportions?

5.6 A parallel beam of light from a sodium lamp passes through a pair of parallel polarizers, with similar orientations, separated by a calcite plate whose optic axis lies in the plane of the faces, at 45° to the axis of the polarizers. Calculate the thickness of the calcite plate required so that one of the sodium D-lines (5890 Å and 5896 Å) is transmitted and the other absorbed. In the region of the D-lines the principal refractive indices and their derivatives are:

$$\mu_e = 1.486, \qquad d\mu_e/d\lambda = -3.53 \times 10^{-6}\ \text{Å}^{-1};$$

$$\mu_0 = 1.658, \qquad d\mu_0/d\lambda = -5.88 \times 10^{-6}\ \text{Å}^{-1}.$$

5.7 A pile of parallel glass plates can be used as a crude polarizer for light, using reflexion at the Brewster angle. If the incident light is unpolarized, estimate how many plates would be needed so that the transmitted light is 90 per cent polarized.

5.8 A plastic strip of rectangular cross-section (thickness t, width w and length l) is supported horizontally by being clamped rigidly at one end. It is observed in monochromatic light between crossed polarizers with axes at 45° to the horizontal. When the free end of the strip is loaded with weight W, show that there appear dark and light bands along the length of the strip, and the number of bands in the width varies linearly along the strip. What is observed in white light?

Chapter 6

6.1 A parallel beam of light of wavelength 5000 Å is incident normally on a circular hole of diameter 1 mm. Find the furthest distance at which the intensity at the centre of the transmitted beam is zero.

6.2 A steel ball of diameter 5 mm is placed in the path of a beam of light, of wavelength 5000 Å, diverging from a point 1 m away. Roughly what irregularity in the surface of the ball must exist for the bright spot in the centre of the shadow not to be seen?

6.3 Investigate spherical and chromatic aberrations in a zone plate, used as a lens. Spherical aberration can be cured by correction of the ring diameters; find these ring diameters for the principal focus. How can an achromatic doublet of zone plates be constructed?

6.4 Calculate the fringe separation in the Fresnel diffraction pattern of a straight edge, as a function of the distance from the edge of the geometrical shadow (Fig. 6.11).

Chapter 7

7.1 Fig. P.1 shows some simple objects and their diffraction patterns. All are correctly oriented and have approximately the same relative dimensions, but the exposures are not identical. Correlate the patterns with their objects.

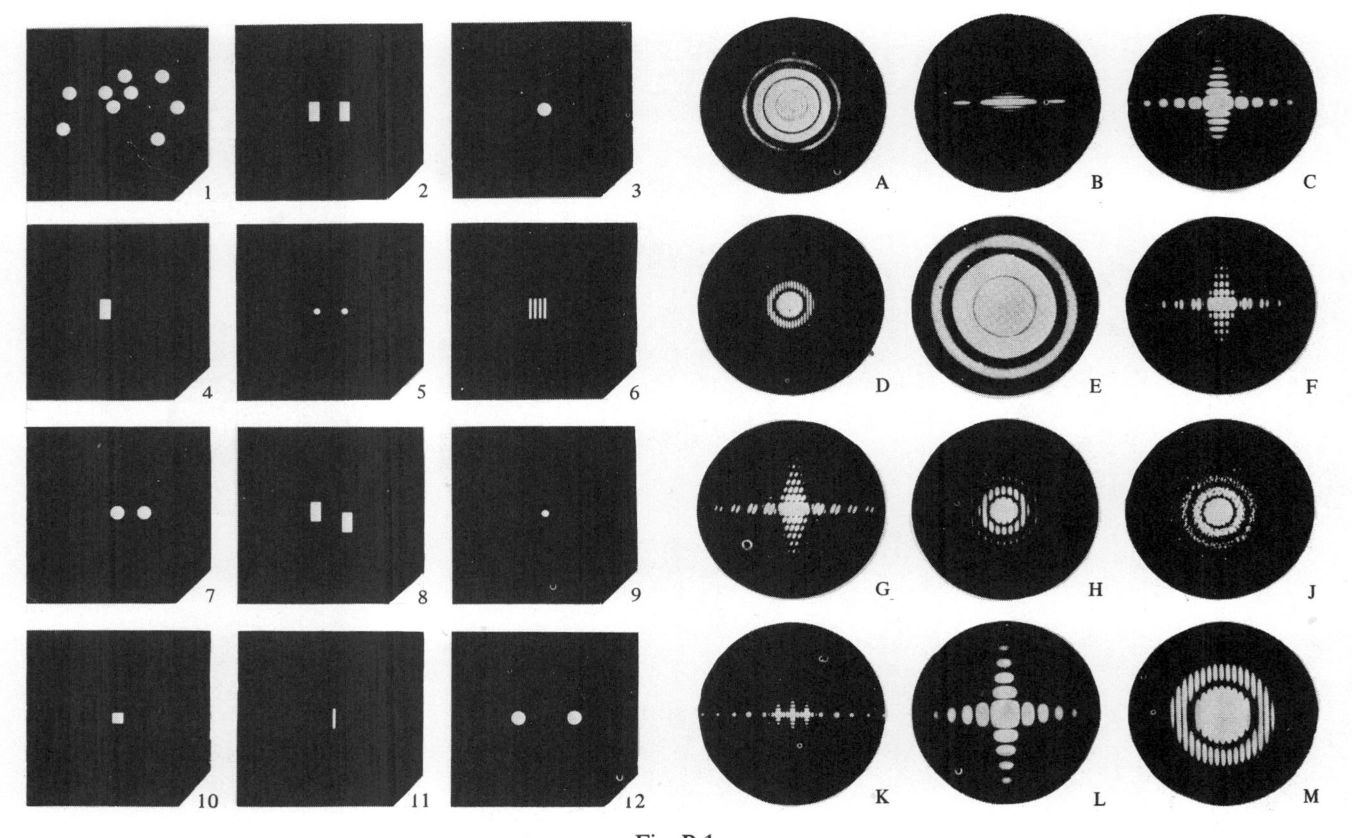

Fig. P.1

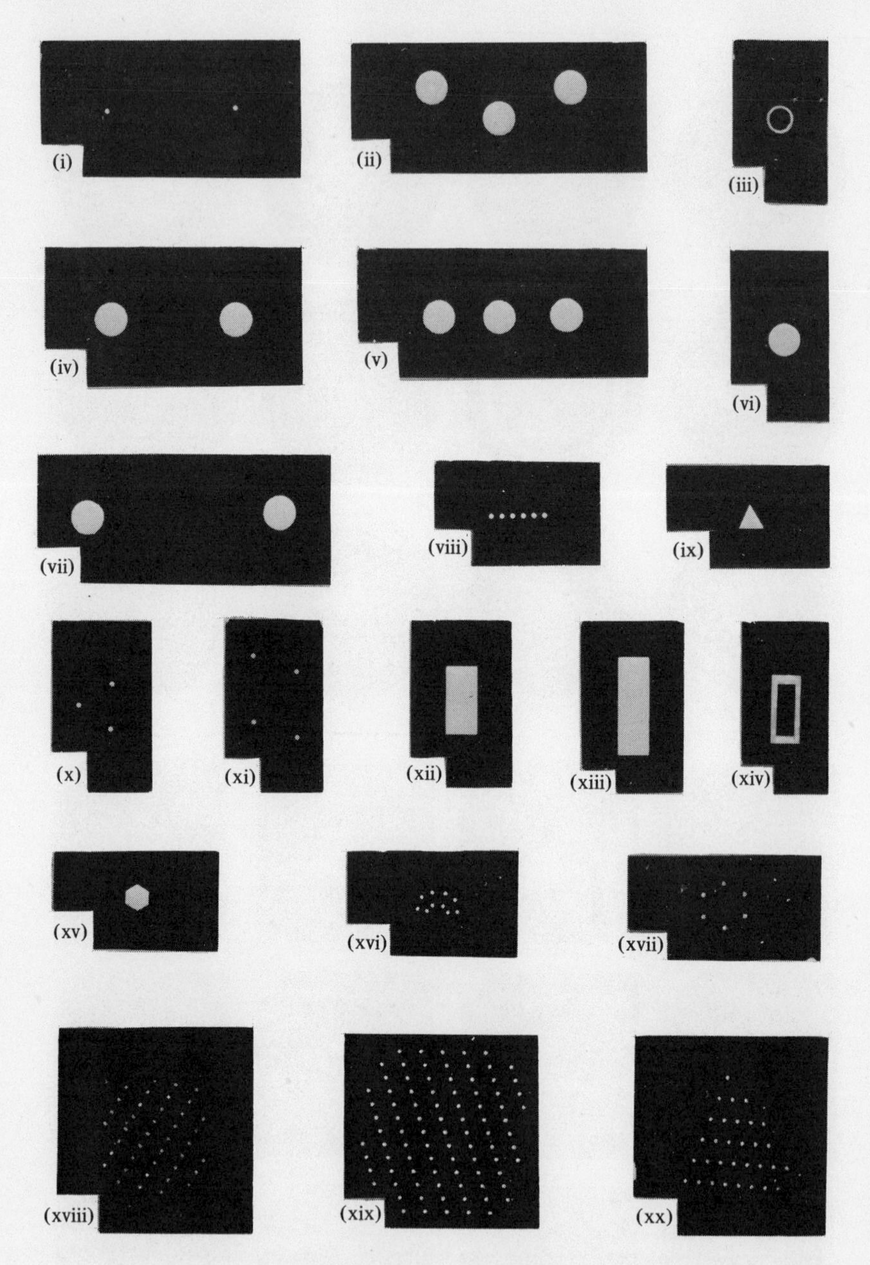

Fig. P.2. (i)–(xx) A set of apertures.

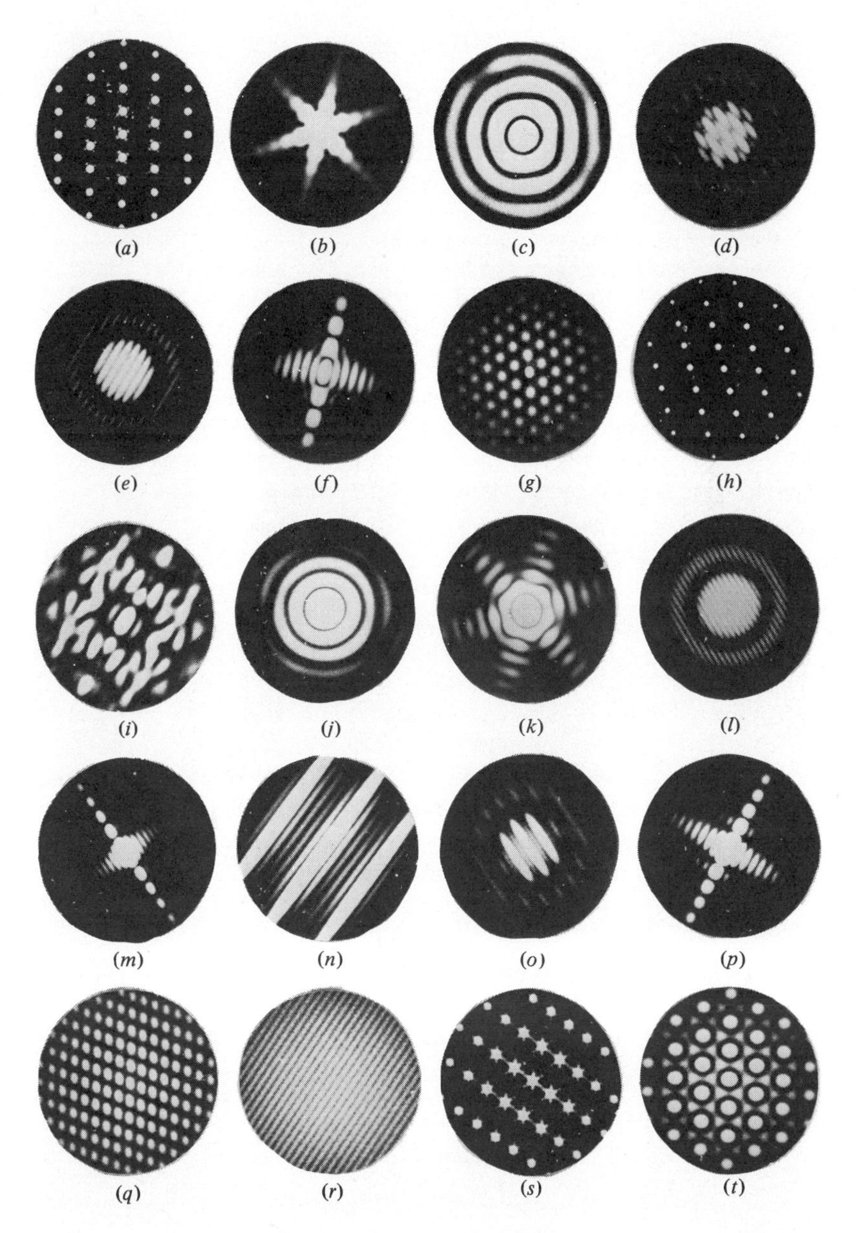

Fig. P.2. (*a*)–(*t*) The diffraction patterns formed from the apertures in (i)–(xx).

7.2 Repeat problem 7.1 with the information in Fig. P.2. Here the patterns are not correctly oriented, and you should deduce the correct orientation for each pattern.

7.3 Deduce the Fraunhofer diffraction pattern of a set of three equally spaced similar slits by adding the transform of the centre slit to that of the two outer ones.

7.4 Deduce the Fraunhofer diffraction pattern of four equally spaced similar slits by considering them as
(a) a pair of identical double slits;
(b) a double slit in the centre, flanked by a double slit of three times the spacing.

7.5 Derive the Fraunhofer diffraction pattern of a double slit by considering it as a rectangular aperture with the centre blocked out.

7.6 A square aperture is half covered by a transparent sheet of mica which changes the phase by $\pi/2$. What is its diffraction pattern?

7.7 A diffraction grating has lines, denoted by the integer p, whose lengths vary periodically according to the expression $a + b \sin 2\pi p/q$, where b is small compared with a. Find the diffraction pattern of this grating. Derive the result also by using the convolution theorem.

7.8 A reflexion grating is blazed for $\lambda = 7000$ Å in the first order, and the zero order is found to be 0.09 as intense as the first order. Treating the grating as composed of flat mirrors, find the relative intensities of the other orders. Find also the relative intensities for $\lambda = 4000$ Å.

7.9 An echelon grating consists of a series of reflecting steps of width c and height b ($b \ll c$). Deduce the observed diffraction pattern assuming that the number of steps is infinite.

7.10 A Michelson interferometer is used in exactly parallel monochromatic light and is adjusted so that the two optical paths SM_1S and SM_2S (Fig. 7.57) differ by exactly $\frac{1}{2}\lambda$. The output intensity is therefore zero. Examine this statement critically and explain where the incident light energy has gone.

7.11 White light is reflected normally from the front and back surfaces of a glass plate. Show how spectral analysis of the total reflected light can be used to determine the optical thickness of the plate.

Chapter 8

8.1 The Young's slit experiment is carried out with two slits 1 mm apart, illuminated by light of wavelength 6000 Å, from a slit 10 cm distant. As the illuminating slit is increased in width, it is observed that the visibility of the fringes becomes less. At what width of slit will the visibility become zero?

8.2 A star has an angular diameter of 0.2 arcsec. What separation of mirrors in the Michelson stellar interferometer is needed in order to measure it in light of wavelength 0.6 μm?

8.3 A strange star consists of a laser with a very long coherence time of the order of seconds. Why would it be impossible to measure its diameter with a Brown–Twiss interferometer, but perfectly possible with a Michelson stellar interferometer?

8.4 Calculate the spatial coherence function in both the x- and y-directions on a table illuminated by a standard fluorescent tube at a height of 10 m. Assume monochromatic light with $\lambda = 0.5\ \mu$m.

8.5 A Fourier-transform spectrometer gives an interferogram of which the positive half is shown in Fig. P.3. What qualitative deductions can you make about the source spectrum?

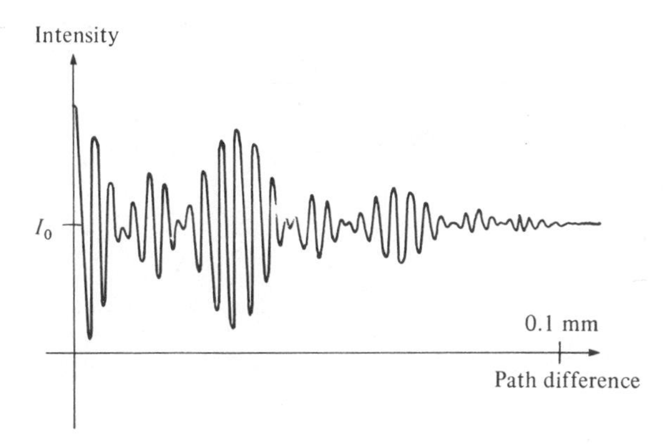

Fig. P.3. Fourier interferogram.

8.6 Monochromatic light ($\lambda = 0.5\ \mu$m) is scattered at 90° from a cell containing 10^{-10} g particles in suspension at 300 K. Estimate the coherence time and linewidth of the scattered light.

8.7 A laser beam is spatially filtered by focusing it through a pinhole (§ 8.7.4). If the laser beam has diameter 1 mm, and is focused with a microscope objective of focal length 5 mm, what size of pinhole is necessary to give a homogeneous beam? Assume $\lambda = 0.5\ \mu$m.

8.8 A very accurate method of measuring the spectral variation of refractive index of a transparent material involves placing a slab of it into one arm of a Michelson interferometer illuminated by white light. An asymmetrical interferogram ($I(d) \neq I(-d)$) results. Develop the analysis of such an interferogram along the lines of Fourier-transform spectroscopy and show how $\mu(\lambda)$ can be calculated from it.

8.9 Calculate the Doppler and collision widths of an emission line from the H_2O molecule at 0.5 μm, at 300 K and atmospheric pressure. Use an estimated size for the molecule. What line-shape would you expect to see when the two linewidths have comparable magnitudes?

Chapter 9 (involving some material from Appendices I, II, III)

9.1 The foci of a mirror in the form of an ellipsoid of revolution are conjugate points. What is the magnification produced if an object is placed at one of the foci?

9.2 An astronomical telescope is used in conjunction with a camera to produce very highly magnified images. If the primary mirror has a diameter of 1 m and a focal length of 12 m, what extra magnification is usable if the photographic film has a resolution limit of 0.05 mm?

9.3 In order to use a microscope on an inaccessible specimen, one can introduce a relay lens between the specimen and the objective lens. The relay lens produces an unmagnified real image of the specimen in the object plane of the objective. Draw a ray diagram of the system, and find the influence of the relay lens on the exit pupil and the field of view.

9.4 Design a periscope having a length of 2 m, and a tube diameter of 10 cm. The angular field of view has to be a cone of semi-angle 30°, and the magnification unity. Ignore aberrations.

9.5 A lens, corrected for spherical aberration, is used to form an image of a distant point of light on its axis. The lens has a diameter of 10 cm and a focal length of 50 cm. How close on the axis to the focal point will it be possible to detect that the image is out-of-focus? (Hint: consider the path difference between an axial ray and a ray passing through the edge of the lens.)

9.6 A telephoto combination consists of a converging lens of focal length 10 cm, and a diverging lens of focal length 2 cm, separated by 8.2 cm. What is its focal length, and where are its principal planes? Given that the two constituent lenses have apertures $f/2$, what is the f-number of the telephoto combination?

9.7 An object consists of two bright points on a dark background. Their separation is 3λ (where λ is the wavelength of light). Calculate the image which is obtained when the object is viewed through a microscope under the following conditions:

(a) coherent illumination, NA = 0.5;

(b) coherent illumination, NA = 0.2;

(c) incoherent illumination, NA = 0.2.

(Treat the problem as one-dimensional.)

9.8 A telescope lens is apodized so as to reduce the prominence of the diffraction rings surrounding a point image. If the radius of the objective lens is R, the apodization is done by reducing the amplitude transmission of the lens at radius r by the factor $\cos(\pi r/2R)$. Treat this problem as one-dimensional, and calculate by how much the relative intensity of the first diffraction ring is reduced.

9.9 A photographic transparency shows a monkey behind a fence consisting of parallel vertical equally spaced bars. How would you use a spatial filtering technique (apparatus as in Fig. 9.10, with a suitable filter) to produce an image of the monkey without the fence?

9.10 For certain types of grey-scale object (containing intensity variations but *no phase variations*) the dark-ground image is the negative of the object, in the photographic sense. What conditions must the object satisfy for this to be true?

9.11 A hologram of a certain object is made using laser light of wavelength λ_1. The hologram is reconstructed with light of wavelength λ_2. How are the size and position of the reconstruction related to those of the object? Assume all angles involved in the problem to be small.

9.12 In Fig. 9.36(*c*) both the real and virtual reconstructions from a hologram are shown. By considering, in the same way, a number of points on an extended object, show that the real image is distorted unless the reference beam is normal to the holographic plate.

Chapter 10

10.1 A Gaussian wave-group of wave-number k_0 and half-width $\sigma(0)$ propagates in a slightly dispersive medium defined by its dispersion relation $\omega(k)$. By expanding $\omega(k-k_0)$ as a Taylor series up to the second order, show that after propagation for a time t the wave-group is centred on position

$$x = v_g t$$

and its half-width has increased to $\sigma(t)$ where

$$\sigma^2(t) = \frac{\sigma^4(0) + u^2t^2}{\sigma^2(0)};$$

$v_g = \partial\omega/\partial k$ and $u = \partial^2\omega/\partial k^2$.

10.2 Calculate the brightness of the 'blue sky' at $\lambda = 0.45\ \mu$m in terms of the intensity of sunlight at the Earth's surface, and in a direction at right angles to that of the Sun. Ignore multiple scattering. (If the atmosphere were compressed uniformly to atmospheric pressure at sea level, it would have a height of 5 km. Brightness is defined as the energy flux emitted by a source per unit area, per unit solid angle. The Sun's angular diameter is 0.5°.)

10.3 A plasma contains two types of free charge carrier, having number density, mass and charge N_1, m_1, e_1 and N_2, m_2, e_2 respectively. Ignore interactions between the two types of carrier, and calculate its plasma frequency and the dependence of its refractive index on frequency.

10.4 A material has a spectral absorption line at wavelength λ_0. Assuming that the absorption line can be represented by a δ-function of strength a_0, use the Kramers–Krönig relations to calculate the refractive index as a function of wavelength.

Solutions

Solutions to problems on Chapter 2

2.1 Writing the argument of the cosine as $(\mathbf{k}\cdot\mathbf{r}-\omega t)$ we see that $\mathbf{k}=(l, m)k_0$, $\omega = k_0 c$. Thus the wavelength is $2\pi/|\mathbf{k}| = 2\pi/k_0(l^2+m^2)^{\frac{1}{2}}$ and the direction of propagation is (l, m).

2.3 As 2.1, but in three dimensions.

2.4 The phase velocity is c/μ and then equation (2.35) can be used to evaluate v_g.

2.5. This example is discussed in detail in books on elementary solid-state physics e.g. Dekker (1967) § 2.8; Kittel (1971) Chapter 5.

2.6 (a) $v = v_g = a(K/m)^{\frac{1}{2}}$; (b) $v = 2a(K/m)^{\frac{1}{2}}/\pi$, $v_g = 0$.

2.7 See § 10.3.2.

Solutions to problems on Chapter 3

3.1 (a) b sinc $(bk/2)$, which has oscillations decaying as $1/k$ at large k;
(b) $\frac{1}{2}\pi b \cos(bk/2)/\{(\frac{1}{2}bk)^2-(\frac{1}{2}\pi)^2\}$, which has oscillations decaying like $1/k^2$;
(c) in this case, the oscillations decay like $1/k^3$.

3.2 See § 7.5.1.

3.3 See Fig. 7.21.

3.4 The auto-correlation function is $f_c(x) = b - |x|$, whose transform is calculated in § 7.2.2.

3.5 See § 7.7.4.

3.6 Since the nth term in the series occurs at time $t = nt_0$, one can write the series as the product of $\Sigma\,\delta(t-nt_0)$ and the function $\exp\{(t/t_0)(\ln G_0 - \beta t)\}$. The latter is a Gaussian function. The transform has Gaussian lines with variance $(2\beta/t_0)^{\frac{1}{2}}$.

3.7 The zigzag line can be considered as the function derived in problem 3.4 convoluted with the series of δ-functions $\Sigma\,\delta(x-2nb)$. The terms have amplitudes 1, 0.4, 0, 0.044, ...

3.8 $2\pi\delta(\mathbf{k})$.

3.9 The transform of the product can be written down immediately as

$$\sum \delta(k-2\pi m/a) * 2b \operatorname{sinc}(kb).$$

Writing this convolution out term by term and combining the terms $\pm m$ gives the sum of two series of the forms

$$\sum_1^\infty \frac{\cos nx}{n^2-c^2}, \qquad \sum_1^\infty \frac{n \sin nx}{n^2-c^2}.$$

These can be summed by analytical methods described by R. E. Collin (1960).

Solutions to problems on Chapter 4

4.1 (a) When μ_2 is complex, absorption occurs in the second layer and the value of μ_2 becomes important. Obviously, if the absorption is large enough, the value of μ_3 becomes irrelevant.

4.3 $(\mu^2-1)/(\mu^2+1)$ for the perpendicular component.

4.4 The amplitude of the wave (perpendicular polarization) is:

$$1\bigg/\left\{\cosh(k\beta d)+\mathrm{i}\left(\frac{1}{\mu\beta\sqrt{2}}-\mu\beta\sqrt{2}\right)\sinh(k\beta d)\right\}.$$

The intensity is the square modulus of this quantity.

Solutions to problems on Chapter 5

5.1 See § 5.5.4.

5.2 Light is transmitted when there is $(2m+1)\pi$ phase difference between the principal components after transmission through the mica. If we define the blue light as 4200 Å we find the thickness of the mica to be 0.21 mm. As the mica rotates, the colour is extinguished whenever the principal axes of the mica are parallel and perpendicular to the polarizer axis, but it remains purple. As the analyser rotates, the colour changes from purple to green (complementary).

5.3 $0.014t$.

5.4 The polarization plane of the incident light lies at 45° to the axes of the quarter-wave plate. The sense of polarization is reversed by turning the quarter-wave plate 90° in its plane.

5.5 Transmit the light through a quarter-wave plate to convert the circularly polarized light into plane-polarized light. Now measure the intensities of light transmitted by an analyser at 45° to the axes of the quarter-wave plate, and at −45°. The difference corresponds to the circularly polarized component in the original light.

5.6 The system rotates the plane of polarization of one D-line by $m\pi$ and that of the other $(2m+1)\pi/2$ when the thickness of the calcite is about 1.8 mm.

5.7 When $\mu = 1.5$, about 7 plates are necessary.

5.8 First solve the problem in elasticity. When (x, y) are measured from the centre of the strip at the loaded end, the local stress is $12\,Wxy/tw^3$. Light and dark bands signify contours of constant stress, and so the number of bands in the width of the strip is proportional to x. With white light, the central band is black, and others are coloured.

Solutions to problems on Chapter 6

6.1 From equation (6.25) the smallest value of d_0 which gives zero intensity is $\rho^2/2\lambda = 0.25$ m.

6.2 We expect the bright spot to disappear if the error in the exponent $k\rho^2/2d_0$ is of the magnitude of π. This is equivalent to an error in ρ of about 0.1 mm.

6.3 If we let $d/\lambda = n_0$, spherical aberration is absent when the ring radii are $r_n = \lambda(n^2 - n_0^2)^{\frac{1}{2}}$. Since the equivalent focal length, $f = a_0^2/\lambda$, the equivalent focal length of any number of such plates in contact is still proportional to $1/\lambda$ and is therefore not achromatic. If the plates are separated, lateral chromatic aberration can be corrected (Appendix II.6, Fig. II.7).

6.4 One fringe corresponds to each turn of the Cornu spiral. After the first few turns, this gives spacing

$$\delta x = \frac{2\pi}{x}\left(\frac{d_0}{k}\right)^{\frac{1}{2}}.$$

Solutions to problems on Chapter 7

7.1

1	2	3	4	5	6	7	8	9	10	11	12
J	*F*	*A*	*C*	*M*	*K*	*H*	*G*	*E*	*L*	*B*	*D*

7.2 The following patterns should be obvious:

e, h, j, l, m, n, o, p, q, r and *t*.

Patterns (*b*) and (*k*) both have hexagonal symmetry and so must correspond to the triangle and hexagon; (*k*) is the closer approximation to the pattern (*j*) of the circular hole, and therefore corresponds to the hexagon (xv).

Both (*c*) and (*j*) have approximately circular symmetry (the apertures are not precisely circular); but (*c*) has the smaller 'Airy disc' and is therefore associated with the annular ring (iii) (§ 9.6.7). Similarly, (*p*) and (*f*) are associated with the complete and annular rectangular apertures respectively.

The pattern (*d*) is clearly that of more than two large holes; its aperture is therefore (ii). The intensity distribution along one diameter is the same as that in (*o*).

Possible uncertainty between the correspondence of (*g*) or (*q*) to (x) or (xi) can be resolved by measuring the angles of the reciprocal lattices.

The patterns (*a*) and (*s*) are produced by limiting the lattice (xix) by a triangular (xx) or rectangular (xviii) aperture; thus the diffraction patterns are the convolutions of these two apertures by the reciprocal lattice (*h*) (§ 7.4.3). The reciprocal lattice is the interference function, and the diffraction pattern of the aperture (e.g. (*b*)) is the diffraction function.

The pattern (i) can be recognized as the diffraction pattern of (xvi) only by elimination. There are two ways of finding the correct relative orientations. First, the central peak in (i) is reciprocally related in shape to the overall shape of the pattern; secondly, although (i) has two planes of symmetry, only one is real (approximately NW–SE) and is perpendicular to the plane of symmetry in (xvi).

There is an ambiguity in the orientation of all the non-centrosymmetric apertures; diffraction patterns all have a centre of symmetry.

7.3 Assume the slits to be δ-functions at $x=0, \pm a$. The two outer slits have transform $2 \sin ua$; the centre one has transform 1.

7.4 Assume the slits to be δ-functions at $x=\pm a/2, \pm 3a/2$. Using the two methods, the transforms are:
(a) $2 \cos (ua/2) \times 2 \cos (ua)$;
(b) $2 \cos (ua/2) + 2 \cos (3ua/2)$.
These two results are equal.

7.5 Assume the slits to have width ϵ centred on $x=\pm a/2$ and to have length b. The result is then:

$$\begin{aligned} b \operatorname{sinc}(\tfrac{1}{2}vb)[(a+\epsilon) \operatorname{sinc}\{u(a+\epsilon)/2\} \\ -(a-\epsilon) \operatorname{sinc}\{u(a-\epsilon)/2\}] \\ = 2b \operatorname{sinc}(\tfrac{1}{2}vb) \operatorname{sinc}(\tfrac{1}{2}u\epsilon) \cos(\tfrac{1}{2}ua). \end{aligned}$$

7.6 The aperture can be represented by a convolution of a rectangular aperture with two δ-functions having phases 0 and $\frac{1}{2}\pi$. The resultant amplitude is

$$a^2 \operatorname{sinc}(\tfrac{1}{2}va) \operatorname{sinc}(ua/4) \cos(ua/4+\pi/4) \exp(i\pi/4).$$

7.7 Consider the problem in one dimension, where the length of the line indicates its scattering amplitude. We write the grating as a product $\sum_p \delta(x-pd) \cdot \{a+b \sin(2\pi x/dq)\}$. The transform is

$$\sum_n \delta(u-2n\pi/d) * \{a\delta(u)+\tfrac{1}{2}b\delta(u-2\pi/dq)+\tfrac{1}{2}b\delta(u+2\pi/dq)\}.$$

Each diffraction order, n, is flanked by two equal ghosts.

7.8 If the angle of specular reflexion from the small mirrors is θ_0, the diffraction pattern amplitude is

$$\sum_n \delta(u-2\pi n/d) \operatorname{sinc}(d'u/2-dk \sin\theta_0/2),$$

where d' is the width of the small mirrors. At the blazing wavelength λ_B, θ_0 is also the angle of the first order; thus $\lambda_B = d \sin\theta_0$. For any wavelength λ, the intensity of the nth order is then shown to be $\operatorname{sinc}^2\{\pi d'(n-\lambda_B/\lambda)/d\}$. The given value of the intensity of the zero order for λ_B shows d'/d to be 0.75. When $\lambda = 4000$ Å, the 0, 1, 2 and 3 orders have intensities 0.040, 0.311, 0.889 and 0.0046 respectively.

7.9 Draw the plane through the centre lines of all the steps. With reference to this plane, the echelon looks like a blazed grating and the result of problem

7.8 applies. The grating has spacing $d=(C^2+b^2)^{\frac{1}{2}}\approx C$, and the angle $\theta_0=2b/C$; also $d'/d=1$. Thus the nth order at wavelength λ appears at angle given by $n\lambda = C(\sin\theta - b/C)$ and has intensity $\text{sinc}^2\{\pi(n-2b/\lambda)\}$.

7.10 Part of the light passing through the interferometer returns in the direction of the illumination source. For any real beam-splitter this light intensity is complementary to that received at E (Fig. 7.57) and their sum is equal to the incident intensity. For example, suppose that the beam-splitter consists of the front surface of a piece of glass, the other surface being non-reflecting. The front surface has reflexion coefficient $(1-\mu)/(1+\mu)$ for light incident from the air, and reflexion coefficient $(\mu-1)/(\mu+1)$ for light incident in the glass. The normal output beam at E contains one wave reflected from each side, and therefore the two have opposite signs and cancel when the optical paths are equal. The beam reflected to the source contains one wave which is not reflected at the beam-splitter and one reflected twice at the front surface; these waves have the same sign and therefore reinforce when the paths are equal. This effect was demonstrated experimentally by Shamir & Fox (1967).

7.11 There is destructive interference between the two reflected waves when the optical thickness is a whole number of half-wavelengths, and these wavelengths appear as dark bands crossing the spectrum at $\frac{1}{2}n\lambda = \mu d$. The difference between values of $1/\lambda$ for adjacent dark bands is therefore $1/2\mu d$.

Solutions to problems on Chapter 8

8.1 0.06 mm.

8.2 62 cm.

8.3 The Brown–Twiss interferometer employs correlation of fluctuations. If the characteristic fluctuation time is of the order of seconds, measurement of the correlation coefficient would be impossibly slow. If the source is spatially coherent itself, the van Cittert–Zernike theorem does not apply, but the source actually produces a diffraction pattern at the Earth's surface. The Michelson stellar interferometer would then give its usual result, although for a different reason!

8.4 If the light source has dimensions 4 cm × 1 m, the spatial coherence function is

$$\gamma(x, y) = \text{sinc}\,(0.002k_0x)\,\text{sinc}\,(0.05k_0y),$$

where $k_0=2\pi/\lambda$.

8.5 The following conclusions can be drawn by inspection. It is convenient to use spectroscopic wave numbers, $1/\lambda$ in cm^{-1}.

There are 30 periods of the basic interference fringes in 0.1 mm, so that the central wave number is $3000\ \text{cm}^{-1}$ ($\lambda = 3.3\ \mu\text{m}$).

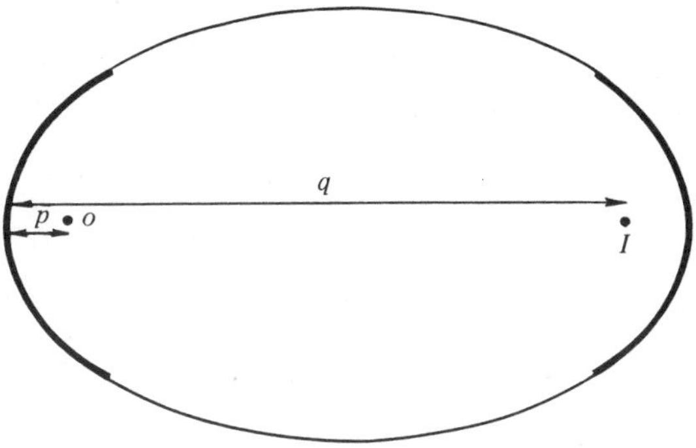

Fig. P.4

The pattern of alternate large and small beats indicates that the spectrum contains three closely spaced components of approximately equal intensity.

The disappearance of the interference pattern when the path difference exceeds 0.1 mm indicates that the components have width of order $1/(0.1\ \text{mm}) = 100\ \text{cm}^{-1}$.

The repeat period of the beat pattern (one large and one small beat) is about 0.03 mm, indicating that the components are separated by $1/(0.03\ \text{mm}) = 300\ \text{cm}^{-1}$.

8.6 The frequency of the scattered light is Doppler-shifted by the thermal motion of the particles. The theory is the same as that for Doppler broadening (§ 8.3.1), and gives a width of 400 Hz, corresponding to a coherence time of 2.5 ms.

8.7 Assuming the beam to have a Gaussian profile with $\sigma = 0.5$ mm, we need a pinhole of about 4 μm diameter to transmit 95 per cent of the light and to block most noise.

8.8 See ch. 8 of Bell (1972).

8.9 The geometrical size of the H_2O molecule is about 3 Å. We then find (equations (8.31)–(8.34)) that both Doppler and collision linewidths are about 0.006 Å. The two line-shapes would be combined by convolution.

Solutions to problems on Chapter 9

9.1 If we consider the system as a symmetrical one the magnification is obviously unity. If we consider the part of the ellipsoid nearest to the object as a small spherical mirror (Fig. P.4) the magnification is obviously q/p, and if we consider the furthest part it is equally obviously p/q. In fact, a finite image is *not* formed; the system has zero spherical aberration but infinite coma.

9.2 When the image at the focus of the mirror is magnified by factor 7, the Rayleigh resolution limit of the telescope becomes compatible with that of the film.

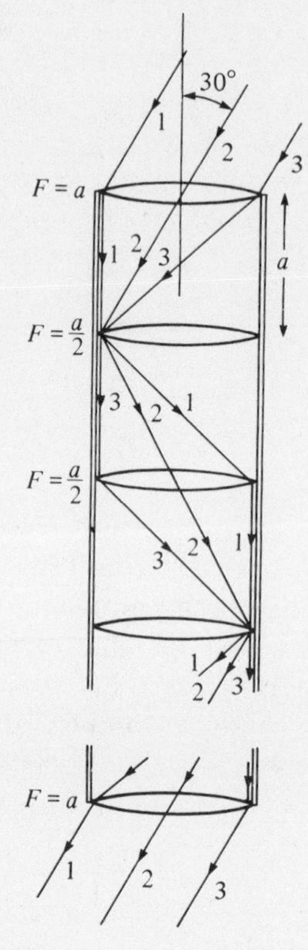

Fig. P.5. A periscope using lenses with apertures equal to the full diameter of the tube. The length a is about 8 cm and 24 lenses would be needed. For clarity, the 45° mirrors of the periscope have been omitted.

9.3 The relay lens replaces the objective as the entrance pupil. The exit pupil is then found to be further from the eyepiece than it was without the relay lens.

9.4 A periscope is essentially an astronomical telescope with the addition of 45° mirrors at each end. To get a large angular field of view, it is necessary to make these mirrors entrance and exit pupils by including relay lenses. The solution must be confirmed by drawing a ray diagram including rays entering at the margin of the field of view (Fig. P.5).

9.5 If we move a distance l along the axis away from the focal point, the axial ray gets longer by l, and the marginal ray by $l/\cos \alpha$, where α is the angular semi-aperture of the lens. The difference between the two elongations becomes comparable with the wavelength when $l \approx \lambda f^2/R^2$. Then l, called the *depth of focus*, is about 50 μm.

9.6 The focal length is 100 cm. The principal planes are 73.8 cm and 410 cm from the converging lens, both on the side further from the diverging lens. The aperture is $f/20$.

9.7 The image intensities are, in the three cases:

(a) $[\text{sinc}\,\{0.5k_0(x-3\lambda/2)\}+\text{sinc}\,\{0.5k_0(x+3\lambda/2)\}]^2$;

(b) $[\text{sinc}\,\{0.2k_0(x-3\lambda/2)\}+\text{sinc}\,\{0.2k_0(x+3\lambda/2)\}]^2$;

(c) $\text{sinc}^2\,\{0.2k_0(x-3\lambda/2)\}+\text{sinc}^2\,\{0.2k_0(x+3\lambda/2)\}$.

These functions must be drawn out or calculated numerically to confirm that the points are resolved in (a) and (c) but not in (b).

9.8 With no apodization, the first ring has intensity 4.7 per cent of the central maximum. With cosine apodization, the ratio falls to 1.95 per cent. The two-dimensional problem has no analytical solution.

9.9 The transparency is used as object. One then inserts a spatial filter which blocks the diffraction spots (equally spaced on the horizontal axis) resulting from the fence.

9.10 The negative must have negligible zero order; thus it must be mainly black. Therefore the object must be mainly white.

9.11 Use the method of Fig. 9.36. The change of wavelength increases the diffraction angles (assumed small) in 9.36(*c*) by factor λ_1/λ_2. Simple geometry then shows that the image moves closer to the hologram by the factor λ_2/λ_1 and is magnified longitudinally by the same amount. The lateral dimensions of the image remain unchanged.

Solutions to problems on Chapter 10

10.2 One calculates the intensity of the polarized light which is scattered at $\theta=\pi/2$. The brightness is 2.5×10^{-6} of the incident intensity.

10.3
$$\Omega=\left(\frac{N_1e_1^{\,2}}{m_1\epsilon_0}+\frac{N_2e_2^{\,2}}{m_2\epsilon_0}\right)^{\frac{1}{2}}.$$

10.4
$$n^2(\lambda)=1+\frac{2a_0\lambda^2}{\pi(\lambda^2-\lambda_0^{\,2})}.$$

Bibliography

Bacon, G. E. (1954). *Research*, London, **7**, 257.

Bacon, G. E. (1962). *Neutron Diffraction*. Oxford University Press.

Bell, R. J. (1972). *Introductory Fourier Transform Spectroscopy*. London: Academic Press.

Binder, R. C. (1962). *Fluid Mechanics*, 4th edn. London: Constable.

Bochner, N. & Pipman, J. (1976). *Journal of Physics*, D**9**, 1825.

Born, M. & Wolf, E. (1964). *Principles of Optics*. Oxford: Pergamon.

Bragg, W. L. (1939). *Nature*, **143**, 678.

Brigham, E. O. (1974). *The Fast Fourier Transform*. New York: Prentice-Hall.

Brillouin, L. (1960). *Wave Propagation and Group Velocity*. New York: Academic Press.

Brown, R. Hanbury & Twiss, R. Q. (1956). *Nature*, **178**, 1046.

Budden, K. G. (1967). *The Propagation of Radio Waves in the Ionosphere*. Cambridge University Press.

Bunn, C. W. (1961). *Chemical Crystallography*. Oxford University Press.

Cole, T. W. (1977). Quasi-optical techniques of radio astronomy. In *Progress in Optics*, vol. 15, ed. E. Wolf, p. 189. Amsterdam: North-Holland.

Collier, R. J., Burckhardt, C. B. & Lin, L. H. (1971). *Optical Holography*. New York: Academic Press.

Collin, R. E. (1960). *Field Theory of Guided Waves*, p. 576. New York: McGraw-Hill.

Cosslet, V. E. (1950). *Electron Optics*. Oxford University Press.

Dekker, A. J. (1967). *Solid State Physics*. London: Macmillan.

Erickson, R. A. (1953). *Physical Review*, **90**, 779.

Ewald, P. P. (1962). *Fifty Years of X-ray Diffraction*. Utrecht: Oosthoek.

Forrester, A. T., Gudmundsen, R. A. & Johnson, T. O. (1955). *Physical Review*, **99**, 1691.

Fröhlich, H. (1949). *Theory of Dielectrics*. Oxford University Press.

Gabor, D. (1948). *Nature*, **161**, 778.

Gloge, D. (1979). *Reports on Progress in Physics*, **42**, 1777.

Grant, I. S. & Phillips, W. R. (1975). *Electromagnetism*. London: Wiley.

Grivet, P. (1972). *Electron Optics*, (second edition, translated from the French). Oxford: Pergamon.

Hartshorne, N. H. & Stuart, A. (1964). *Crystals and the Polarizing Microscope*. London: E. Arnold.

Hoffman, R. & Gross, L. (1975). *Applied Optics*, **14**, 1169.

Hooke, R. (1665). *Micrographia.*

Jeans, J. (1960). *An Introduction to the Kinetic Theory of Gases*. Cambridge University Press.

Labeyrie, A. (1976). High resolution techniques in optical astronomy. In *Progress in Optics*, vol 14, ed. E. Wolf, p. 71. Amsterdam: North-Holland.

Lamb, J. (1946). *Transactions of the Faraday Society*, **42A**, 242.

Landau, L. D. & Lifshitz, E. M. (1960). *The Electrodynamics of Continuous Media*. Oxford: Pergamon.

Landau, L. D. & Lifshitz, E. M. (1963). *Statistical Physics*. Oxford: Pergamon.

Leith, E. N. (1976). In *Advances in Holography*, vol. 2, ed. N. Farhat. New York: Marcel Dekker.

Lipson, H. (1961–2). *Memoirs of the Manchester Literary and Philosophical Society*, vol. 106.

Lipson, H. & Beevers, C. A. (1936). An improved method of two-dimensional Fourier synthesis for crystals. *Proceedings of the Physical Society*, **48**, 722.

Lipson, H. & Cochran, W. (1966). *The Determination of Crystal Structures*. London: Bell.

Lodge, O. J. (1920). *Pioneers of Science*. London: Macmillan.

Mandel, L. (1958). *Proceedings of the Physical Society*, **73**, 1037.

Margenau, H. & Murphy, G. M. (1964). *The Mathematics of Physics and Chemistry*. New York: Van Nostrand.

Michelson, A. A. (1927). *Studies in Optics*. Reprinted by University of Chicago Press, 1962.

North, A. C. T. (1963). *Reports on Progress in Physics*, **26**, 105.

Siegman, A. E. (1971). *An Introduction to Lasers and Masers*. New York: McGraw-Hill.

Shamir, J. & Fox, R. (1967). *American Journal of Physics*, **35**, 161.

Smith, F. G. (1962). *Radioastronomy*. Harmondsworth: Penguin Books.

Smith, W. J. (1966). *Modern Optical Engineering*. New York: McGraw-Hill.

Squires, G. L. (1978). *Introduction to the Theory of Thermal Neutron Scattering*. Cambridge University Press.

Taylor, C. A. & Lipson, H. (1964). *Optical Transforms*. London: Bell.

Thewlis, J. (ed.) (1973). *Concise Dictionary of Physics*. Oxford: Pergamon.

Thomas, D. H. (1948). *Applied Electronics*. Glasgow: Blackie.

Thomas, W., Jr (ed.) (1973). *SPSE Handbook of Photographic Science and Engineering*. New York: Wiley.

Thompson, B. J. (1958). *Journal of the Optical Society of America*, **48**, 95.

Tolansky, S. (1960). *Surface Microtopography*. London: Longman.

Vinen, W. F. (1978). *Physics Bulletin*, **29**, 511.

Wood, R. W. (1934). *Physical Optics.* Macmillan; reprinted by Dover Publications, New York, 1967.

Yariv, A. (1975). *Quantum Electronics.* London: Wiley.

Index